Illustrated Encyclopedia of Building Services

Other Titles From E & FN Spon

European Construction Costs Handbook
Edited by Davis Langdon & Everest

Spon's Mechanical and Electrical Services Price Book 1993
24th edition
Edited by Davis Langdon & Everest

Combustion Engineering and Gas Utilization
3rd edition
British Gas

Noise Control in Industry
3rd edition
Sound Research Laboratories Limited

Ventilation of Buildings
H. B. Awbi

Effective Speaking
Communicating in speech
Christopher Turk

Good Style for Scientific and Technical Writing
John Kirkman

For more information about these and other titles please contact:
The Promotion Department, E & FN Spon, 2–6 Boundary Row, London,
SE1 8HN
Telephone: 071-522-9966

Illustrated Encyclopedia of Building Services

EurIng DAVID KUT
BSc (Eng), CEng, FIMechE,
FCIBSE, MinstE, MConsE, FCIArb
Consulting Engineer for
Building Services

E & FN SPON
An Imprint of Chapman & Hall
London · Glasgow · New York · Tokyo · Melbourne · Madras

Published by E & FN Spon, an imprint of Chapman & Hall, 2–6 Boundary Row, London SE1 8HN

Chapman & Hall, 2–6 Boundary Row, London SE1 8HN, UK

Blackie Academic & Professional, Wester Cleddens Road, Bishopbriggs, Glasgow G64 2NZ, UK

Chapman & Hall, 29 West 35th Street, New York NY10001, USA

Chapman & Hall Japan, Thomson Publishing Japan, Hirakawacho Nemoto Building, 6F, 1-7-11 Hirakawa-cho, Chiyoda-ku, Tokyo 102, Japan

Chapman & Hall Australia, Thomas Nelson Australia, 102 Dodds Street, South Melbourne, Victoria 3205, Australia

Chapman & Hall India, R. Seshadri, 32 Second Main Road, CIT East, Madras 600 035, India

First edition 1993

Typeset in 9.5/12pt Times by Graphicraft Typesetters Ltd., Hong Kong
Printed and bound in Hong Kong

ISBN 0 419 17680 2 0 442 31653 4 (USA)

A catalogue record for this book is available from the British Library

Library of Congress Cataloging-in-Publication data available

This book was commissioned by Maritz Vandenberg on behalf of E & FN Spon

Acknowledgements

The author is pleased to acknowledge the assistance of the undermentioned who have provided illustrations for this encyclopedia:

Contramec Installations Ltd.

Greenwood Airvac Ltd.

G. Lakmaker Esq.

Silentair Ltd.

Spirax–Sarco Ltd.

Terrain Ltd.

Trox Brothers Ltd.

Waterloo–Ozonair

Woods of Colchester Ltd.

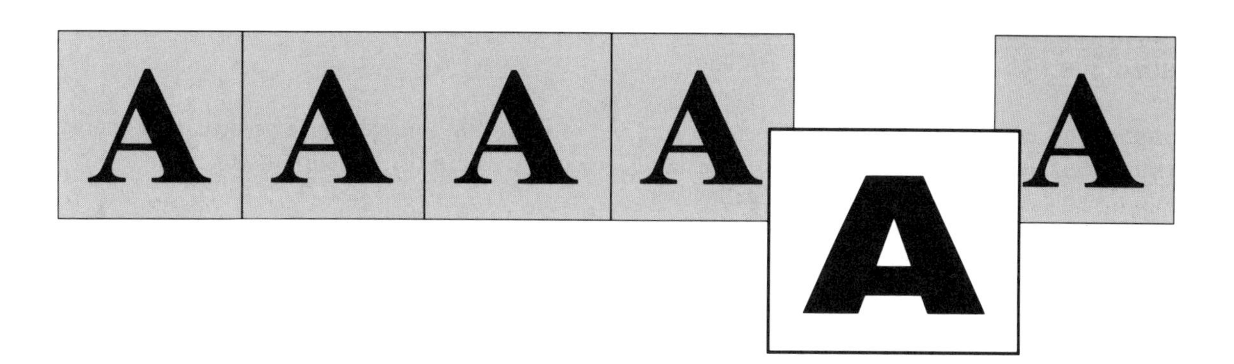

ACE agreements A series of agreements published by the Association of Consulting Engineers, for use between clients and their consulting engineers, which specify duties to be performed for a range of options from the preparation of full designs to performance specification. The fee scales are no longer mandatory.

ADR Alternative dispute resolution. This involves mediation in lieu of action in the law courts with the object of reducing delays and costs.

APC Air pollution control. Whilst the most polluting processes are controlled by HMIP under the IPC scheme, lesser polluting processes are regulated by the air pollution control (APC) system. The enforcing authorities are the Local Authorities in respect of releases into the air. Where substances are released into water, NRA and/or the sewerage undertakers must authorize the process.

ASHRAE The American Society of Heating, Refrigeration and Air-conditioning Engineers, a professional body known widely for the annually updated handbooks issued by them which are the guides to which the building services engineers in the USA conform.

abrasive A substance used for rubbing or grinding down surfaces. Emery is widely used for such purposes.

absolute air filter A filter suitable for an ultra-high degree of air filtration, e.g. in hospitals, laboratories, and the atomic and food industries. It usually has a filtration medium of glass or asbestos (substitute), can filter out particles to 0.01 micron, has a fairly high (fine) dust-holding capacity, and operates at a face velocity of about 1.3 m/s, and at an efficiency in the range of 95–99.9%. It must be fitted with a pre-filter at the inlet side to reduce the burden of coarser dust affecting the filtration efficiency.

absolute humidity See **humidity, absolute**.

absolute pressure Minus 760 mm Hg, i.e. zero pressure or –1 bar (compared to gauge pressure which is 760 mm Hg under standard conditions).

absolute temperature That temperature at which a perfect gas, kept at constant volume, exerts no pressure. It equals zero on the Kelvin scale or –273.15°C.

absorption The intrusion of one substance or force into another substance, e.g. capillary action within the pores of a solid may attract, and hold a liquid by cohesion.

absorption capacity The quantified ability of one substance to absorb another into itself.

absorption chiller Equipment comprising a generator, condenser, absorber and evaporator built into a single or twin-shell assembly, together with refrigerant and solution circulating pumps. Commonly water is the refrigerant and lithium bromide the absorbent. The heat source may be direct gas firing, steam or medium-pressure hot water. The operating costs are low.

Application: for cooling air-conditioning in locations where gas, steam or medium-pressure hot water is available, or where adequate electric power does not exist.

Disadvantages: high equipment cost and space requirements for a chiller and associated cooling tower.

absorption refrigeration [Figs 1, 2] Substitutes a heat source (gas firing, steam or medium-pressure hot water) for the compressor in the conventional chiller. The heat vaporizes the absorbent, usually lithium bromide; it is then taken through a cycle of operation which includes concentrating and condensing the absorbent, evaporation of the liquid and spraying it into an absorber from where the dilute solution is returned to the concentrator. This is an alternative to a compressor-operated refrigeration system. It reduces operating costs where cheaper heat sources are available (possibly surplus steam during the summer air-conditioning season). However, it is, relatively costly and space-consuming with an associated cooling tower plant.

Application: general use; in particular for energy recovery where there is surplus heat available.

acceptance test A test conducted on an installation or item of equipment when it is offered as completed by the manufacturer before being handed over to the customer or user. For example, a lift installation is subjected to such a test by the customer's insurer or consulting engineer before it is certified as having been satisfactorily completed and ready for operational use.

access cover A cover in one part or in sections placed above equipment to which access is or may be required. It is sometimes embodied within floor screed and clearly marked.

access cover, sanitary installation A tightly secured removable cover on pipes and fittings providing access to the interior of the pipework for purposes of inspection, testing and cleansing. Required locations are at junctions, changes of direction and all other places in the installation where obstruction to free flow might occur. All these must be easily accessible to maintenance personnel.

accumulator A vessel which stores a commodity, e.g. steam, at times of surplus capacity and discharges it when demand exceeds the supply.

accumulator (electric) A device for storing electricity in a storage battery. An electric current is

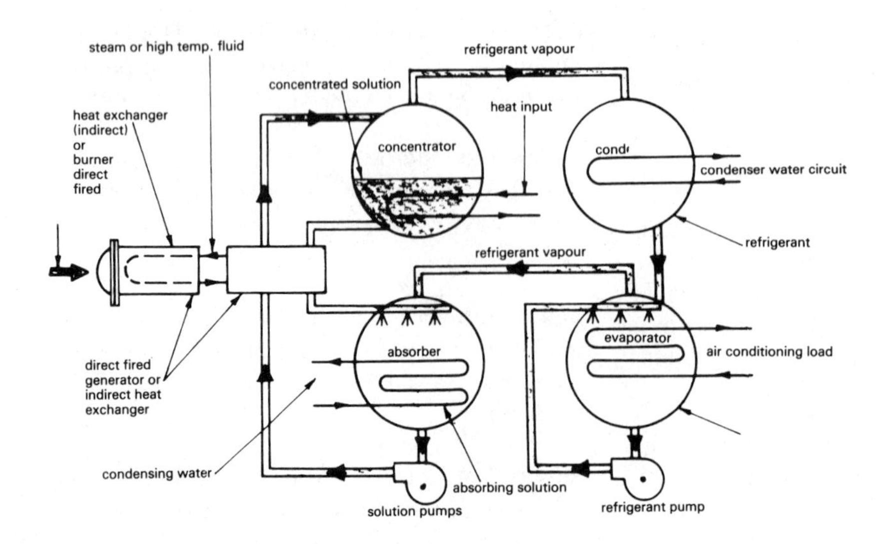

[Fig. 1] Absorption refrigeration: diagrammatic arrangement of process

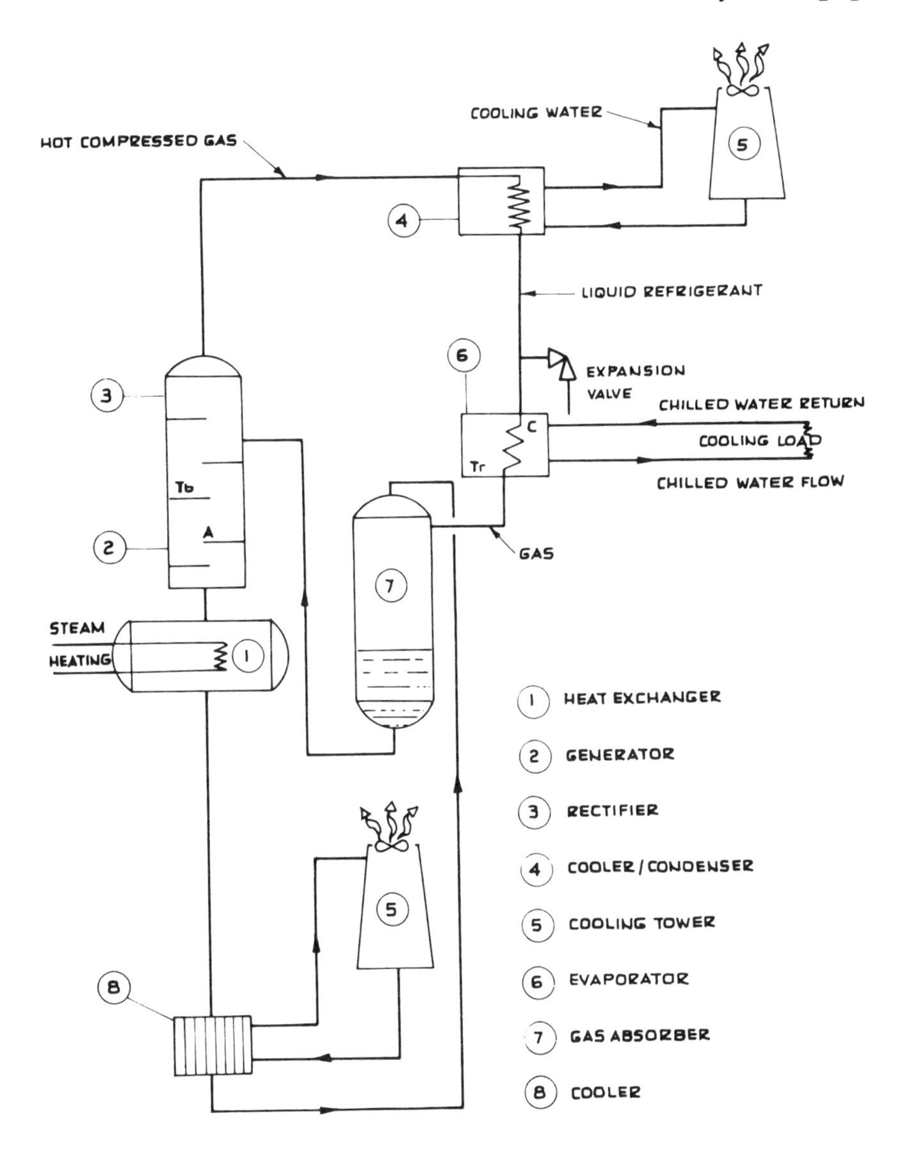

[Fig. 2] Schematic operation of an absorption refrigerator

passed between two plates in a liquid. The plates are commonly of lead and the liquid is sulphuric acid of specific gravity 1.20–1.28. This causes chemical changes due to electrolysis in the plates and liquid. When the changes are complete the accumulator is fully charged. When the charged plates are joined externally by a conductor of electricity, the chemical changes are reversed and a current flows through the external circuit until the reversal is complete and the accumulator has been discharged to a specific gravity of the liquid of 1.15. The process is then repeated.

acetylene C_2H_2 Colourless, poisonous, gaseous fuel produced commonly by the chemical reaction of calcium carbonate and water or from methane, heavy oil and naphtha. It is a major industrial gas used in conjunction with oxygen as oxyacetylene. It burns with the highest temperature of any common combustible gas and is therefore widely employed with an oxyacetylene torch for welding and cutting metals. It is stored within a high-pressure steel vessel or cylinder and is discharged to one port of the oxyacetylene torch via a regulating valve.

acid condensation The deposition of acid out of acid-containing gas or vapour when cooled below the dewpoint of the acid. It relates particularly to chimney flue gases. Acid condensation can cause major damage to a chimney and its associated installations and must be avoided.

acid dewpoint Refers particularly to chimney operation in which the flue gases have an acid content (e.g. sulphur). Cooled below the acid dewpoint, the acid will condense out of the flue gas.

acidic Water with a pH value below 7 is termed acidic. Waters with low pH values are generally corrosive.

acoustic louvre [Fig. 3] An air transfer grille arranged to reduce the transfer of sound from one room to another or to or from the outside of a building or plant room. When located in an external wall, it is fitted with weather-resistant slats.

[Fig. 3] Sound-attenuated louvre (Trox Brothers Ltd.)

acoustics The science and practice of controlling noise, based on an understanding of the physical nature of sound and on the way it behaves under certain conditions.

activated carbon filter [Fig. 4] A filter utilizing granules of activated carbon which are located in position by a glass fabric on either side of a filter panel. The filter assembly is placed into the air stream to be filtered, offering an adequate filtration area to the flow of air. Activated carbon adsorbs vapours or gases; such filters are used widely for the adsorption of body odours, tobacco smells, fumes and cooking smells. Only a relatively brief contact period is required to be effective. Such filters are also suited to an air-conditioning installation where a high proportion of the air must be recirculated, where cooking smells are likely to cause offence, etc.

activated carbon filter, monitoring The condition of such a filter is monitored on a continuous basis by a test panel which is fitted to each section of the filter. This is removed after a specific period of service and undergoes laboratory tests which assess the further life expectancy of the main filter medium in the circumstances of the particular installation.

activated carbon filter, pre-filter When an activated carbon filter handles fairly dirty air, such as would be the case in cities or in industrial areas, it is essential to install a pre-filter upstream to limit the burden on the activated carbon filter. The pre-filter is usually of a coarse type, such as fibre-glass throw-away panel.

activated silica A highly effective coagulant aid prepared from sodium silicate 'activated' by various chemicals.

active fire protection Equipment such as sprinklers, drenchers, etc., which are activated to protect designated areas in the event of a fire.

actuator A mechanism which causes a controlled device to obey the control command; e.g. operate a control damper or motorized valve.

adhesion Sticking to a surface; an effect which is produced by forces between molecules.

adhesive bonding Bonding which may be applied to the jointing of pipework. Metal pipes with

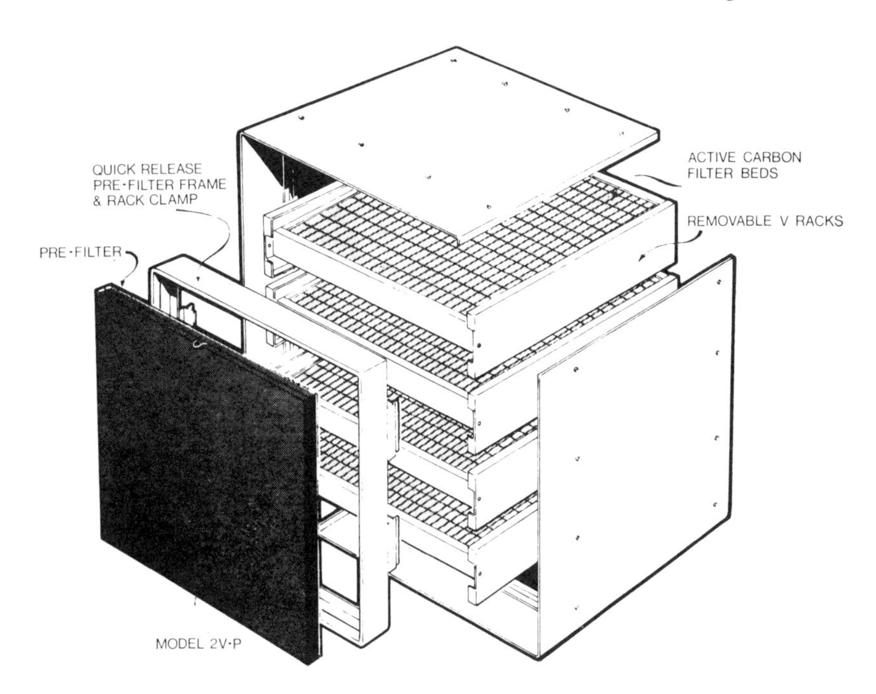

[Fig. 4] Activated carbon filter

capillary joints can be bonded using anaerobic adhesives developed for this purpose. These only cure when air is excluded and they are in contact with a metal surface. They provide an acceptable alternative to the traditional hemp and PTFE tape which were commonly used for joints with screwed fittings. Solvent welding techniques also use suitable adhesive bonding methods.

adhesive A substance used for sticking (binding) surfaces together, e.g. cement, glue or a special formulation for specific applications.

adiabatic process A process conducted without gain or loss of heat; a theoretical concept which cannot be fully realized in practice.

adsorption The process, caused by mass migration of the vapour molecules from the air to the adsorbent because of the existence of a higher vapour pressure in the air than in the pores of the adsorbent substance. Some adsorbents are selective relative to the vapours they will attract to themselves; e.g. silica will inevitably adsorb the water vapour and charcoal odour-bearing vapours.

a/e ratio The ratio of the absorption coefficient for solar radiation to the emission coefficient at operating temperature: it is a measure of the selectivity of absorber surfaces.

aerate Induce air or oxygen into a liquid or operation. For example, an aerated gas burner is one in which the gas induces primary air immediately before the burner ports. Air can be drawn into the water of a heating system through continuous overflow from an open vent pipe into the feed tank.

aerated gas burner A burner in which the gas induces primary air into the burner immediately before the burner ports.

aerosol A particle of solid or liquid matter which is of such small size that it can remain suspended in the atmosphere for a long period of time. Aerosols can be classified into categories of dust, fumes, mists and smoke.

aeration The exposure of small droplets of water to air to encourage the absorption of oxygen from the atmosphere or the release of other gases.

aerobic Relates to the action of micro-organisms in the absence of oxygen.

aerogenerator A machine for converting wind energy into electric power. There are a growing number of applications in Europe and the USA of projected outputs up to 4 MW per unit. The UK Central Electricity Generating Board plans wind farms with individual outputs of up to 1 MW fed into the electricity grid. Modern aerogenerators of large output are mounted atop towers up to 100 m high and may have blades of up to 40 m long.

after-burner A gas or oil burner which is located within the path of the exit gases from a combustion process, e.g. an incineration. It supplies the heat required to destroy smoke and odours.

after-cooler A device associated with air compressor systems. It is fitted between the air compressor and the air receiver and cools the compressed air to ambient temperature after compression; this permits most of the water vapour in the air input to condense and be removed before it enters with the air into the receiver. The usual cooling medium is water. A common design basis is as follows: a 5.5°C temperature rise of the cooling water; the air temperature cooled within 11°C of the water inlet temperature. Oil and moisture must be drained off periodically. The presence of these pollutants can be a major factor in the failure and breakdown of the control equipment of the compressed air system.

after-heater A heater installed within a ducted air system downstream from the main air heater to boost the air temperature or to overcome heat losses from the duct.

agrément certificate A certificate issued by British Board of Agrément for equipment approved by it, following submission for inspection.

air admittance valve A valve which relieves negative pressure in sanitary discharge systems, whilst also preventing foul air from entering the building if positive pressures occur. Care must be exercised in their selection and use as some designs may be prone to being tampered with when installed in easily accessible locations and freezing may cause malfunction.

air-blast cooler A cooler which dissipates heat from refrigeration machines by blowing air over the heat exchange coils. It is an alternative to a cooling tower.

airborne infra-red survey Equipment using an infra-red camera system to study, monitor and accurately locate heat emissions from buildings or equipment. It is a valuable tool in energy surveys for identification of major heat losses and energy wastage.

airborne thermal infra-red heat loss survey Airborne thermal infra-red line-scan equipment for locating energy losses from the external surfaces of buildings, storage vessels and pipelines due to structural or material defects. Surveys are carried out during the winter heating period.

Applications: energy conservation schemes for manufacturing plants, chemical works, hospitals, public buildings, etc.

air brick A perforated brick of various standardized dimensions built into an external wall and used for natural ventilation in, for example, small boiler houses, larders, etc.

air change The replacement of the air content of a space x.

air change rate The number of times the air content of a space is replaced, expressed in air changes per hour.

air classifer [Fig. 5] A device employed in waste separation and recycling to direct the different separated materials into their respective discharge streams. Basically, it comprises a blower, a rotating drum fitted with an internal helix and a large engagement chamber. In use, a current of air is directed along the drum and the sized feed is allowed to fall through the airstream, so that any paper and plastic in the mixture is carried along the drum and into the disengagement chamber. The dense objects fall through the airstream and are conveyed to the opposite end of the drum by the internal helix.

air compressor A machine for compressing air and delivering it into a compressed air system.

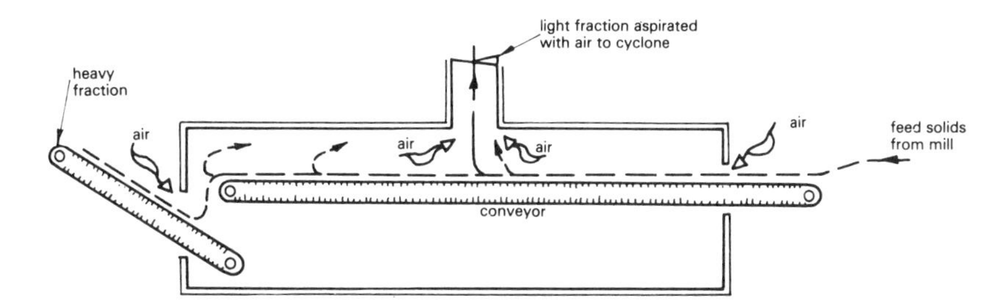

[Fig. 5] Principle of the enadimsa horizontal air classifier

air compressor, oil free A compressor which handles air being compressed without contamination by lubricating oil. See also **oil-free air compressor**.

air-conditioner, split see **split air-conditioner**.

air-conditioning by cold ceiling A system by which the cooling effect is transmitted by radiation from a construction which comprises sinuous pipe coils which are supported off the ceiling and through which chilled water is circulated. Panels or strips of pressed steel/aluminium or extruded aluminium sections are clipped closely to the pipes and provide positive heat transfer.

The cold ceiling system is supplemented by a low-velocity balanced diffused fresh air and extract ventilation system to provide humidity-controlled and filtered air via low-level air diffusers at a rate of about two air changes per hour to the climate-controlled space. It is essential to integrate the temperature and humidity control of the air system to obviate condensation problems which could otherwise occur.

Such systems are particularly popular in the Scandinavian countries and the claimed advantages are: greatly reduced structural space for the air-conditioning elements; design flexibility for varying loads; ease of maintenance; elimination of draughts and noise; effective control of odours; good and even room temperature gradient; enhanced safety factors in the event of a fire; reduction in Sick Building Syndrome complaints.

air-conditioning cassette unit A comprehensive packaged assembly complete with supply air grilles ready-made for mounting above ceiling. It only requires electrical, condenser and drain connections.

air-conditioning, comprehensive Provides control of temperature (heating and cooling), relative humidity, air condition (filtration) and air movement within the treated environment.

air-conditioning compressor protection A mechanism comprising a current/temperature overload cut-out which will restore the electricity supply automatically when conditions have returned to normal. Three-phase compressors usually embody single-phasing protection which will reset automatically when the fault has been rectified and the power has been restored.

air-conditioning, dual duct See **dual duct air-conditioning system**.

air-conditioning, final connection A connection between the distribution duct and the item of equipment, e.g. a fan coil unit, with a flexible spiral tube or fibreglass duct.

air-conditioning, four-pipe system See **four-pipe air-conditioning system**.

air-conditioning, high-velocity system See **high-velocity air-conditioning system**.

air-conditioning, mixing box A box provided for the mixing of air at different conditions; e.g. recirculated air and fresh air supply.

air-conditioning panel cooling A system which incorporates coils or panels laid over a suspended ceiling. The cooling effect is achieved by circulating chilled water through these coils or panels.

air-conditioning, primary air That air which is delivered directly from the central air handling plant to individual terminal units.

air-conditioning, secondary air That air which is induced from the air-conditioned space over the heat exchanger of the individual induction unit by the action of a bank of high-velocity nozzles.

air-conditioning system A system which correctly conditions and distributes air into a space to achieve predetermined conditions in that space of air temperature, humidity, filtration and air flow.

air-conditioning system, eutectic salts See **eutectic salts air-conditioning system.**

air-conditioning system, fan coil A system which utilizes fan-assisted room-located casings, each with heat exchanger(s), circulating fan, coarse filter, condenser tray, and controls. It can be floor or ceiling mounted and there is the option of two-pipe or four-pipe systems. The former heats during winter and cools during summer; the latter has two heat exchangers and heats or cools as required by controls. Two-pipe systems are suited mainly to climates where there is a sharply defined weather difference between summer and winter. Fan coils are served with hot water and with chilled water from a central plant. The pipes must be thermally insulated and chilled water pipes require a vapour barrier.

air conditioning system – thermal storage See **eutectic salt air-conditioning system.**

air-conditioning system, VRV See **VRV air-conditioning system.**

air-conditioning system, wet A term commonly referring to those systems which function with chilled water at low temperatures and which thereby tend to cause condensation.

air-conditioning terminal unit A device which delivers the conditioned air into the air-conditioned space.

air-cooled condenser [Fig. 6] A condenser which utilizes air as the cooling/condensing medium and which incorporates fan(s) to circulate the air.

air cooler Equipment for cooling of air.

[Fig. 6] Air-cooled condensing unit of 30 TR capacity

air curtain [Fig. 7] A cold or warm air curtain with adjustable blast direction control. Single side, top blow or double side blow units are available. Wind velocities of up to 15 m.p.h., depending on conditions, can be resisted. Warm air curtains are available from which warm air from a door heater is discharged directly into the work area as a prime source of heating. A door-actuated switch directs the air blast across the door only when the door is open.

Applications: to prevent warm air escape from heated space.

air diffuser [Figs 8, 9] A terminal air supply fitting which comprises a number of metal cones (circular, square or rectangular) with the air flowing between them; each configuration offers a particular mode of air distribution/pattern.

air diffuser, circular pattern See **circular pattern air diffuser.**

air diffuser, jet nozzle See **jet nozzle air diffuser.**

air diffuser, swirl pattern See **swirl pattern air diffuser.**

air eliminator A device fitted at the highest point of a pipe loop or system; usually in form of a air bottle to which is attached a manual air release cock or automatic air vent.

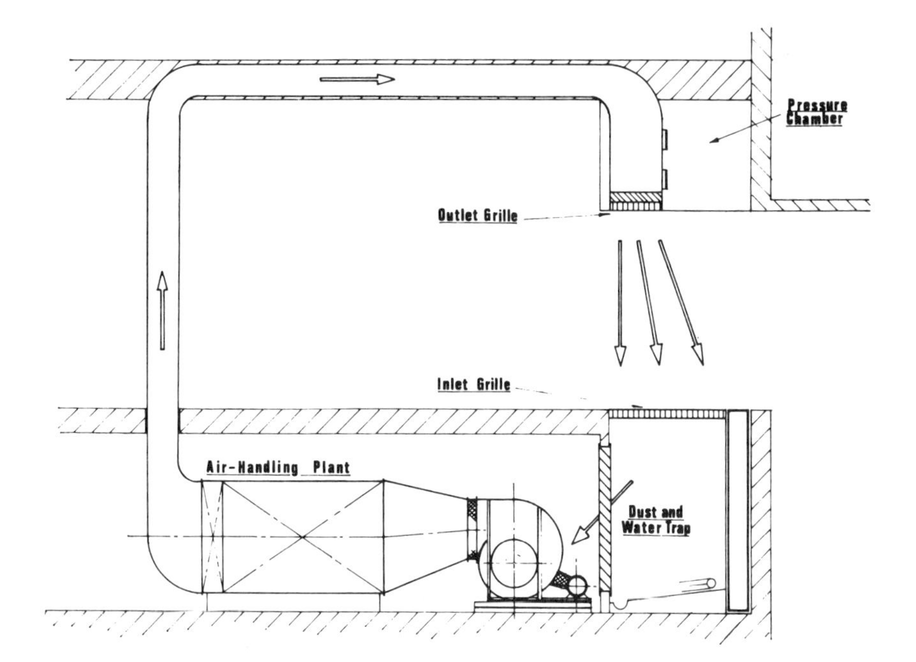

[Fig. 7] Arrangement of a typical warm air curtain plant

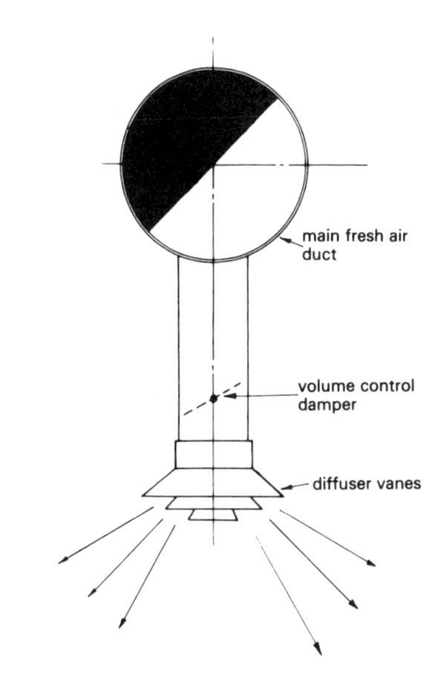

[Fig. 8] Circular air supply diffuser

air entrainment Air drawn into a jet of air.

air filter, absolute See **absolute air filter**.

air-filter, bag-type See **bag-type air filter**.

air filter, dry metal wool See **dry metal wool air filter**.

air filter efficiency blackness test See **blackness test**.

air filter, electrostatic See **electrostatic air filter**.

air filter gravimetric efficiency test The dust-holding capacity of an air filter. It relies on the accurate weighing of the filter being tested before and after collecting the test dust. It is unsuitable for filters which cannot be accurately weighed.

air filter, grease filter See **grease air filter**.

air filter methylene blue efficiency test See **methylene blue air filter efficiency test**.

air filter, panel type [Fig. 10] Replaceable or washable, usually square, air filter element capable of only modest filtration efficiency. Used as a pre-filter and as a main filter in situations where low to moderate cleanliness is required.

air filter, roll-type See **roll-type air filter**.

air filter, throw-away-type See **throw-away-type air filter**.

air filter, vee-mat-type See **vee-mat-type air filter**.

air filter, viscous See **viscous air filter**.

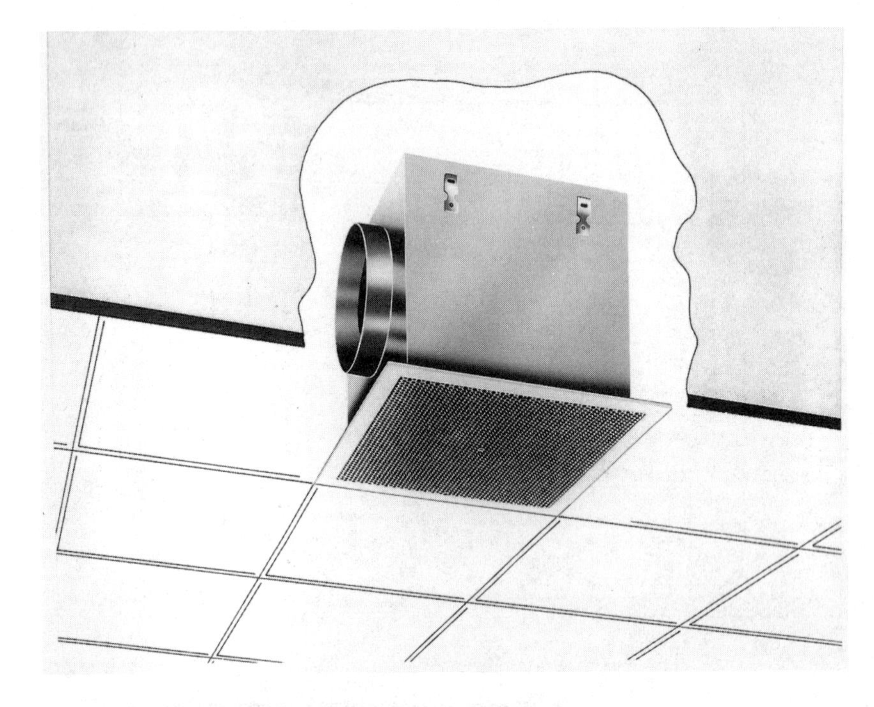

[Fig. 9] Air distribution: diffuser for installation in perforated ceilings (Trox Brothers Ltd.)

[Fig. 10] Air filter: dry panel disposable module (Waterloo–Ozonair Ltd.)

air filter, viscous-coated See **viscous-coated air filter**.

air filtration [Figs 11, 12] Any method of removing impurities from the air.

air flow in ducts The rate of air flow in ducted systems which depends on the following parameters:

(a) fan outlet pressure;
(b) resistance offered to air flow by ducts and fittings;
(c) density of air being conveyed;
(d) internal dimensions of ducts and fittings;
(e) shape and air flow characteristics of system components;
(f) roughness of duct walls;
(g) velocity of air.

airflow pattern The manner in which the air supplied into a space flows through that space. The designer plans for a specific pattern of air flow to conform to the design parameters of the air distribution system. The airflow pattern can be observed by the use of smoke bombs or capsules.

air governor The automatic control of an air compressor at predetermined pressure.

air governor, automatic Used in conjunction with air compressors. It provides for the continuous

[Fig. 11] Air distribution: terminal incorporating filter (Trox Brothers Ltd.)

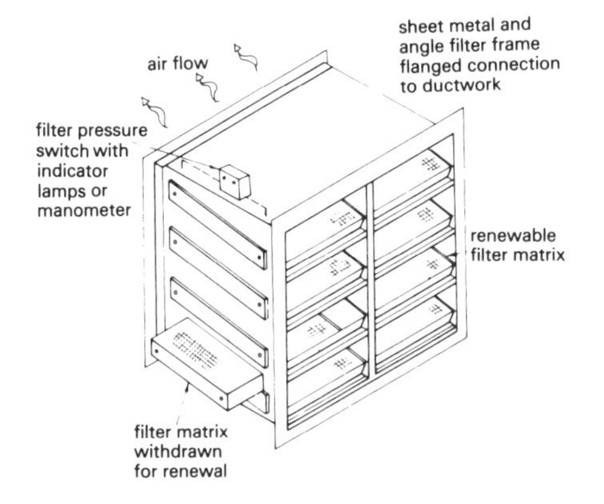

[Fig. 12] Air filter assembly showing withdrawal facility

running of the compressor but allows the plant to run under conditions of light load when a predetermined pressure has been reached in the air receiver. The driving unit (e.g. the electric motor) will then be running on light load with a corresponding reduction in power consumption. When the demand for air increases, the pressure in the air receiver drops and the automatic air governor allows the compressor to resume operation. A hand-lifting device is often incorporated and enables any air under pressure from the air receiver to lift the governor valve and so pass into the unloading cylinder of the compressor. This ensures that the air compressor can be started up against frictional load only, even if the air receiver and pipelines are full of air under pressure.

air handler Short for 'air handling unit'.

air handling luminaire [Figs 13, 14, 15] A luminaire which incorporates air extraction and a light fitting within one compact assembly, withdrawing at source the heat generated by the light fitting. A cooler environment improves the life of the light source.

air handling unit [Fig. 16] The assembly of air treatment equipment within one casing. It may include a pre-filter, main filter, air heater battery, humidifier, fan plant, cooler battery, reheater and associated controls.

air heater A heat exchanger which utilizes steam, hot water, electricity or gas to warm a stream of air flowing across it. It is commonly associated with a fan plant which propels the air at the required velocity.

air infiltration The random inflow of air into a building via cracks around windows and porosity or gaps in the building fabric. It varies with external weather and wind conditions and is a factor in heat loss calculations for a building.

[Fig. 13] Combined modular light and air supply outlet

air lift A method of raising water from a low to a higher level by the injection of compressed air.

air movement patterns The movement of air discharged from air supply terminals as established by design and adjusted following smoke tests. The correct pattern of air movement is essential to the success of any air distribution scheme.

air pre-heater See **preheater, air**.

air preheater, combustion A mechanism which transfers heat from the flue gases of a boiler or furnace plant to the air which is supplied for combustion of the fuel. It is usually located between the boiler economizer and the entrance to the chimney. Recovery of heat from the flue gases recycles waste energy and thereby raises the overall efficiency of the combustion process, reduces the temperature of the flue gases entering the chimney and raises the flame temperature in the combustion chamber, increasing the rate of heat transfer and reducing the requirement for excess air.

[Fig. 14] Air distribution: ceiling diffuser with integral centre light fitting (Trox Brothers Ltd.)

[Fig. 15] Air distribution: ceiling diffuser with light fitting (air handling luminaire) (Trox Brothers Ltd.)

[Fig. 16] Air handling unit: access door to fan section removed for servicing

air purity Specifies the dust or contamination content permitted in the air supply to a space.

air receiver A pressure vessel for the storage of compressed air pending its consumption by the distribution system.

air receiver A mechanism associated with air compressor equipment. It dampens out the pulsations of the air delivered by the compressor, stores the air to provide a supply 'cushion', cools the air and removes the oil and moisture which is contained in the input air (usually in association with ancillary equipment). Each air receiver must be permanently marked with: the maker's identification number, date of test, specification number, hydraulic test pressure, and maximum permitted working pressure. Air receivers must be installed, maintained and periodically inspected under the provisions of the Factory Act. BS 429 and 487 relate to air receivers.

air register Controls or regulates the amount of air permitted to enter a combustion appliance, e.g. a boiler or incinerator.

air register, combustion A component of combustion apparatus which permits the adjustment of the amount of air being fed to the combustion process. It is installed to control/regulate primary and secondary air ports.

air relief valve A valve which manually or automatically exhausts or releases air from a fluid system.

air splitter See **duct splitter**.

air stratification See **stratification**.

air supply diffuser [Fig. 17] An air supply terminal best suited to ceiling mounting. It may be circular, square, rectangular or of the slot type and permits good control over air flow pattern. A variety of dampering methods are available, e.g. by a screwdriver inserted from the front, quadrant control at the rear, or local adjustment at each slot diffuser. Air diffusers are engineered to suit particular circumstances and require careful selection.
Application: for all types of air-conditioning and ventilation systems. Very large diffusers are available for use in tall and large spaces.

air supply nozzle [Fig. 18] A nozzle-shaped air supply fitting suitable for applications which require a long air throw from the discharge. They tend to be noisy.
Application: air supply into large and tall areas, such as exhibition halls or factories.

air terminal device A fitting at the termination of an air duct into a ventilated space. There is a wide choice available to suit the design intent of the air distribution.

air test of sanitary installations The most common method of testing the soundness of a newly

[Fig. 17] Air distribution: one- or two-way discharge linear or finite diffuser particularly suitable for installation in false ceilings (Trox Brothers Ltd.)

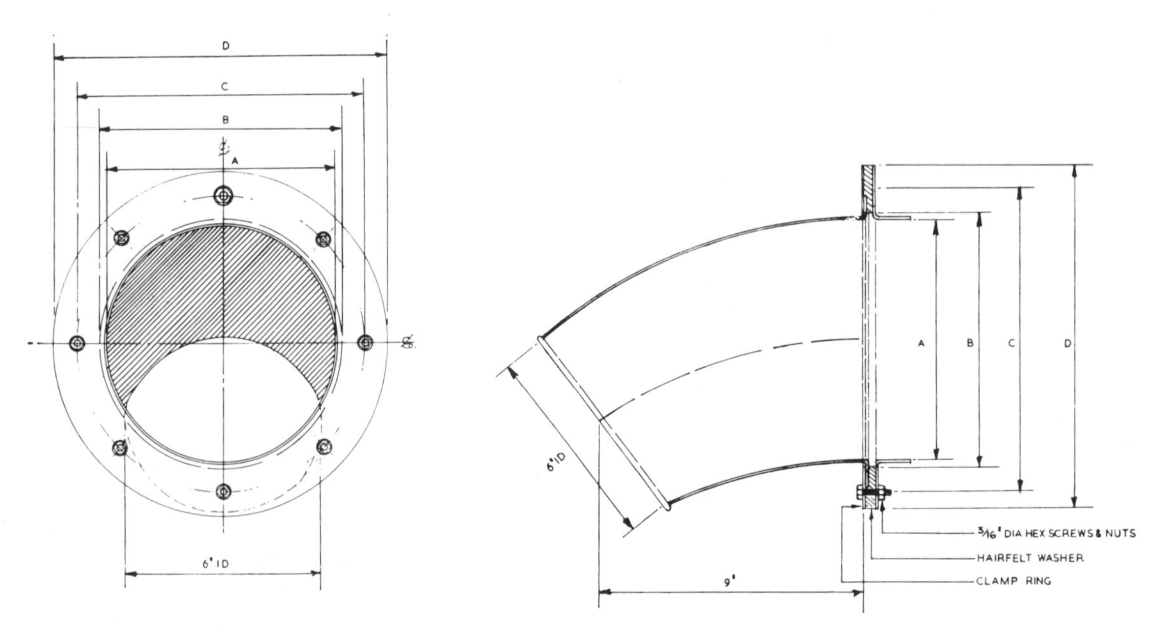

[Fig. 18] Six-inch diameter air supply nozzle. A, 8" i.d. nozzle and connections. B, $8\frac{1}{4}''$ i.d. clamp ring and hairfelt washer. C, $9\frac{1}{4}''$ i.d. flange to nozzle. D, $10\frac{1}{2}''$ o.d. clamp ring hairfelt washer and flange to connection

installed sanitary installation. To apply it, all open ends are sealed using test plugs or bags and one test plug takes the form of a tee piece. A manometer is fitted into one branch and air is pumped into the pipes through the other branch of the tee piece until a pressure of 38 mm is indicated by the manometer. The air inlet is then closed and the manometer reading observed over a period. The pressure in an airtight system will remain constant over a period of not less than three minutes.

air throw The projected distance of an air jet issuing from an air terminal device into the ventilated space. The actual throw can be demonstrated by means of a smoke bomb or cartridge.

air treatment The heating, cooling, filtration, humidification, dehumidification, ionization or sterilization of a quantity of air to suit specification or requirements.

air valve A single or double valve for water mains with a small orifice for automatically releasing air accumulating at the working pressure and with a large orifice for discharging the air during pipe filling and admitting air during pipe emptying. The design of the valve ensures that the controlling ball will not lift until venting is complete. A kinetic adaptor may also be fitted in applications where very high air velocities are encountered during filling and emptying of the pipeline.

air vent [Fig. 19] A manual or automatic device for removing accumulated air from a fluid system.

air vents for steam systems Vents which open automatically to discharge air when receiving air/steam mixture and close against live steam.

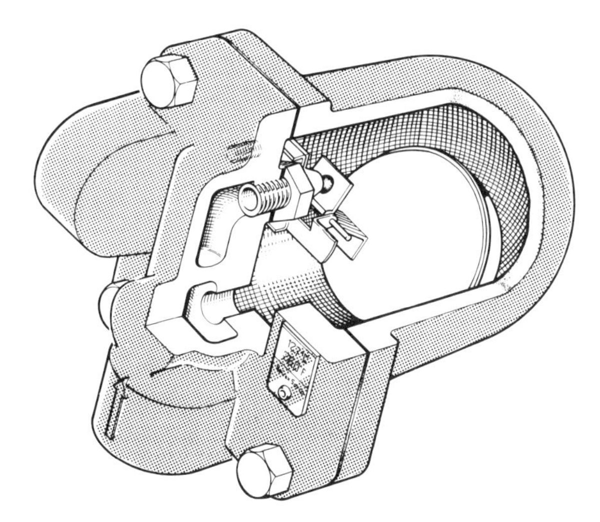

[Fig. 19] Automatic air eliminator for water systems (Spirax–Sarco)

air vessel A closed chamber used in connection with cold water services, which utilizes the compressibility of the contained air to either promote a more uniform flow of water when connected to the delivery or suction pipe of a reciprocating pump or to the delivery pipe of a hydraulic ram. It is also installed for the purpose of minimizing shock caused by water hammer in a pipe when connected to a high-pressure system.

air washer [Fig. 20] A device which sprays water into the air stream of an air-conditioning or ventilation system to humidify, dehumidify or cool. It may rely for its effect on evaporative cooling or it may operate with chilled water; the equipment incorporates pump(s), a pre-filter and controls.

air washer, capillary See **capillary air washer**.

air washer, recirculating An air washer which relies on the recirculation of the spray water between the washer tank and the spray nozzles. The use of such air washers is regarded as a possible way of generating legionnaire's disease because of the water recirculating feature.

air–water storage vessel A vessel which stores water under the pressure of air which is compressed in its upper portion, forming part of a pumped water supply system. Variations in the quantity of the

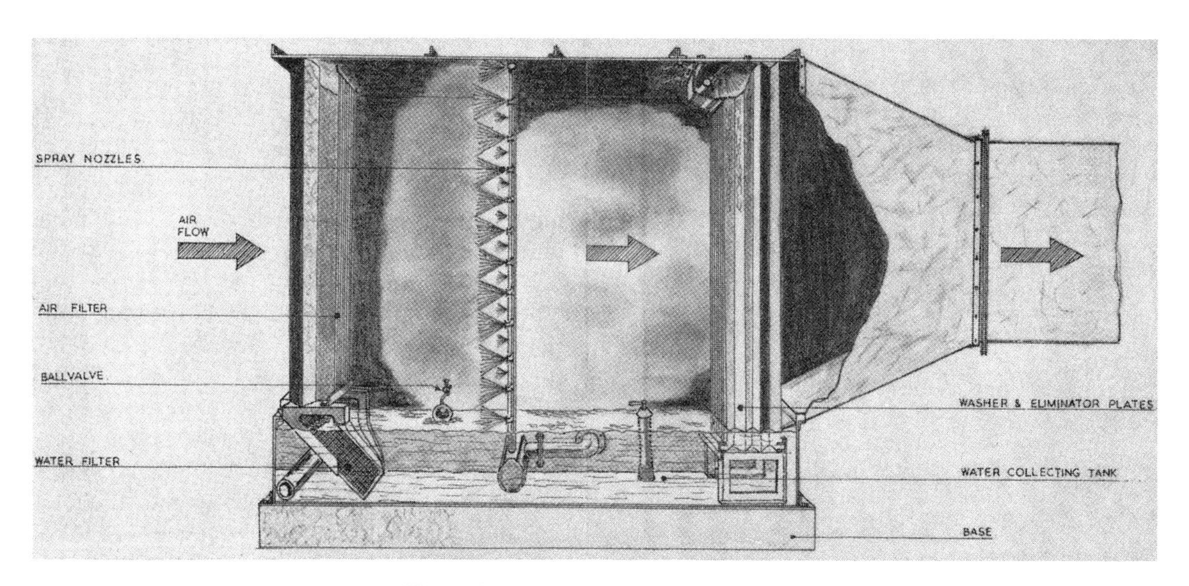

[Fig. 20] Typical spray-type air washer

stored water lead to changes in the pressure of the water and air; these can be utilized to activate and control the starting and stopping of related pumps.

algae Unicellular or filamentous plants, usually fast growing in fresh or sea water and in recirculating water systems. Algae can cause fouling of equipment and must be controlled by methods such as the use of bactericides or ultraviolet sterilization.

algaecide A chemical solution formulated to inhibit and or remove algae in cooling ponds, water tanks, etc. It is commonly applied via manually or automatically controlled dosing apparatus.

algebra That branch of mathematics which concerns itself with the properties and relationships between quantities by means of generalized symbols.

algorithm The system of mathematical procedure(s) which enables a problem to be solved in a finite number of steps. Problems for which no algorithms exist require heuristic solutions. See also **heuristic**.

alkaline A substance with a pH value above 7.

alkathene A proprietary make of low-density polyethylene pipe.

all-air-conditioning system A system which utilizes conditioned air, with associated air exhaust, to handle all the air-conditioning functions.

alloy The composition of two or more metals. It may be a compound, a solid solution, a heterogeneous mixture or any combination of the metals.

alternative energy A non-fossil fuel energy source such as solar, wind, wave, hydraulic, geothermal or refuse resources.

altitude The angle which the rays of the Sun make with the horizontal plane at a given point.

alum Aluminium sulphate, a widely used coagulant.

aluminium cladding External chimney insulation with a skin of aluminium sheet over an air space or over the thermal insulation of a vertical steel chimney. The method of applying aluminium cladding is laid down in British Standards Specification BS 4076: 1978.

aluminium foil, self-adhesive Aluminium foil 0.002 inches thick with a self-adhesive coating and protected by a peel-away backing paper for easy application. It is supplied in easily handled rolls.

Applications: wide width material is used primarily for backing radiators. Narrow width material is used for sealing joints in foil-faced duct and pipeline insulation.

Energy-saving potential: applied to the wall behind radiators (reflective side out), it will show a substantial decrease in heat loss through a wall. As an insulation joint sealer it makes insulation efficient.

ambient Surrounding (of temperature).

ambient air The state of the general atmosphere at a particular location and time.

ambient noise Existing background noise in an area which can be compounded from many sound sources, near or far.

amortize To apportion the cost of capital plant over a given/stated number of years.

ampere A unit of electric current.

analogue data The presentation of a variable quantity by signals displayed on a computer which may be a portable device.

anchor, bottom An anchor located at the bottom of a riser pipe.

anchor, directional An anchor which restricts the movement of the pipeline in one direction only.

anchor, force on When an installation is cold and not subject to pressure, the elasticity force is inwards due to the cold draw on the bellows. When an installation is cold and subject to pressure test, then the elasticity force is inwards and the pressure force outwards; the result is a net pressure force acting on the anchor. When a pipe is hot and not subject to pressure, then the elasticity force on the anchor is outwards and the bellows joint is then in compression.

anchor, intermediate An anchor which divides a pipeline into a number of separate sections, each being self-contained as regards thermal movement compensation. It is subject to lighter loads than main anchors.

anchor, main The anchor located at each end of a straight run of pipe to contain the pipe movement and permits same only towards the thermal movement compensation device.

anchor point [Fig. 21] The location of a pipe anchor. It must have adequate strength to withstand the stresses induced by the thermal movement of the pipe and it commonly comprises a steel girder or reinforced concrete block to which the anchor is fastened with nuts and bolts.

anemometer A hand-held vane (windmill) operated instrument for measuring the velocity of air movement.

aneroid chamber A sealed metal chamber with flexible sides which expand or contract with changes in the pressure difference between the inside and outside of the chamber, together with a mechanism which magnifies the movements of the chamber for the purpose of initiating control action related to pressure changes or to record such changes.

angle of incidence The angle at which solar radiation strikes a horizontal plane or solar collector.

angle of repose The natural running angle or natural angle of slope of a material (e.g. coal) at which a bulk store of the material will settle. The natural angle of repose of major coal types are as follows: dry broken coal – 33 degrees; dry slack coal – 30 degrees; dry coke – 35 degrees; coke breeze with 5% moisture – 40 degrees. The angle of repose is important in the design of coal storage hoppers and bunkers; a minimum angle of 50 degrees is recommended to establish a smooth flow of the fuel to the handling equipment.

angle valve A valve which generally has a spherical body in which the ends are at right angles to

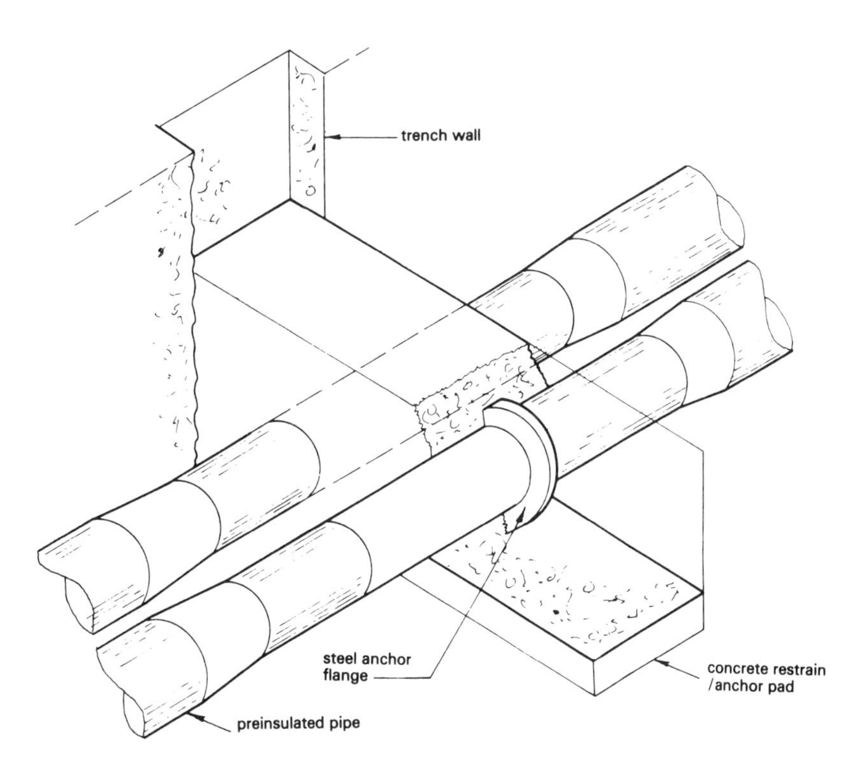

[Fig. 21] Pre-insulated pipe system anchor point

each other and in which the valve stem is in line with that of one body end.

annular flow Fluid flow within the space formed by an inner and outer cylindrical configuration.

annular heater A closed annular fitting located inside a hot water calorifier and connected to the primary heating circuit. It heats water inside the calorifier shell.

anode The positive electrode of an electrolytic cell at which oxidation occurs. In a corrosion process, the electrode has the greater tendency to go into solution; this property is utilized in the design of corrosion prevention systems in which the anode is allowed to waste away, to counteract the corrosion effect.

anthracite Hard, dense coal which contains only 5–9% of volatile matter of high calorific value of order 8500 kcal/kg. It is hard to ignite and burns with a short intense flame in the virtual absence of smoke.

anti-flood valve A valve inserted as protection into a drainage system which is subject to surcharge backflooding.

anti-freeze That which prevents freezing, usually in water, by means of various additives or by heating trace elements to keep the fluid temperature above freezing-point (0°C or 32°F).

anti-reflection coating (AR) A coating applied to a surface (bloomed) to increase the amount of light penetration. It may be applied to the surface of solar cells or to glass/plastic covers of solar collectors.

anti-siphon A vent pipe inserted into a circuit prone to siphoning to prevent formation of an active siphon.

anti-splash shoe A discharge fitting fixed at the lower end of a rainwater pipe and shaped to reduce splashing when rainwater is discharged into the open air.

anti-static flooring A highly durable seamless epoxy resin system which helps to control industrial hazards caused by electrostatic discharges by reducing static build-up. The static charge decay curves for this product show rapid charge dissipation on both new floors and floors already in industrial use.

anti-vacuum valve An automatic type of air valve installed for the prevention of the formation of a vacuum or for the elimination of vacuous conditions in a large-bore pipeline.

anti-vacuum valve at terminal A device through which water does not flow. It is fitted on the vertical branch pipe at the end of heating or plumbing system. It has air entry ports which close mechanically when the system is under positive pressure; in the presence of negative pressure in the connecting pipework it opens the air ports to limit the vacuum pressure, thereby preventing back siphonage.

anti-vacuum valve in line A device installed in a pipe through which water normally flows. It is equipped with one or more air inlet ports which are closed by a flexible seal under normal flow and pressure conditions. When below atmospheric pressure occurs at the device, the air ports are unsealed to admit air and thereby prevent backflow.

anti-vibration mounting [Fig. 22] A device placed below vibrating machinery to isolate the vibration and prevent its transmission via connected pipes or adjacent structures.

apparent electric power That power in an alternating current circuit which is obtained by multiplying the circuit voltage by the current flowing to the load or by adding together the active and reactive components of the current. Translation of apparent power into usable energy requires multiplication by the power factor of the circuit.

appliance rating A manufacturer's specified output for a particular appliance, e.g. for a hot water boiler, usually in kilowatts per hour.

arbitration The process whereby the parties to a contract in which a dispute arises elect to appoint a tribunal of their own choice to determine the outcome of that dispute. Most forms of contract

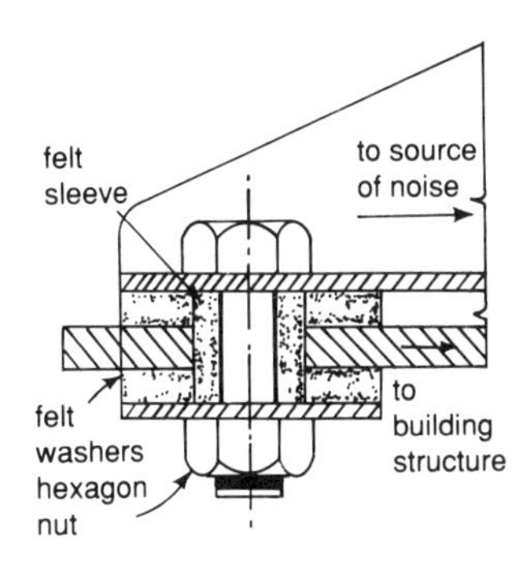

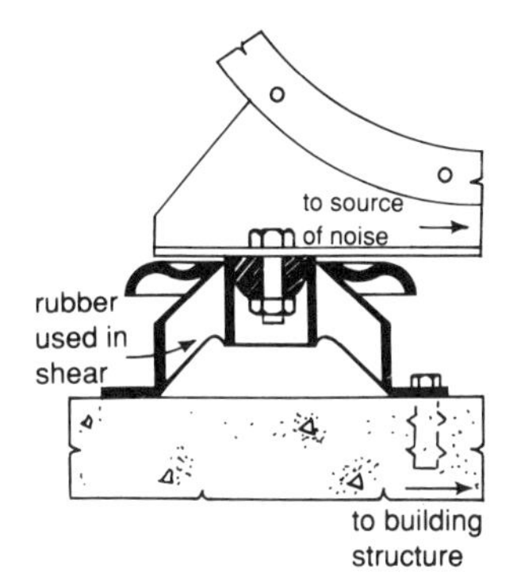

[Fig. 22] Anti-vibration mountings

related to building works incorporate an arbitration clause.

Arbitration Act The conduct of arbitration is based on the Arbitration Act of 1979 which supersedes previous Arbitration Acts.

arbitration award The arbitrator's findings issued in writing and dated. Unless the parties agree otherwise, it states the reasons upon which the award is based.

arbitration, claimant The party initiating the arbitration.

arbitration, clause The terms of a contract which specify that disputes between the parties to the contract are to be settled by the process of arbitration.

arbitration, costs The amount of money specified by the arbitrator in his award covering the total amount of his fees and costs. Unless the parties agree otherwise, the arbitrator determines the proportions in which the parties shall pay such fees and expenses. The arbitrator has the power to order in his award that all or part of the legal or other costs of one party be paid by the other party and to tax these costs, if so requested by the parties.

arbitration, direction An instruction by the arbitrator to the parties to pursue a specified course of action, e.g. submit a list of documents or attend a hearing.

arbitration, hearing The occasion when all parties to the arbitration present their case to the arbitrator at one selected venue.

arbitration, preliminary hearing The meeting of parties to the arbitration to prepare preliminary moves, and the scope and details of the arbitration hearing.

arbitration, real evidence Visual evidence, e.g. defective boilers or malfunctioning air-conditioning, available for inspection by the arbitrator.

arbitration, respondent The party that has to respond to the arbitration claim.

arbitration rules The rules issued in pamphlet form by the Chartered Institute of Arbitrators, 1988 edition.

arbitration, Scott schedule The summarized costed statement of claims in the arbitration, including the claims and the response to them.

arbitration, 'unless' provision The provision activated by the arbitrator in the event of one party failing to implement his directions. He will then advise that unless his directions are carried out as specified he will proceed in the arbitration without that party.

arbitrator A person (or persons) appointed with the consent of all parties to the dispute to conduct the arbitration.

arc light An electric lamp which utilizes an electric arc between carbon electrodes as a source of light.

arc welding The welding of metal by an electric process which establishes an arc between the welding electrode and the material being welded, causing the metal surface to be heated to melting-point and material of the electrode to be deposited to form a joint.

architect A designer of buildings. He commonly heads the design team.

arcing The short-circuiting of electricity across switch contacts. This may lead to the melting of contact faces and is indicated by burn marks.

armature That part of an electric machine in which, in the case of a generator, the electromotive force is produced; in the case of an electric motor, the torque. It includes the winding through which the current passes and the portion of the magnetic circuit upon which the winding is placed.

armouring A metal covering, usually in form of tape or wire, which is applied to a cable to protect it against mechanical damage.

array See **solar array**.

asbestos A group of some 30 or more minerals of fibrous crystalline structure, only six of which are of economic significance. In the order of importance these are chrysotile, crocidolite, amosite, anthophylite, tremolite and actinolite. Chrysotile is a fibrous form of serpentine; the others comprise the amphibole group. The common name 'asbestos' does not distinguish between its natural types, the two main groups being serpentine and amphibole asbestos. The chemical basis of all types of asbestos is magnesium silicate combined with lime or alkalines in varying proportions. All grades are fire resistant and have a varying degree of acid resistance coupled with good mechanical strength.

asbestos, blue The crocidolite group of asbestos, mainly mined in Bolivia, South Africa and Western Australia. It can be identified by its rich lavender blue colour. It is held that blue asbestos presents a greater health hazard to workers engaged in its manufacture and to users than the other grades. Blue asbestos has been very widely used in the past for thermal insulation (largely within asbestos magnesia plastic site-applied insulation). Because of the risk to health, this asbestos is now only rarely used in the UK, but many installations still exist which incorporate this material.

asbestos, chrysotile The most abundant form of asbestos belonging to the serpentine group of rock-forming minerals and the most extensively used asbestos for industrial applications. Mainly mined in Canada, Russia and Zimbabwe. Grades with relatively long fibres are spun into a yarn, which is used in weaving asbestos cloth for protective clothing, thermal insulation, resin-impregnated pads, and wire braiding and ropes.

asbestos, crocidolite See **asbestos, blue**.

asbestos, health hazard Asbestos is a dusty substance. It has been proved that this dust can be highly injurious to persons who inhale it in sufficiently large quantities. Asbestos can cause a specific fibrosis of the lungs, now termed asbestosis, largely associated with workers who have been exposed to heavy concentrations of asbestos dust over a long period of time. A relationship also appears to have been established between exposure to certain types of asbestos dust and the occurrence of mesothelioma, a form of cancer, after only relatively short periods of exposure. Regulations have therefore been introduced in the UK aimed at greatly minimizing this health hazard.

asbestosis A type of lung cancer caused through the inhalation of asbestos dust over a prolonged period. See also **asbestos, health hazard**.

asbestos, regulations Regulations applying to every process which involves asbestos and to any article which is composed wholly or partly of asbestos, excepting only processes in which asbestos dust cannot arise. The regulations dated 1969 supersede the earlier 1931 regulations. For the purpose of the regulations, asbestos is defined as any of the minerals crocidolite, amosite, chrysotile or fibrous anthophyllite, as well as any mixture containing any of these materials.

The purpose of the regulations is to lay down strict rules for the manufacture, handling and removal of asbestos materials in order to minimize any health hazard. The regulations have resulted in the emergence of specialist firms, licensed to carry out asbestos removal under the stringently controlled conditions laid down in the asbestos regulations. On completion of the removal, a certificate of analysis is issued, following the taking and analysis of air samples in the areas from which the asbestos has been removed. The certificate specifies the names of the independent testing authority (or laboratory) and the sampler (usually by a code reference), the sample location, the date, the sample number and the associated fibre count (fibres per cubic centimetre) which must be well within the permissible limit.

asbestos use Guidance for the use of asbestos is given in the 'Approved Code of Practice and Guidance Note – work with asbestos insulation and asbestos coating' by the Health and Safety Commission, effective as from 1 October 1981. This gives practical guidance with respect to the Asbestos Regulations 1969 (S.I. 1969 No. 690) and the Health and Safety at Work Act. It relates to precautions to be observed in any work concerned with asbestos-based thermal and acoustic insulation in compliance with current regulations.

as-built drawings See **record drawings**.

ash The residue of the combustion of solid fuel.

ash, bulk storage Commonly, an overhead silo raised sufficiently to permit the ash collection vehicle to park directly beneath it. A skip hoist raises the bin ash containers along sloping guides and discharges them into the top of the silo. An effective seal must be fitted at the hopper entry to prevent the ash being blown about during tipping.

ash content The ash content of a fuel determines partly the calorific value and the required rate of ash removal. The fusion temperature of the ash is important in combustion practice; also the knowledge of whether ash solidifies and forms clinker or whether it falls freely.

ash dump A covered enclosure adjacent to a coal burning boiler house of output up to 9000 kg/hr for the storage of ashes awaiting disposal. It is provided with water sprays to quench hot ashes and to lay dust.

ash fusion The melting of ash to form clinker.

ash fusion temperature That temperature at which ash melts and begins to flow. It is a critical parameter in the selection of grade of coal and of a combustion appliance grate; ash with a low melting temperature is likely to clog the spaces between the grate bars and links.

ash handling Large coal-fired installations employ pneumatic ash and clinker handling systems. Ash and clinker must be cooled before being transported, and it is wetted in an ash conditioner before discharge into the ash collecting vehicle. Other systems use conveyors submerged in water and may comprise belt conveyors, vibratory conveyors and drag-line conveyors.

ash pit That part of a solid fuel combustion appliance into which the ash drops and accumulates pending removal.

as-installed drawings See **record drawings**.

aspect ratio The ratio of internal length to width of a duct.

aspirated hygrometer A dry bulb and wet bulb mercurcy-in-glass thermometer together with a small fan which cools the wet bulb.

asset management A high standard of organized plant maintenance aimed at preserving the assets invested in it.

assisted circulation Air movement which is generated or assisted by a fan, or water which is circulated or boosted by a pump.

associated builders' work That building work which is associated with the engineering services, e.g. plant bases or holes for pipes and ducts.

associated electrical works Those electrical installations which directly connect items of equipment commonly included within the mechanical services contract and which encompass electrical connections between the equipment and the adjacent isolator, switchfuse or distribution board.

atmosphere, standard Atmosphere at 'normal' pressure. One standard atmosphere at 0°C equals 760 mm Hg or 1 bar. The zero pressure mark on a steam pressure gauge indicates pressure of 1 bar.

atmospheric gas boiler A boiler with a chimney-induced operation (without a fan), with an integral draught diverter, copper heat exchanger and burner assembly designed for easy cleaning and maintenance. A typical modular design boiler output in 60, 90, 120 and 150 kW sizes.

atomization The separation into fine particles of a liquid fuel to facilitate intimate mixing with the air supplied for combustion.

atomize To break a liquid fuel into fine particles for the purpose of ensuring intimate air-to-fuel contact during combustion. Also to break a liquid, e.g. water, into fine particles to generate a mist such as is used to humidify air.

atomizing humidifier An air washer in which the circulating water is atomized and sprayed into the passing air stream.

atmospheric cooling tower A cooling tower which functions with the natural circulation of air, i.e. does not incorporate a fan.

atria, active fire protection The use of equipment designed to prevent the spread of fire within a multi-storey atrium. One favoured method relies upon long-throw sprinklers which are located at ground level and spray water horizontally. In use such sprinklers would cover a predescribed atrium floor area at a given application density in accordance with the rules of the Loss Prevention Council for automatic sprinkler installations. An infra-red detection system locates the source of the fire and, when the fire reaches a predetermined size, causes the nearest long-throw sprinklers to operate and discharge water on to the fire. Infra-red fire detection may also be used during the early stages of a fire to activate alarms or to operate other fire safety equipment.

atrium The large central court of a building which provides light and ventilation to adjacent areas. It requires special attention in smoke and fire control.

attenuation Transmission loss or reduction in magnitude of a signal between two points in a transmission system. Also the reduction in sound level by the use of sound attenuation devices.

autoclave A pressure vessel used for the sterilization of foods and pharmaceutical articles. Larger sizes are fed with steam from a boiler plant; smaller units have integral gas or electric steam generation. Process autoclaves usually operate on a preset programme of pressure and vacuum operation. They are major consumers of steam. For optimum fuel efficiency, correct steam trapping and pressure controls must be applied and maintained in good order. An autoclave may be front-operated or be arranged for front loading and rear discharge to best suit its use.

automatic control, floating control A control in which the final control element is moved gradually at a constant rate towards either the open or closed position, depending on whether the controlled variable is above or below the neutral zone. The valve (or other controlled element) can assume any position between its two extremes so long as the controlled variable remains within the values which correspond to the neutral zone of the controller. When the controlled variable is outside the neutral zone of the controller, the final control element moves towards the corrective position until the valve of the controlled variable is brought back into the neutral zone of the controller or until the final control element reaches its extreme position.

Major advantages: gradual load changes can be compensated for by a gradual adjustment of the valve position.

automatic control, two position A control where a valve moves between two fixed positions such as high or low, or open and closed.

automatic controls Controls which may be self-activating, electric, electronic, pneumatic or a mixture of these. Automatic controls are an integral part of a comprehensive environmental control system.

automatic differential pressure condensate controller A device which quickly expels condensate air and other insulators directly they become present. Where these must be lifted to the drain, the unit combines the advantages of blow-through and syphon drainage without the self-imposed disadvantages of either system.

Applications: to improve the efficiency of heated rollers and roller dryers in the paper and corrugating industry.

automatic doors Electromechanical and electro-hydraulic sliding and swing doors with automatic operation.

Applications: for entrance and internal doors in airports, hotels, supermarkets, hospitals, offices, warehouses, old people's and handicapped persons' establishments.

Energy-saving potential: typical installations give a 50% increase in traffic flow allowing for either a smaller opening to be used or resulting in the door being in the closed position more often.

automatic ignition A method of initiating the combustion of a fuel by an electric spark, thereby avoiding the use of a pilot flame or manual ignition by match, taper, etc. It must incorporate a comprehensive set of fail-safe equipment which will cut off the incoming fuel if satisfactory ignition (flame) has not been established within a preset time (a few seconds). See also **flame-failure device**.

automatic lighting control A device which detects photoelectrically when the natural daylight has fallen to a predetermined level and switches the lights on. The lighting is switched off when the natural daylight increases above the level of the artificial lighting.

Applications: interior lighting installations in factories, offices, public buildings, etc.

Energy-saving potential: energy savings of up to 30% are possible.

automatic voltage stabilizer and load shedding transformer An auto-transformer with on-load tapping switch which automatically controls the voltage to a set level and incorporates an overload facility to reduce the voltage as dictated by a maximum-demand controller. It stabilizes the voltage supply and automatically sheds a percentage of load if the maximum demand setting is exceeded.

auto-transformer A single-winding transformer in which the primary voltage is applied to the whole winding and the secondary (output) voltage is taken from suitable tappings on the winding.

auxiliary burner An oil or gas burner which is fitted as back-up or standby, e.g. oil burners on gas-fired boilers which are served by an 'interruptible' gas supply, or an oil burner fitted to a wood waste incinerator for firing at times when the wood waste supply is inadequate or interrupted.

auxiliary energy The use of an alternative energy source, such as solar energy, lessens the amount of primary energy (from gas, electricity, etc.) otherwise required. The primary energy provision required is referred to as the auxiliary or back-up need.

axial fan, bifurcated A fan in which the electric motor driving it is located outside the air stream which passes through the fan casing. It is suitable for handling hot or contaminated air which might damage the motor when in contact with it. It is suitable for kitchen extract systems, laboratory fume cupboards and similar applications. The fan can handle hot air at temperatures of up to 340°C.

axial fan multi-stage, contra-rotating An axial flow fan which incorporates two or more impellers fitted within one casing which rotate in opposite directions. The arrangement permits the development of high fan pressures of up to 50 mb.

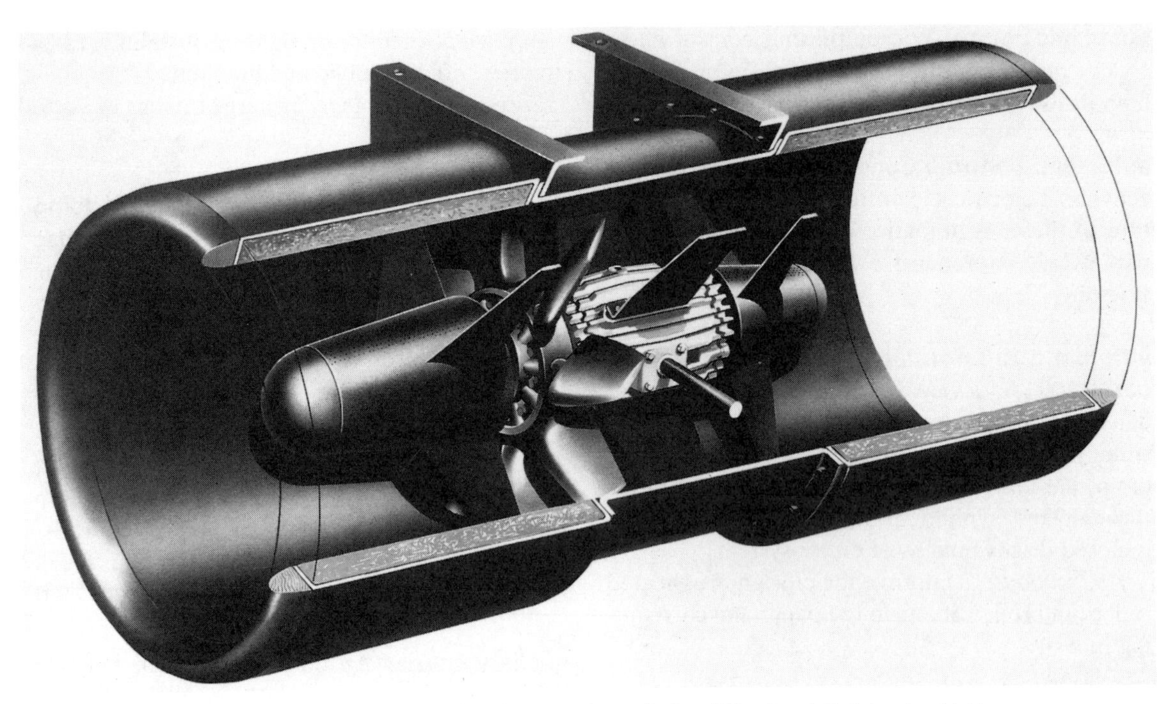

[Fig. 23] Jetfoil axial flow fan for tunnel ventilation (Woods of Colchester Ltd.)

[Fig. 24] Axial flow fan

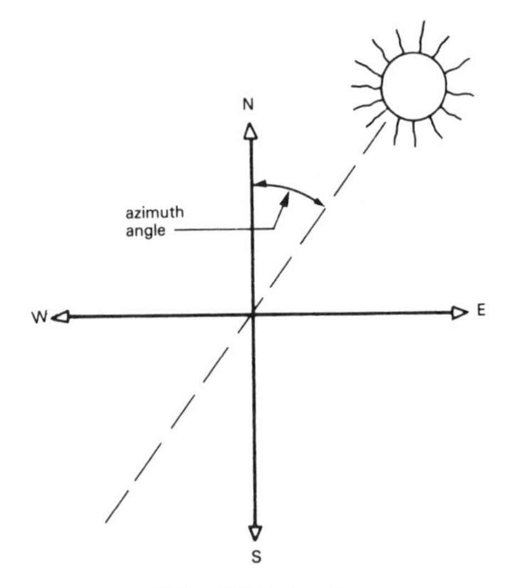

[Fig. 25] Azimuth

axial fan, single impeller An axial flow fan which is fitted with one impeller. It is commonly specified for situations where the fan is intended to operate against only moderate air resistance; e.g. in applications of warm air heating.

axial flow compressor A type of compressor in which the flow of air is essentially parallel to the rotor axis.

axial flow fan [Figs 23, 24] A fan in which air enters the fan inlet axially and is discharged axially. It is arranged for direct insertion into ducting and comprises an impeller with an aerofoil section and with a directly coupled electric motor rotating inside a circular diameter casing which is fitted with flanges for attachment to the adjoining ductwork. It offers a choice of blade angle to provide different outputs from the same fan size.

axial flow fan, stall point The point at which the characteristic curve for the fan begins to dip steeply. The use of axial flow fans of high pitch angles to handle air volumes less than that volume at which the fan will stall must be avoided.

azimuth [Fig. 25] The angle between the horizontal component of the rays of the Sun and the true south. Azimuth angle is usually measured in degrees east (morning) and degrees west (afternoon) of south.

B

BATNEEC A term derived from the Environmental Pollution Act 1990; an abbreviation for best available techniques not entailing excessive cost for:

(a) preventing or, where practicable, reducing to a minimum and rendering harmless, releases of substances prescribed for any environmental medium into that medium; and
(b) rendering harmless any other substances which might cause harm if released into the air, water or land.

'Harm' has a wide meaning within the EPA as it covers not only harm to the health of living organisms or other interference with the ecological system of which they form part and harm to property, but also includes the concept of offence caused to any man's senses; 'harmless' has a corresponding meaning. 'Technique' covers both the process and its relating aspects, not merely the technology. 'Available' means that the technology is generally accessible. 'Best' means most effective in preventing or minimizing or rendering harmless polluting emissions.

BPEO A term derived from the EPA; it denotes the best practicable environmental option. It is that option which provides the most benefit or the least damage to the environment as a whole at acceptable cost, in the long term as well as in the short term. HMIP decides on what is the best practical environmental option in any particular case.

BSI kitemark An award of the British Standards Institute signifying the independent verification of the performance of a particular item of equipment.

BSRIA The Building Services Research and Information Association. The Association researches specific aspects of building engineering services; issues reports and overviews on them; maintains a comprehensive library of publications; and advises and informs members concerning new developments in the building engineering services industry. It also undertakes sponsored research and performance tests.

BTU British thermal unit, an imperial unit of heat. It is that quantity of heat which is required to raise the temperature of one pound of water through one degree Fahrenheit. It has been superseded by the SI unit of watt, one BTU being equivalent to 3.14 W.

BTU meter Equipment designed to integrate and record BTU (or other units) measuring hot water flow and flow return temperature.

Applications: designed for use within industrial boiler plant.

Bacharach number A number indicating density of smoke on a scale of 1 to 10. See **Bacharach smoke scale**.

Bacharach smoke scale A scale comprising ten shades of smoke staining, ranging from white to dense black, employed in the assessment of the smoke density of flue gases.

backbone system, telephones Multi-core risers from a building telephone (frame) room which cross-connect the patch panels on different floors or areas of a building.

back draught damper A device fitted into a ducted air or flue gas system to prevent the air flow being reversed from the intended direction.

back-drop connection A vertical or steeply sloping connection to, or near, the invert level of a manhole from a drain or sewer which is situated at a higher level.

back-drop manhole [Figs 26, 27] A manhole which accommodates a difference in the level of the connected drains. This difference may be accommodated within the manhole or external to it. Where the difference in levels is considerable, the external solution is usually easier and much cheaper.

back flow The flow of a fluid in a direction which is contrary to the intended or natural direction of flow.

background central heating A space heating installation which provides only partial comfort heating either by limiting the installed heating surfaces in the various spaces or by providing space heating only to certain areas, such as hallways and living-rooms. This form of space heating maintains a minimum set temperature in a commercial building during unoccupied periods when the main plant is switched off, as a means of frost protection.

background heating Space heating which provides less than a full standard of heating, e.g. a radiator only in the entrance hall and the living-room.

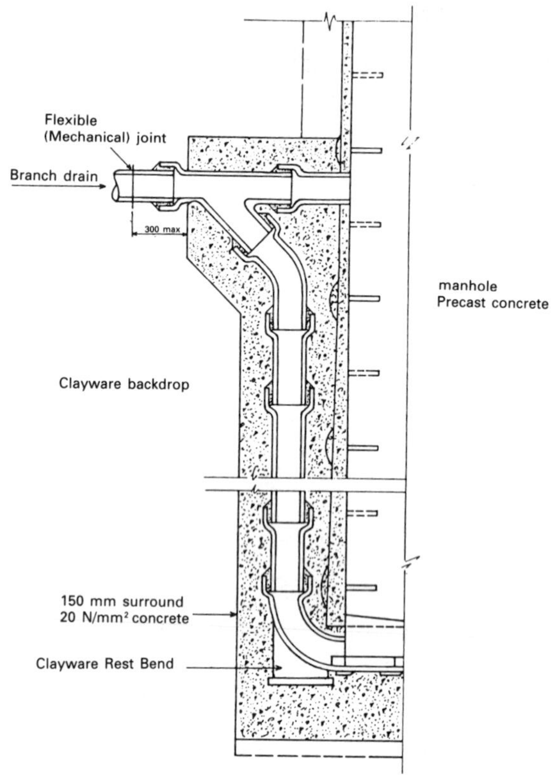

[Fig. 26] Back-drop manhole (precast concrete)

background noise Sound other than the wanted signal. In room acoustics, background noise is the irreducible sound field measured in the absense of any building occupants.

back-pressure steam generator A steam-driven electricity generator which utilizes for its operation the low-pressure steam discharged from turbines; this is a valuable method of energy recovery.

back-pressure turbine A steam turbine forming part of a combined heat and power system in which the exhaust steam is not condensed but is piped away and utilized at the exhaust pressure and temperature for process or water heating.

back siphonage All types of back flow whereby water intended for drinking and domestic use, whether in the home, canteen or commercial premises or elsewhere, may become contaminated either by contact or by mixing with water which

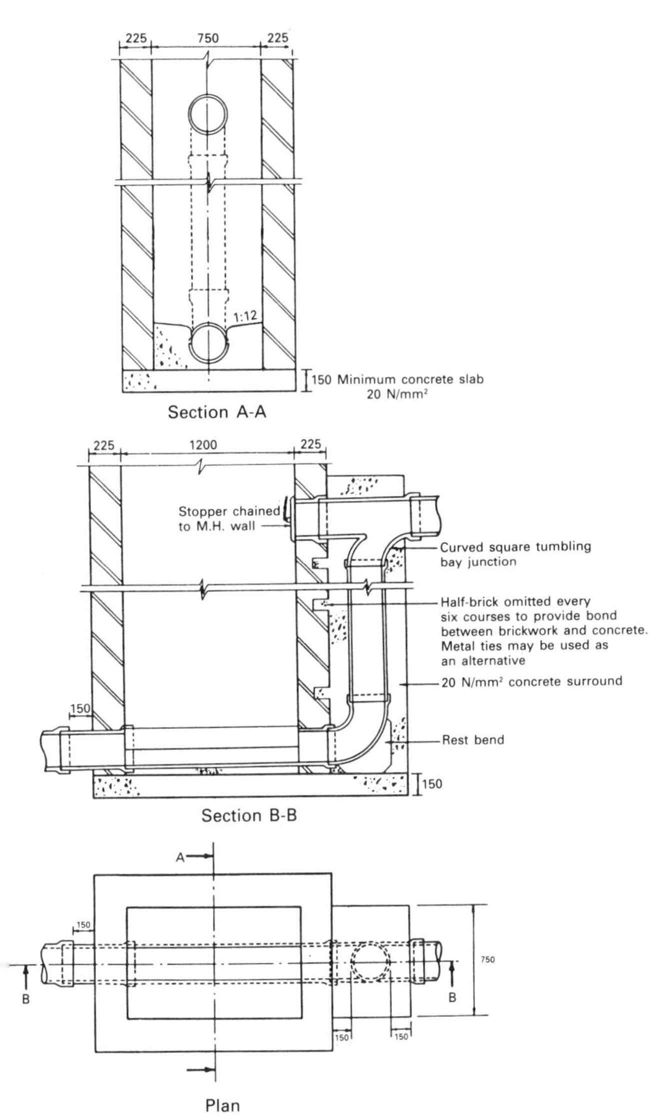

[Fig. 27] Back-drop manhole (brick)

could contain matter that is potentially hazardous to health or which could give rise to complaints from consumers. It must be avoided.

back siphonage, classification of risks The classes of risk arising from potential back siphonage situations are as follows.

Class 1: risk of serious contamination likely to be harmful to health (continuing or frequent occurrences); e.g. bidets, WC pans, dialyser washing equipment associated with haemodialysis machines.

Class 2: risk of contamination by a substance which is not continuously or frequently present, but which may be harmful to health; e.g. taps at sinks, baths and washbasins, domestic clothes and dishwashing machines, hose union taps in gardens or garages, flushing cisterns, flexible shower fittings at bath or basin, boiler primary circuits and indirect heating systems with their feed cisterns.

Class 3: risk of contamination by a substance which is not continuously or frequently present but which could give cause for complaint by consumers; e.g. mixer valves, single outlet taps, domestic water softeners or cold water storage cisterns.

back siphonage, conditions promoting this Back siphonage may be caused by the connection of a fitting or appliance in such a manner that the water outlet is submerged or close to the water surface. The valve or tap controlling outflow from the installation to or through the appliance or fitting must be in the open position. The pressure in the pipe connected to the appliance or fitting must fall below the equivalent of that of the free water level at the fitting or appliance.

Reduced mains pressure in a public utility supply is not uncommon and may be due to heavy draw-off at peak times and in fire-fighting or even by a serious leak in the mains. A significant reduction in the mains pressure translates into a corresponding pressure loss in a consumer's supply pipes connected to a pressurized installation. A sustained reduction in mains pressure entails the risk of contaminated liquid reaching the mains.

backwashing The standard method of cleaning rapid sand filters whereby water is passed upwards through the filter media to dislodge, by viscous drag, the dirt accumulated during the filtering process. This action is sometimes preceded by air also being blown upwards through the filter media to loosen the dirt.

bacteria Unicellular or filamentous microscopic organisms occurring in air, water, animals, plants and decaying organic materials which multiply rapidly by simple fissure. They attack metals, thereby causing corrosion and the formation of harmful gases. For example, they form hydrogen in the radiators of heating systems and generate black corrosion products.

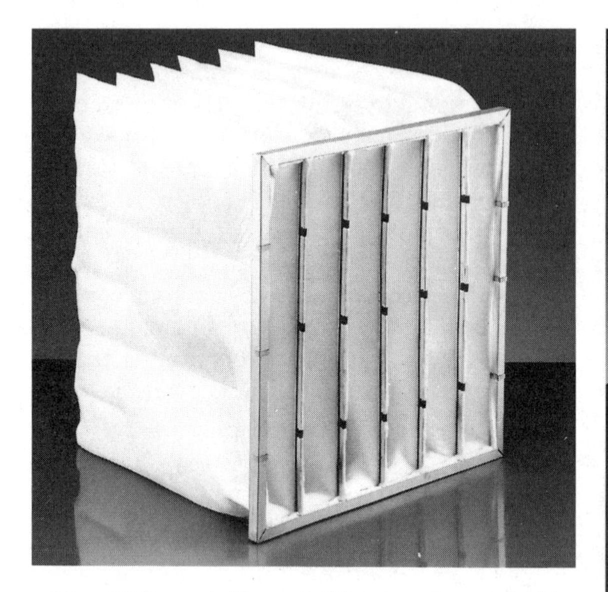

[Fig. 28] Bag air filter unit (Waterloo–Ozonair Ltd.)

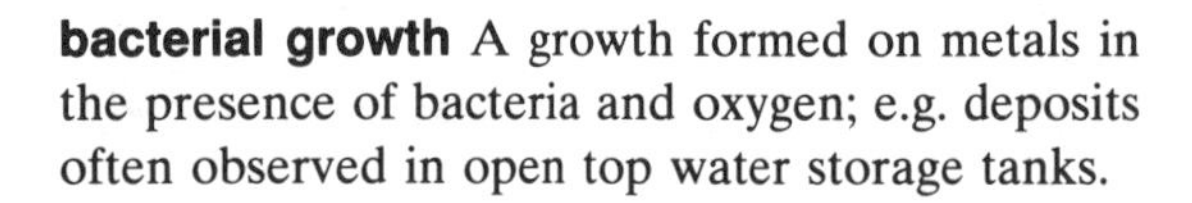

[Fig. 29] Bag air filter modules (Waterloo–Ozonair Ltd.)

bacterial growth A growth formed on metals in the presence of bacteria and oxygen; e.g. deposits often observed in open top water storage tanks.

baffle A device installed within air, gas or fluid systems for the purpose of changing the direction of the flow of air, combustion gases, etc.

baffle plate A plate constructed of metal, timber or refractory (lined) material and placed in the path of a gas or fluid to change its direction of flow.

bag-type air filter [Figs 28, 29] An air filter which employs a filtration medium of scrim-protected glass wool or similar in the form of a bag. It is a fairly coarse filter and can filter out particles to about 1 micron at face velocity of 2.6 m/s. It has a large dust-holding capacity and filtration efficiency of about 80% (blackness test). It is a useful pre-filter in dusty atmospheres. On shutdown of the air system, each filter bag collapses on loss of pressure and can then be easily removed and replaced by a new bag. Bags should be replaced on a pressure increase of about 1.24 mb.

Bailey wall A wall formed of plain tubes in the combustion chamber of a water tube boiler, faced with metal blocks.

balanced draught See **chimney draught, balanced**.

balanced-flue boiler/heater A domestic range of gas-fired boiler which operates with a balanced combustion air intake and flue discharge terminal placed behind or alongside the boiler. It obviates the requirement for a conventional chimney. The terminal should be fitted with a protective guard. The maximum boiler output is about 35 kW.

balanced flue gas convector An individual gas-fired unit flued to the outside via an exhaust grille with built-in safety devices, for installation through walls.

balancing, portable test set Equipment for measuring and pressure differentials commissioning of heating and chilled water systems.

balancing valve system A system employed in the commissioning of heating and chilled water systems through mass flow measurement. The method uses two valves in the circuit to be balanced; that in the flow pipe for isolating and as a fixed orifice across which the pressure drop is used as a metering signal; that in the return pipe is of the double-regulating pattern and is used for isolating and regulating purposes, the double-regulating feature avoiding the necessity for subsequent rebalancing.

Pressure differential signals are indicated at the orifice valve which, being a fixed orifice, has only one flow curve and its accuracy is influenced by a minimum of manufacturing tolerances. The flow characteristics of the double-regulating valve are of no consequence, other than that it must provide fine regulation and effective isolation.

A given pressure differential signal at the orifice valve results in a positively known mass flow of water. The resistance offered by the circuit being balanced does not matter and may vary from the designer's estimate. When the double-regulating valve has been adjusted to provide a specific signal at the orifice valve, the required (balanced) flow of water will be passing through the circuit.

Each orifice valve should be (and usually is) supplied with a pair of PSA/DOE approved test plugs, complete with captive blank caps fitted to each orifice. In commissioning use, probe units are connected to these plugs via plastic tubing to a portable pressure differential test set.

ballast A device used with discharge lamps to stabilize the current.

ball check valve A valve whose check mechanism is a ball.

ball float valve A valve which controls the automatic water replenishment or make-up in a water tank or similar container. A ball valve usually comprises a metal or plastic float linked by a lever arm to a slide valve mechanism of a (mains) water valve.

ball valve A control valve of the piston and disc type which is operated by a ball float and lever mechanism to maintain a specified liquid level in a tank or other container. It is available in different types and modes to suit particular applications.

ball-valve silencing pipes Ball valves equipped with screw threads for the attachment of silencer pipes. These have been prohibited since 1973 as they present a potential back siphonage risk.

banking A very slow rate of combustion which is only just adequate to retain combustion in readiness for a change to full boiler output. It is equivalent to switching off a gas or oil burner pending heat demand.

bar A unit of pressure in the c.g.s. system. 1 bar (b) = 10^5 pascals (Pa) = approx. 760 mm of mercury (Hg).

barometric damper A pivoted, balanced plate placed within a flue system between the furnace and the chimney stack. It is actuated by changes in chimney draught.

barometric pressure Pressure exerted by the prevailing atmosphere.

barrel (oil) A commercial measure for a quantity of oil. One barrel is 159 litres. The term is used mainly in the oil exploration industry.

baseboard Timber board attached to the skirting of a wall.

baseboard radiator A low-level heater attached to a baseboard. It may be of the convector type with extended heat transfer fins or comprise a plain radiating surface, of steel or cast-iron construction.

base exchange water treatment A method of softening hard waters in which a natural or synthetic material introduces sodium-based salts to replace those of calcium and magnesium, thus softening the water which is passed through it. The material has to be regenerated at frequent intervals by washing with a brine solution.

base load That loading of a network or plant (electricity, heat, etc.) which is constant. Loadings above the base load are subject to fluctuations.

baths for the elderly Baths which commonly incorporate recessed soap discharges and a slip-resistant base. Many different types and sizes are available manufactured from porcelain, enamelled pressed steel, cast-iron, cast acrylic sheet or other suitable plastics. Installed in homes for the elderly, these baths commonly incorporate special gripping attachments and are of lower height. The waste

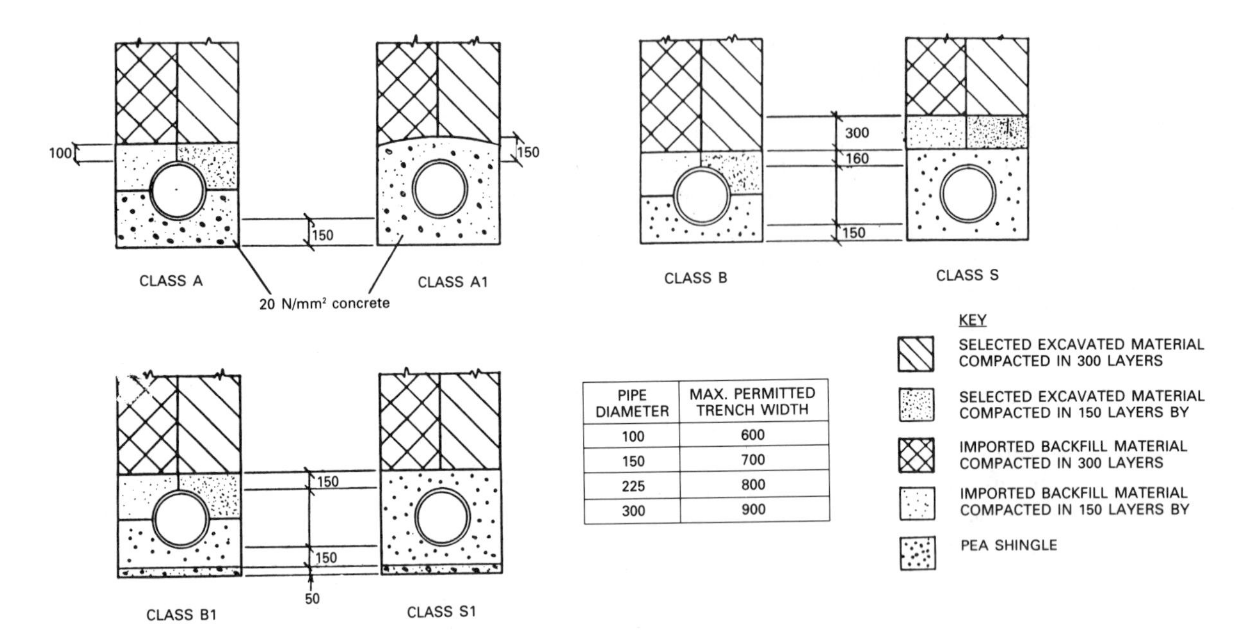

PIPE DIAMETER	MAX. PERMITTED TRENCH WIDTH
100	600
150	700
225	800
300	900

[Fig. 30] Drain: bedding

outlet is a trapped outlet and the overflow is usually fitted into or before the water seal in the trap. Water services may be of the pillar, mixer or wall-mounted type.

bearing A support for a rotating shaft.

bearing support A means of supporting a rotating shaft and its bearing.

bedding, drainage [Fig. 30] The base on which buried drain-pipe is laid. The underlying excavation must be compacted before the bedding material is introduced; such material will be pea gravel, shingle or other granular fill.

bedplate A support for an assembly of equipment such as a fan, pump, motor, diesel engine, air compressor, refrigeration compressor, etc., enabling the equipment to be attached rigidly by holding-down bolts to a concrete or masonry base.

bellows Where pipe movements, which may be due to thermal movement or vibration, cannot be taken up by the natural integral flexibility of a pipe layout the use of bellows compensators is commonly considered. Such compensators are versatile and, in addition to compensating for pipe movements of up to 50 mm taken up by a single such compensator, can attenuate the low frequency vibrations at oil burners, air compressors, circulating pumps and similar installed equipment items. They are also used to compensate for minor misalignments of pipes caused by earth movements or subsidence. The bellows are the flexible core element of a bellows compensator. They are made with a specified number of concertina type convolutions by one of a number of patented manufacturing processes. For general purposes use, most bellows are manufactured of chrome–molybdenum or stainless steel. Special applications may require to incorporate neoprene or rubber materials.

bellows, angular type Bellows which compensate for angular movement; e.g. placed into a expansion loop to take up (compensate for) major expansion.

bellows compensator, axial Bellows that compensate for axial movement only. It is the most basic type and has limited compensation requiring firm anchors and guides.

bellows compensator, axial with tie bar [Fig. 31] Reinforced standard axial bellows to withstand higher pressure and thrusts without buckling.

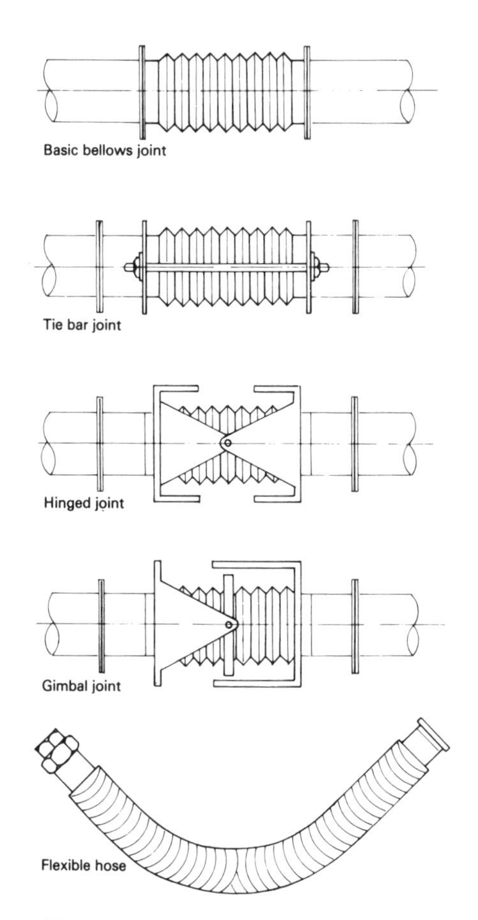

[Fig. 31] Pipeline expansion joint types

bellows compensator, cold draw See **bellows compensator, CPU**.

bellows compensator, CPU Standard type bellows designed to take half the total traverse as a cold pull-up (CPU). Allowance for cold pull-up (or cold draw) is made in the layout of the pipeline by making the gap between the flanges of the pipe section adjacent to the bellows joint equal to the free length of fitting plus a length allowance of half the listed traverse. When the bolts have been drawn up and tightened, the bellows joint will have been extended as required.

bellows compensator, gimbal joint A variation of an ordinary hinged joint. It is hinged at its quadrant points. It can absorb expansion in any plane. Gimbal joints are commonly used in pairs to absorb the pipe movement in planes other than those directed by the pipe itself.

bellows compensator, hinged Bellows that allow free angular movement in a plane. The two halves of the hinge are each attached to one end of the joint and are pinned at the centre to permit free angular movement restricted in one plane about the centre of the joint. They are commonly applied in sets of two or three, the extent of lateral movement which can be accommodated by a double-hinged compensator. Specification details relate to the amount of angular movement the bellows can handle and on the length of pipe between them.

bellows compensator, specification Detailed tables published by manufacturers generally listing the following information:

(a) size of connection;
(b) overall free length;
(c) total axial traverse from bold pull-up;
(d) extended length with 50% cold pull-up;
(e) spring rate;
(f) effective area;
(g) safe working pressure.

bellows expansion compensator A device that employs the action of stainless steel corrugated bellows to permit unstrained expansion and compression of pipelines. It must be part of a carefully designed compensatory arrangement which will include anchors and pipe guides.

bellows joint, articulated A joint suitable for compensation of larger thermal movement under great pressure. It comprises an assembly of two bellows between a short length of pipe, held together by two steel struts pinned at the joints so that the whole joint assembly can move.

bellows joint, externally pressurized A joint where the bellows are fully enclosed within a pressure casing and pressurized from outside offering the advantage of large movement under great pressure.

bellows joint, oval flanges A bellows joint incorporating tie bars which are usually supplied with oval flanges, the oval shape accommodating the fixings for the tie bars.

bellows joint, rubber Effectively a length of reinforced rubber hose with a small bellows shape in the middle. The parallel hose part is reinforced with heavy wire and the 'convolution' with textile or nylon plies to resist internal pressure whilst still allowing flexibility.

Advantages: the ability to offset movement at right angles to the axis of the bellows; very compact and flexible; particularly suitable for noise and vibration absorption; standard neoprene bellows has good corrosion resistance against the more usual hazards of sea water, oils, aggressive waters, mild chemicals, etc.; relatively inexpensive, except in very large diameters.

bellows joint, spring rate The force which results from resistance of the bellows to the movement (compression of expansion). The spring load is substrated from the main anchor loads due to pressure (cold) and added to the anchor loads due to pressure load (hot).

bellows, wall penetration sealing bellows Bellows that seal off, neatly and safely, the gap between a wall and a moving pipe that passes through it.

belt-drive [Fig. 32] An arrangement for connecting the pulley of a driving motor or engine to the pulley of a driven item of plant (fan, pump, etc.) by means of a flat leather belt.

benching A construction inside a manhole with sloping surfaces on either side of the channels at the base of the manhole to discourage the accumulation of drainage deposits.

benchmark The datum level established for a particular site of building operations and displayed throughout the site by means of a number of prominently displayed survey poles. All site installation levels are then related to that benchmark. Applies, for example, to invert levels for drainage manholes and sewers, crawlways, floor levels, etc.

Bernoulli's theorem The theorem that states that for any mass of fluid in which there is a continuous connection among all the particles, the total head of each particle is the same. It is most important to the solution of problems relating to fluid flow. The venturi meter is based on this theorem.

[Fig. 32] Centrifugal fan belt drive assembly

bib tap A water outflow control valve designed to connect to the female fitting of the water connection. It is operated by a simple two-prong lever secured to the top of the valve shank by a screwed nut. The tap may incorporate a garden hose union connection.

bibplus tap A bib tap which incorporates an integral double check valve to protect the connected system from back syphonage when using hose-pipes with substances such as liquid fertilizers and car shampoos.

bidet A low-level basin that may be one of two types: the douche type with ascending spray, hot and cold water controls and a diverter to direct warm water to the rim, or the over-rim type which has valves mounted on the rim and taps to fill the bowl or to supply a hand spray, usually with a thermostatic valve similar to shower valve. In dwellings, bidets are fitted with and next to the WC suite; in hospitals, they are sometimes located within separate wash rooms.

Water supplies to a bidet are commonly 15 mm with a 32 mm trapped waste outlet. The arrangement of water supplies must avoid any possibility of back syphonage and Water Authority regulations are strict in that respect. Except in the case of over-rim bidets fitted with pillar taps, the hot and cold supplies should be from a tank or break tank, depending on the regulations. No other water supply branches or other appliances may be connected off the supply feeding a douche type bidet.

bifurcated fan A type of axial flow fan with the fan casing containing the built-in motor compartment and a venturi inlet to a shrouded compound flow impeller with the electric driving motor situated outside the air stream. Such fans can handle hot air or gases at temperatures of up to 340°C. Bifurcated fans are widely used in situations where fans must handle contaminated, corrosive or grease-laden air which is likely to damage the electric motor. They are often used in kitchen extract systems.

binary-coded decimal (computer) A numeric coding system. Each decimal numeral is represented by a group of four binary digits.

bin discharger A mechanism that is usually fitted within the coned section of a cone-bottomed storage silo to rotate in an angular or radial fashion, in either a clockwise or anticlockwise direction, to agitate the stored waste in the bottom of the silo and thereby encourage the free flow of the material into the associated rotary valve. A manual override is usually incorporated into the bin discharger controls to permit the operation of the discharger independently of the rotary valve, in the event of bridging conditions when fuel starvation arises.

biochemical oxygen demand (BOD) The weight of the quantity of oxygen absorbed as a result of the oxidation of the constituents of a specific weight of water through biological action, expressed in parts per million or grams per litre. It is an indication of the extent of biologically degradable organic material in polluted waters. BOD is used by Local Authorities in assessing the performance of sewage plants which discharge into waterways.

biodegradable The property of a material, mixture or specific waste which permits it to decompose. Anaerobic bacteria digest the decaying putrescible organic material and, in the process, generate methane (landfill) gas.

bioenergy process A process that transforms effluents from food and other factories into a fuel gas. The plants are fully enclosed and do not produce bacterial aerosols, smells or noise.

Applications: the dairy industry, fermentation (yeast, beer, cider, pharmaceuticals), distilling and paper and pulp industries.

Energy-saving potential: the process transforms over 40% of organic pollutants into gas with each tonne of organics yielding the equivalent of half a tonne of fuel oil. Large distilleries can be made independent of fossil fuels and cheese factories with a whey disposal problem can be made energy independent of both electricity and heating fuels.

Availability: survey and laboratory services are available to ensure reliable design data. Full-scale plants take nine months to build and put into operation.

biogas [Fig. 33] A gas generated in anaerobic methane fermenters which convert the organic acids and alcohol into methane and carbon dioxide, their relative proportions depending on the detention time and on the process temperature.

biogas plant [Fig. 34] A plant that transforms waste effluents into methane gas.

bio-reactor Municipal refuse landfill in which the accumulated micro-organisms produce a biogas which is composed of approximately 50% carbon dioxide and 50% methane. Biogas can be extracted from 'gas wells' through a network of perforated plastic tubes laid within the accumulated refuse.

birdnesting The adhesion to boiler tubes of fused ash which cements together with unburnt carbon, causing reduction in heat transfer and dangerous overheating of boiler tubes.

bit (computer) The smallest unit of computer storage. It is represented by one of the digits used in the binary notation, i.e. 0 or 1.

bituminous coal A widely used solid fuel with a volatile content of 20–35%, fixed carbon of 45–65% and inherent moisture of 3–10%. The calorific value ranges between 26 kJ/g and 32 kJ/g.

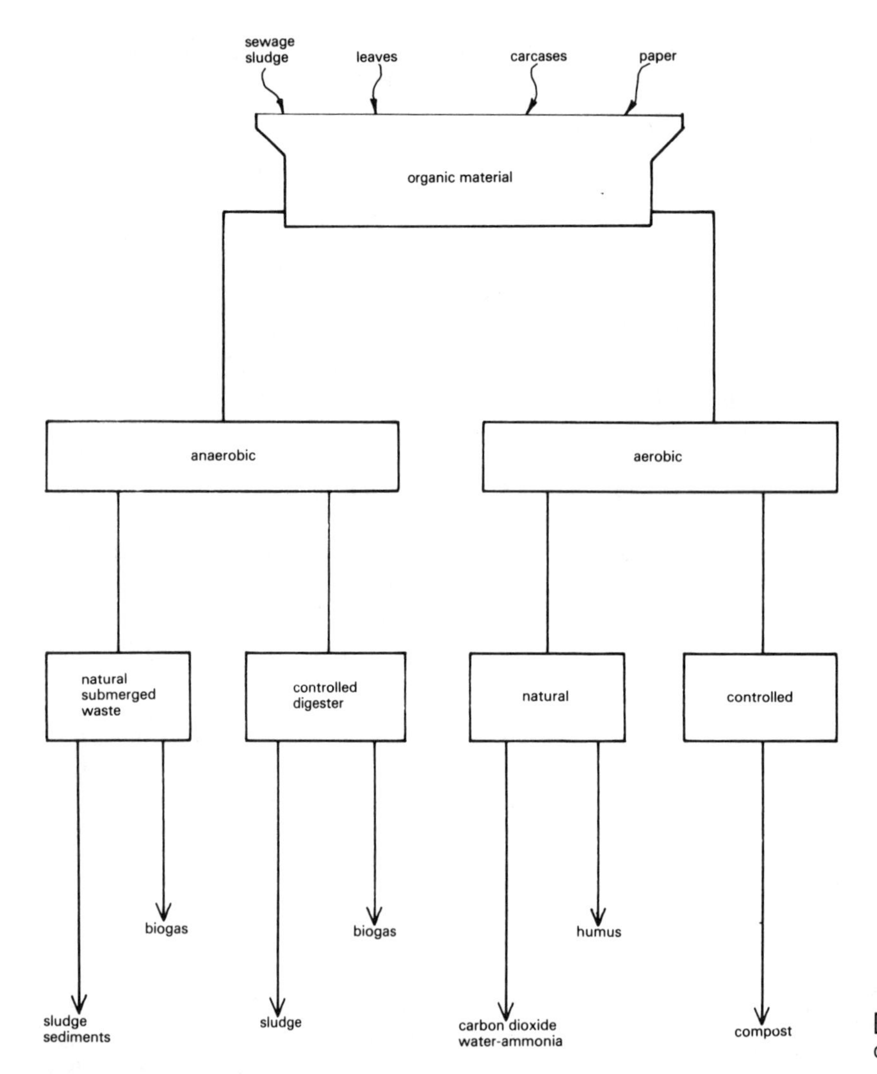

[Fig. 33] Flow diagram of organic decay

black heat This relates to air heating applications (mainly electrical), e.g. duct-located electric warm air heaters, in which the temperature of the heater element(s) is controlled to remain below the red (visible) spectrum, mainly for reasons of safety.

blackness test A test to determine air filter efficiency by injecting a test dust sample upstream of an air filter and then drawing the dust-laden air at the inlet and outlet of the air filter through special filter papers. The efficiency is assessed by comparing the relative blackness of the stains made on the filter papers. The colour of the test dust is unimportant.

blade angle The angle between a blade and an impeller. Pressure developed by a fan rises with an increase of blade angle. For an impeller of given proportions, the volume then also tends to increase.

blank flange See **flange, blank**.

blast cooler A radiator through which warm water is recirculated and a fan plant which blasts cool air at high velocity across the radiator.

blast heater Similar to a blast cooler but operates with heated air being discharged at high velocity.

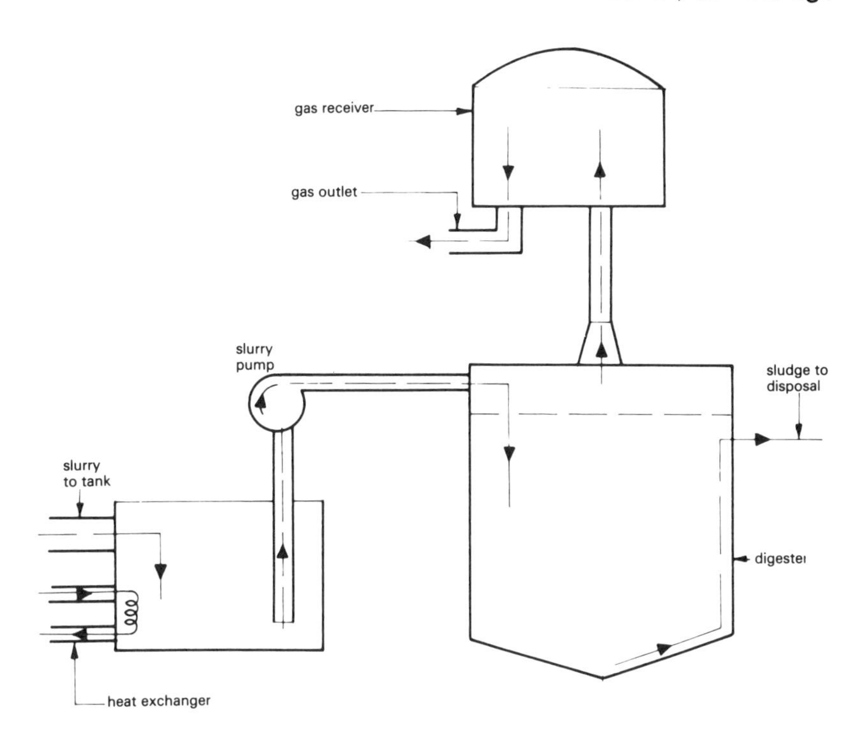

[Fig. 34] Biogas digester system

bleed valve A valve that permits a small preset quantity of fluid to flow from an item of equipment, such as from a cooling tower pond.

bleeder valve A valve that enables the air in the delivery pipe line from an air compressor to be exhausted to atmosphere when the pressure switch contacts open, causing the compressor driving the motor to stop. It may incorporate an adjustable pressure differential and may be centrifugally, magnetically, mechanically or oil operated.

blender A hot water or steam or water mixing device, set to one specific mixed water temperature.

blind flange See **flange, blind**.

block tariff An electricity tariff in which a fixed number of electric units is charged and consumed at a comparatively high price before the unit price is reduced to a lower charge.

block wiring The central telephone wiring system to which a consumer connects his telephones.

blowdown The controlled discharge of water from the lowest part of a steam boiler to reduce the dissolved solids content in the boiler water and to dislodge accumulated sludge and other foreign matter.

blowdown pit A pit that provides for safe discharge of steam or boiler hot water blowdown into the drains. It is arranged to store the hot blowdown water to permit cooling before discharge. It must be fitted with a substantial cover to safeguard boiler plant personnel from steam emission.

blowdown valve A valve adapted for the control of the high-temperature water being discharged from steam boilers during blowdown. It must be strongly constructed to withstand high pressures and temperatures.

blown fibre loft and wall insulation Loose glass fibre granules used for cavity wall and roof space insulation.

boiler, air leakage The uncontrolled leakage of air into the boiler via gaps in the brick-setting, defective doors and inspection cover seals, unsound joints in the flue system, etc. Air leakage must be obviated by careful workmanship during the assembly of the plant and by subsequent good monitoring and housekeeping. The ingress of unwanted air into the boiler unit depresses the thermal efficiency as this air must be wastefully heated. Ingress of air into the flue system will reduce the buoyancy of the draught and, in extreme conditions, can overload the chimney flue gas-carrying capacity.

boiler, automatic damper A damper that isolates the boiler from a connected flue pipe or chimney. It typically comprises a flanged steel casing which houses a hinged blade fitted with a stainless steel sealing strip to give tight closure. It is fitted with motor drive for open/close or modulate control and has an integral explosion relief facility. Its use prevents the cooling effect on a hot idle boiler of chimney-induced air flow through the boiler.

boiler availability The actual boiler plant capacity which is available for use, expressed as a percentage of the maximum installed capacity.

boiler blowdown The process of discharging a quantity of boiler water from a steam boiler to maintain the dissolved solids in it within specific limits, to avoid carryover, foaming or priming. The frequency and quantity of blowdown must be carefully monitored and controlled to ensure minimum safe blowdown quantities. Excessive blowdown will waste heat and fuel, possibly in meaningful quantities. The blowdown process must be carried out with care and to a specific procedure to safeguard the operator.

boiler blowdown, automatic A blowdown controller, for use with steam boilers, which is activated by the conductivity of the water in the boiler. It can incorporate sequence control for multiple boiler installations and can be used in conjunction with heat recovery equipment. It enables boiler(s) to operate continuously at the maximum dissolved solids level, thereby reducing the quantity of required blowdown to a minimum, yielding major energy saving.

boiler blowdown controller A device that automatically and continuously monitors the boiler water density and adjusts the rate of blowdown to maintain the density within preset limits to minimize the amount of the water blown down and consequently the heat loss therein. There is some evidence that the replacement of manual blowdown methods by automatic ones in the average boiler installation reduces the blowdown quantity by about 20%.

boiler, burner unit The heat source for a boiler. The burner characteristics are matched closely to those of the boiler which it fires.

boiler capacity The amount of steam or heat obtainable from a boiler, expressed in terms of kg/h or kW/h. The information is usually supplied by the manufacturers of the boiler. See also **boiler rating**.

boiler casing A metal enclosure placed around a boiler to protect the thermal insulation below and to provide the boiler assembly with a neat appearance. It is also an insulated tailored metal covering placed over a sectional boiler.

boiler cleaning powder Dry powder for application to the dry side of boilers and furnaces for in-line cleaning and the removal of combustion deposits in oil, coal, gas and waste fired boilers.

boiler, combustion efficiency The energy conversion by the boiler (or other heat generator) firing equipment, expressed as a percentage:

$$100\% \times \frac{\text{Heat output (to combustion chamber)}}{\text{Heat input (calorific value of fuel)}}$$

boiler, condensation See **condensation boiler**.

boiler damper air sealing A device that prevents unnecessary heat loss while the boiler is not being fired. Fitted to the burner combustion fan inlet, the air sealing damper prevents cooling air being drawn through the boiler by natural draught during stand-by or shut-down, helps maintain boiler pressure and increases boiler availability.

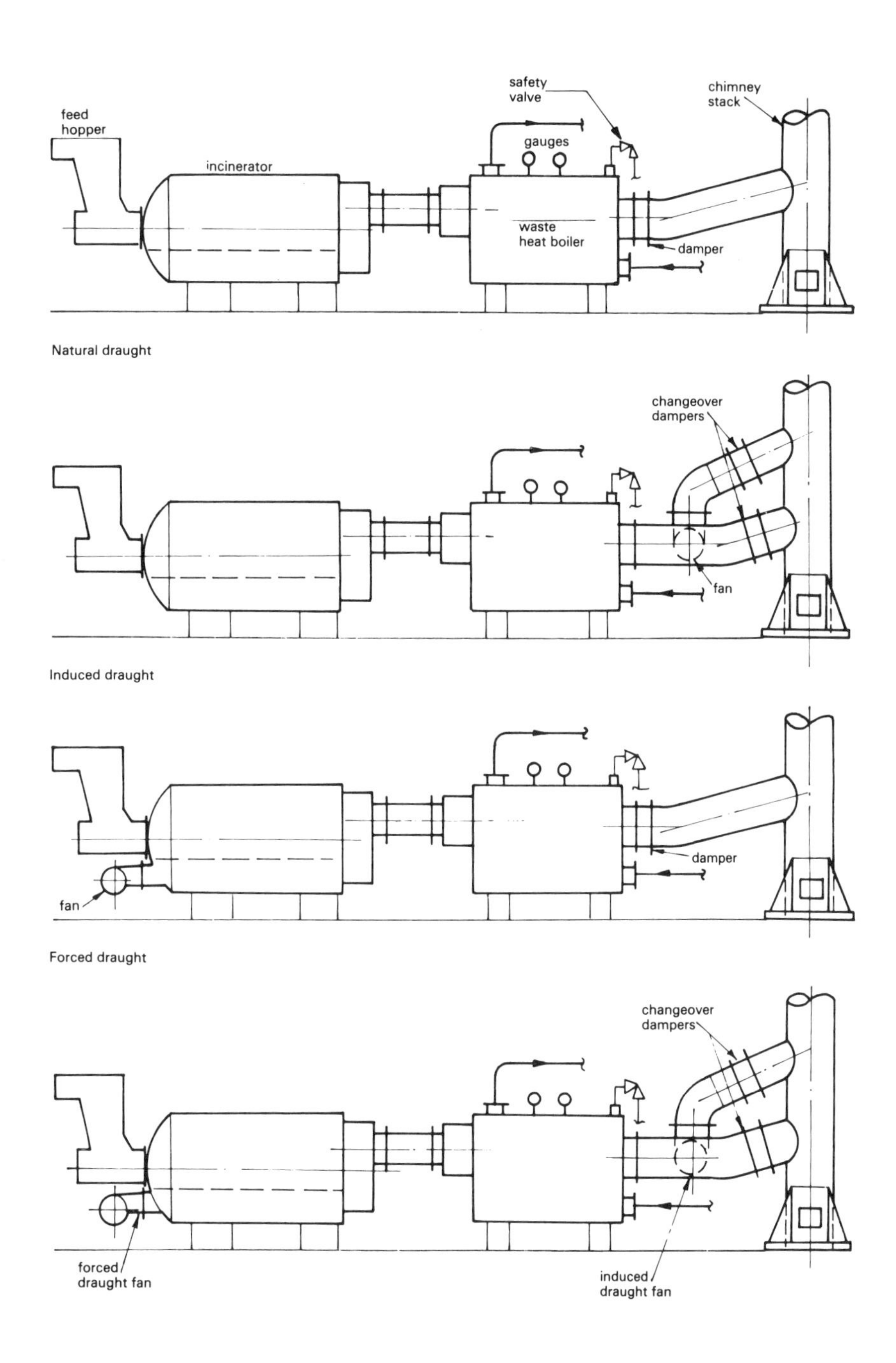

[Fig. 35] Boiler draught conditions

Construction is of heavy galvanized steel complete with servo-motor and interlock. A butterfly damper and synthetic rubber seal design can achieve 99.95% sealing.

boiler draught [Fig. 35] The difference in pressure (positive or negative) which moves the combustion flue gases through the boiler and the attached chimney. See also **chimney draught**.

boiler drum A cylindrical vessel which provides a reserve of water for a water tube boiler. It is used as a collecting space for the steam that is generated.

boiler economizer A stack-mounted device designed to recover energy from the exhaust gases for preheating boiler feed water or process usage.
Energy-saving potential: savings of 3–7% on boiler operating costs.

boiler efficiency The ratio of the available heat output at the boiler flow tapping to the heat input from the fuel, expressed as the full load combustion efficiency less the percentage losses from the boiler casing including the plate work. It can also be derived from direct measurements of the thermal input and the output of the boiler plant. Boiler efficiency can vary with the following modes of operation:

(a) full load, rated output, continuous operation;
(b) part load, continuous operation;
(c) part load, cyclic operation.

boiler escape valve See **three-way boiler escape valve**.

boiler feed pump A pump that feeds water into a steam boiler, overcoming the boiler pressure to do so. The pump may be a steam-driven reciprocating or centrifugal type with an electric motor.

boiler feed water The piped water supply to a steam boiler fed from a hot well. It may be pre-heated, deaereated or conditioned. The condition of feed water to a steam boiler must be closely monitored on a routine basis to conform with specified quality. Most boiler rooms include appropriate test kits with their maintenance tools.

boiler foaming See **foaming**.

boiler, gravity feed A solid fuel fired boiler which incorporates an inclined fuel bunker which permits the fuel to flow to the fire bed under action of its weight.

boiler head The pressure of water which may be safely applied to a boiler. It is the pressure under which the boiler operates.

boiler heating surface The heat transfer surface between the combustion flue gases and the water being heated, expressed in square metres.

boiler, lead See **lead boiler**.

boiler, minimum waterflow The flow specified by manufacturers of compact small water content boilers to prevent the formation of local hot spots on heat transfer surfaces. This requirement is met by pumping of the circulation.

boiler mountings The assembly of safety, control and indicator fittings attached to a boiler such as gauge glasses, safety valves, altitude gauge, boiler feed float controls, blowdown valve, feed valve, main steam valve and operating and high limit safety thermostats.

boiler, multi-pass See **multi-pass boiler**.

boiler, multi-fuel [Fig. 36] A boiler that can be easily adapted for firing with different fuels to take advantage of availability and/or changes in fuel prices.

boiler oxygen trim control system A system designed to recognize any inefficiencies in boiler/burner combustion and to adjust automatically the various parameters to keep the plant operating at optimum efficiency while consuming the minimum fuel. The use of such control is likely to yield a worthwhile improvement in the combustion efficiency in the order of $2\frac{1}{2}$%.

boiler, packaged The assembly of all the components required for the functioning of the boiler and the combustion system mounted on a common base. It is brought to the site in one piece and only requires local connection to chimney, water, fuel and electricity supplies.

boiler priming See **priming**.

boiler, radiant tube type A boiler that incorporates the heat exchange surface in the form of banks of tubes located within the boiler combustion chamber. This permits quick-response operation.

boiler rating The heat or steam output from a boiler specified by the manufacturer for a particular boiler type and model, usually given in terms of kW/h (hot water) or kg/h (steam). In the case of a

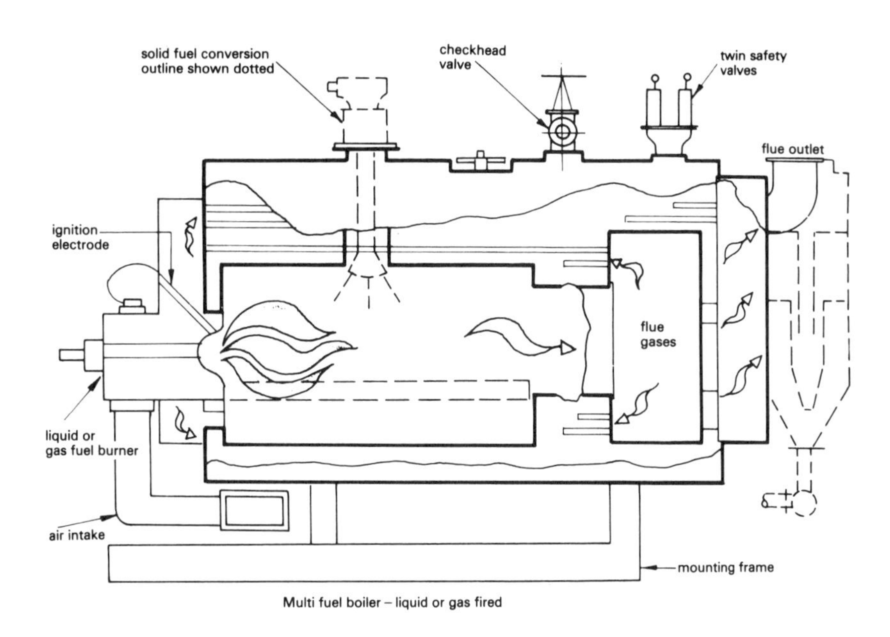

Multi fuel boiler – liquid or gas fired

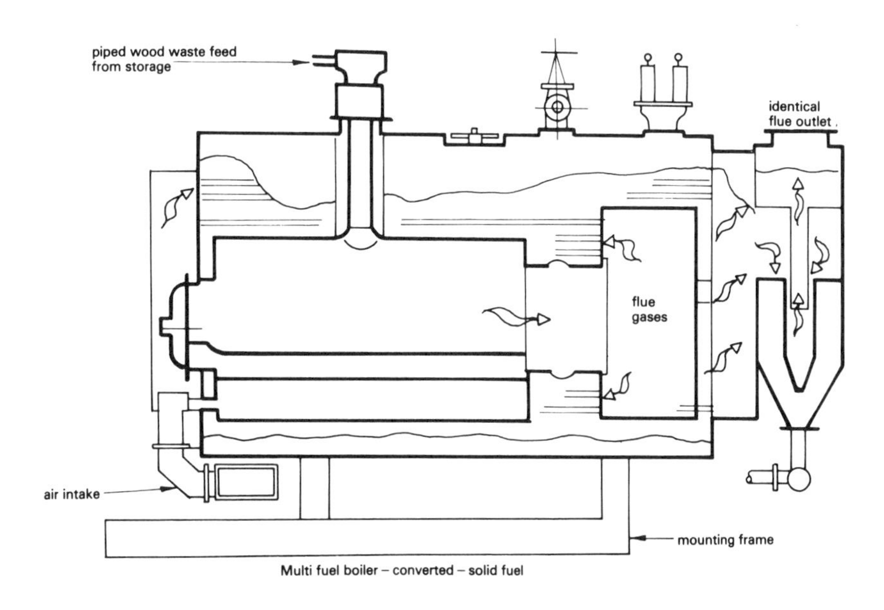

Multi fuel boiler – converted – solid fuel

[Fig. 36] Multi-fuel boiler

steam boiler, the rating is tied to a particular output pressure and feed water temperature.

boiler, sectional The assembly of a number of cast-iron or steel sections with integrated flue and water ways and heat transfer surfaces around a central combustion chamber. It is joined together by means of nipples and tie-rods. The front plate is arranged to suit the required combustion equipment, e.g. underfeed stoker, or oil or gas burner. Flue gases exit via a smoke-hood.

boiler sequence control A control that ensures that boilers within a group of boilers are switched on and off in a manner which is strictly related to the requirements of the connected heat load.

boiler sequence controller An electrical mechanism for effecting sequence control of multiple boiler installations. See also **boiler sequence control**.

boiler sequencing control system A system consisting of a control console, housing all the required electrical components. It may be arranged for series or parallel modulation, automatically controlling two or more boilers and their rates of firing in response to steam or hot water requirements. It is suitable for commercial and industrial steam and hot water boilers. Fuel savings arise from optimal boiler usage and savings in electrical power.

boiler setting The arrangement of the refractory brickwork within a boiler or furnace.

boiler, shell type A horizontal or vertical mild steel shell which incorporates combustion equipment and heat transfer surfaces.

boiler siphon A U-shaped pipe fitting mounted between the boiler steam space and the pressure gauge to interpose condensate and prevent direct contact between the live steam and the gauge mechanism. It is commonly fitted with an isolating cock to permit removal of the gauge under operating conditions.

boiler, thermal efficiency The energy conversion by the overall boiler unit (or other heat generator), including losses, e.g. via chimney, radiation, or infiltration.

$$100\% \times \frac{\text{Heat output from boiler to connected system}}{\text{Heat input (calorific value of fuel)}}$$

boiler tube cleaning equipment Electric, air-driven or manual tube cleaning machines. Tools and brushes for all types of tubular boilers and process plant.

boiler water line A line that indicates the maximum height to which a steam boiler may be filled with water to leave adequate space for steam generation. It is commonly indicated on the boiler gauge glass.

boiling-point The temperature at which liquid changes into vapour, e.g. water to steam. It depends on the prevailing pressure; see **steam tables**. High temperature heat transfer fluids possess higher boiling-points than water. See also **heat transfer fluid**.

bonding The connection of pipes and other metalwork by a conducting metal bond to ensure common electric potential. Such bonding is mandatory for reasons of safety.

bonding and bonds A term used in connection with boiler or furnace refractories. The stretcher bond is mainly used in smaller furnace construction, the header bonds are used for furnaces which operate at high temperature, and the English bond comprises alternative courses of header stretchers. The Dutch bond is similar to the English bond, but additionally ensures that alternative stretcher courses are not coincident.

Boost compressor A compressor that boosts the pressure of the gas supply to that required at the gas burner equipment.

boost heater See **after-heater**.

booster pump A pump installed within a pumped circulation to provide a localized pressure boost to overcome frictional losses in a pipeline or to compensate for poor regulation of circuits.

bottle trap A trap in which the division between the inlet and outlet legs is formed by a dip tube or vane within the body of the trap. The lower part of it is removable for access and there must be no reduction in flow through the trap. Bottle traps are often used in conjunction with wash-basins where the trap is exposed or where there may be difficulty in fitting a tubular trap.

boundary layer The stagnant layer of a fluid or gas immediately adjacent to the wall of a boiler or pipe.

Bourdon gauge A pressure gauge for steam boilers. It comprises a coiled tube with one sealed end which is connected to the pointer mechanism of the gauge. The admission of steam tends to straighten the tube which results in the movement of the sealed and attached gauge pointer. A siphon U fitting must be inserted between the gauge and the steam space to interpose condensate and thereby prevent live steam coming into direct contact with the gauge mechanism.

Bower barffing The process for rust-proofing cast-iron or mild steel in which the metal's temperature is raised to red heat and then treated with live steam.

bowl and trough urinal A type of urinal. Such urinals are adequate for use in offices and in public buildings. They can be securely installed to avoid damage through misuse and offer the advantage that any leakage on an outlet is visible rather than concealed within the floor. Trap outlets and sparge pipes may be fitted as exposed or concealed.

Boyle's law A scientific law stating that the absolute pressure of a given mass of any gas varies inversely as the volume, provided that the temperature remains constant.

braiding The plaited protective covering of a cable, generally made of fibrous or metallic material.

branch circuit A circuit that connects between a main circuit and terminal(s).

branch discharge pipe A pipe that connects a sanitary appliance to the discharge stack.

branch duct A duct that connects between the main duct of a system and the terminal fitting.

branch ventilating pipe The ventilating pipe connected to a branch discharge pipe.

brazed joint A joint that unites the parts with molten brass.

break pressure cistern A cistern interposed between a potentially contaminated liquid and the water supply system. It is provided with piped overflow, and float-operated or another automatically controlled inlet. The covered inlet should be fitted with a protection device.

breakout noise Airborne noise radiated from a duct system to the outside.

breech fitting A symmetrical pipe fitting in which two parallel pipes combine into one pipe. The fluid flow may be in either direction according to the purpose for which the fitting is used.

breeching piece An inlet fitting to a fire-fighting dry riser.

bridge (or bridge wall) Refractory material formed as a partition wall between parts of a combustion chamber, generally at changes of direction of the flue gases.

brine A solution of inorganic salt in water, usually sodium chloride or calcium chloride salts. Brine is used as a secondary heat transfer fluid in refrigeration systems. Sodium chloride may come into contact with foodstuffs; calcium chloride may not. Brine permits operation at lower temperatures than chilled water (about 2°C) but tends to be corrosive.

briquettes storage Storage for housing wood-waste briquettes pending use. It must be dry and safeguard the stored material from crumbling and reverting to its original stage.

briquetting The process of compressing wood-waste chips and sawdust into briquettes of a convenient size for storage, prior to use. A briquetting press operates at high pressure and is noisy, so requires careful location.

British Board of Agrément The body that issues agrément certificates for approved equipment which is then so labelled.

bronze welded joint A joint that unites the parts with molten brass. It differs from brazing by the local building up of the welding material.

brush draught excluder Strips of brush-type fibre (e.g. polypropylene) bonded to a non-corrosive flexible carrier which can be cut to length at random. It may include a (polypropylene) fin run integrally through the brush material to improve draught proofing and to act as a water barrier. The excluder is cut to the required length and is fitted at the bottom and sides of external doors to exclude draughts.

Applications: doors to warehouses, factories, hospitals, loading bays, hotels, etc.

brush strip draught excluder A brush strip made by looping nylon filaments over a core wire. The assembly is locked into a metal channel and the resultant brush strip is housed in an aluminium and PVC carrier. It is available in a variety of depths.

Buchan trap An intercepting trap with a horizontal inlet which incorporates a vertical socket on the inlet side of the trap for the purpose of connecting an access pipe. It is also available with means of access on the trap outlet.

bucket elevator An enclosed, inclined or vertical elevator which conveys material by scoops or buckets. It is used for conveying coal in boiler plants.

bug (computer) An error in a program or an equipment fault.

builder's work drawing A drawing related to services installation indicating the builder's work that is associated with it, e.g. holes or plant bases.

building allergies Allergies caused by a wide range of construction materials and other substances found within buildings which may affect the health of some individuals. These include cements, adhesives, brickwork cleaners and protectants, metals, wood preservatives, treatments and finishes of various kinds, grouting materials, insulants and or sealants containing chemicals which may produce their unpleasant effects after inhalation or skin contact. Certain metals, such as chromium, may cause delayed reactions. See also **Sick Building Syndrome**.

building contract The documents forming the tender for building works and the acceptance thereof.

building energy management systems Computer-based systems for the supervision and automatic operation of building services and systems. Programs are available for peak electrical demand limitation, optimum start/stop enthalpy optimization, efficiency monitoring, planned maintenance, actual energy consumption and cost, direct digital control of plant, access control, fire detection, security patrol, flexitime, lighting, etc.

Applications: large buildings or complexes, chain stores, etc.

Building Services Research and Information Association See **BSRIA**.

bulk density, waste The density of different wastes stored in bulk. This varies greatly, particularly if the waste is wet. For example, the density of wood waste may vary from about 80 kg/m^3 to over 160 kg/m^3, depending upon whether it is hard wood or soft wood, whether it is in the form of wood dust, chips or off-cuts and whether it is moist, wet or dry. The variation of bulk density in other forms of waste may be just as great. The variation of waste fuel volume within a combustion chamber has a significant effect on the combustion air requirements.

bund The impervious masonry construction around the perimeter of a storage area built to an adequate height to contain, in the event of leakage or rupture, all the contents of the storage vessel(s). Bund is generally designed with a 10% margin of over-capacity. It is required as a safety measure around oil storage tanks which are placed outside buildings.

bunker A store for solid fuel.

bunker-to-boiler stoker A feed mechanism located inside a fuel store to convey coal into the boiler via a screw feed tube.

buoyancy, driven ventilation See **displacement ventilation**.

burner purge See **purge**.

burner purge period See **purge period**.

bursting disc A machined disc which is inserted in the shell of a pressure vessel and designed to burst at a predetermined excess pressure.

bus-bar A short solid uninsulated conductor to which several circuits are connected, usually protected inside a bus-bar chamber.

busbars A continuous set of separate solid copper conductors extending the full length of a switchboard or busbar system, to which all the circuits are connected individually via some form of switch or circuit breaker.

bush A fitting that joins pipes of different internal bores. The outer thread screws into the larger bore fitting and the smaller bore pipe or fitting is screwed into the reduced central bore.

butt joint A method of joining pipes by laying the two prepared ends adjacent to each other and welding them.

butterfly control valve A valve that has a control disc rotating in axial trunnion bearings. It may have resilient seating in the body or disc, and resilient lining in the body or metal seatings. It is used for tight shut-off service, low leakage rate service, for fluid flow regulation or for a combination of these. It is suitable for use with thermostatic controls.

butterfly damper A regulating device inserted into ductwork to control gas (air) flow. It comprises a single-leaf control from a damper quadrant. It is a crude method of control due to leakage around the damper leaf.

butterfly valve A regulating device fitted into pipework to control fluid flow. It comprises a single-blade control. It offers crude control as it is susceptible to leakage, and it is used for isolation purposes in a circulation with thermosyphon effect and is fitted within the flow pipe for the best effect.

butt welding Welding where pipes are placed to touch each other and are then welded together by oxyacetylene or arc welding. Abutting surfaces may be bevelled to give greater depth of weld penetration.

bypass A pipe or duct which serves to divert the fluid or gas being conveyed. One application whould be to divert the flue gases from an incinerator around the waste heat boiler directly into the chimney or through a dump heater at times when there is no demand for the waste heat.

bypass control See **compressor by-pass control**.

bypass damper A damper that directs the gas flow around the main duct.

bypass valve A valve that diverts a proportion of the fluid flow directly between the flow and return pipe for the purpose of balancing or pre-heating the mains.

byte (computer) A fixed number of bits. Often these correspond to a single character.

BZ system Short for British Zonal system, it is a system that classifies light fittings according to their distribution of downwards light.

C

CAD Computer aided design.

CEDR The Centre for Dispute Resolution. This body promotes alternative dispute resolution (ADR) in the UK.

CFC refrigerants Refrigerants (Freons) are manufactured from chlorine, fluorine and carbon. They are a group of synthetic organic compounds known as fluorocarbons in which some or all of the hydrogen atoms have been substituted by atoms of fluorine. Such compounds are usually non-inflammable, chemically resistant, non-corrosive, non-explosive, immiscible with water or air and low in toxicity. Given their favourable thermal characteristics they are suitable and widely used as refrigerants in air conditioning and refrigeration apparatus. Further use of Freon refrigerants is discouraged as their release into the atmosphere via leaks and through dismantling of equipment damages the ozone layer which protects the earth.

CHP database Comprehensive information on combined heat and power (CHP) schemes available to the general public through the Office of Electricity Regulation (OFFER) database.

CHP installation with gas turbine A combined heat and power installation which incorporates gas turbine prime movers with alternators, a gas compressor or pump and exhaust heat recovery. Combined cycle installations incorporate gas turbine prime movers with a driven unit together with a steam generator and a condensing or back-pressure turbine.

CIBSE The Chartered Institution of Building Services Engineers, a professional body of engineers concerned with engineering services in buildings. The Institution advances the technology of this industry by means of lectures, seminars, a monthly magazine and the publication of authoritative guides to good practice. It possesses a Royal Charter and is recognized by the Engineering Council for the conferment of the title of Chartered Engineer on suitably qualified members of the Institution.

COP Coefficient of performance. See **heat pump**.

COSHH Regulations Abbreviation of 'Control of Substances Hazardous to Health' regulations.

CSA Commissioning Specialists Association. This association looks after the interests of firms which specialize in the commissioning of building engineering services and offers training courses to their operatives.

cable trunking A box-type conduit for housing cables, usually complete with removable cover, couplers, screws, washers and earth links.

calibration The process of adjusting the readings of an instrument against those of another (standardized) instrument.

call-out A request from the user of an installation, e.g. a lift system, for a service mechanic to attend the installation to rectify an operational fault.

call-out charge The invoice raised by a service company in respect of time and materials expended during a call-out visit. Some forms of annual service contracts include call-out visits being made without additional charge.

calorie A unit of heat quantity. A calorie is the heat required to raise the temperature of one gram of water through one degree Centigrade. It has been replaced by the SI unit of the joule (J), one calorie being 4.19 J.

calorific value The quantity of heat which is released during combustion per unit weight or volume. The intrinsic value of any waste as a fuel, therefore, is directly related to its calorific value. Calorific values of different fuels depend largely on the amounts of carbon and hydrogen they contain, the heat being derived from the combustion of these elements to carbon dioxide and water.

calorifier A heat exchanger which comprises two independent heat circuits: a primary one which introduces the heat into the equipment from an external heat source; and a secondary one which accepts its heating from the primary circuit. Thus, there are two sets of circuit connections to a calorifier, the primary which services the primary heater and the secondary via which the heated medium (water or other liquid) circulates between the calorifier and the points of heat use.

calorifier bundle An assembly of tubes on a tube plate to provide immersion heating to a liquid (most commonly water) storage vessel. It may be suitable for steam or liquid heat transfer medium. It is arranged with flow and return or steam and condensate connections for connecting up to primary heating medium pipes. The storage vessel should be fitted with a suitable access/inspection manhole for viewing the condition of the bundle. The heat transfer surface must be specified to suit the required heat output with reference to the available primary heating medium.

calorifier chest That part of a calorifier which embodies the primary pipe connections and which is assembled with the flanged ends to the section which contains the heat exchange configuration.

calorifier, cradles Purpose-made supports for horizontally mounted calorifiers, each usually being cast in one piece to suit the specified calorifier diameter.

calorifier, glazed A calorifier that offers a solution to the problem of hot water storage of aggressive waters, avoiding the use of galvanized surfaces and of copper materials. One such arrangement is the Thermoglazed Calorifier Vessel in which the heating coils are made from special steel. A patented vacuum heating process ensures that the thermoglazing fuses with the steel and forms an even covering over the secondary heating and internal calorifier surfaces of about 0.5 mm thickness. The finished surfaces then have a green glass-like appearance. The smaller sizes (up to 600 litres) are offered as standard packages incorporating a fixed heating coil, access hand-hole, insulation, control module, magnesium anode, lifting lugs and steel protective jacket. The external surface is painted with rust-resistant paint. Larger capacity calorifiers are usually purpose-made.

calorifier, interface A heat exchanger which is interposed between two different systems to enforce physical separation. It is commonly interposed between an old pipe system and new boilers to protect the latter from scale and corrosion products which may reside or be formed in the old system.

calorifier, mountings Devices such as safety values, altitude gauges, thermometers, combined thermometer/altitude gauges, thermostat immersed stems, etc. depending on the selected heating medium.

calorifier, non-storage A calorifier that does not incorporate meaningful storage capacity. The hot water flows rapidly across the calorifier to the heat load and transfers heat from a higher temperature medium to one being circulated at lower temperature.

calorifier, specification A detailed description of a calorifier which must include the following particulars:

(a) available primary heating medium – steam, hot water or other heat transfer fluid;
(b) pressure or temperature of heat transfer fluid;
(c) required heat output per hour;
(d) static pressure to which the calorifier will be subjected;
(e) required bosses for connecting pipes, mountings and associated controls;
(f) integral thermostatic control, if required;
(g) integral steam trapping set, if required;
(h) integral drain cock, if required.

calorifier, steam-to-water A calorifier in which the primary heating medium is steam. This condenses within a heater battery or coil and transmits heat to the secondary hot water circuit.

calorifier, storage A calorifier that incorporates volume for the storage of hot water.

calorifier, water-to-water A calorifier in which the primary medium is hot water (commonly at high-pressure at a suitable temperature) which heats the secondary domestic or process hot water. The latter must always be at a lower temperature than the primary hot water.

candela The SI unit of luminous intensity.

capacitance The particular property of a system of electric conductors and insulators which allows them to store an electric charge when there is a potential difference between the conductors. Capacitance is measured in farads (F).

capacitor An arrangement of electric conductors in the form of sheets or foil separated by thin dielectric material, serving as 'capacity' for the system by storing a small electric charge. Any system or part of an electric system or a component which possesses appreciable capacitance. See also **capacitance**.

capillarity The rising or falling of a liquid in a narrow tube. When a tube (pipe) of small diameter is inserted into a liquid, the liquid within that tube can be observed to rise or fall, this being caused by the fall in pressure on the underside of the meniscus due to the surface tension. The pressure at the centre of the meniscus is subatmospheric as the meniscus is sagging and this causes the elevation of the liquid in the tube. This phenomenon is termed capillarity. With mercury in the tube, the meniscus is reversed in curvature, thus causing a capillary depression in the tube.

capillary action The elevation or depression of a liquid within a small tube inserted into a liquid due to capillarity.

capillary joint A spigot and socket joint with fine clearance which draws molten solder into the space by capillary action.

capillary tube A restricting flow control device that causes the pressure drop of the refrigerant and regulates the refrigerant flow according to the load. It is used with dry expansion evaporators. A capillary tube is a very small diameter tube of considerable length which thus causes the required pressure drop. It is used mainly in small units, such as in domestic refrigerators and unitary air conditioners.

capillary air washer [Fig. 37] An air washer assembly comprising a number of capillary cells which provides more efficient water-to-air contact than can be achieved in a conventional air washer by dividing both the circulating water and the air into fine streams flowing together through low resistance channels. A capillary air washer consists of a sheet metal housing containing a number of capillary cells, all-glass eliminator plates, water spray header, atomizer nozzles and a water collecting tank. Each cell consists of a casing entirely filled with some 60 000 fine glass strands offering a gross area of close to 12 m^2 which are installed in a direction

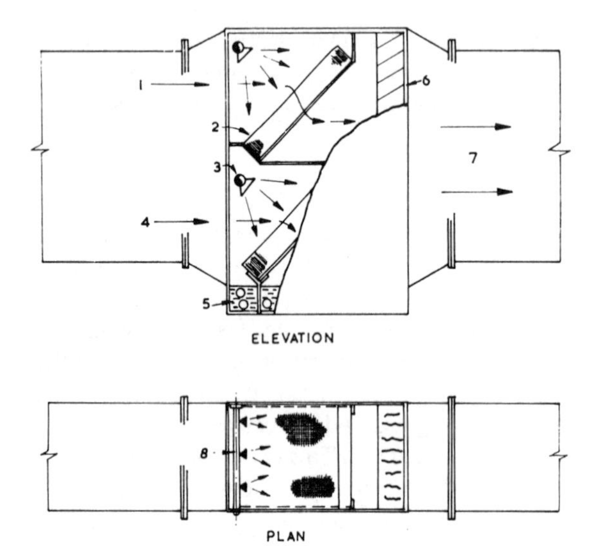

[Fig. 37] Capillary air washer: diagrammatic arrangement. 1, Air in. 2, Capillary cells. 3, Water atomizing nozzles. 4, Air in. 5, Water-collecting tank. 6, Eliminator plates. 7, Air out. 8, Sparge pipe

parallel to the air flow. A water-to-air contact surface of less than 0.25 mm in thickness is established. An acceptable face velocity is 2.5–4.0 m/s. The water is distributed by low pressure nozzles over the upper face of the capillary cell and then gravitates along the many orientated strands to the lower surface where it accumulates until large drops are formed which eventually fall into the collecting tank below. Saturation efficiency of over 95% is claimed when recirculated water is used, so that the air leaves the washer virtually saturated at the wet bulb temperature of the entering air. It is capable of eliminating from the entering air most gaseous and vaporous contaminants and it can be chemically dosed to sterilize the air. Its main use is for humidification of the air; the filtration action is coarse. It requires a pre-filter; dosing with a wetting agent can improve filtration efficiency.

carbon (C) A chemical element and the major combustible component of a fuel. The greater the carbon content, the more available the fuel.

carbon dioxide (CO_2) A gas which is the predominant product of the combustion process. The percentage of CO_2 in a flue gas, coupled with the flue gas temperature, indicates the efficiency of combustion. For example, for fuel oil, 11% CO_2 at 260°C (500°F) yields a combustion efficiency of 80%. The same conditions, but with CO_2 of 8%, yields a combustion efficiency of only 75%.

carbon dioxide, fire protection Carbon dioxide extinguishes fires by reducing the oxygen content and/or disrupting the air/fuel ratio to a stage where further combustion is prevented. Reduction of the normal proportion of oxygen in the air from 20% to 15% will extinguish inflammable liquids. It is well suited for the protection of electrical switchrooms, computer rooms (but only when taking precautions to protect the personnel), substations and electrical equipment in general.

carbon dioxide, gas analyser A device designed to continually aspirate from one or two boiler outlets, analyse flue gas and transmit to a separate indicator or recorder the percentage of carbon dioxide in the flue gas.

carbon dioxide, properties Carbon dioxide is a non-corrosive, odourless, colourless, inert gas with good dialectric strength which has virtually no adverse effect on most materials and foodstuffs. At normal pressures and temperatures, its density is about 50% greater than that of air, so that it will accumulate at low level. It penetrates into otherwise inaccessible areas and disperses without leaving a trace. Whilst it is non-poisonous, it can cause suffocation if breathed in quantity, due to its property of excluding oxygen from the lungs. Hence, its use in fire-fighting requires great care to obviate the possibility of injury to personnel.

carbon dioxide, storage When highly compressed, carbon dioxide liquifies and it is therefore generally stored in liquid form within steel cylinders which are supplied in three standard sizes, having capacities of approximately 23 kg, 36 kg, and 46 kg of carbon dioxide content, respectively, under a pressure of 46–56 bar. There is a risk of explosion when carbon dioxide containers are heated; storage areas are therefore protected by means of water sprays to act as coolant in the event of a fire. Also available are fully insulated low-pressure carbon dioxide storage vessels which are equipped integrally with suitable refrigeration plant to control the pressure and consequently the temperature of the liquid carbon dioxide.

carbon dioxide, total flooding system A system that comprises the fixed installation of permanently connected CO_2 cylinders with piping and discharge nozzles, used where a hazard area is within a permanent enclosure and permits build-up and maintenance of CO_2 concentration.

carbon filter See **activated carbon filter**.

carbon monoxide (CO) A gas that is the product of incomplete combustion. Its presence indicates a deficiency in the energy process.

carbon monoxide monitor A device that continuously monitors carbon monoxide levels in industrial boilers, kilns, large furnaces, etc. using an on-line non-sampling infra-red absorption technique.

carbonization The destructive distillation of coal to produce coke, gas and liquid residue.

Carnot cycle The ideal cycle for a heat engine, giving maximum theoretical efficiency. Heat is added at constant temperature T_a and is rejected at a constant lower temperature T_b.

$$\text{Thermal efficiency} = \frac{\text{Work done}}{\text{Heat added}} = \frac{(\text{Heat added} - \text{Heat rejected})}{\text{Heat added}} = \frac{T_a - T_b}{T_a}.$$

carryover Droplets of water and associated impurities carried by steam from a boiler to the superheater.

cartridge fuse A protective device in an electrical circuit in which the fuse element is totally enclosed. It may be suitable for fixing in the fuseholder of a fuseboard or switch-fuse; alternatively, it may be placed inside a plug which connects a flexible cable to an appliance or be in the form of a contact pin which mates with a socket contact. A cartridge fuse consists of a barrel or cartridge of heat-proof insulating material with metal end caps, sometimes incorporating fixing lugs. The fusible element is made from silver wire and is stretched between the end caps inside the cartridge. The space around the element is filled with treated and graded quartz. In the event of an overload, the element is vaporized and the resulting heat causes the quartz filling to fuse into a solid non-conducting core which will not permit arcing.

cast-iron sectional boiler A sectional boiler made of cast-iron. See also **sectional boiler**. Such a boiler has a limited working pressure and thus is unsuitable for use in tall buildings where the boiler is placed on the lowest floor level. This problem may be overcome by locating the boiler in a roof-top boiler room. Cast-iron boilers do not generally suffer severely from corrosion at the metal surfaces and thus offer a longer working life than boilers manufactured of steel.

catalyst A substance which accelerates or retards the rate of a chemical reaction without itself undergoing permanent change in its composition. It is normally recoverable when the reaction is completed.

cat and mouse contents gauge A gauge in which a float suspended inside a vessel floats on top of the liquid to be measured and is connected over a pulley to a weight suspended from a strong wire. As the liquid rises and falls, the weight travels over a vertical calibrated contents scale.

catastrophe theory The analysis of major mishaps in organizations. The five stages in catastrophe theory are:

(a) the formation of an error chain;
(b) the identification of a responsibility gap;
(c) information difficulties;
(d) poor control;
(e) rigid and limited perception.

Most catastrophes are the consequence of a long chain of small events coincidentally occurring at the same time. If the small events can be managed properly, then major ones are unlikely to arise.

catchment areas Those areas from which rainwater will run off into a particular surface water drain.

catchpot trap A type of bottle trap with a removable lower part which is sufficiently voluminous to retain waste for recovery and examination.

cathode A negative electrode. In building services it applies to corrosion protection systems.

cathodic protection [Fig. 38] A commonly used method of protecting a material subject to corrosion (cathodic metal) by using a more electronegative metal, the most common of these being either magnesium, aluminium or zinc, in the form of sacrificial anodes which are connected electrically to the component(s) requiring protection. These sacrificial anodes will then slowly dissolve instead of the protected metal. Another such method uses an impressed electric current from a generator in conjunction with auxiliary anodes of iron, steel, graphite, lead or platinized titanium.

cathodic water storage tank protection by impressed current A method that prevents corrosion of the inside surfaces of steel tanks. In a typical installation, equipment for each tank consists of corrosion-resistant silicon iron anodes which are suspended from the roof with polypropylene ropes and supplied with low-voltage d.c. current from a transformer/rectifier. The anodes are suspended on either side of the centre line of the tank and at such a height to ensure that they will always remain submerged. Electric cables from each anode pass through a steel roof plate in watertight glands to a junction box situated on the roof. The positive feeder from the transformer/rectifier also terminates in this box and supplies current to all the anodes in parallel. The circuit is completed with a negative cable which is connected between the rectifier and a stud on a tank wall.

The d.c. current flowing from the anodes through the water and into the tanks prevents electrolytic corrosion at areas where the protective tank coating has been damaged or has deteriorated. Silver/silver chloride electrodes, suspended from the tank roof, monitor the electrochemical potential of the steel. This has to remain within a band of –850 mV to –1400 mV to ensure protection over a wide range of varying water conditions. Controls can be devised which enable the current to be varied automatically to maintain the metal of the tank at a preset potential. Such a system can protect water storage tanks over very long periods of time with minimal maintenance requirements.

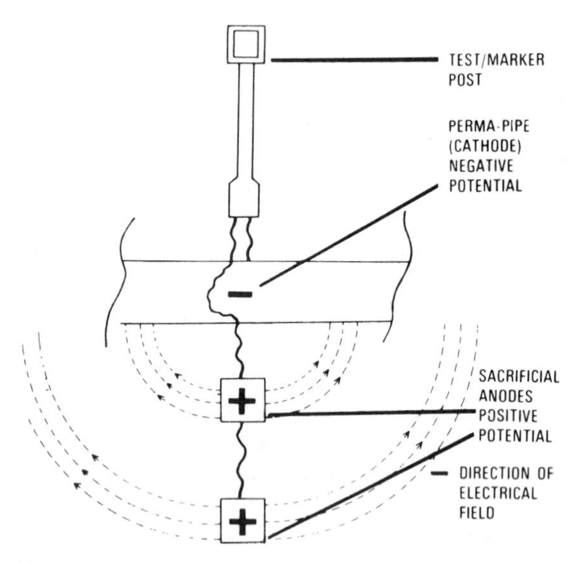

[Fig. 38] Sacrificial anode used for pipeline protection

caulking A method of jointing cast-iron pipes whereby the jointing material is forced between pipe and socket by means of a caulking tool.

caustic embrittlement The embrittlement of metal caused by impurities carried by water. It is a cause of metal failure in steam boilers at riveted joints and at tube ends.

cavitation The generation of localized air pockets within a liquid caused as a result of a reduction of pressure in equipment such as pumps. High-pressure hot water systems are particularly prone to cavitation which may be caused by loss of pressure due to pipe friction allowing the high temperature water to flash into steam. In such systems, a common precaution is to install a fixed-position bypass valve which always blends a fixed quantity of return water into the main flow pipe of the system to maintain a predetermined temperature margin between the temperature in the boiler and that in the system to obviate cavitation. The effects of cavitation may include irregular pumping, water hammer and damage through erosion at pipe fittings.

cavity wall insulation Water-resistant polyurethane foam inserted into cavity walls which does not shrink or crack and which adds to the strength of the building.

ceiling fan economizer A thermostatically controlled ceiling fan in a casing which blows hot air downwards. It is suitable for heights of up to 16 m.

ceiling heating, embedded The arrangement of space heating in which the heat emitters are serpentine pipe coils which are directly embedded in the ceiling construction.

ceiling heating, suspended The arrangement of space heating in which the heating elements comprise pipes with attached heat reflectors overlaid with thermal insulation and are laid within a suspended ceiling space.

cellular air filter See **capillary air washer**.

cellular glass insulation A totally inert insulation material formed of glass fibres. It is water vapour, fire, rot and vermin proof, and is dimensionally stable. It is supplied in slab form or fabricated to fit around pipework.

cellular offices The division of office space into a number of partitioned-off rooms which are usually entered from a common corridor.

cement grouting with pistons A well-tried method of repairing a rigid drainage system, comprising two rubber pistons separated by a slug of cement grout which are pulled through the drain run from one manhole to the next. The pressure of one piston against the other squeezes the cement grout into any cracks and gaps and seals them. This procedure is repeated until the drain satisfies the appropriate water test.

centistokes An international unit for measurement and expression of all kinematic viscosity values, measured in m^2/s – 1 centistoke = 10^{-6} m^2/s.

central air-conditioning An arrangement in which a number of air-conditioning terminals are served from a central plant, usually including, chiller, boiler plant, air- or water-cooled cooling tower/condenser and air handling/treatment equipment.

central battery system A system of emergency lighting which has a battery room in which the charger, batteries and switching devices are located.

central control panel A panel that houses an assembly of switchgear, time switches, meters, indicator lights, etc. It is suitably sited to facilitate central monitoring and operation of the controlled installation from one convenient location.

central heating A space heating arrangement in which one boiler plant serves a number of dwellings, rooms or offices.

centrifugal chiller See **chiller, centrifugal**.

centrifugal compressor A compressor that incorporates vaned impellers which rotate inside a casing. These increase the velocity of the gas which is then converted into a pressure increase by decreasing the velocity. It is suitable for large capacities. The impellers can be rotated at speeds up to 20 000 r.p.m., enabling it to handle a large quantity of refrigerant. It may be of the hermetic or open type.

centrifugal fan [Fig. 39] A fan that consists of an impeller running inside a casing of spiral-shaped contour. The air enters the impeller in an axial direction and is discharged at the periphery, the impeller rotating towards the casing outlet. The pressure development of the fan depends primarily on the angle of the fan blades with respect to the direction of rotation at the periphery of the

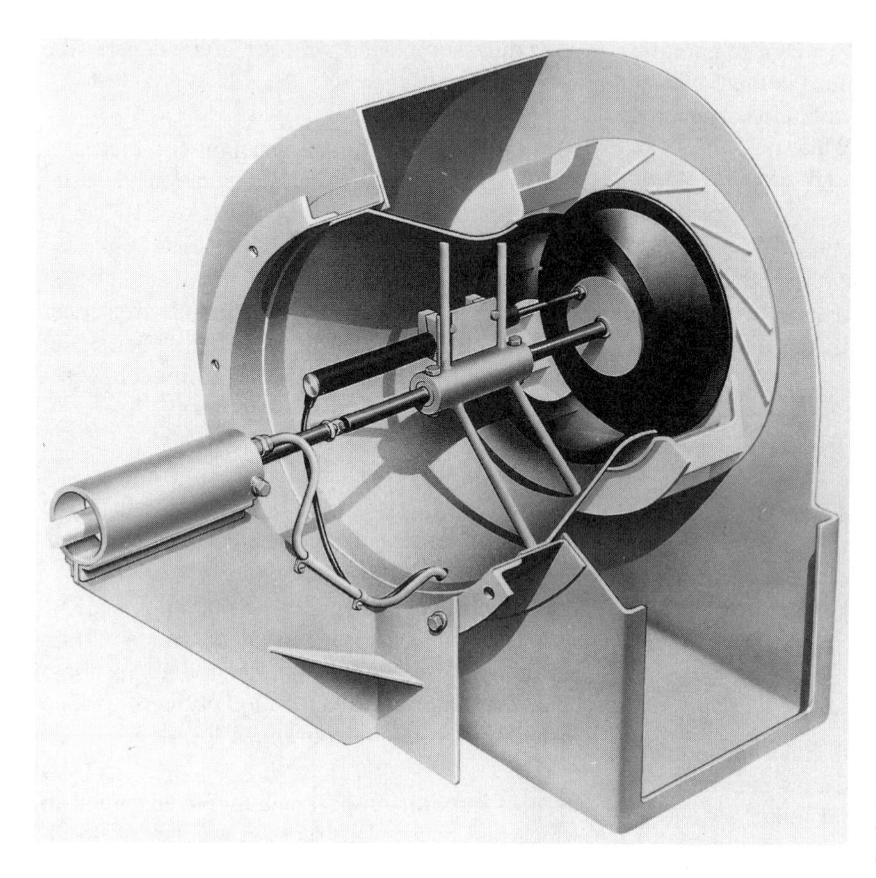

[Fig. 39] Disc throttle centrifugal fan: variable air volume control (Woods of Colchester Ltd.)

impeller. Three main forms of fan blade are available as follows.

(a) Backward blade: the blade tips incline away from the direction of rotation; the blade angle is less than 90 degrees.
(b) Radial blade: the blade tips (or the whole blade of a paddle blade fan) are radial; the blade angle is 90 degrees.
(c) Forward curved: the blade tips incline towards the direction of rotation; the blade angle is greater than 90 degrees.

The pressure development of the fan rises with an increase of blade angle; for an impeller of given proportions, the volume flow then also tends to increase. Centrifugal fans may be of the single-inlet type in which the air enters from one side only or of the double-inlet type with the air entering through both inlets. The latter can handle about twice the amount of air.

centrifugal fan, double inlet A fan of double width in which the air is drawn into the fan via the two inlet ports. It can handle twice the volume of a single-inlet fan.

centrifugal fan, single inlet A fan in which the air enters via one inlet port.

centrifugal in-line duct fan An extract fan designed for in-line ducted ventilation systems where centrifugal fans are considered essential to create sufficient air flow along lengths of ducting. Generally robustly constructed with a steel outer casing which is coated with pressure-cooked epoxy resin, it is manufactured for long operating life, minimum maintenance (because of its intended location) and rust resistance. One manufacturer offers a range of seven models in standard diameters ranging from 97 mm to 312 mm, arranged to fit commercially available ducting.

centrifugal pump A pump that essentially comprises an impeller which rotates within a volute casing which must be completely filled with liquid when the pump is in operation. The impeller throws the liquid being pumped to the outside of the volute casing thereby imparting kinetic energy and generating a certain head (pressure) which is a function of the pump speed and which causes the pumped liquid to flow against the resistance of the pipe system.

ceramic heat wheel A heat recovery system incorporating a ceramic heat wheel for applications of up to 1200°C. The system package supplied includes combustion air and exhaust fans, interconnecting ducting, safety controls, etc.

ceramic membrane Alcan's Ceramesh ceramic membrane is a composite ceramic/metal-mesh membrane which is uniquely flexible and formable. It can be retrofitted into conventional membrane filtration equipment. The mesh material is a high-temperature and corrosion-resistant alloy, such as Iconel 600. The ceramic phase is usually zirconia.

certify To issue a certificate to confirm or instruct. For example to issue a certificate for payment of monies to a contractor, certify that a section of works has been satisfactorily completed, certify that a particular welder is suitably qualified for the job in hand, etc.

cesspool A chamber below ground provided for the reception and storage of sewage or foul water installed in locations which are without main sewer facilities. The contents are periodically removed by special vehicles for disposal.

The term is also used for a box-shaped receiver which is fitted in a roof or gutter for collecting rainwater which then passes into a rainwater pipe connected to it.

cesspools, etc., connection to The velocity of entry of the discharge must be limited to that which will provide an adequate flow into the cesspool without disturbing the contents. Where the incoming drain is of size up to 150 mm, it is usual to provide a maximum gradient of 1:50 for an upstream pipe length of at least 12 m and adequate provision must be made to permit effective rodding of the incoming drain and of its connection to the cesspool.

chain grate A continuous chain of fire bars which are linked together to convey a solid fuel bed through a furnace.

chain grate stoker [Fig. 40] A type of mechanical stoker which comprises a continuous chain of cast-iron fire bar links driven through sprockets and moving from the front of the furnace to the rear of the combustion chamber. The fuel is fed to the grate by gravity chute and is spread over the grate to a bed thickness which is regulated by the fire door to suit the fuel being burnt. The fuel burns as the grate moves along. Ash and clinker are removed at the rear end by ash plates. To stabilize ignition by radiation, the front section of the chain grate is shrouded by a refractory arch which extends down to the level of the grate and is heated by the burning fuel below.

chair Applied to sanitation, a chair is a metal frame for building into a thin wall partition and the floor to support a wash-basin, W.C. pan or other sanitary appliance clear off the floor.

change of state The melting or formation of ice, or the changing of liquid into vapour at constant pressure and without change of temperature.

change-over switch A switch that changes operation from one unit to its standby or alternative one. It may be manually or automatically actuated.

change-over valve system A valve system that operates to change the functioning of a plant from one mode to another; for example, changing over a two-pipe fan-coil or induction air-conditioning system from cooling to heating or changing the heat pump mode from cooling to heating.

characteristic The specific property of an item of equipment or of a material, e.g. a fan or a pump. See **fan characteristic**.

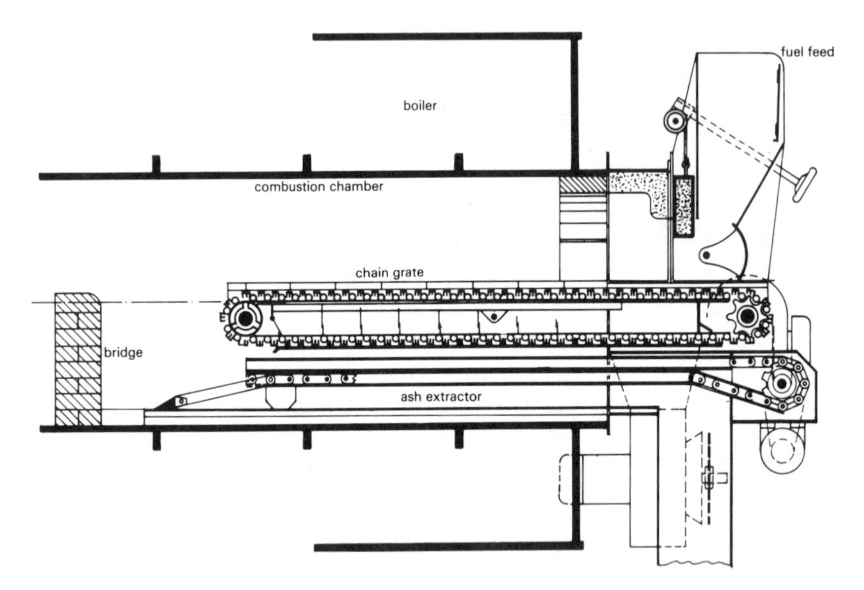

[Fig. 40] Chain grate stoker

charging door A hinged or sliding door/gate through which a solid fuel fired combustion chamber is fuelled.

Charles's law A scientific law which states that a given mass of a gas expands by a constant fraction of its volume at 0°C when its temperature is raised by one degree, provided that the pressure remains constant.

chartered building services engineer An Associated Member, Member or Fellow of the Chartered Institution of Building Services Engineers.

chartered engineer A registered title conferred on qualified engineers by the Engineering Council.

Chartered Institution of Building Services Engineers See **CIBSE**.

Chartered Institution of Electrical Engineers See **IEE**.

Chartered Institution of Mechanical Engineers See **IMechE**.

check meter A meter interposed between a meter at a consumer's boundary and a section or department to measure only the consumption of that section.

check valve A valve that prevents the reversal of fluid flow by means of an integral check mechanism, the valve being opened by the flow of the fluid in one desired direction and closed when the flow ceases or by back pressure on the valve.

check valve, ball See **ball check valve**.

check valve, disc See **disc check valve**.

check valve, featherweight See **featherweight check valve**.

check valve, lift See **lift check valve**.

check valve, piston See **piston check valve**.

check valve, screw-down and stop See **screw-down and stop check valve**.

check valve, swing See **swing check valve**.

check valve, vertical See **vertical check valve**.

chemical cleaning See **cleaning and descaling of sanitary installations**.

chill beam A device in the form of a structural beam which has been developed from the radiant cooling panel to overcome its major drawbacks. Essentially, there are two types of chill beam; one that acts purely as a cool radiant surface and one that includes heating and/or ventilation. Room air must be allowed to circulate freely on all sides of the beam.

Chill beams are best incorporated within a system of displacement ventilation which moves the ventilation air upwards from a low level. The chill beams are located at the upper level to counteract heat gains from the lighting and from the rising air. The output from a chill beam is essentially self-balancing; i.e. when there is no heat gain, its convective output (some 85% of the output) approaches zero, whilst the radiative effect (some 15%) continues to cool the occupants regardless of the air temperature.

chilled water system A system that circulates water which has been reduced to a specified temperature by passage over a water cooling device, such as cooling coils, and connects to cooling units, e.g. fan coils or air handlers.

chiller A machine for air or liquid cooling.

chiller capacity The refrigeration output available from a chiller expressed in terms of tons refrigeration. One ton equals a capacity of 3.15 kW (SI) or 12 000 BTU/h (imperial).

chiller, centrifugal A centrifugal compressor and prime mover, water-cooled condenser and flooded evaporator assembled as a complete unit with interconnecting pipes and controls. Capacities vary between 175 kW and 17 500 kW.

chiller, reciprocating A reciprocating compressor and prime mover, water-cooled condenser and a flooded or direct-expansion evaporator assembled on to a common base, complete with pipes and controls. The capacity can be up to 350 kW.

chimney The assembly of all sections, fittings and accessories necessary for the conveyance of flue gases from the point of discharge to atmosphere. It may be constructed of masonry, concrete, steel, etc.

chimney access ladder A tall chimney should be provided with permanent access facilities to the inspection doors and to the chimney terminal comprising hooped safety ladders.

chimney by-pass A section of the flue system which is installed in parallel with a heat exchanger, waste heat boiler, etc. and functions to pass flue gases directly to atmosphere. The operation of the by-pass facility is activated by manually or electrically actuated tight-fitting flue gas dampers suitable for high-temperature use.

chimney cladding A metal sheath placed about the outside of a steel chimney for insulation or protective purposes. The space between the metal of the chimney and the cladding may be taken up with thermal insulating material.

chimney condensation Condensation caused by the cooling of the chimney gases below their dewpoint and the consequent deposition of moisture which generally has corrosive properties. Condensation can be averted by thermal insulation of the chimney to minimize the cooling of the flue gases. Some condensation will occur when the plant is started up from cold but this is usually re-evaporated when the operating temperature is reached. A trapped discharge pipe should be connected to the base of the chimney to convey the water of condensation to a gully or drain.

chimney downwash The descent of flue gases down the outside of a chimney, caused by the excessively low efflux velocity of flue gases which cannot discharge into the atmosphere and tend to flow down along the outside of the chimney terminal. A similar effect may arise from high wind velocities acting on a gas plume.

chimney draught The difference in pressure between a location at the base of a chimney and the chimney terminal which motivates the evacuation of the flue gases through a chimney system.

chimney draught, balanced A chimney draught system that utilizes two fans; one to supply the air for combustion and the other to induce draught. The arrangement is termed a balanced-draught

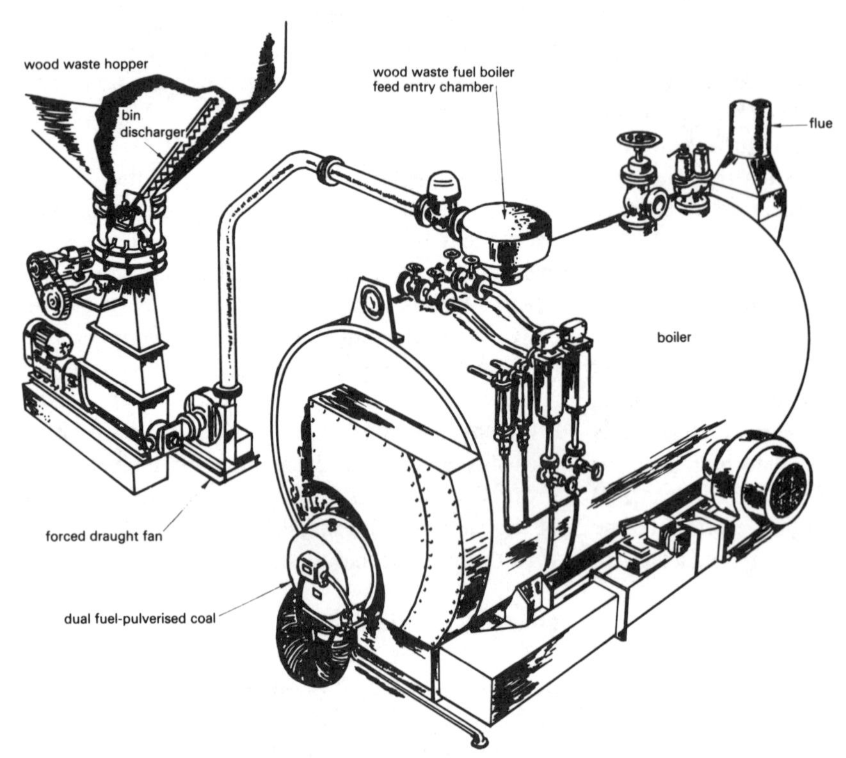

[Fig. 41] Forced draught boiler

system and it offers the most sophisticated draught arrangement, as it permits close control over the combustion air supply and chimney draught conditions.

chimney draught, forced [Fig. 41] A chimney draught system where a fan supplies the air for combustion, overcomes the frictional resistance of the boiler or furnace and associated chimney system and maintains the required efflux velocity. In some plants the fan pressure is calculated to overcome only the resistance of the boiler or furnace system, the natural chimney draught being utilized for the necessary suction to draw the flue gases to atmosphere.

chimney draught, induced A chimney draught system where a fan is inserted between the flue gas outlet from the boiler or furnace and the entry to the chimney. Usually this fan is located in a bypass to the main flue connection into the chimney. The latter is fitted with a tight shut-off damper which remains closed whilst the induced draught fan operates. Pressure generated by the fan discharges the flue gases against the resistance of the chimney system and maintains the required terminal efflux velocity. Fan suction must be adequate to draw the flue gases through the incinerator, boiler or furnace, overcoming friction to the flow of the flue gases.

chimney draught, mechanical A system in which the chimney draught is generated by the use of fans to motivate the evacuation of the flue gases. It confers control of the chimney function regardless of wind, weather, chimney height and boiler conditions and permits the choice of chimneys of a smaller cross-section and height, subject only to regulations concerning chimney heights. The assembly of the chimney system must be designed to minimize the loss of draught and pressure to economize on the electrical consumption of the fan(s).

chimney draught, natural Natural buoyancy forces caused by the difference in densities between the gases at the base and the terminal due to the temperature differential between the flue gases and the ambient air. This depends on the height of the chimney and on the friction in it to the flow of flue gases. It varies with wind conditions but can be controlled automatically by the admission of ambient air into the chimney, near the base, by means of a draught stabilizer.

chimney, draught stabilizer A device that adjusts the temperature of the gases travelling up the chimney by diluting them with colder ambient air. There is commonly a combined chimney clean-out door and draught stabilizer within one common frame built in or attached to the chimney just below the entry of the flue pipe. The stabilizer comprises a weight-loaded steel damper which is adjusted to maintain a certain pressure at the base. The damper swings to a more open position when the draught is excessive, admitting more ambient air and closes when the draught is low.

chimney effect Air movement caused by a vertical temperature gradient.

chimney efflux velocity The speed at which the flue gases exit from the chimney terminal. This should be about 6 m/sec for natural draught operation and 7.5–15 m/sec for mechanical draught systems. Incinerators usually operate at not less than 6 m/sec when at full output with forced draught, not less than 7.5 m/sec with induced draught for a connected load of up to 9000 kW, increasing to 15 m/sec for connected loads up to 135 000 kW. Lower efflux velocities tend to promote conditions for downwash and temperature inversion.

chimney, flexible liner A flexible steel tube inserted into a masonry chimney to protect it from the effects of conveying boiler flue gas. It is secured by closure/support plates at top and bottom.

chimney foundation A reinforced concrete block sunk into the ground, of suitable dimensions and materials to support the weight of the chimney. The bottom fins of the chimney are attached to a metal plate which is secured to the foundation block by means of stainless steel holding-down bolts and nuts.

chimney, free-standing A metal chimney which is supported off a concrete base without the use of guy ropes or brackets.

chimney inspection doors Doors provided to permit the periodic inspection of the condition of the inside of the chimney and of any lining installed therein. It must be substantially constructed with stainless steel fastenings.

chimney, lined A chimney whose internal surfaces are covered with a material which is suitable for the chimney operating conditions to maintain the temperature of the flue gases above the acid dewpoint and/or to protect the material of the chimney construction from contact with corrosive gases.

chimney lining The insertion of an inflated rubber bag into a chimney after which a suitable insulation compound is pumped between the bag and the masonry to provide a permanent protective chimney lining.

chimney loss The residual heat content of the flue gases discharged into a chimney and then lost to atmosphere.

chimney loss measurement kit A kit typically containing thermometer, oxygen analyser, probe, reference table, soot pump and papers and battery charger within an attache case, which can quickly determine approximate dry stack losses and make suitable adjustments to the air/fuel ratio of the burner by obtaining a direct digital readout of percentage oxygen. It is suitable for use with boilers and furnaces.

chimney, multi-compartment [Fig. 42] Two or more separate compartments housed within one chimney construction, each connected to a separate boiler, incinerator, etc. Such a configuration subjects the separate compartments to unequal thermal stresses which tend to promote fractures of the chimney construction and therefore should be avoided.

chimney, self-supporting A chimney construction which is supported directly off a foundation block without requiring brackets or guy ropes.

chimney strake Metal strips in the form of a helix which are wound around the top section of a tall

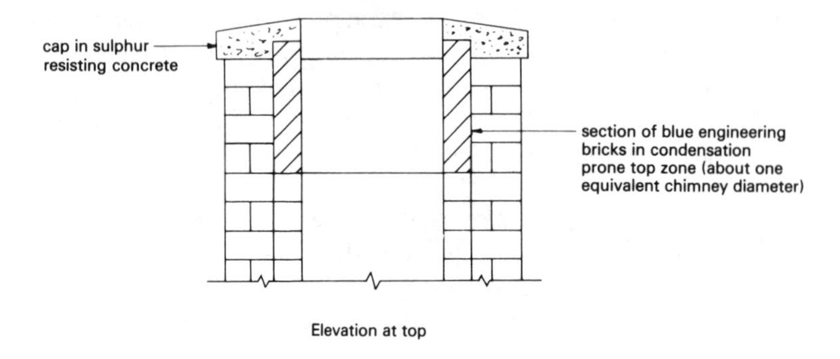

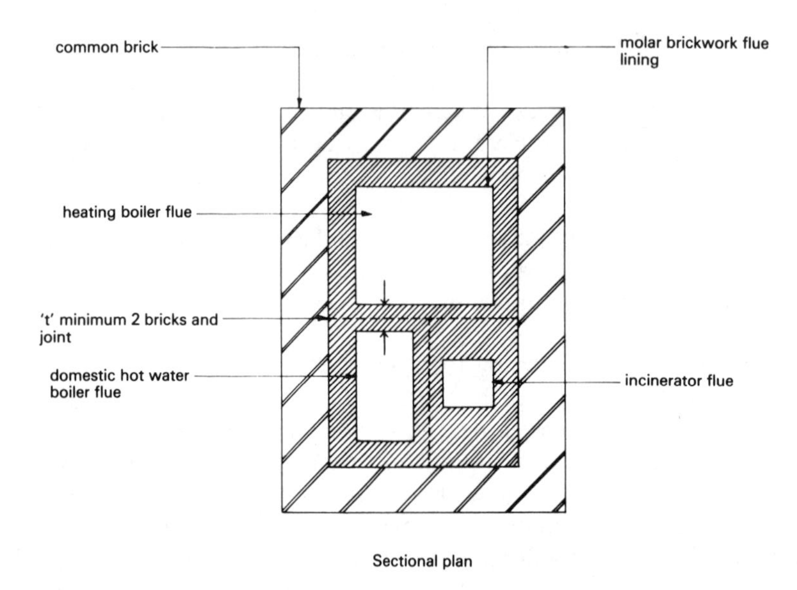

[Fig. 42] Multi-compartment chimney construction (above), and section through it (below)

chimney for the purpose of preventing it from oscillating in high winds.

chimney support [Fig. 43] A device for supporting safely the load imposed by the chimney. It may be in the form of a foundation base, wall-attached brackets, guy ropes and strainers, etc. Chimneys may be free-standing, bracketed off walls or secured by guy ropes with supports to suit these applications. Local Authorities in the UK insist that all such supports are safe and monitor their design.

chimney terminal [Fig. 44] The point in a chimney at which the flue gases exit to atmosphere. This must be designed to promote the required efflux velocity of the flue gases and to exclude the ingress of water. The terminal can be of conical shape to increase the efflux velocity. It is usual to construct the final courses of a brick chimney in engineering brick to withstand the corrosive effect of flue gases condensing close to the exit. The terminals of metal chimneys are commonly pointed black to comouflage the effect of smut emission.

chimney, TV survey The process of inspecting the internal condition of a chimney by lowering a television camera and surveying all otherwise inaccessible parts. The developed film of the survey provides the result and records of the survey.

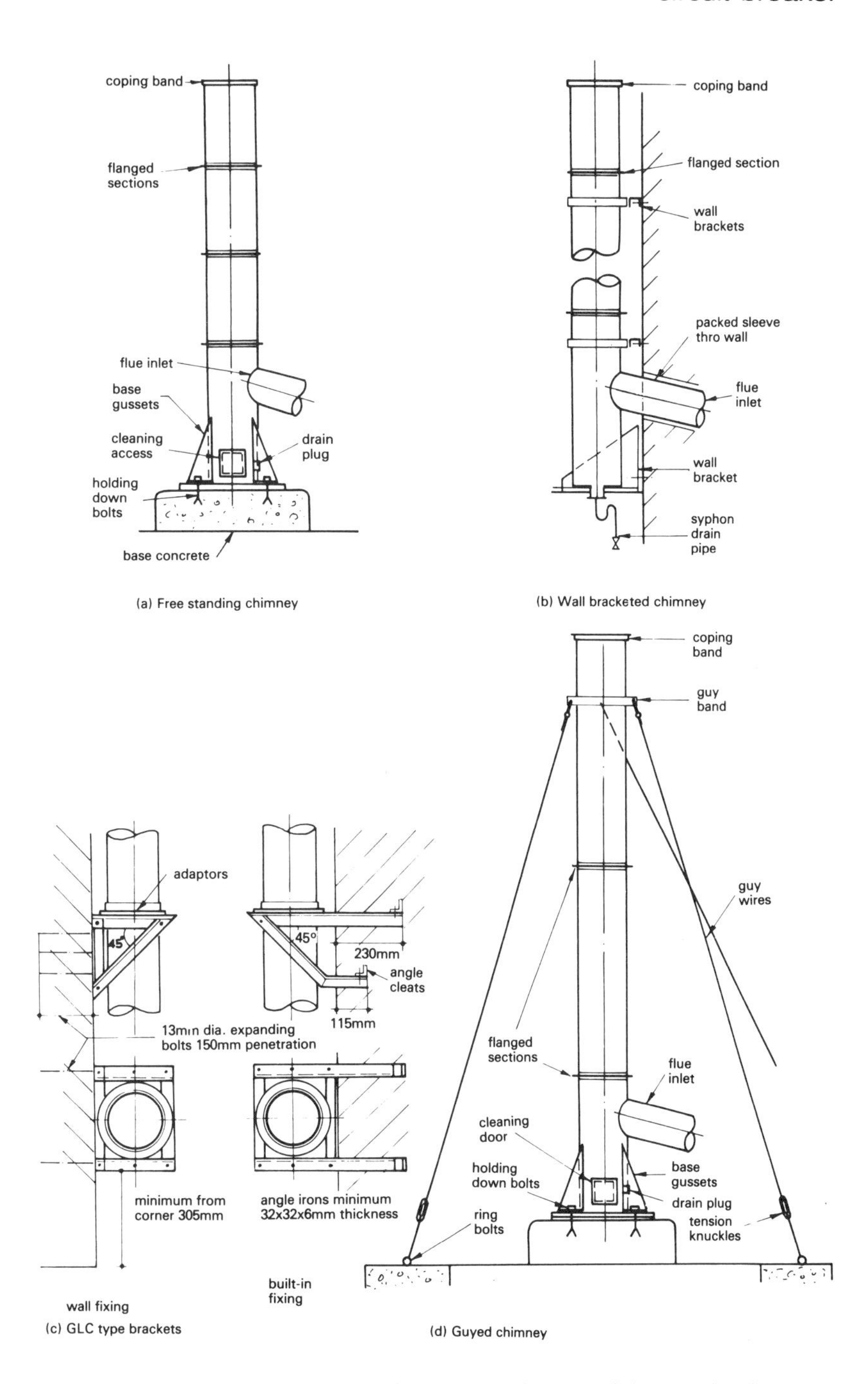

[Fig. 43] Various types of chimney support

chimney windshield A concrete or masonry shell protecting chimney compartments rising therein.

chipper knives Knives incorporated with hoggers which operate to reduce chunks of wood to manageable chips. The action wears out the knife edges rapidly; the harder the wood being hogged, the greater the wear. It is essential to maintain spare sets of knives for speedy replacement and to have a facility for sharpening the knives.

circuit breaker A device capable of making, carrying and breaking electric currents under normal operating conditions and also under specified

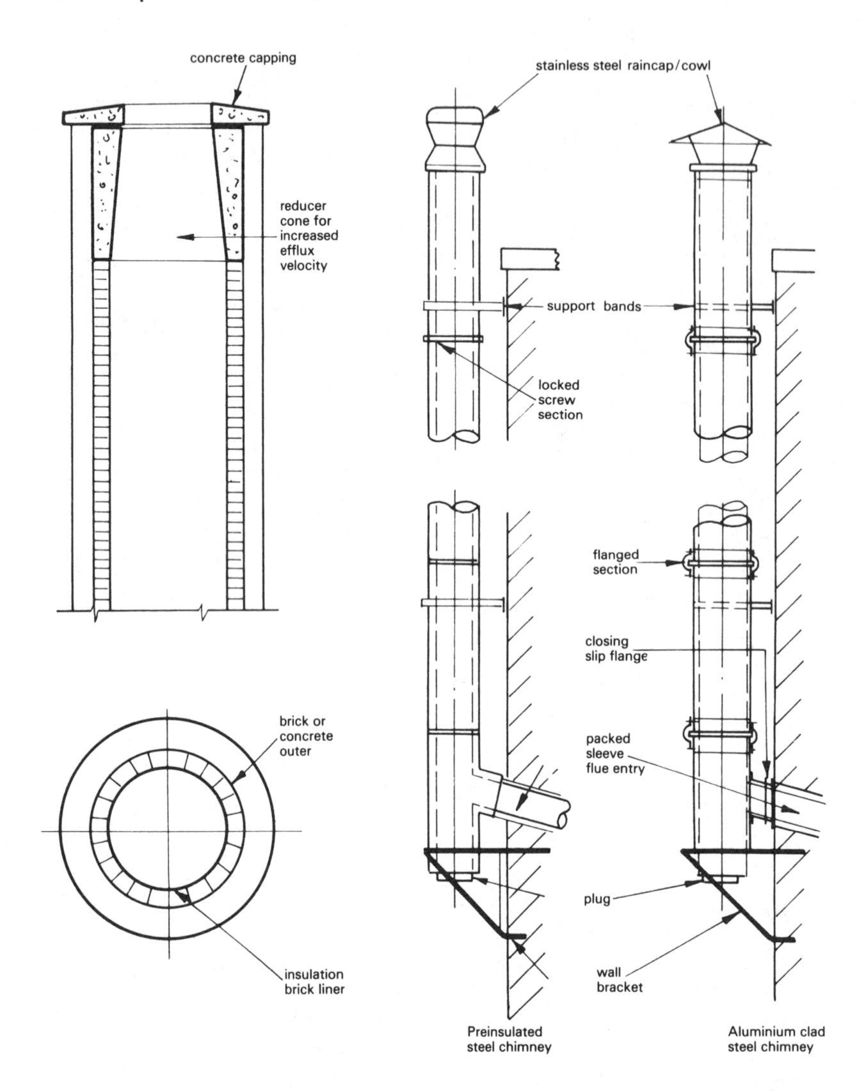

[Fig. 44] Chimney terminals for mild street, masonry and pre-insulated packaged chimneys

conditions or under abnormal circuit conditions such as a short circuit.

circular pattern air diffuser [Fig. 45] A diffuser that discharges the supply air in a circular movement.

circulating pressure That pressure available within a liquid system to overcome frictional resistance to the velocity of flow and to provide the required outflow (the latter in water supply systems).

circulating pressure, gravity That circulating pressure which is generated within an unpumped heating or hot water supply system by the thermosiphonic effect which depends on the difference in density between hot and cooler water in a circuit and on the circulating height.

circulator Short for circulating pump.

cistern Usually a small tank; e.g. a flushing cistern or feed and expansion cistern.

cistern flushing mechanism This should be fully compatible with the toilet pan and must be specified as a total component.

[Fig. 45] Air distribution: circular pattern air diffuser (Trox Brothers Ltd.)

cistern flushing mechanism, ball valves Ball valves must be selected to suit the available water pressure, i.e. whether mains or tank water. Incorrect selection is liable to reduce the water filling effectiveness and to generate noise during filling.

cistern flushing mechanism, syphonic Operation of the flushing lever lifts the water in the bell of the syphon over the invert and discharges it into the flushpipe to the water closet. The action continues until the water level falls low enough to allow air to enter the bell of the syphon.

Cisternmiser The trade name of a device for reducing water wastage in the use of automatic flushing systems in men's toilets. It consists of a non-electric mechanical control valve which automatically shuts down the water supply to such systems when they are not in use and offers considerable scope for reducing water consumption.

cladding A metal or plastic layer applied over vulnerable components such as thermal insulation as protection against damage or to enhance appearance.

clean room An area designated for a high standard of air filtration. It generally uses high-efficiency particulate air filters capable of removing 99.97% of all particles in the supply air of 0.3 micron diameter or more. Air distribution is commonly achieved by downward displacement of the air from a plenum above a ventilated (perforated) suspended ceiling.

Applications: electronics, pharmaceuticals, microbiology and similar industries.

clean zone ceiling A stainless steel structure designed to promote clean air techniques in hospital operating theatres to protect the operation field and other sterile zones from bacterial contamination and to promote thermal comfort for the patient and for the operating team. It incorporates supporting air flow nozzles which are integrated into the ceiling design to stabilize the downwards sterile air flow even alongside low cooling loads and high-performance suspended particle filters on the air admission side in conjunction with pre-filters.

cleaning and descaling of sanitary installations Such installations will be periodically required to be cleared of an accumulation of limescale, soap, blockages, etc. The following methods are in common use.

(a) Plunger: a simple method of clearing minor blockages at the traps of sinks, basins and even VCs.
(b) Rods: used for manual clearance of blocked drains. Rods are coupled together and are progressively introduced into the drain via access manholes or gullies. They are manipulated backwards and forwards until the blockage is cleared. Mechanically rotated versions of rod systems are available.
(c) Kinetic ram: a kinetic ram gun can be used to clear obstructions in branch pipes, but its function and limitations must be understood as misuse can prove dangerous. On plumbing installations the use of the kinetic gun should be restricted to removal of blockages which consist of compacted soft material, such as grease, soap residue and saturated paper.
(d) Coring and scraping: coring of pipes can be considered in pipes of 100 mm bore and greater, where the pipe bore is severely restricted or even completely obstructed by hard lime and/or similar material. Before contemplating such an operation, it must be ascertained that the

pipe material cannot be damaged in the process. This method involves the use of a rotating steel cutter on a flexible drive which is pushed into the pipe to clear the blockage. The peripheral accumulations of grease and other deposits in these large-bore pipes can generally be periodically removed using profile scrapers which are attached to ropes and are pulled through the pipes.

(e) Chemical cleaning: this method uses acid-based chemicals which must be applied with due regard to the safety of the operator who must follow the manufacturer's detailed instructions. On completion of the treatment, all the pipework must be thoroughly flushed out with clean water.

clerk of works The supervising officer appointed by an employer to oversee his interest at the site of the works. The clerk of works monitors compliance with specification, checks works before these are covered up, attends site meetings and maintains a log of progress of the works.

clinker Solidified non-combustible residue separated in the fused state. It is also referred to as slag and must not be confused with ash.

clorius heat meter See **evaporation heat meter**.

chlorometer (Palintest Ltd) Direct-reading electronic photometer, suitable for laboratory analysis or field testing. It is calibrated to provide fast, accurate readings of residual chlorine in drinking water samples. The reagent system enables free, combined and total chlorine levels to be determined.

closed-circuit television (CCTV) A reliable method of inspecting and recording the internal condition of sewers and drains on black and white or colour video film, the latter being the more popular. The eye of the modern compact camera is of postage stamp size and can negotiate 90 degree bends and branch pipes. Most Local Authorities insist on being handed a video record of a sewer or drain which has been photographed by an independent survey firm before adopting the system.

closed-cycle gas turbine [Fig. 46] A turbine in which the working fluid, usually air, is continuously recirculated, the heating required for expansion being supplied by an air heater.

closed-loop control system A system that measures a controlled variable and compares it with the designed performance. Any deviation from the desired standard is fed back into the control system which acts to correct the deviation.

closed system Describes that part of a heating circuit that circulates between the heat generator, usually a boiler, and an interface heat exchanger. The output distribution circuit(s) connects only to the heat exchanger and does not come into contact with the water that is conveyed between boiler and the heat exchanger, with the object of preventing the contamination of the boiler by any scale, sludge, corrosion products or other foreign matter which has accumulated in the distribution circuit.

Closed systems are essential in cases where a new boiler with small waterways is installed onto an old pipe installation. Provides similar function in steam systems.

coal, brown The commercial terms for lignite, a coal of relatively low energy content in the order of 14 MJ/kg. It is available in large deposits of wide quality variations.

coal bunker, access provision A bunker must be fitted with suitable access doors and hatches for safe operation and maintenance and to permit (in the case of a pneumatic system) delivery to be made by tipper truck in an emergency.

coal bunker, air filter When discharging coal into a bunker air is displaced and must be vented via a suitable filter to protect the environment. A commonly used filter medium is disposable fibreglass. The filter must be placed in such a location that it cannot be coated with coal inside the bunker.

coal delivery, pneumatic [Fig. 47] Coal is conveyed pneumatically between the delivery vehicle and the store. The delivery pipe terminates with a pipe coupling close to the parking area of the delivery truck. The delivery pipe must be as straight

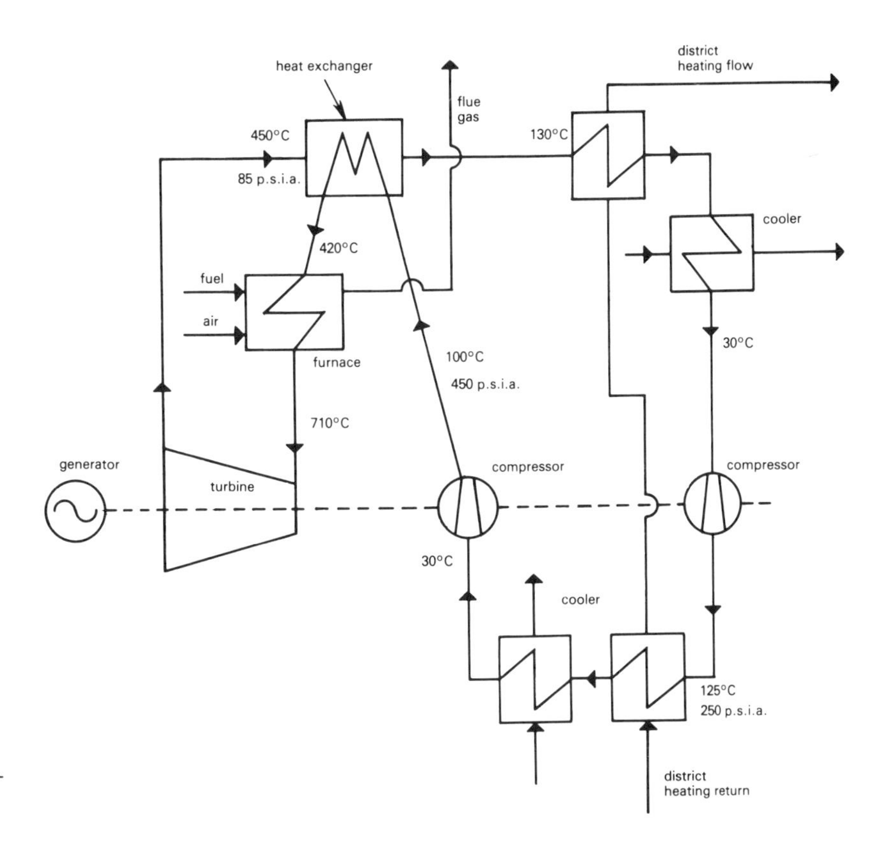

[Fig. 46] Diagrammatic representation of a closed-cycle gas turbine

as practicable with only a few long sweep bends. The final leg of the pipe must be enlarged at a point about 2.2 m from the bunker entry to reduce the exit velocity. In the NCB area the enlargement is from 127 mm to 178 mm.

coal, hard A grade of coal with a high energy content – about 60% carbon by weight.

coal slurry A mixture of crushed coal with water. In this form, coal may be efficiently transported in pipelines over long distances.

coal storage bunkers. These may be underground for delivery by tipper truck or above ground for delivery by pneumatic tube system.

coal tar A by-product of the manufacture of coke. It provides coal tar viscous fuel of calorific value from 38 to 40 kJ/g.

coefficient of discharge A term that expresses the ratio or proportion of the actual flow obtained from a terminal discharge fitting to that theoretically possible, the difference being due to the effects of velocity head and friction.

coefficient of expansion The increase in the dimension of a material or substance for each degree rise in temperature, expressed as a fraction of the original dimension. It may relate to linear or cubic expansion.

coefficient of linear expansion, pipes The increase in length by unit length of pipe when its temperature is raised by one degree Centigrade.

cogeneration See **combined heat and power**.

coke A solid fuel obtained by the heat treatment of suitable coal. It is commonly a by-product of gas production (town gas). It used to be widely employed in space heating boiler plant and industrial users are steelworks which require this fuel for smelting, reduction, etc. Coke is a high-energy fuel containing about 80% carbon with a calorific value of 28 MJ/kg. It has a relatively high ash content.

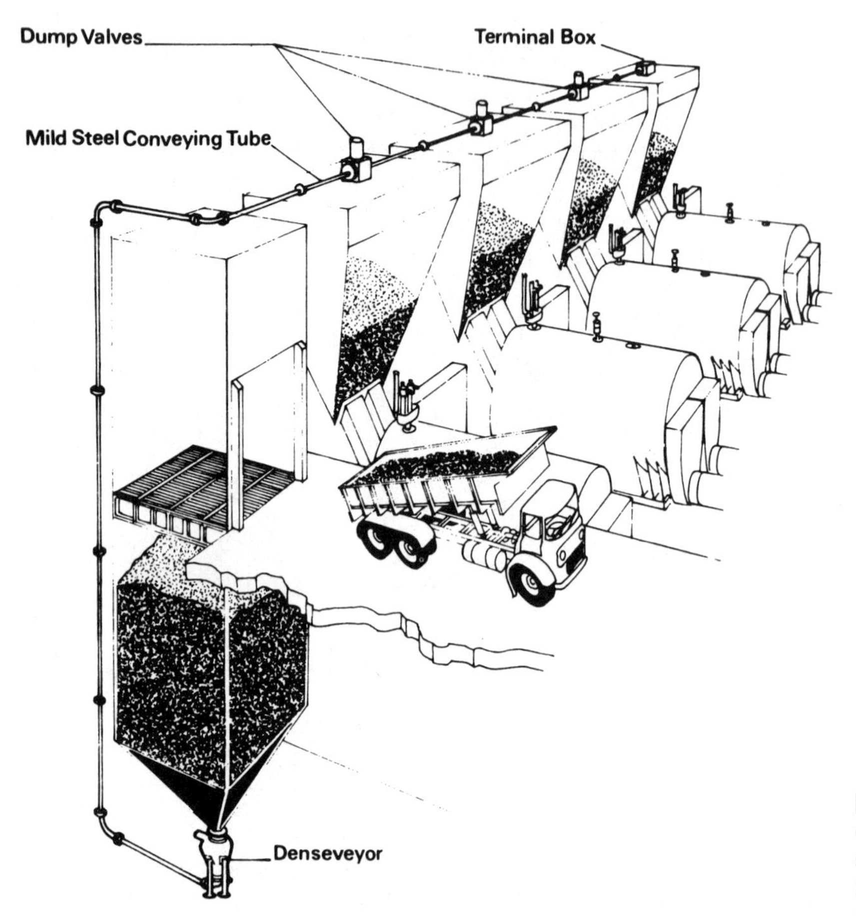

[Fig. 47] Pneumatic distribution of washed small (coal) by Macawber Denseveyor system

coking stoker A type of mechanical stoker for solid fuel in which the coal is partially carbonized by the heat of combustion before spilling over on to the bottom coking or deadplate which holds sufficient coal to ensure rapid ignition as the coal spills over from the top plate and thereby creates a condition for both overfeed and underfeed combustion. The movement of the fuel caused by the reciprocating fire bars and by the fresh coal being introduced over the top plate by the ram causes the burning fuel to drop off the bottom coking plate and to travel along the grate.

cold ceiling air conditioning See **air-conditioning by cold ceiling**.

cold feed pipe A distribution pipe which conveys cold water from a cistern to a hot water apparatus, e.g. a calorifier.

cold water systems, sterilization On completion of a new water supply or refurbishment of an existing one, the whole of the water system should be sterilized to eliminate possible traces of bacteria.

collatoral warranty An agreement intended for use by employers and specialist contractors in situations where they wish to create a separate contractual link with each other, in addition to the principal contractual route via the main contractor. Model forms of collatoral warranty are available and are commonly used. These provide the employer with reasonable protection but without imposing unduly onerous liabilities on to the specialists.

collector A terminal for the collection of solar energy.

colour code The marking of different piped services or other materials by specific coloured bands in accordance with relevant standards. See also **identibands**.

colour rendering The ability of a light source to reveal the colours of an object.

column radiator A space heating heat emitter constructed most commonly of cast-iron comprising an assembly of a number of cast sections or loops connected together by means of nipples. A column radiator offers greater output per unit length than an equivalent single panel radiator. These radiators are also available constructed of aluminium, copper or steel.

combi boiler Short for combination boiler.

combination boiler A boiler that incorporates within one assembly the facility for space heating and domestic hot water provision.

combination boiler, industrial A boiler that incorporates within one common casing a boiler and a fast-recovery close-coupled hot water calorifier. The choice of fuel is optional. It may be controlled to accord priority to hot water supply requirements. It optimizes the floor space in the boiler room, offers a fast response to hot water draw-off and improves the appearance of the boiler room.

combination wall-hung boiler A compact gas-fired boiler assembly which provides domestic-scale space heating and hot water supply. It incorporates a combustion chamber, heat exchanger, expansion vessel, circulating pump, all controls and electric wiring. It may also embody a fan-assisted flue which obviates the need for a chimney and which can be installed through any convenient wall (subject to being set a minimum distance from window(s)). It can be operated in 'winter' or 'summer' mode; the former yields priority to space heating, the latter to domestic hot water supply. It operates in conjunction with a programmer and water and room thermostats.

A closed system operates without an open vent and feed and an expansion cistern. It is pressurized off the public water main and is fitted with all necessary associated safety devices. Connection to the water main is via two stopcocks, one on either side of a temporary pipe connection. When commissioning or servicing, the connection is inserted between the valves and mains pressure is applied until the set pressure (as shown on an integral pressure gauge) is reached. The valves are turned off and the connection is removed and stored safely for the next service. The connection must not be left in place, as the accidental opening of the valves is likely to over-pressurize the system and cause flooding via the safety valves or the fracture of components. The safety valves are commonly set to lift at a pressure of 3 bar.

combined heat and power (CHP) [Figs 48, 49] The generation of heat in the course of the production of electric power and utilization of the heat for the energy supply to a district heating system. It can be demonstrated that the sacrifice of only one GJ of electricity produces about 5–6 GJ of heat, so that the process of obtaining heat as a by-product of electricity generation can be considered to be 500–600% efficient; hence the popularity of CHP schemes in the USA, Germany, Russia, Scandinavia, etc. The main technical problem of efficiently operating CHP systems is to establish continuously the correct balance between the ongoing generation of electricity and heat use.

Combined heat and power with back pressure turbines In back pressure turbines, steam is fed in at high pressure, but instead of being exhausted at the normal low temperature of about 40°C, as is the case with conventional condensing turbines, the steam is taken off at a higher temperature which is still useful for heating purposes. The combined effective output of such turbines is therefore all the heat input in the fuel, less the unavoidable losses (by radiation, from bearings, etc.).

A back pressure turbine can be operated at a combined heat plus power efficiency of about 80%. Such an arrangement is best suited to an industrial process which offers constant heat demand, but it is too inflexible for use with urban district heating as, in practice, it gives far too much heat for its output of electricity.

combined heat and power with ITOC (intermediate take-off condensing) turbine Intermediate

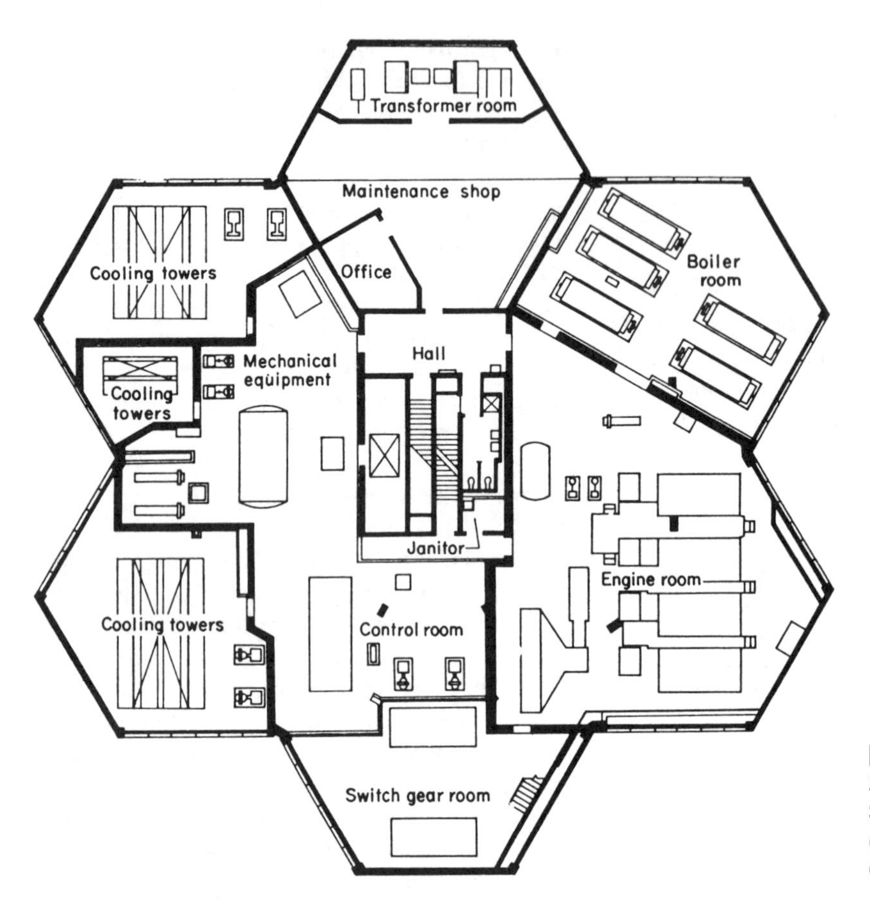

[Fig. 48] Plan view of the 24th storey of the Ohio State University Hostel (USA) showing location of CHP system

take-off condensing systems of combined heat and power are all characterized by being able to vary the ratio of electricity to heat extracted; i.e. they are flexible to meet varying demands. One or more such turbines are essential features of all major combined heat and power district heating schemes as offering the only acceptable method of balancing the demand with heat supplied in a system. Basically, ITOC turbines can be operated within the following modes.

(a) When little heat is required by the district heating system, the ITOC turbine is switched to the production of electricity at full condensing operation, the relatively small heat load being supplied from back-up plant (e.g. an incinerator waste heat boiler).
(b) As the demand for heat rises, the ITOC turbine is switched gradually to partial back pressure operation and, in some designs, even to full back pressure operation.

Sophisticated electronic control and valve systems are incorporated for this purpose, and it is essential that these are under the control of the district heating operating authority which has to decide, second by second, the manner of turbine operation and the balance between electricity and heat output.

combined heat and power (CHP), packaged unit The concept of CHP incorporated within a compact assembly to provide energy savings in small installations. This supplies heat and electric power outputs and for optimum efficiency it must have balanced heat/electricity loads.

combined sewer system A system that collects and conveys foul sewage and surface water within one common pipe installation.

combustion air requirement The quantity of air which must be continuously supplied to support

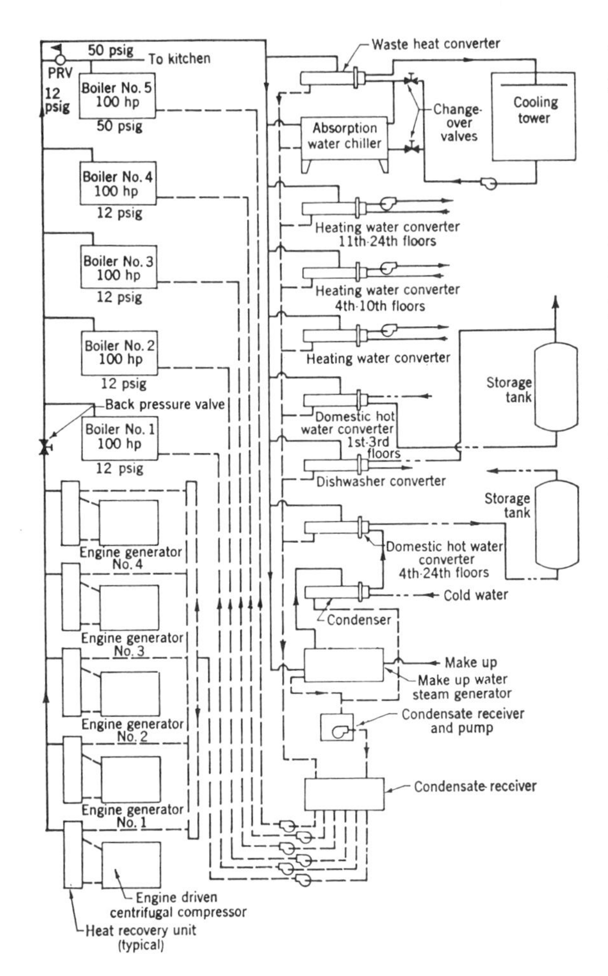

[Fig. 49] Flow diagram of CHP system used at students' hostel, Ohio State University (USA)

and maintain the combustion of a material. It can be theoretically calculated, based upon element molecular weight principles, which yield the following statements:

(a) 12 kg of carbon plus 32 kg of oxygen produce 44 kg of carbon doxide.
(b) 4 kg of hydrogen plus 32 kg of oxygen produce 64 kg of sulphur dioxide.

Air contains only 23% oxygen. The theoretical air supply quantity for the complete combustion of 1 kg of fuel containing these three combustible elements can be mathematically stated as follows:

$$\frac{1}{0.23}\left(\frac{32C}{12} + 8H + 1S\right) \text{kg}$$

C, H and S are the carbon, hydrogen and sulphur fractional constituents of 1 kg of the fuel.

combustion air preheater See **air preheater, combustion**.

combustion chamber That space within a furnace or boiler in which the combustion of the fuel takes place.

combustion, complete The burning of a fuel (or other material) in the presence of adequate air supply. The nitrogen in the air takes no part in the process. The combustible substances must reach a temperature at least equal to their individual ignition temperatures before they can ignite with the oxygen for complete combustion.

combustion efficiency See **boiler, combustion efficiency**.

combustion efficiency monitor A microprocessor-based instrument which automatically monitors and calculates boiler or furnace efficiency via temperature and oxygen measurements. Readings are presented as digital readouts. It can operate in conjunction with automatic combustion regulating devices to materially improve operating efficiency.

combustion, incomplete This is usually associated with an inadequate air supply to the furnace and occurs when the carbon component is burnt to carbon monoxide instead of to carbon dioxide.

combustion products The gaseous products resulting from the process of burning fuel in a combustion chamber, consisting in the main of carbon dioxide, nitrogen, water vapour, oxygen, oxides of sulphur, as well as some carbon monoxide (indicating incomplete combustion) and oxide of nitrogen, together with some particulate matter.

combustion, spontaneous See **spontaneous combustion**.

combustion test kit, portable Portable testing equipment for measuring carbon dioxide, oxygen, carbon monoxide, smoke, draught, flue temperature and gas pressure in the combustion process, supplied individually or in kits. A combustion analyser can give a digital display for oxygen readings and combustibles or temperature.

combustion trim unit A microprocessor-based oxygen trim (adjustment) unit incorporating an on-line combustion efficiency display and plant/transducer alarm monitoring. It is capable of storing two combustion characteristics for dual fuel firing.

comfort air-conditioning The provision of air-conditioning to a space to achieve the comfort of the occupants, as opposed to air-conditioning for process or industrial purposes.

comfort assessment system A system, using five transducers, which can indicate air, surface, asymmetric radiant temperatures, humidity and air velocity within a space, thereby summarizing the true overall comfort conditions affecting the occupants of that space.

comfort cooling The cooling of premises to achieve the comfort of the occupants, as opposed to cooling for industrial purposes.

comfort heating The heating of a space to achieve the comfort of the occupants, as opposed to heating for industrial purposes.

commissioning [Fig. 50] The adjustment of the various controls of a new installation to meet the design intent and the recording of the results. This is an essential activity before handing a new installation to the user. Adequate time must be built into the building programme to permit unhurried commissioning.

commissioning codes A series of publications which stipulate established procedures for commissioning building services and aim to provide the engineer with a rational procedure based on good current practice. Such codes are issued in the UK by CIBSE.

commissioning engineer A specialist in the setting up and regulating of new installations.

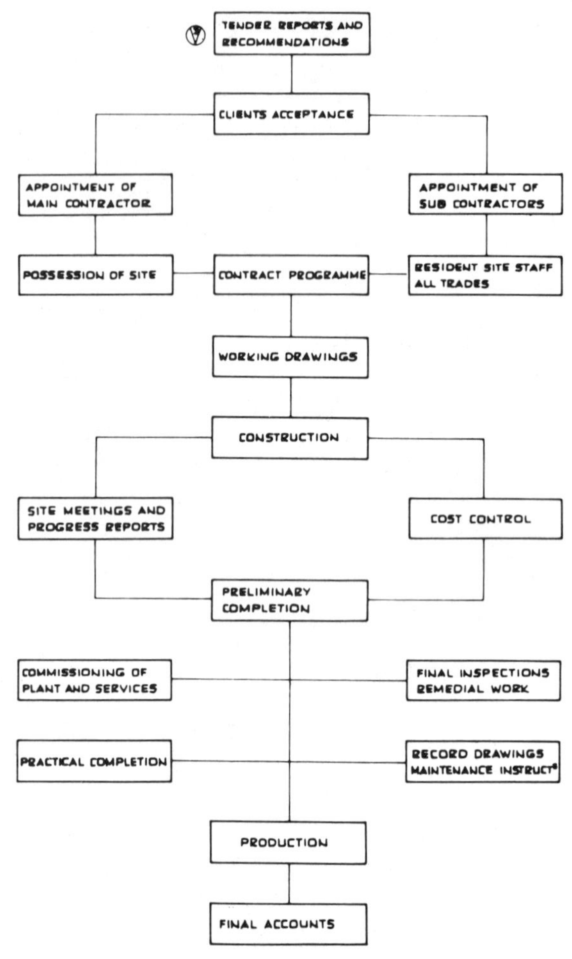

[Fig. 50] From analysis and reports on tenders to the final accounts

commutator A device for reversing the direction of an electric current.

compact Fischbach fan A fan that embodies a disc rotor motor. The high slip characteristic of this arrangement endows these fans with a steep pressure/volume performance characteristic. The performance of the fan adjusts automatically as system pressure increases. Thus, on free air, the fans operate well below normal speed and then adjust as the pressure increases until reaching a maximum speed of about 1400 r.p.m. Constant volume air flow can therefore be maintained throughout the life of the associated air filter.

Since the fan shaft does not rotate, it can be secured in rubber mountings, isolating the rotating part of the fan from the casing and doing away with the need for anti-vibration fittings. The motor/impeller assembly is electronically balanced in two planes statically and dynamically. Also available are fans with a twin drive motor and two impellers which can be driven separately or together, as well as a tandem motor which can be integrated in a fan with a tandem drive. Two stators on one shaft drive the rotor, which is fixed between the two stators. See also **disc rotor motor**.

compact fluorescent lamp (Thorn 10 W 2D type) A lamp that can cut lighting costs by as much as 80% compared with a 60 watt GLS tungsten lamp. It is 94 mm square with a light output of 650 lumens. The nominal circuit wattage is about 18. The life claimed is 8000 hours, about eight times that of an ordinary light bulb. It is available with four-pin cap only and in two colour temperatures. It offers good colour rendering conferred upon it by high quality polylux phosphors.

companion flange That flange at the end of one pipe which matches with the flange on another pipe to form a bolted flanged joint between them.

compensator [Fig. 51] A fitting inserted into a pipeline to absorb the thermal expansion movements. It must be part of a system which includes guides and anchors.

complete combustion See **combustion, complete**.

compressed air, leakages Leakages commonly occur at air receiver mountings, hose unions, pipe connections, drain-off points, process and control equipment, quick-release couplings, etc. The associated power loss can be a serious drain on the capacity of the system; 30% of total requirement is not untypical. Compressed air systems must be routinely tested for leakages to avoid major power losses and the leakages must be rectified.

compressed air, mains These must be laid to falls of about 1 in 40 to permit water to drain to drain points from where it is periodically removed. Branch pipes must be taken off the top of the main to avoid water entering the pneumatic equipment.

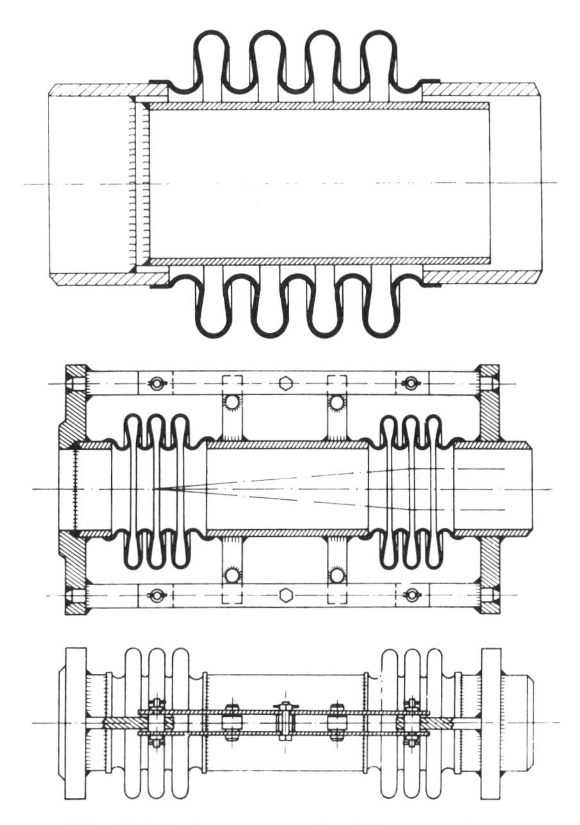

[Fig. 51] Various types of axial compensator

Adequate drain traps must be provided in accessible locations. A connection to a pneumatic tool must be taken off the side of the drop pipe before the drain connection.

compressibility The ability of a substance to suffer reduction in volume when placed under pressure.

compression joint, manipulative A joint that requires the end of the pipe to be shaped outwards before assembly into the joint to enable the pipe fitting to securely grip the pipe.

compression joint, non-manipulative A joint that grips the outside wall of the plain pipe by means of a compressing ring or olive.

compressor, after-cooler A device sometimes installed after the last stage of a compressor to remove moisture from the air or gas. It cannot influence the work done in compression.

compressor air displacement The actual volume of air swept out per minute by the low pressure pistons or during the suction strokes.

compressor bypass control A method of capacity control applicable a to multi-cylinder compressor assembly in which the refrigerant gas is bypassed around the compressor when reduced capacity is called for.

compressor clearance volume To allow for wear, and to give mechanical freedom, a space must be left between the cylinder end and the piston of a compressor. Provision must also be made for the reception of valves. The sum of these two spaces is the clearance volume.

compressor cooling The cooling required to limit the temperature at air compressors to permit adequate lubrication and to avoid excessive thermal stresses. Such cooling may be either by air or by water circulation. Air cooling is acceptable for duties up to 350 m^3 or for continuous duty at pressures up to 14 bar. To provide enhanced cooling, the compressor external surfaces are finned to give an extended heat transfer surface and additional cooling is provided by a fan directing a stream of air at the compressor. The air-cooled compressor must be so located that cooling air can circulate freely over it. Water cooling is by circulating cool (but not chilled) water through an integral water jacket. The heated water is often piped away to a heat exchanger where heat energy is recovered for space, process or domestic hot water supply.

compressor displacement The reduction of the volume of gas in the confined space of the compressor, thereby raising its pressure.

compressor, hermetic See **hermetic compressor**.

compressor inlet guide vanes A set of adjustable vanes fitted in the compressor suction. The vanes are gradually closed to reduce the volume of refrigerant gas compressed, thereby reducing the capacity. When fitted to centrifugal chillers, they tend to obviate surging by curving the direction of the gas in an efficient manner that permits capacity reduction to 15%.

compressor, intercooler A cooler (heat exchanger) placed between the stages of a compressor.

compressor, multi-stage See **multi-stage compressor**.

compressor, oil-free See **oil-free compressor**.

compressor, open See **open compressor**.

compressor, positive displacement See **positive displacement compressor**.

compressor, screw See **screw compressor**.

compressor, semi-hermetic See **semi-hermetic compressor**.

compressor, single vane rotary See **single vane rotary compressor**.

compressor suction The side of a compressor connected to low pressure.

compressor surging Surging caused by the throttling of the gas flow rate with a butterfly type discharge damper when an unstable condition can be reached in which the gas is constantly surging back and forth through the compressor. This is a serious event that can damage the machine. It can be alleviated/avoided by fitting inlet guide vanes to the compressor suction. See **compressor inlet guide vanes**.

compressor, turbo See **turbo-compressor**.

compressor, twin cylinder See **twin cylinder compressor**.

compressor, vertical See **vertical compressor**.

computer floor A false floor. It is particularly convenient in computer rooms where it permits short service connections between electrical and mechanical services run via the floor space directly to the computer equipment.

computer room, fire protection Three major methods are in use for protecting the valuable and delicate equipment housed within computer rooms.

(a) Carbon dioxide (CO_2): stored under pressure within cylinder CO_2 is discharged and fills the computer room in the event of a fire alarm. This is an unsatisfactory method, as the gas is injurious to personnel and these must be evacuated before the discharge and until the gas has been cleared.
(b) Halon gas: stored under pressure within cylinders this gas is discharged and fills the computer room in the event of a fire alarm. Halon is a CFC (chlorofluorocarbon) which is largely banned because of its adverse effect on the Earth's ozone layer. Where its use continues to be permitted, legislation requires that the gas is discharged at roof level – a difficulty where the computer is housed in the basement of a 30-storey building.
(c) Water: the water is discharged over the computer equipment via a sprinkler system. Such water protection used to be unacceptable because of the damage the water would inflict on sensitive computer equipment but modern computers function with printed circuit boards which are not easily damaged by water. Even the flooding of the area below the computer floor is no longer a major nuisance, as vacuum suction plant is available which can suck up the water rapidly.

Given the above choices, opinion is veering towards the use of sprinkler systems in computer rooms as being the least of three evils, combined with rigorous preventive maintenance routines to eliminate accidental water discharge.

computerized optimizer See also **optimum start control**. Computerized form of optimum start control using microelectronic components, including a microprocessor.

concentrating solar collector A collector which incorporates a reflector system to increase the intensity of solar energy input on a given area.

conciliation The process of reconciling opposing views without recourse to legal action or arbitration.

condensation [Fig. 52] Water that separates from the air at a cold surface. This occurs when moisture-laden air comes into contact with a cold surface which is at a temperature below the dew-point of the air.

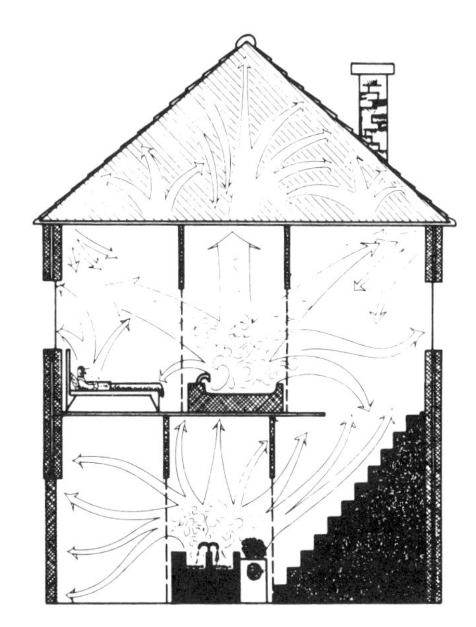

[Fig. 52] Sources of condensation in a typical domestic dwelling

condensation channel An integral guttering formed around the perimeter of a kitchen exhaust canopy to collect and drain off condensation water.

condensate pump, steam/air motivated A pump powered by steam or compressed air which is self-contained and automatic. It pumps water, oils and other liquids as high as 25 m or can discharge against the equivalent back pressure. It can also drain a closed vessel under vacuum or pressure.

condensate recovery unit A packaged unit consisting of a receiver vessel, one or two close-coupled pumps and all interconnecting pipe work and valves.

condensate return pump A pump installed as part of the condensate return system of a steam installation to pump the condensate into the boiler room hot well. It must be suitable to handle water at the operating temperature of the system. See also **condensate recovery unit**.

condense Hot water which arises from the condensation of steam after the latent heat has been abstracted from the steam. It is usually the discharge from steam traps.

condense meter [Fig. 53] A device that measures the consumption of steam by the weight of condensate discharged from a steam-using installation.

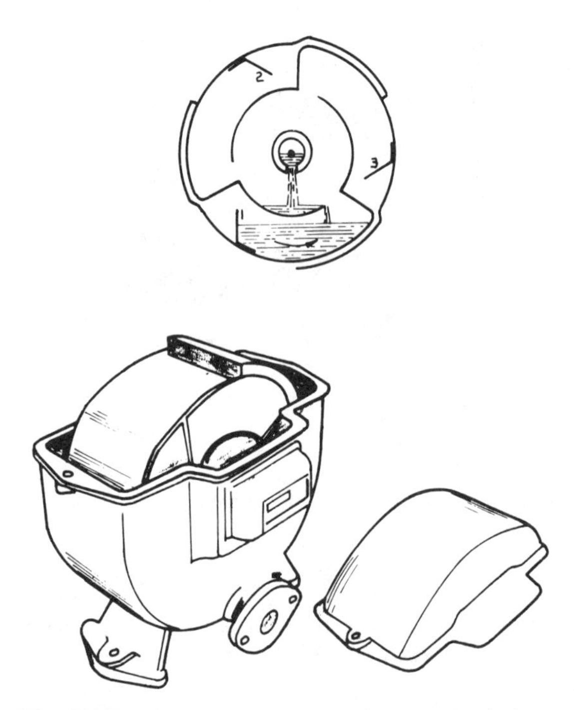

[Fig. 53] Condensate meter measuring supply of steam

condense pump A light-duty pump incorporated with an air-conditioner to lift accumulated condensation into a higher level condense drain-pipe.

condenser A heat exchanger in which a vapour (e.g. steam, refrigerant gas) is condensed to a liquid, giving up its latent heat in the process. Liberated heat may be used for heat recovery purposes. A condenser may be water-cooled or air-cooled.

condenser, air-cooled See **air-cooled condenser**.

condenser coil A heat exchanger within a condenser which serves to condense gas into a liquid.

condenser coil duty The available performance from the condenser installed within a refrigeration or air-conditioning system. It is usually expressed in terms of tons refrigeration or kW/h.

condenser, evaporative See **evaporative condenser**.

condenser, reflux See **reflux condenser**.

condenser, shell and coil See **shell and coil condenser**.

condenser, shell and tube See **shell and tube condenser**.

condenser, water-cooled See **water-cooled condenser**.

condensing boiler A boiler that uses the latent heat in the combustion gases. When applied to 'wet' space heating, it can achieve an appreciably higher thermal efficiency than the conventional heating boiler as it utilizes this latent heat, in addition to heat transfer through the boiler sections. In one typical such boiler, the combustion gases transfer their maximum heat to the boiler block and they are then directed to the inlet side of the condensing unit where they pass over the bank of finned heat exchanger pipes and then over the final plain condensing pipes. The final pass of the combustion gas is through an induced-draught fan.

When the return water from the heating distribution is sufficiently cool, the temperature of the combustion gas is reduced to its dewpoint of 54°C, at which temperature and below, the water vapour in the combustion gas will condense and thereby release its latent heat. Maximum efficiency is available from the condensing boiler when the return temperature is sufficiently low for complete condensation. The condensation is routed to the bottom of the boiler unit through a water trap, from where it must be piped to drain.

It is essential to design the space heating to extract most of the heat, so that a low return temperature can be achieved. At 70°C return water temperature, the efficiency will be about 85%, whilst with the low return water temperature of only 40°C, it will approach 92%. Since the water of condensation is acidic, the materials of construction of both boiler and drain must be acid resistant.

condensing temperature That temperature at which the fluid or gas changes state from vapour or gas into a liquid at the prevailing pressure.

condition monitoring An organized system for continuously monitoring the wear of moving parts which warns of the need for some specified action to avoid plant breakdown. It aims to keep plant in continuous efficient operation.

conditioned air Air being conveyed in an air-conditioning or ventilation system which has been subject to air treatment, such as filtration, heating, cooling, ionization, humidification and/or dehumidification.

conduction The transfer of heat within a, say, solid material by passage of the heat from the portion closest to the heat source through the layers of the material towards the colder portion of the material.

conductivity See **thermal conductivity**.

conduit valve A valve that incorporates a gate which has parallel faces and has a circular aperture of the same diameter as the valve bore, thereby offering an uninterrupted fluid flow when the valve is fully open.

console air, conditioner An air-conditioning cabinet suitable for room installation.

constant flow valve A factory preset control device to give a constant fluid flow, irrespective of variations in upstream and downstream conditions.

consulting engineer An independent adviser on engineering matters.

consulting engineer, building services An engineer concerned with one, several or all aspects of the engineering services for the project.

consulting engineer, civil An engineer concerned with civil works, commonly including structures and main drainage.

consulting engineer, electrical An engineer concerned with all electrical services, including facilities for data processing, telecommunications, etc.

consulting engineer, mechanical An engineer concerned with the mechanical engineering services only. These commonly comprise space heating, ventilation, air-conditioning, water and gas installations, piped fire protection and other piped and ducted systems as well as the electrical installations associated with them.

consulting engineer, vertical transportation An engineer concerned with passenger and goods lifts, escalators, document hoists, car transporters, etc.

consumer unit An electrical isolator and distribution board provided in a package at the electrical intake of a consumer.

contactor A power-operated switch for making and breaking a circuit. It utilizes the electric field generated by the contactor coil which is connected to allow the sensing circuit to activate the 'on' or 'off' action of a much larger power capacity.

contactor starter A motor starting device which incorporates a coil by which the starter can be activated by a remote device, such as a time clock, thermostat, etc.

contingency sum A sum of money defined and set aside within a contract to meet unforeseen expenditures.

contra-rotating The operation of two impellers within one casing of an axial flow fan which rotate in opposite directions to each other for the purpose of developing maximum fan pressure.

contract A legally enforceable agreement between parties.

contract, accident records clause A clause in the contract that obliges a contractor to maintain at site records of all accidents which occur at site during the contract.

contract, Health and Safety at Work clause A clause in the contract that obliges a contractor to comply with the provisions of the Health and Safety at Work Act.

contract, insurance clause A clause in the contract that requires a contractor to arrange specified insurance cover for works and personnel.

contract maintenance An arrangement whereby a firm of maintenance specialists undertakes the routine maintenance of plant under a specific contract and specification.

contractor A firm appointed to carry out the contract works.

contract, site accommodation clause A clause in the contract that obliges a contractor of provide temporary housing for administration, meetings, staff canteen, washrooms and toilets.

control agent That process, energy or material of which the manipulated variable is a condition or characteristic.

control deviation A departure of the system from the intended controlled condition.

controlled device A device that performs a controlling function, such as opening or closing a valve or damper.

controlled medium That process, energy or material in which a variable function is controlled.

controlled variable That quantity or condition which is measured and controlled.

controller/data logger A microprocessor-based controller with interfaces for digital and analogue values and serial communication to printers, monitor screens or control computers employing three-wire transmission.

control panel [Fig. 54] A centralized assembly of controls which may include a mimic diagram and indicator lights.

control panel, cubicle type [Fig. 55] A control panel that houses controls within purpose-made compartments secured by doors.

control point That value of a controlled variable which, under any set of conditions, the automatic controller operates to maintain.

convection The transfer of heat from one place to another by actual movement of the hot body. In

[Fig. 54] Central control panel serving a swimming pool complex: doors open showing switchgear and wiring (Contramec Installations Ltd.)

[Fig. 55] Air-cooled water chiller: close-up of control panel

fluids, the portions adjacent to the heat input source become hot and expand, accompanied by a lowering of density. Motion is then set up under the action of gravity, with the lower density liquid moving away from the heat source, permitting the colder, denser liquid to approach to become heated in turn. Thus, heat transfer by convection takes place by reason of the movement within the volume of fluid. This process is described as thermosyphonic.

convection, natural See **natural convection**.

convector radiator A radiator that combines a flat front surface with convector features. It offers greater heat output per unit length and some types offer lowered surface temperatures for use in areas where there is a risk of scalding to occupants; e.g. in nurseries, mental homes, etc.

conveying velocity The velocity of air required to convey given solids in an exhaust system; e.g. 10.4–15.3 m/s for light, dry sawdust to 20.4 m/s for heavy wet or green wood blocks or edgings.

cooler battery [Fig. 56] An assembly of cooling coils within a casing, usually with flanged ends for attachment to a ducted system.

cooling coil The heat exchanger of a cooling system. It may be plain or have fins (extended surface) or be formed of heat pipe(s).

cooling load The amount of heat which is removed by the cooling system, expressed in tons refrigeration or kW/h.

cooling pond A cooling system where water is stored in an open pond. Cooling progresses by contact between the pond surface and the air passing over it, causing evaporation of some of the water. Usually, the cooled water is drawn off at one end of the pond and recirculated at the other end. It suffers from a low rate of heat transfer, the likelihood of contamination, the requirement for a large area and uncertainty of specific performance but it is useful where an ornamental pond is required. It can also act as a heat sink for an associated heat pump system.

cooling, sensible See **sensible cooling**.

cooling system, condensation tray A pan fitted below the heat exchanger to collect water or condensation dripping off the exchanger and immediate connections. It requires a condense outlet for

[Fig. 56] Water heat exchange coil showing pipe headers

connection into a central condensate drain with trapped discharge. The outlet should be flush with the bottom of the tray to avoid the accumulation of standing water which may collect bacteria.

cooling system, thermosyphon effect See **thermosyphon cooling**, **thermosyphon cooling effect** and **thermosyphon cooling system**.

cooling system, wet A cooling system which operates below the dewpoint temperature of the ambient air, causing condensation off the heat exchangers and any uninsulated cold pipes and fittings.

cooling tower [Figs 57, 58, 59] Equipment for cooling water by evaporation in the ambient air. Air may be induced by natural or mechanical means (a fan). The purpose of a cooling tower is to economize on water used for condenser cooling by recirculating the water between the cooling tower and the condenser. The water being cooled may be in direct contact with the air flow or it may be cooled indirectly within a closed circuit.

cooling tower, anti-vortex plate A metal plate fitted close to the inlet and outlet of a tank to improve the distribution of water flow.

cooling tower, approach The termperature differential between the water leaving (off) the tower and the design (specified) air inlet wet bulb temperature. For example, for a water outlet at 27°C and a wet bulb air inlet at 20°C the approach is 7°C.

cooling tower, bleed The quantity of water which must be removed from the evaporative water circulation to control the content of the dissolved solids, to prevent an excessive build-up of these and/or other possibly harmful atmospheric constituents. Bleed may be by intermittent or continuous drainage and manual or automatic activation to suit a water treatment schedule.

cooling tower, blowdown As **cooling-tower, bleed**.

cooling tower, carryover Spray which is entrained in the air as it flows through the tower and

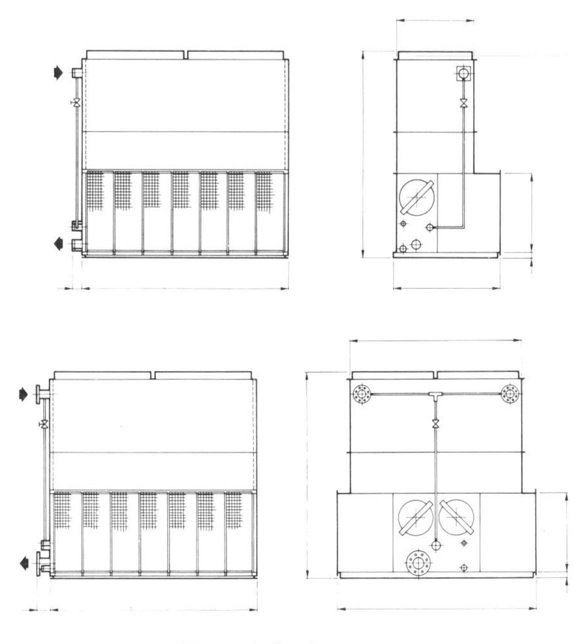

[Fig. 57] Cooling tower

which it carries out of the tower. Excessive carry-over can constitute a nuisance.

cooling tower, casing The envelope which encompasses the packing and the water system within the tower.

cooling tower, cladding As **cooling-tower, casing**.

cooling tower, counterflow The arrangement by which the fan (usually induced draught) induces vertical air flow up the tower and across the packing in the opposite direction to the flow of water streaming down the tower so that the coldest water is in contact with the driest air. This provides the best cooling tower performance.

cooling tower, counterflow See **counterflow cooling tower**.

cooling tower, cross draught The arrangement by which the fan creates a horizontal air flow as the water streams across the air movement (usually an induced draught fan). This permits the design of cooling towers with a low silhouette (which may be required under certain circumstances of the site). It must be protected against the recirculation of saturated vapour which is likely to occur in confined locations.

cooling tower, cross flow See **cross flow cooling tower**.

cooling tower, direct A cooling tower in which the circulating system cooling water comes into direct contact with the cooling air. This achieves optimum evaporative cooling, but inevitably causes contamination by atmospheric pollutants of the water which circulates through the cooled equipment. It is unsuitable for installations where foreign matter (even though strained and filtered) is likely to choke small waterways, valves, etc.

cooling tower, distribution header An arrangement that directs the water flow into the tower for the best evaporative effect.

cooling tower, dosing pot A device through which the water treatment chemicals are introduced. This commonly incorporates a metering device which controls dosing at a specified rate.

cooling tower, drift The evaporative loss from water surfaces. It is taken as being in the order of 1% of the water being circulated per 5.5°C drop in the temperature of the water passing through the tower. See also **carryover**.

cooling tower eliminator [Fig. 60] A device that removes droplets of water from the air before this is discharged from the tower. Eliminator blades may be constructed of timber, metal or plastic. Metal eliminators are vulnerable to aggressive atmospheres; exposed to these they are likely to waste away within a short space of time.

cooling tower fill [Fig. 61] See **cooling tower packing**.

cooling tower, filter screen Wire mesh panels fitted in front of a cooling tower sump to trap coarse particles.

cooling tower, forced draught See **forced draught cooling tower**.

[Fig. 58] Cooling tower: assembly of large capacity output

cooling tower, frost protection The means of providing protection to obviate the freezing of the water in the tower (and coils). This is usually done by electric immersion heater(s) or through bypassing some boiler heat through the tower in cold weather under thermostatic control.

cooling tower, indirect A cooling tower in which the water which circulates through the primary (cooled equipment) circuit is kept out of contact with the air by circulating the water through pipes over which the cooling air is directed. This protects the cooled equipment from contamination. This type of cooling tower is of larger dimensions than a direct cooling tower. Cooling coils must be protected to avoid frost damage.

cooling tower, induced draught A cooling tower in which the air is drawn into the tower by the action of a fan which is fitted to the air outlet of the tower.

cooling tower, make-up [Fig. 62] The supply of water required to compensate for losses of water from the system through evaporation, bleed, pump glands or overflow. On the larger installations make-up is supplied from a treated water reservoir; smaller installations may draw directly off the water main, possibly incorporating a water treatment dosing pot. Make-up water flow is usually controlled by a float and ball valve arrangement associated with the tower.

cooling tower, natural draught (atmospheric) A cooling tower in which the cooling air is drawn over the tower by the chimney effect of the arrangement.

Drawback: it depends on wind conditions, external temperatures, etc.

cooling tower packing The material placed inside the cooling tower which functions as the heat transfer surface over which the water is directed and distributed in its passage through the tower. This may be timber, glass fibre or plastic.

cooling tower, performance This is expressed in terms of water range and approach. The former is the difference between the entering and leaving water temperatures; the latter refers to the difference between the leaving water temperature and the wet bulb temperature of the entering air.

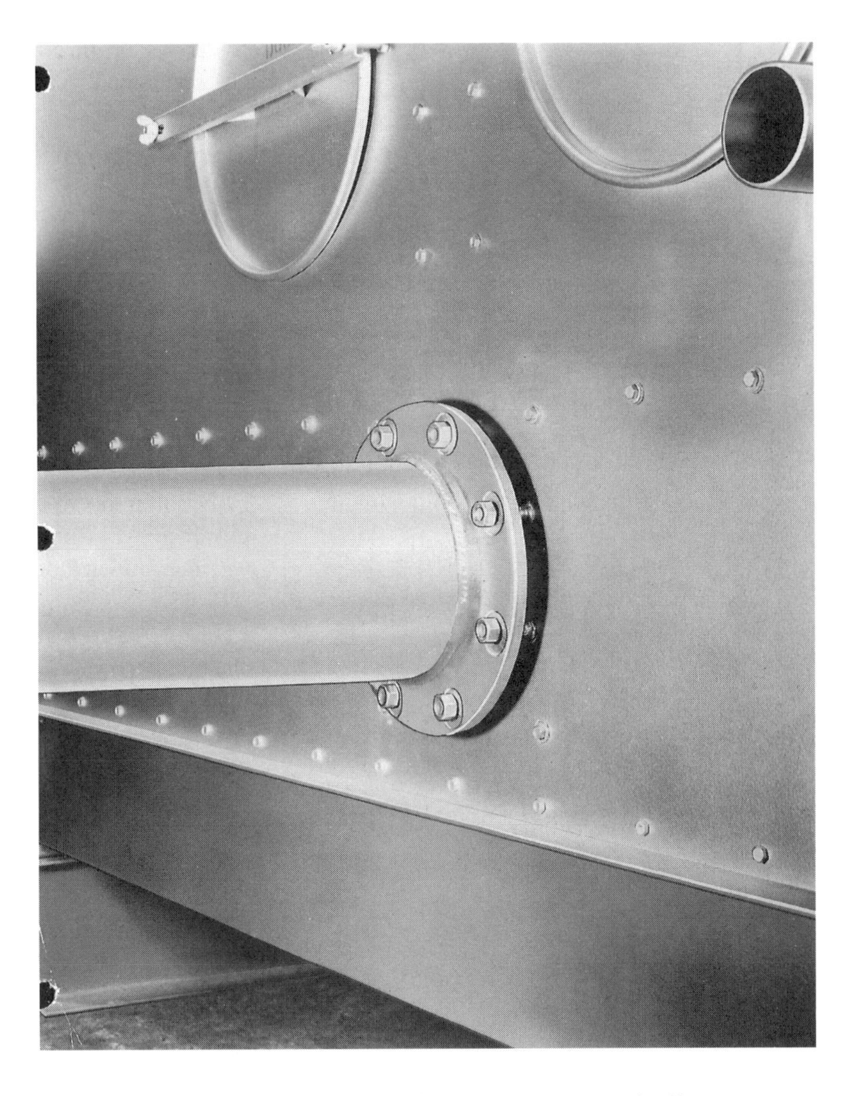

[Fig. 59] Cooling tower: close-up view of pipe connections

cooling tower, plastic Plastic offers natural resistance to aggressive atmospheres and operating conditions. Plastics commonly encountered in packings and casings are polystyrene, polypropylene, high-density polythene, PVC and reinforced glass fibre.

cooling tower plume The visible discharge of water vapour from a cooling tower into the atmosphere.

cooling tower, purge As **cooling tower, bleed**.

cooling tower, range The temperature differential between the circulating water entering the tower and leaving the tower. For example, if water enters at 40°C and leaves at 30°C, the range is 10°C.

cooling tower, spray tree See **spray tree**.

cooling tower, spray See **spray cooling tower**.

cooling tower, sump A depression at the outflow from the cooling tower tank to trap foreign matter.

cooling tower, UV system A cooling tower that incorporates ultraviolet lamps to disinfect cooling water and to eradicate bacilli.

cooling tower, water treatment Water treatment required/provided to obviate fouling of the cooling

[Fig. 60] Eliminators

[Fig. 61] Fill

[Fig. 62] Compartment with water feed level control and strainer

tower system through the growth of algae, bacteria, scale and/or corrosion.

cooling tower, windage loss The amount of small water droplets carried away by the prevailing wind. This is taken as being in the order of 0.1%–0.3% of the water circulated in mechanical draught towers and up to 0.1% in evaporative condensers. Some loss of dissolved solids is also due to windage.

cooling tower, winter operation Equipment is subject to frost damage when it contains water. Precautions must be taken, either by draining the tower before winter, by using antifreeze solution in circulation, activating an electric immersion heater or bypassing hot water through the tower at all times under thermostatic control.

cooling water Water circulated through heat exchangers to reduce the temperature of warmer fluids.

cooling water system A system that circulates water between refrigeration condensers and heat dissipators such as cooling towers, blast coolers, etc.

co-ordinated design A design that joins together the design of all the various services and disciplines forming a project.

co-ordinated design drawings Drawings that indicate all the elements of a project design in their relationship to each other. This is an invaluable construction tool as it obviates services clashing with each other.

copper pipe, end feed capillary fittings A range of capillary fittings suitable for end feeding with solder. The jointing utilizes solder wire which is fed into the mouth of the fitting after it has been fluxed and heated. At the correct temperature, the solder will flow freely and be drawn by capillary action into the space between pipe and fitting.

copper pipe, integral ring soft solder fitting Pipe fitting that utilizes the principle of capillary attraction to place a continuous film of solder between the fitting socket and the pipe to ensure a perfect joint. During manufacture, a ring of solder is positioned in the wall of each capillary socket; this solder is released as heat is applied externally to the fitting. A slight protrusion of solder around the joint indicates satisfactory jointing.

copper pipe, making a capillary joint

(a) Clean the outside of the pipe and socket with fine sandpaper or with steel wool.
(b) Apply adequate (but not excessive) flux to both the outside surface of pipe and the inside surface of the fitting.
(c) Heat the fitting and tube until a complete ring of solder appears at the mouth of the fitting.
(d) Allow the joint to cool without disturbing it.
(e) Carefully wipe off all surplus flux.

corbel A unit cantilevering from a wall surface to form a bearing.

coring and scraping See **cleaning of sanitary installations**.

cork The outer bark of the cork tree which grows principally in countries bordering the Mediterranean. Bark and cork products incorporate minute air cells, each completely sealed by a strong membrane from the adjoining cells. When the tree has reached the age of 20 years, the outer bark is removed, and since it is unsuitable for the manufacture of cork products, it is broken into granules of various sizes suitable for thermal insulation purposes. The granulated cork from the first stripping contains a natural resin which, under the combined action of heat and pressure, is used to bind the granules together. Cork is used mainly for the insulation of cold surfaces, but has now been largely superseded by less costly alternative materials.

corkboard thermal insulation Insulation material manufactured from granulated cork pressed into moulds and baked at a temperature of about 316°C. This liberates the resin in the cork which cements the granules firmly together, no other binding agent being necessary. The moulded cork insulation is coated with a waterproofing material to seal it against air and water. Thermal insulation boards are supplied in slabs to specified thickness for application to ducts. Joints are sealed with waterproof cement. The density of corkboard with bitumen and asphalt binder varies between 235 and 1040 kg/m^3, the limiting temperature is about 82°C and the thermal conductivity lies between 0.028 and 0.055 W/m °C. The main use is for thermal insulation of cold surfaces but is being largely superseded by less expensive insulating materials.

corner tube boiler [Figs 63, 64] An adaptation of the water tube boiler. It consists basically of an all-welded cage system of corner water tubes fitted together with longitudinal and transverse headers. It incorporates welded-in heating surfaces such as furnace walls and convection heat banks. The corner tubes ensure that the boiler cooling system is self-supporting. For the smaller size of boiler, the furnace is made completely gas-tight by being enclosed within a sheet metal casing which expands and contracts in unison with the tubes. The larger size models utilize integral fins on the furnace tubes. This type of boiler contains no refractories, except the burner quarls, thereby saving weight and reducing maintenance. A gas-tight membrane wall avoids low temperature corrosion problems arising within the furnace. The boiler is compact and is used widely in European group and district heating schemes. A corner boiler may be designed to function both as a hot water boiler and as a steam boiler. One popular design is a combined steam/hot water boiler in which 30% of the output is available in the form of line steam with the balance as high temperature hot water. When using the standard hot water boiler of this type the steam generated in the drum of the boiler, which is mounted directly above the boiler, can be used for the purposes of oil fuel heating, soot blowing and or oil jet atomizing.

Whilst it is practicable to use corner tube hot water boilers for direct connection into a group or district heating system, it is better to circulate the boiler water in a closed loop by a separate circulating pump and to transfer the heat to the network via a heat exchanger as such an arrangement protects the boiler from any scale or debris which may

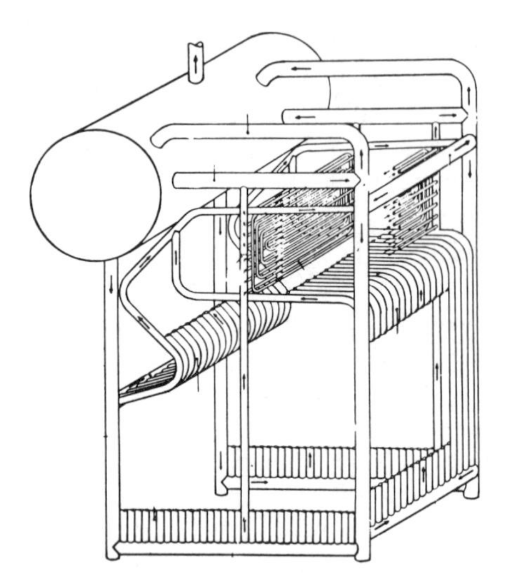

[Fig. 63] Schematic arrangement of corner tube boiler

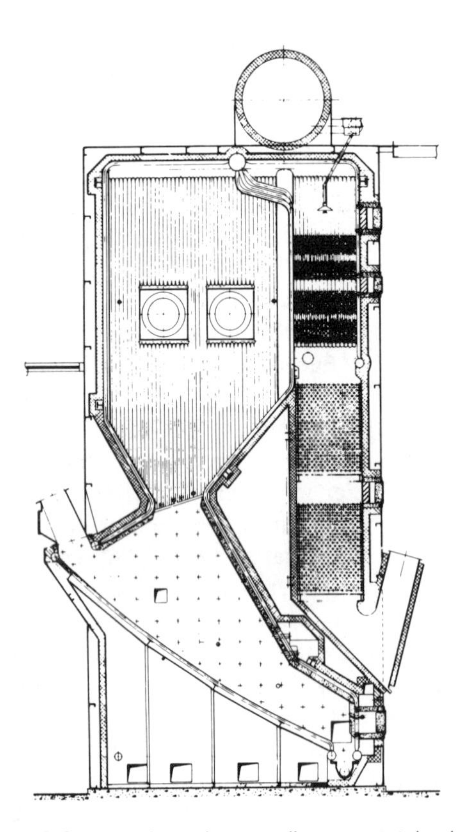

[Fig. 64] Section through a smaller corner tube boiler

circulate within the system network. A dual circuit must be maintained when the corner tube boiler is utilized to supply steam and hot water.

Cornish boiler A horizontal shell boiler which incorporates one single furnace tube with a brick flue setting. Flue gases leaving the grate travel to the back of the furnace tube down into a bottom flue situated immediately below the boiler. The gases return to the boiler front and divide into two streams which pass through the left and right side flues respectively and then combine to enter the main flue which connects to the chimney.

corrosion [Fig. 65] The reaction of a metal with its environment, resulting in damage which impairs the stability, service life and function of a component or system. It is one of the major causes of premature failure or breakdown of mechanical services systems.

corrosion by bacteria The presence of anaerobic bacteria, in the absence of oxygen, may promote the cathodic reduction of sulphate which is present in most soils and natural waters, as these bacteria can use cathodic oxygen in their living process and convert the sulphate to sulphide. This type of corrosion is generally responsible for the smell of bad eggs, which emanates from air vented at central heating radiators.

corrosion by stray electric current Stray earth currents, particularly direct current, can generate cathodic and anodic areas where they enter and leave buried pipelines or other metal structures, causing severe corrosion.

corrosion inhibitor A chemical introduced into a piped system to inhibit internal corrosion. Some forms of corrosion generate hydrogen gas within a heating system which causes the symptoms of air locking. There are inhibitors available to combat this. Small systems may require periodic manual addition of chemicals via a dosing pot; larger systems require an automatic dosing system incorporating metering pump and dosing pot.

cost plan An itemized budget for the works or any part thereof prepared for the employer by a quantity surveyor and/or consulting engineer.

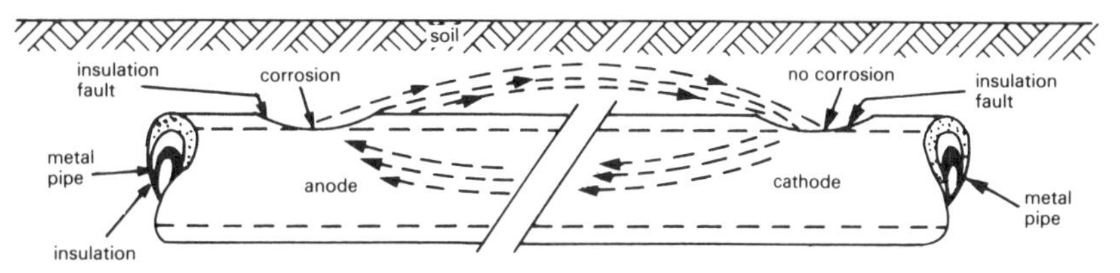

Drawing shows buried pipework without cathodic protection. Note the area of corrosion (anode) when the current leaves it.

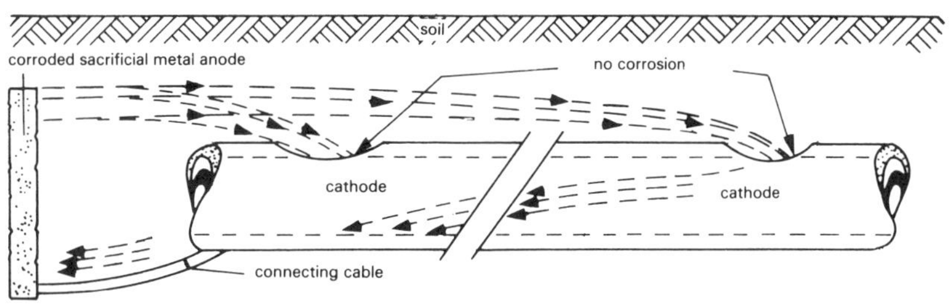

Buried pipework with cathodic protection. Note the corroded state of the sacrificial anode. There are other and more sophisticated forms of cathodic protection.

[Fig. 65] Corrosion effects with buried pipelines

cost-in-use The total cost of operating a plant, system, etc.

cowl A fitting which prevents ingress of rain and snow into the open terminal of a duct or pipe.

coulomb A unit quantity of electricity; that which is transported in one second by a current of one ampere.

counterflow The flow of fluids in opposite directions to each other.

counterflow cooling tower A tower in which the cooling water and the condenser water flow in opposite directions.

counterflow heat exchanger The arrangement of heating or cooling a heat exchanger in which the primary and secondary fluids flow in the opposite direction to each other to achieve an enhanced rate of heat exchange.

counterflow shell and tube heat exchangers Shell and tube counterflow heat exchangers of U-tube and spiral tube configurations. Offer improved heat transfer as hottest primary water is in contact with coolest secondary water.

coupling nipple A metal fitting which joins together two adjacent sections, e.g. cast-iron boiler sections.

crankcase The casing of a compressor in which the crank rotates.

crankcase heater Fitted to a compressor to maintain the lubricant at free-flowing conditions at low ambient operation. An air-conditioner operating under programmed control in low ambient conditions must incorporate a separate electrical circuit to the crankcase heater to maintain the same operational conditions at all times.

crawlduct See **crawlway**.

crawlway [Fig. 66] A service duct big enough to permit an average-size person to crawl through it with adequate space for him to work on the services which are run through the crawlway.

creep The slow plastic deformation or movement of a material under stress.

criterion of satisfactory service in sanitation That percentage of time during which the design discharge flow loading will not be exceeded.

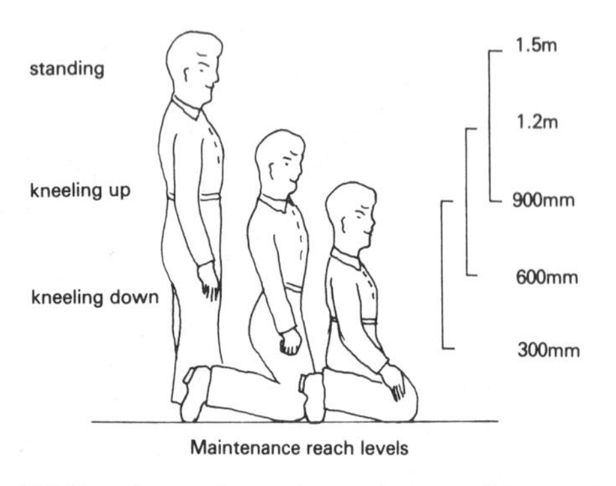

[Fig. 66] Crawlway: dimensions of accessible manholes and inspection pits

critical path The method of arranging the execution of a programme of works in such a manner that all critical operations are defined and their relative priorities established.

critical path diagram A chart which displays the programme of work and indicates specifically all those operations which are critical to the fulfilment of the programme.

critical pressure That steam pressure below which latent heat must be applied to convert a liquid into a vapour.

critical temperature That temperature above which it is impossible to liquefy a gas by application of pressure alone.

cross connection Incorrect pipe connection between circuits: e.g. connection between a potable water supply and a downstream or process water supply.

cross flow Where opposed small discharge pipes without swept entries connect to a stack in a sanitary system. Such entries must be arranged to enter the stack in such a manner that cross flow cannot occur.

cross-flow cooling tower A cooling tower in which the air and the water move at right angles to each other.

cross-flow fan See **tangential fan**.

cross tube boiler A boiler that incorporates heat exchanger tubes which cross through the combustion chamber to achieve optimum heat transfer between the flue gases and the water being heated.

cross vent As **yoke vent**.

cross wire thread insert A locking wire thread insert that maintains bolt tension under conditions of severe vibration.

crosstalk The transfer of airborne noise from one area to another via secondary air paths, such as ventilation ducting or ceiling voids.

crosstalk attenuator [Fig. 67] An attenuator that prevents transmission of conversational sound between rooms or offices across partitions.

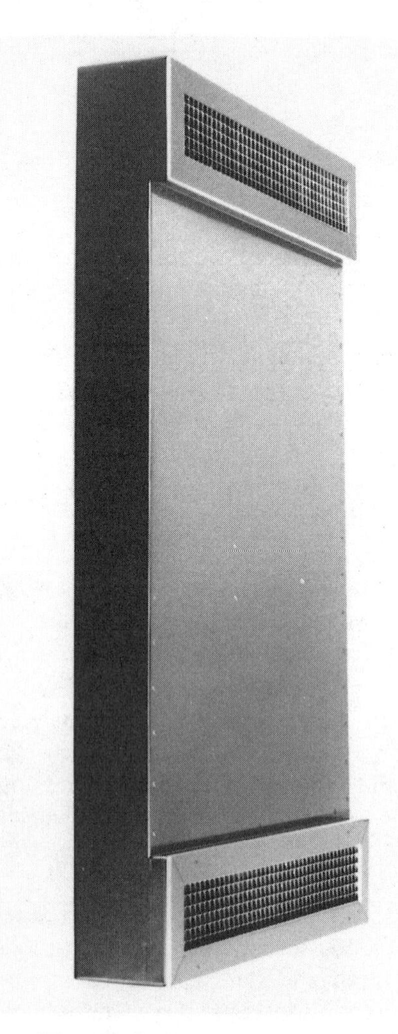

[Fig. 67] Crosstalk sound attenuator

[Fig. 68] Multi-section cubicle-type central control panel for major installation (Contramec Installations Ltd.)

crown of trap The topmost point of the inside of a trap outlet.

crude oil Petroleum as it occurs naturally.

cryogenics The science and technology of substances used at very low temperatures, say, –250°C (–418°F).

cubical expansion The increase in volume which unit volume undergoes when its temperature is raised through one degree.

cubicle-type control panel [Fig. 68] A substantial control panel which is constructed with a number of sections, each housing its own dedicated switchgear, door and locking device.

culm An anthracite coal mining waste product with an average ash content of 70%. It has been found suitable for successful combustion in a fluidized bed boiler (Pennsylvania).

cupro-solvency The property of some waters to dissolve copper. This can have a deleterious effect on brass fittings incorporated with pipe systems conveying such water.

curing A term used in connection with refractories relating to the time required for the settling down of cast refractories before the application of heat. In no circumstances should the curing time be less than 24 hours; it may well be longer, depending on the specification of the refractories used.

customized tariff structure A tariff that can be negotiated with the utility supplier to suit the load and pattern of energy use of major consumers who use in excess of 1 MW (averaged over a three-month period) and can benefit from a non-franchised tariff structure.

cut-off frequency That frequency below which a travelling sound wave in a duct cannot be maintained.

cycle The complete sequence of a value of a periodic quantity which occurs during a period.

cycling Achieving a closed sequence of specified operations.

cycling control The method of programming a sequence (cycle) of operations.

cyclone [Fig. 69] A means of removing particulate matter from dust-laden air or flue gases. A cyclone basically comprises a cylindrical upper section and a conical bottom section. The dust-laden air or flue gases enter the upper section tangentially, centrifugal forces throw the particulates against the outside walls causing them to fall by gravity to the dust outlet at the bottom, whilst the cleaned air or gases leave through a centrally situated tube within the upper section of the cyclone.

Particles down to 40 microns can be removed in this relatively coarse filtration device. Cyclones may be installed singly or in groups; high efficiency cyclones are available to achieve a higher targeted dust collecting efficiency. Cyclones are used widely for the collection of waste from wood working plant and are installed ahead of incinerators. They are also used to clean flue gases.

cyclone furnace [Figs 70, 71, 72] A furnace that utilizes a cyclonic (see **cyclone**) system of combustion, burning waste products to generate heat from the resultant waste gases as steam, hot water or warm air via a waste heat boiler or heat exchanger to which it is connected. Incineration occurs in the cyclone chamber. It is therefore

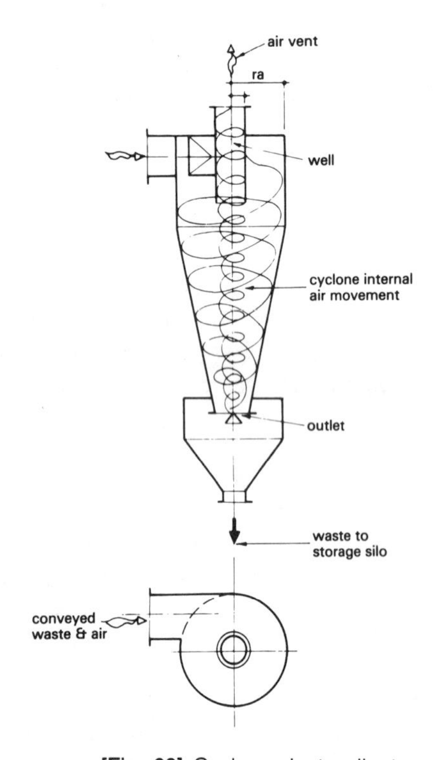

[Fig. 69] Cyclone dust collector

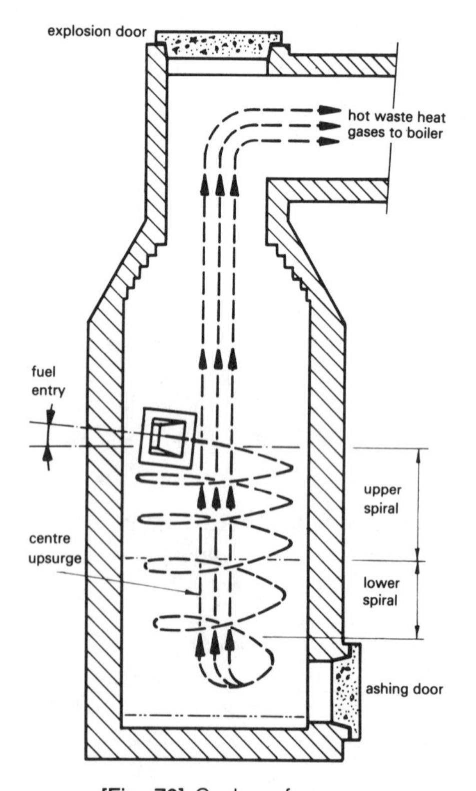

[Fig. 70] Cyclone furnace

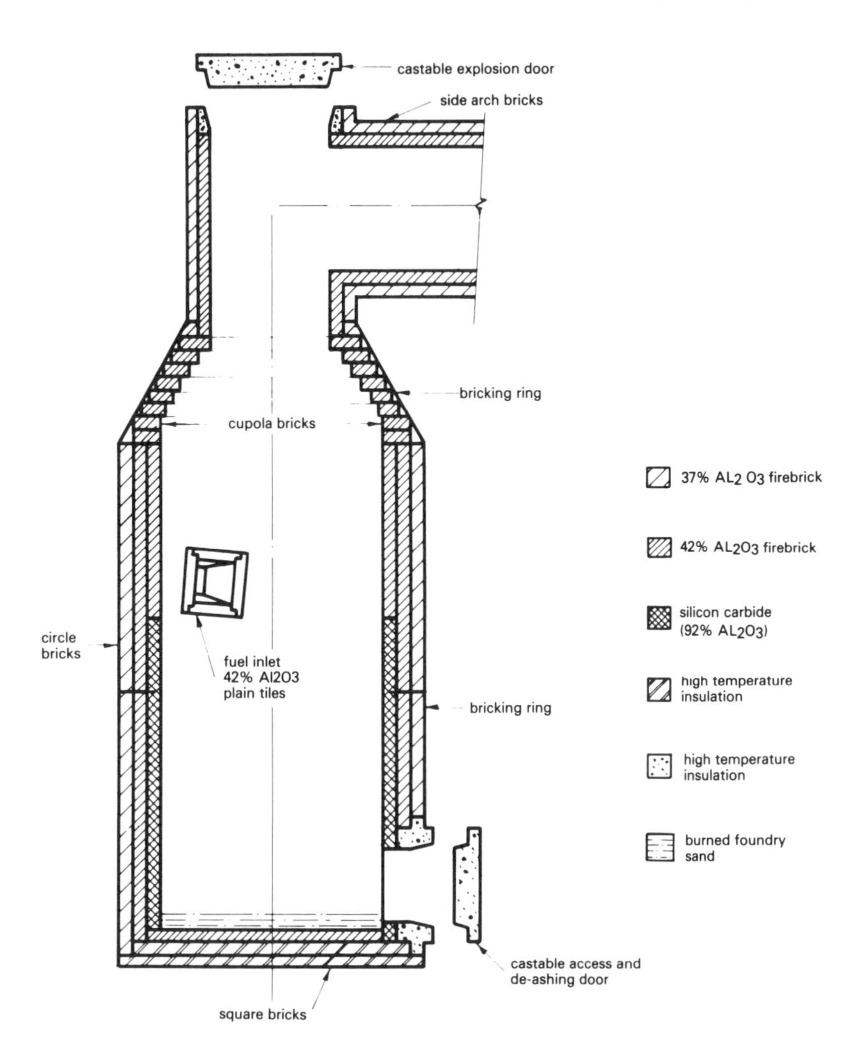

[Fig. 71] Cyclone furnace refractory composition

practicable to utilize existing boilers in conjunction with cyclone furnaces. They are suitable for the combustion of difficult fuels or wastes, such as sludges, wet timber, scrap tyres, etc.

cylinder and pipe thermostat A thermostat that functions with a changeover snap action switch supplied with a fixing strap for cylinder or pipe mounting.

cylinder jacket A covering for industrial and domestic hot water cylinders and tanks. Usually manufactured in wrap-around quilt with zip fasteners and vents to accommodate pipework.

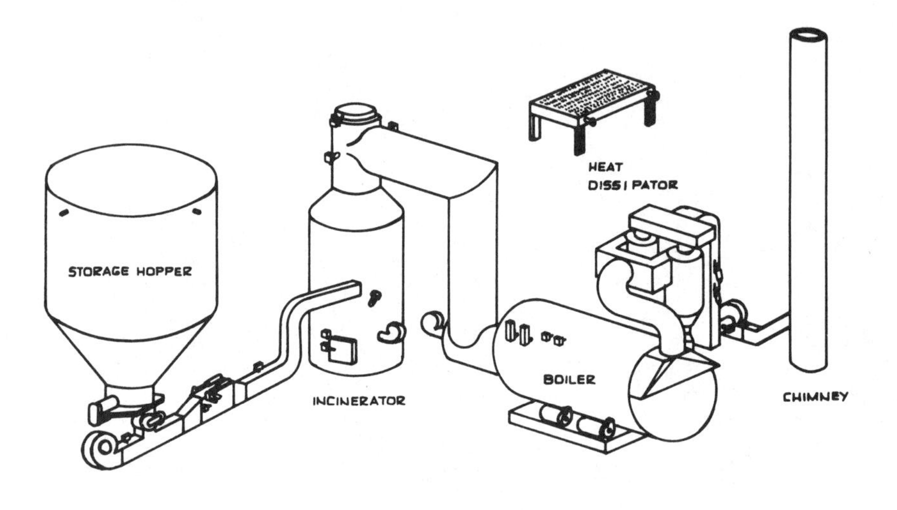

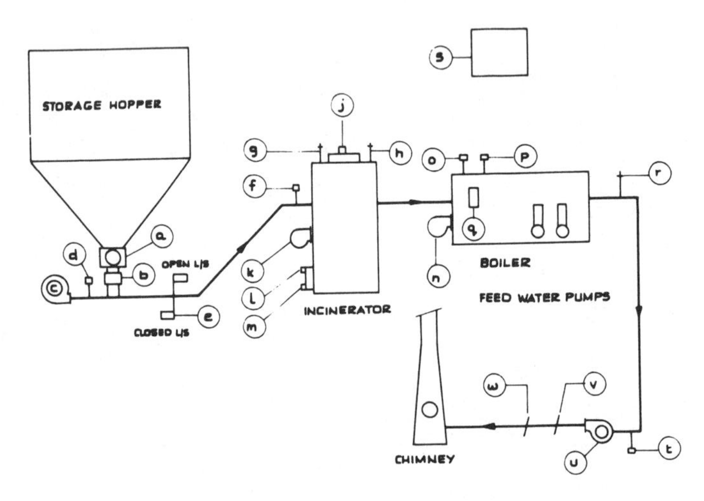

a BIN DISCHARGER (VARIABLE SPEED)
b ROTARY VALVE (" ")
c PRIMARY AIR FAN
d PRIMARY AIR PRESSURE SWITCH
e SAFETY GATE (WITH OPEN . CLOSED LIMIT SWITCHES)
f FURNACE INLET PRESSURE SWITCH
g FURNACE TEMP (T_1)
h " " (T_2)
j EXPLOSION RELIEF LIMIT SWITCH
k IGNITION OIL BURNER
l FURNACE DOOR LOCK CONTROL LIMIT SWITCH
m " " CLOSED LIMIT SWITCH

n MAIN OIL BURNER
o BOILER LIMIT PRESSURE SWITCH
p " CONTROL " "
q BOILER WATER LEVEL ALARMS
r BOILER BACK TEMP.
s HEAT DISSIPATOR
t I.D. SUCTION SWITCH
u I.D. FAN
v I.D. PRIMARY DAMPER
w I.D. SECONDARY "

[Fig. 72] Cyclone furnace operating and safety controls

D

DX coil [Fig. 73] A direct-expansion cooling coil.

daily service tank A tank of limited capacity – commonly about 500 litres – which feeds an oil burner or diesel generator by direct gravity flow. It is a safety feature and may be filled automatically under level control in one single operation. The next replenishment requires a manual start.

Dalton's law The scientific law which states that the pressure which a vapour exerts in a mixture is very nearly independent of the pressure exerted by any other gas or vapour in that mixture.

damper actuator [Fig. 74] A device which operates a motorized damper under the control of a detecting device such as a thermostat or pressurestat.

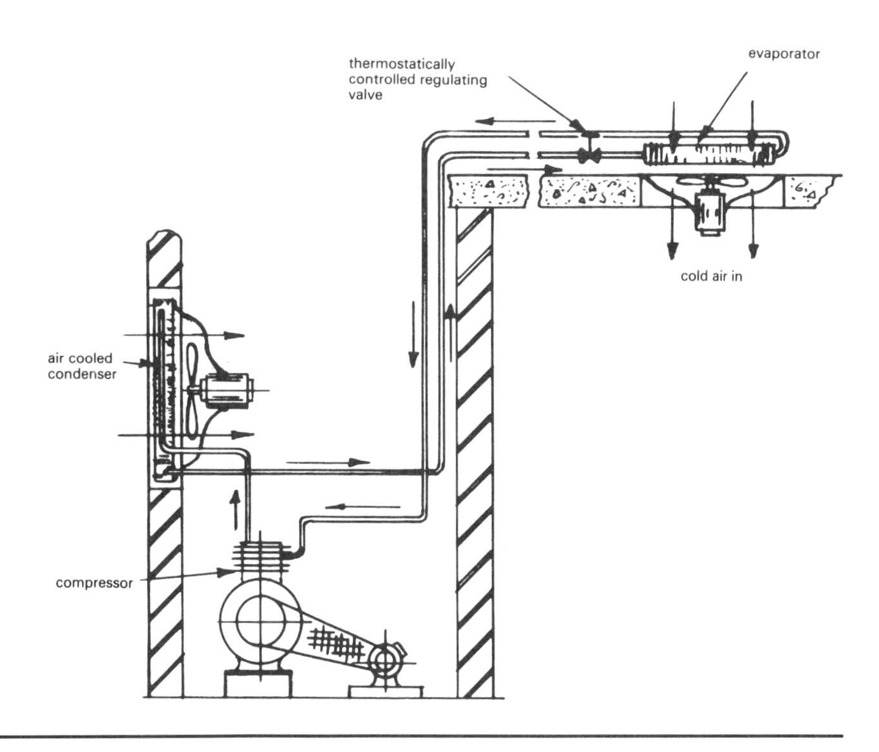

[Fig. 73] Basic direct expansion (DX) refrigeration system

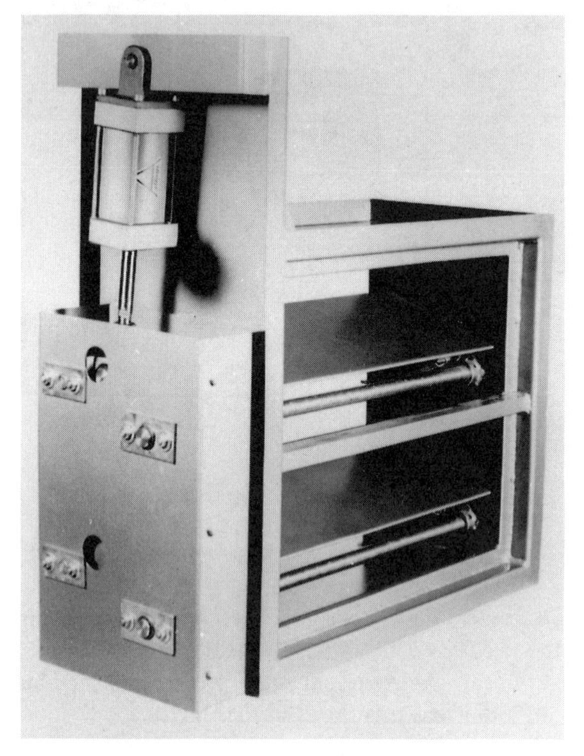

[Fig. 74] Shut-off damper with automatic actuator control (Trox Brothers Ltd.)

damper, barometric See **barometric damper**.

damper, bladed A damper used in larger size ducts comprising a number of airofoil-shaped metal blades moving on a common spindle.

damper, dish A damper that incorporates a metal dish which is forced against a rubber gasket to give a very tight air seal.

damper, multi-leaf [Fig. 75] A damper comprising an assembly of interlocking blades on a common shaft. It is used to balance/regulate air flow with minimum loss of air pressure.

damper, opposed blade See **opposed-blade damper**.

damper, pneumatic See **pneumatic damper**.

damper quadrant A quarter circle frame fitted outside the damper in which the damper spindle moves and is secured in position.

[Fig. 75] Assembly of a multi-leaf damper suitable for manual or automatic control modes (Trox Brothers Ltd.)

damper setting The fixed position of a damper inside a duct, set following commissioning to regulate, e.g. permanently, the proportion of outside air entering a ventilation system.

damper, shut-off type See **shut-off damper**.

damper spindle A rotating rod on which a damper section or blades are mounted.

damping The process of smoothing out vibrations or shock waves.

damping plate A plate that is immersed in the water of a ball-valve supplied water storage tank/cistern in such a position as will offer the greatest resistance to vertical motion. It is fitted to a float or float arm with the object of preventing the oscillation of the float valve assembly with attendant water hammer.

database A structured collection of long-lived data conforming to a data model and managed by a database management system which gives ready access to the data contained in it.

dayworks Works (commonly extra works) instructed to be carried out on the basis of time and materials expended on them. The contractor will be expected to keep comprehensive records of dayworks and to obtain a signature to them from the supervising officer whilst these works are progressing or very shortly after completion. Neglect of such procedures will cause difficulties with the agreement to the final account.

deadleg The length of hot water supply pipe connecting to a draw-off point and not forming part of a circulating loop or circuit. The maximum length of deadleg is restricted by regulation to avoid undue waste of water and discomfort to the user.

dead storage That section or area of a fuel store from which the fuel is not recovered during normal operation. Storage must be designed to minimize wasteful dead storage.

dead zone That part of a control range in which the controls are ineffective.

deaeration The process of removing air (oxygen) from the feed water to a steam boiler to minimize the risk of corrosion.

deaerator [Fig. 76] Device that removes the dissolved gases (mainly oxygen) from boiler feed water.

decibel A unit of the international weighted scale of sound levels.

decibel A (dBA) scale The international weighted scale of sound levels. 0 dBA represents the threshold of hearing; sounds above 140 dBA are uncommon. An increase of 10 dBA indicates that the noise perceived by the listener has doubled in loudness; e.g. a passing car sounds twice as loud

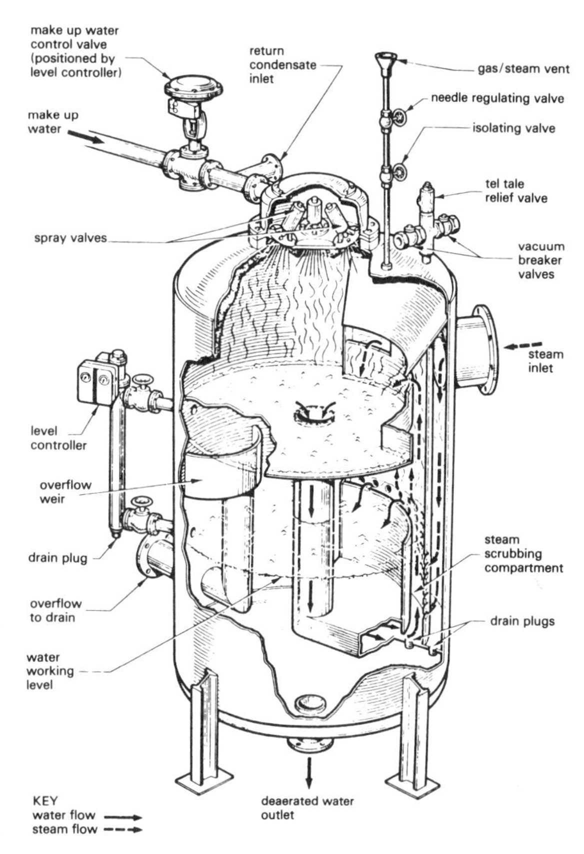

[Fig. 76] Section through an industrial water deaerator

at 70 dBA as one at 60 dBA. The scale is used as a basis for noise legislation.

decibel scale A logarithmic scale for the measurement of levels of sound intensity, pressure, etc. The units denote equal ratio, not equal absolute changes. The decibel scale is related to acoustics.

declination of Sun The angle between the plane of the Earth's orbit and the equatorial plane. Also the angular distance of the Sun from the equator.

dedicated circuit An electric cabling circuit which is arranged to serve a specified load only; e.g. a computer suite.

deep seal trap A trap that provides a water seal 75 mm or deeper.

defects period The period following practical completion of the works during which any defects found in the works must be rectified under the terms of the contract. This commonly varies between six and twelve months. A twelve-month period is usually applied to environmental services to ensure that their performance may be checked over all the seasons of the year.

defrost To remove accumulated ice from a heat transfer surface.

defrost control A control that initiates the defrosting cycle only when the outside temperature is likely to cause freezing on the coil. It terminates the defrost operation when the coil has defrosted and drained.

defrost cycle A programmed arrangement for the periodic removal of accumulated ice from the heat transfer surface of a refrigerator, heat pump, etc.

defrosting The removal of ice which has formed on the heat transfer surfaces of an evaporator within a refrigerator system. This is required periodically to maintain the heat transfer (cooling) capacity. It is usually achieved by automatic means using a timer and defrost heater.

degree-days method A means of comparing, over different periods, the variations in the seasonal heat loads of heating installations in various geographical locations; it is essentially a climatic parameter. It assesses for monthly periods the daily difference in temperature (in degrees Centigrade) between a base of 15.5°C and the mean external temperature over 24 hours. The monthly totals are then used to compare monthly changes in the weather factor, or they may be added together for the complete heating season, to permit comparison between one year and another and one location and another as regards the severity and duration of the winter.

dehumidification The process of abstracting moisture from the air.

dehumidification, by cooling A common method of removing moisture from the air adopted in air-conditioning, by which the air is cooled below its dewpoint when it will shed moisture through condensation.

dehumidification, by heat pump A process of dehumidification whereby the evaporator of the heat pump is placed within the air stream with the object of depressing the air temperature below the dewpoint, when the moisture will condense out of the air stream.

dehumidification, chemical A process of dehumidification in which the moisture content of the air is reduced by passing it over a bed of absorbent moisture, such as silica gel.

dehumidifier Apparatus for achieving the dehumidification of air. The operating principle may rely on a dehydrating agent (such as silica gel) cooling the air below its dew point to shed moisture.

dehydrate To remove moisture from a material or substance by one of several means.

de-ionized water Pure and ultra-pure water approaching the theoretical H_2O. It is used in processes which require water which is of the highest biological and chemical purity and is produced in single or mixed-bed de-ionizers. The purity is measured by a conductivity monitor which measures the conductivity.

delayed action float-operated valve In some water storage systems a float rides on the surface of the water in an open-topped chamber which is fitted within a cistern. The opening of the inlet valve is delayed until the water level in the cistern has fallen through a fixed distance, when the chamber empties itself into the cistern through a non-return valve. With the rising water level, the chamber is refilled by flow over its top edge, thus quickly closing the float-operated valve.

delivery filter and silencer, air compressor A device fitted at the outlet side of the air receiver of an air compressor close to where the pipeline branches off in various directions. It functions to exclude dirt and scale from pipelines entering the delivery of compressed air lines. It may be a felt or fabric, centrifugal or chemical type.

deluge system A method of fire protection which incorporates open sprinklers which are controlled by a quick opening valve (deluge valve) operated by a system of approved heat detectors or conventional sprinklers installed in the same area as the open sprinklers. It is designed to meet special hazards such as polyurethane foam making machines, drying sections of hardboard manufacturing plant, fireworks factories, aircraft hangers, etc. where very intensive fires with a fast rate of fire propagation can be expected and it is desirable to apply water simultaneously over a complete zone in which a fire may originate by admitting water to open sprinklers or to high velocity sprayers. See also **drencher system**.

deluge valve A quick opening type of control valve which is incorporated in a drencher system of fire protection. It is controlled from heat detectors or conventional sprinklers installed in the same area.

Demco sectional slip liners A method of lining sewer pipes or drain-pipes with lengths of polypropylene screwed pipes. See **preformed new sewer replacement pipes**.

demineralized water Water that has been purified by a process of ion exchange in which almost all the dissolved solids and gases (excepting nitrogen and oxygen) have been removed from the raw water supply.

density The mass of a unit volume of a substance. In SI units density is expressed in kilograms per cubic meter.

depth of water seal of trap That depth of water which would have to be removed from a fully charged trap before air could pass freely through the trap.

derate The planned reduction of one of the originally specified parameters of an operating system or equipment, e.g. the lowering of the permitted working pressure of a steam boiler because of age or deterioration of the boiler (sometimes such derating is demanded by the boiler insurers); a reduction in the output of a fan to meet changed conditions or to overcome noise and/or draught complaints; a change of firing nozzle of a gas- or oil-fired burner to derate boiler output or to overcome complaints of noise and/or vibration.

desalination The process of producing pure water from brackish or salt (sea) water. This is an energy intensive operation which results in pure water condensed out of water vapour.

design condition(s) Those parameters laid down for the execution of a particular services design; e.g. the minimum external temperature (heating), the maximum external temperature (cooling), the noise level, or the standard of air filtration.

design criteria Specified parameters to which an installation or plant is intended to perform, e.g. temperature, relative humidity, air purity, or sound level.

design team [Fig. 77] A team usually comprising architect, consulting engineers, quantity surveyors and other specialist advisers as may be required.

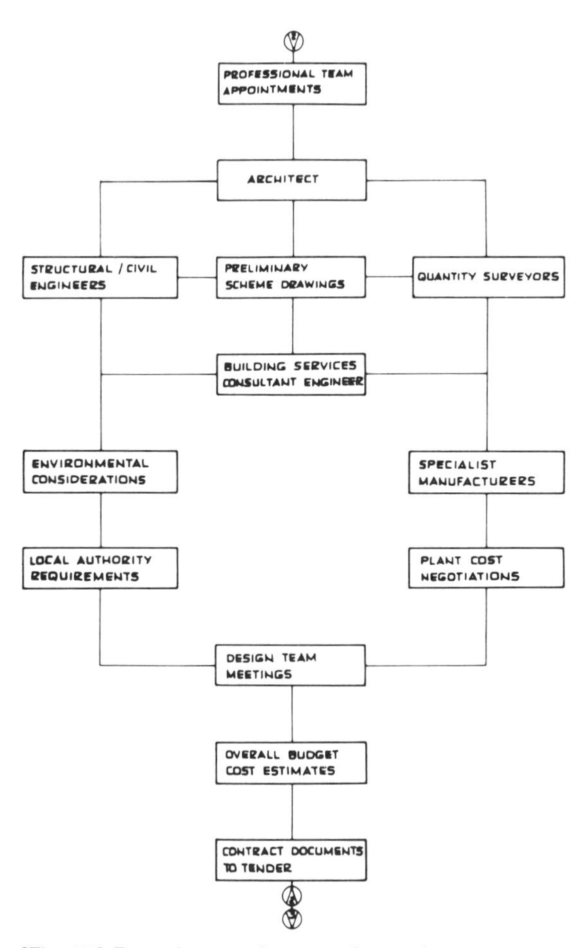

[Fig. 77] From the appointment of a professional team to the completion of the contract documents for tender

dessicant A material (e.g. silica gel) used for removing moisture from a material or substance.

desuperheater A device that reduces superheated steam in temperature by the controlled direct injection of water, aimed at promoting the complete absorption of all cooling water into the steam in the shortest possible time and over the widest possible load range.

detail drawing A drawing that indicates specific detail; bracket, mixing box, etc.

detectatape See **detector tape**.

detector tape Tape which, in conjunction with a signal generator and a signal locator, locates the presence of buried pipes and cables. It is placed in the ground at a specified distance above the buried pipe and provides the only means of locating non-conducting (e.g. plastic) pipes. One model, 'Detectatape' (Boddingtons Ltd.) is available in three alternative types to suit different applications and cost parameters.

(a) Ultra-strong; a wide stripe of foil bonded between a thick reinforced base and a fully sealing covering layer.
(b) Wavelay: two stainless steel wires laminated between two layers of rot-resistant polythene.
(c) Aluminium foil: polythene-coated aluminium foil tape.

determination of contract The unilateral termination of a contract by the employer because of allegedly inadequate performance by the contractor.

dewpoint The temperature of the air at which the saturation vapour pressure equals the actual vapour pressure of the water vapour in the air. If moist air is cooled below its dewpoint, it must shed moisture to maintain the saturation vapour pressure at the new reduced temperature of the air.

dewpoint, acid See **acid dewpoint**.

dezincification The corrosion of brass products, usually indicated by heavy white deposits on the brassware which is in contact with water. Brass is an alloy of copper which contains zinc. Whilst zinc is widely used for protection against corrosion, it can itself be attacked by waters which contain free carbon dioxide or which have a high pH (in excess of 8.2) and a high ratio of chloride to carbonate hardness. In the latter, for a low chloride content, the problem arises when the ratio of chloride to carbonate hardness is greater than 1, and for chloride greater than 20 mg/l; dezincification can occur when the ratio is lower.

Remedy: avoid hot-pressed brass and/or increase the carbonate hardness in applications where dezincification is likely or suspected.

dial thermometer A thermometer that indicates the temperature on a graduated circular dial. It is often screwed directly into the temperature well of a vessel containing heated or cooled fluid.

diaphragm A membrane adapted for the control of valves and pressurization units. It is selected to suit the specific fluid being handled and it is essential that the correct material is specified. A diaphragm is replaceable in the event of failure.

diaphragm-operated valve A valve extensively used for automatic flow control in piped services, usually operated by pneumatic (air) pressure on the diaphragm. The operating medium may also be water, oil or steam. In such valves, the conventional gland stuffing box is eliminated and replaced by two composition rings which require no adjustment, promote easy operation and give a perfect seal under exacting conditions, e.g. for high vacuum, where the smallest back leakage or inspiration of air cannot be tolerated. In a standard arrangement, the operating medium, acting on the diaphragm which is commonly of oil-resistant neoprene, opens the valve; it is closed by means of an adjustable spring when the pressure is released.

diaphragm valve A method of controlling the flow of fluid through the valve body by the action of a replaceable diaphragm, available in different materials to suit the liquid being controlled by the valve. It is essential to specify the correct diaphragm which is generally colour-coded.

diesel engine An engine which is fuelled by diesel oil. In building services it may serve as a stand-by power unit for use in the event of mains power failure to maintain essential services such as ventilation, lifts or computers.

diesel oil Oil used as fuel in diesel and other compression/ignition engines. High-speed engines require a special grade of oil. See also **gas oil**.

differential aeration A type of corrosion which is caused by the breakdown of the natural oxide film of a metal surface due to lack of oxygen. Corrosion is localized and can occur particularly underneath surface deposits such as mill scale and in crevices.

differential controller A controller that measures variations in temperature, pressure or humidity and changes its control pressure accordingly.

differential pressure control A control that functions by sensing pressure differences and actuating controls to maintain a specified pressure.

diffuse reflection Light reflected from a surface scattered equally in all directions.

digester A component of a biogas production facility. It is a container in which the biological material (or biomass) is allowed to ferment and generate methane or alcohol to be used as fuel.

digital display A display indicating the value of a measured function on a light-emitting diode (LED) display screen.

digital temperature indicator An electronic thermometer providing an accurate digital readout of temperature from a thermocouple or resistance thermometer input.

dilution flue An exhaust serving a diluted flue gas system in which the flue gases discharged by a gas-burning appliance are diluted with air with the object of permitting discharge of the gases at a low level and thereby avoiding the need for a chimney.

diode An electronic valve.

direct-acting controller A sensing element which, by the thermal movement (expansion or contraction) of a liquid or vapour, transmits an activating power directly to a bellows or diaphragm, which then actuates the controlled device (e.g. a damper, motorized valve or pneumatic controller).

direct-acting reducing valve (regulator) One in which the valve is operated directly by the combined action of a spring and of the outlet pressure applied to the underside of the valve diaphragm. This type is used for applications involving smaller flows of fluids or in cases where some variation of the outlet pressure can be tolerated.

direct-acting thermostat A thermostat that is directly coupled to the controller and moves it under action of thermal expansion or pressure differential; e.g. a thermostatic radiator valve or immersion heater control.

direct current (d.c.) Unidirectional electric current.

direct drive A drive in which the drive shaft of the motor or engine is directly coupled to the driven machine which may be a pump, fan etc. Driven machine speed must equal that of the motor or engine.

direct expansion coil [Fig. 78] A heat exchange coil which is placed into the airstream of an air-conditioning system and in which the condensed liquid refrigerant evaporates into a vapour or gas. The latent heat is extracted from the airstream, thereby cooling it. The coil may operate dry or flooded.

direct expansion cooling Cooling in which the refrigerant evaporates inside coils which are placed directly into the airstream being conditioned, thereby cooling the air. Its performance is difficult to control within close limits.

direct gas-fired heater A heater in which exhaust (flue) gases are discharged into the space being heated.

direct hot water supply When the hot water discharged from the taps has passed through the boiler

[Fig. 78] Direct expansion (DX) coil

water jacket and not through a heat exchanger. This practice is unsuitable for use with hard water as the inevitable scale deposits on the hot boiler surfaces will cause thermal stress and boiler failure.

directly flued appliance An appliance in which exhaust gases are discharged to atmosphere via a flue pipe directly connected to the appliance.

direct steam heating See **steam injection**.

dirt pocket A pipe pocket fitted ahead of a steam trap within the line of steam flow designed to intercept heavy scale and sludge.

disc check valve A valve in which the check mechanism incorporates or is a disc.

disc feed A method of regulating the flow of water into a flushing cistern using a thin disc with a calibrated orifice fitted within a union connector in the feed pipe to the cistern.

discharge lamp [Figs 79, 80] A lamp in which the electric current passes through a mixture of gas and metallic vapour, giving rise to a luminous electric discharge which is utilized for illumination two and a half to five times greater than that available from filament lamps of equivalent electricity consumption.

discharge stack, connection to drain The bend at the foot of the stack should have a minimum centre line radius of 200 mm or should preferably be a long radius bend to facilitate smooth flow and ease of clearing blockages.

discharge unit A unit employed in the design of sanitary systems and is chosen so that the relative load-producing effect of sanitary appliances can be expressed as multiples of that unit. The discharge unit rating of an appliance depends on the rate and duration of discharge, on the interval between discharges and on the chosen criterion of satisfactory

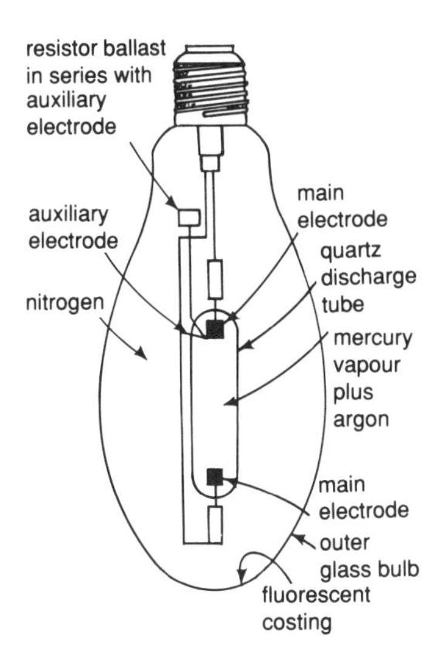

[Fig. 79] Mercury-vapour discharge lamp

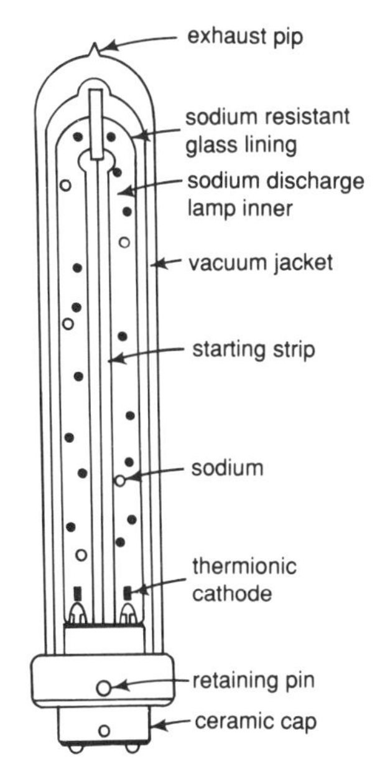

[Fig. 80] Sodium-vapour discharge lamp

service and not merely on the multiple of a rate of flow.

discharging capacity, thermal storage The rate at which a hot water accumulator of a thermal storage system can discharge heat into the system. It is controlled by a motorized valve system.

disclaimer A clause in a specification which states that no claim by the contractor for additional payment will be allowed on the grounds of any misunderstanding or misinterpretation or on the grounds that he did not or could not foresee any matter which might affect or have affected the execution of the works.

discoloration The discoloration of water discharged at taps is due to impurities in the water supply or the presence of corrosion products within the system.

discrimination (between protective devices) Discrimination between two or more electric circuit-interrupting devices occurs when, following the incidence of a short-circuit or of an overload, only that protective device which is required to operate, does so, whilst the other protective device(s) remain(s) unaffected (intact).

disc rotor motor A motor developed by Fischbach for their range of compact fans. It incorporates a motor shaft which is static and the disc-shaped rotor revolves around it. The forward curved impeller is attached to the disc-shaped stator. The motor is mounted in the airstream and thereby draws heat away from the windings, this heat exchange being assisted by a large surface area. The windings are embedded in a special resin developed by Fischbach. In consequence, the motor need not be taken out of the airstream, even in clean room applications. If the motor was earthed, the resin would prevent smoke escaping from the motor.

displacement ventilation [Fig. 81] Ventilation that functions via a plug of air entering the space across a large input area, displacing the hot or contaminated room air with minimal mixing. The principle is applied to high-level input (e.g. laminar downflow clean rooms), floor supply (e.g. computer rooms), sidewall supply for cross ventilation and low-level supply with horizontal discharge and high-level exhaust (e.g. hospital operating theatres).

dissolved gases Mainly oxygen, ammonia and carbon dioxide which are present in boiler feed water. Their presence is undesirable as they may cause corrosion of boiler feed equipment. They may

[Fig. 81] Air distribution: displacement flow pattern diffuser (Trox Brothers Ltd.)

be removed in deaerators or their presence may be inhibited by dosing with suitable chemicals.

dissolved solids Dissolved mineral salts which are present in water supplies. They increase in concentration as the water is evaporated into steam and promote undesirable foaming and instability in the boiler water.

distillation A method of purifying water by the process of boiling and condensation of the resultant vapours.

distilled water Condensed water resulting from the process of distillation.

distribution board (dis. board) A panel or board from which electrical connections are taken off for the distribution of electricity to various electrical circuits, usually embodying fuses or circuit breakers to protect the circuits served off the board.

distribution pipe A pipe, other than an overflow or flushing pipe, which conveys water from a storage cistern or from hot water equipment supplied from a feed cistern and under pressure from that cistern.

distributor A device installed with direct expansion coils to equally divide the mixture of liquid and vapour refrigerant passing the expansion valve amongst the various circuits. The main circuit comprises a refrigerant feed connection from the distributor through the coil to the suction header, the path being designed to offer optimum heat transfer, effective return of lubricant and an acceptable pressure drop. Individual liquid connections from the distributor to the coil inlets are commonly made of small-bore tubing, all of the same length and diameter, in order to impose similar friction between the distributor and the coil.

district cooling [Figs 82, 83] A method of distributing the cooling medium, usually chilled water, from a centralized refrigeration plant to individual operators of air-conditioning or chiller equipment. Another method utilizes transfer of the cooling effect by heat pipes.

district heating, once-through system A system in which a single pipe connects from the boiler or heat exchanger plant to the consumer of heat. The design of the system encourages the maximum abstraction of heat from the water before it is discharged to waste; there are no closed circulations.

district heating station An assembly of boilers and associated equipment located within plant room or energy centre supplying steam or hot water to a district heating system.

district heating system [Fig. 84] A heating system that serves a large number (say, 1000 plus) consumers of heat which are independently connected to it, usually via a consumer terminal which includes a heat metering facility.

diurnal Describes a process or cycle which recurs daily.

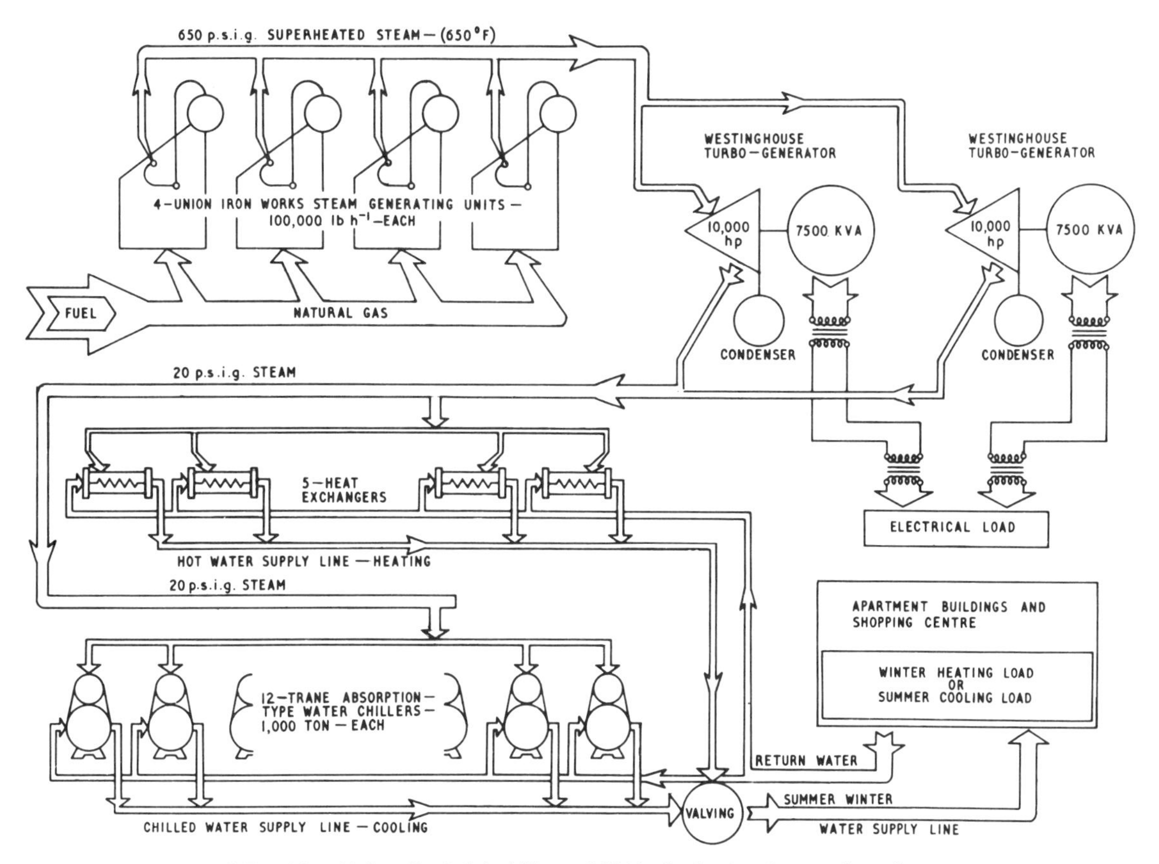

[Fig. 82] Brooklyn Union Rochdale Village (USA) district heating and cooling system

diversity factor The probable rate of flow divided by the total possible rate of flow (e.g. of water or electricity). It is applied in the design of a system which serves a multiplicity of outlets, e.g. water or electric, to determine the maximum rate of flow of water or electricity in a pipe or cable, as it is unlikely that all the outlets will be active at one and the same time.

domestic hot water supply The provision of hot water for domestic use.

domestic hot water supply demand A computation based on numbers and types of draw-off water fittings multipled by an appropriate diversity factor.

domestic subcontractor A specialist firm appointed by the main contractor and included within his applications for payment and accounts. The main contractor carries total responsibility.

Doppler effect A distortion of sound waves brought about by the velocity and direction of a sound-emitting object. The effect is utilized in flow measuring instruments.

double check valve assembly Two separate check valves with an intervening draining tap required for certain water supply applications.

double disc gate valve A valve incorporating a gate consisting of two discs which are forced apart by a spreading mechanism at the point of closure against both parallel body seats, thereby ensuring effective sealing of the valve without the assistance of the fluid pressure.

double-pole switch A cut-out, circuit breaker, fused switch or similar in which the circuit is broken at both poles simultaneously.

double regulating valve A valve that embodies features which permit maintenance of the

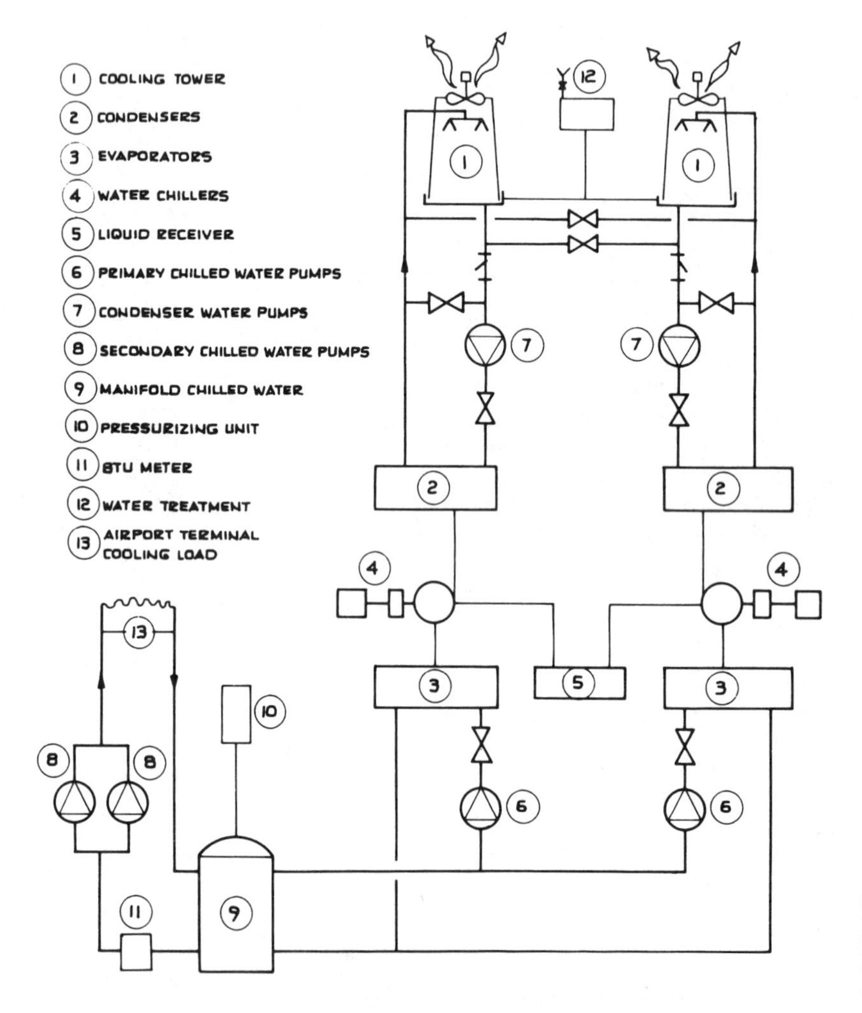

[Fig. 83] District cooling scheme at London's Heathrow Airport

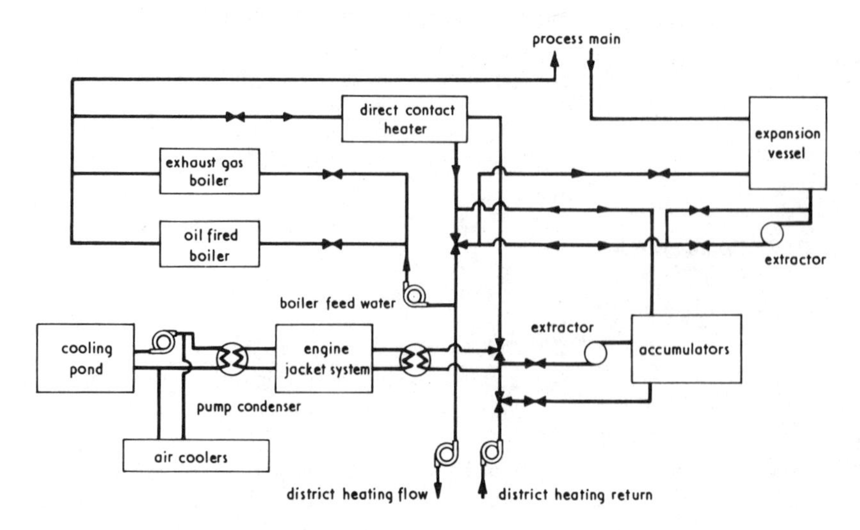

[Fig. 84] General layout of Aldershot district heating system

permanent valve adjustment when the valve is used to isolate the circuit.

double-skin chimney A chimney that has an inner and an outer steel shell. The annular space between the skins may be filled with a suitable insulation material such as mineral wool. The construction must avoid any direct contact between the inner and outer shells.

downlighter A luminaire which directs the light towards the floor.

down service water supply A water supply that is drawn from storage tanks.

downwash The downwards movement of flue gases existing from a chimney into eddies which form in the lee of the chimney when a wind is blowing. This causes discoloration of the chimney terminal and may drive the gases towards the ground. It can be avoided by adopting a sufficiently high discharge velocity using a correctly designed chimney terminal.

drainage pipe flexibility This depends on the pipe stiffness and on the soil stiffness. The drainage design is based on the extent to which deformation of the pipe may be allowed to proceed.

drainage pipe, flexible A pipe that under soil and surcharge loads can deflect to a significant extent without showing structural distress. Such pipes derive their load-bearing capability from their bedding or surround. Under soil and superimposed loading, such pipes deflect vertically and horizontally thereby developing passive support at the sides of the pipe. They may be of ductile iron, GRP, pitch fibre or uPVC.

drainage pipes, flexible joints [Fig. 85] Joints that accommodate changes in the subsoil, such as the seasonal variations in moisture content, bedding of the drains and compaction conditions, as well as the nature of the subsoil itself which may lead to differential settlement and consequent stresses in buried drains. Flexible joints are used to provide flexibility, unstressed longitudinal travel and ease of jointing.

drainage pipes, rigid These may be of asbestos cement, clay, concrete or grey iron and must be bedded in/on granular material or selected fill appropriate to the ground conditions.

drainage pipe, stiffness The stiffness of a pipe depends on the material of manufacture and on the extent of deformation which is allowable as the pipe stiffness increases.

drainage system, suspended See **suspended drainage system**.

drain chute A tapered drain fitting attached to the inlet or outlet of a manhole to facilitate rodding through.

drain cock A cock fitted at the low point of a liquid's distribution pipe system, commonly with a square valve shank operated by a key or spanner. The outlet port should be plugged when not in use.

drain, fin See **fin drain**.

drain liners Either *in situ* liners for renovating an existing drain-pipe with a new liner or a preformed pipe system for the replacement of an existing drain-pipe.

drain outfall The discharge of a surface water drainage system into a water course, water sewer, combined sewer soakaway or storage tank, as appropriate to the installation.

drain saddle A fitting attached to an existing drain at which a branch drain may be connected.

draught sealers A range of devices designed to cope with most draught problems associated with doors and windows.

drencher head Terminal discharge fitting of drencher system.

drencher system A method of fire protection/fighting which relies on a deluge of water to contain/fight a fire.

drinking fountain Equipment connected to mains water which ejects a controlled vertical stream of

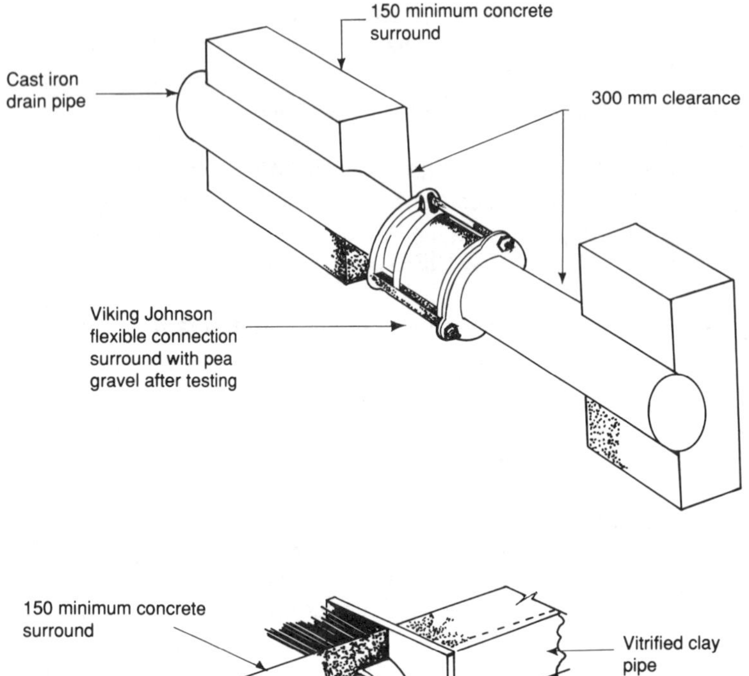

[Fig. 85] Flexible joints.
(a) On cast-iron drain.
(b) On concrete surround to vitrified clay drain

clean drinking water when activated by hand or foot pressure. For reasons of hygiene, it should not be installed within toilet or lavatory accommodation. It may include a water chilling facility.

drop connection A length of drain installed vertically immediately before its connection to another drain or to a sewer.

drop-pipe manhole A manhole built to accommodate significant differences between invert levels by building the manhole in the lower drain and installing a vertical, or nearly vertical, drop-pipe from the higher drain. This pipe may be fitted either inside the manhole, supported on adequate brackets, or be run outside the manhole; in the former case, the chamber must be enlarged correspondingly. A drop-pipe on a branch drain should terminate at its lower end with a bend turned to discharged its flow at 45 degrees or less to the direction of flow in the lower drain.

drum A horizontal cylindrical vessel fitted in the upper part of a water tube boiler to which the boiler steam tubes connect and direct the steam being generated.

drum pump A hand-held electric pump used to transfer liquids from containers. There is a wide range of materials available for the construction of pump tubes to suit the properties of the liquid being handled.

dry air Air which is without moisture content, generally after passing through a process of air drying.

dry bulb temperature The temperature of air as measured by an ordinary (dry bulb) thermometer.

dry core boiler A boiler that comprises a core of heat-retaining bricks in a well insulated casing. The bricks are heated overnight during the cheap off-peak periods. When heat is required it is extracted by means of a fan which drives air around a closed loop within the hot bricks and over a heat exchanger through which water is pumped. The heated water is fed to radiators in the conventional manner. The boiler incorporates a boost facility which can be activated outside the off-peak period to supplement the stored heat.

dry evaporation The process that ensures that only gas (usually with some superheat) reaches the refrigeration compressor.

dry evaporator See evaporator. In the dry evaporator, the water flows through tubing and there is no liquid storage of refrigerant in the evaporator.

dry metal wool air filter A pre-filter used in high temperature filtration applications at a face velocity of about 1.5 m/s. It is cleanable, fire and weather resistant, and there is a choice of metal wool types.

dryness fraction The weight of a quantity of steam in a sample compared to the total combined weight of the steam and of the entrained moisture.

dry riser A pipe installed in tall buildings from ground to top level with an outlet valve and connection at each floor landing. In the event of fire, the fire brigade connects its supply hose to the inlet connection at the lowest level.

dry riser box A valve and capped connection off a dry riser located in an accessible box at each floor landing.

dry riser, fire-fighting An empty (dry) pipe which rises vertically through a building and incorporates a hydrant outlet at each floor and at roof level. An inlet fitting (known as a breeching piece) is provided to the pipe at ground floor level, terminating at an external wall to enable the fire-fighters to pump water into the riser from the closest suitable water hydrant. Dry risers are installed solely for the use of the fire brigade and avoid the need for canvas hose to be run up the staircase of a building under fire-fighting conditions. The fire on individual floors is fought by coupling canvas hoses to the floor hydrants.

dry saturated steam Steam which contains neither free moisture nor superheat; for practical reasons, this is seldom achievable.

dual duct system A system that comprises one duct carrying warm air and another carrying cooled air, mixing taking place near points of use.

dual duct air-conditioning system An air-conditioning system that comprises ducts conveying heated and cooled air streams, respectively, together with plenum mixing boxes and terminal outlets, usually in conjunction with a separate air extract system.

dual gas burner A burner that provides immediate fuel change from natural gas to butane (or propane) or vice versa, even when the burner is firing. Changing fuel according to price advantage, or where interruptable natural gas supplies exist, is possible. The different characteristics of natural gas and, for example, butane, are dealt with economically by a basically conventional gas train incorporating two sets of gas pressure regulators, one for each gas. The two gas inlet sections are connected by short and joined levers in the form of a handle. The interlocked lever assembly operates the gas cocks so that as one opens the other closes. With dual-fuel gas and oil firing, the burner controls are necessarily different for each fuel, and the changeover from one fuel to another is less immediate than with dual gas firing.

duckfoot bend Cast-iron or stoneware bend which has a supporting foot formed integrally in its base.

duct [Fig. 86] A conduit for the conveyance of air, gas or other services. It may be constructed of masonry, concrete, steel, glass fibre, asbestos, etc.

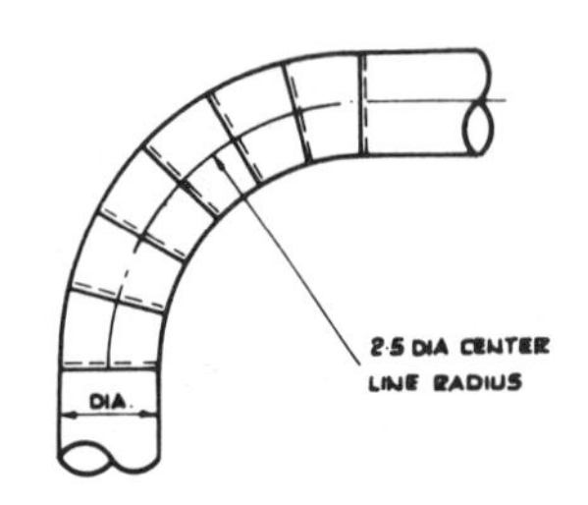

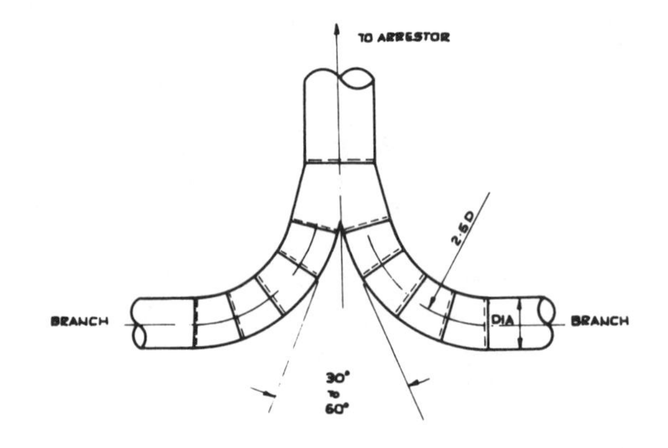

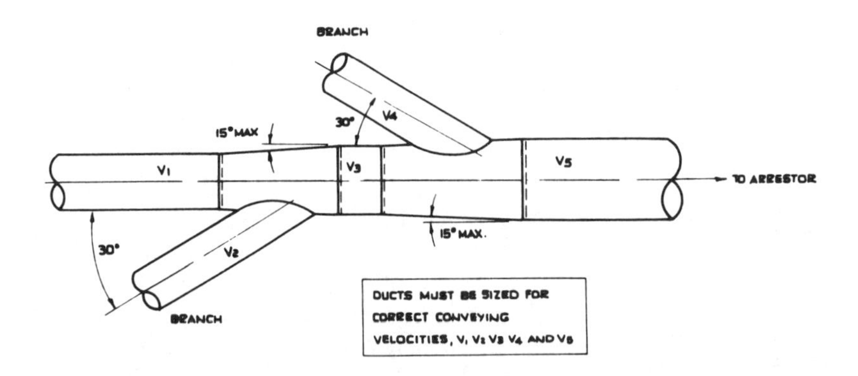

[Fig. 86] Methods of duct construction

duct access opening A hole cut into a duct to permit access to duct-mounted equipment, e.g. a fire damper linkage or thermostat stem. It should be covered by a well fitted access cover to obviate air leakage. A somewhat larger access opening is required for the purpose of cleaning access to the duct. This is necessary for all types of ducts, but particularly essential for ducts conveying exhaust air from kitchen extract hoods and similar equipment.

duct, angle iron joint A means of jointing duct sections. Two angle iron frames are attached to each section of duct and secured by placing a jointing compound between the flanges and tightening with nuts and bolts.

duct breakout noise See **breakout noise**.

duct construction, standards The recommended specifications for duct construction and for duct thicknesses relative to duct dimensions, published by the Heating and Ventilating Contractors Association and by the Chartered Institution of Building Services Engineers.

duct, flexible A duct made from a fabric component interlocked mechanically with its supporting metal spiral and commonly coated with butyl coated cotton or neoprene coated glass fibre where fire resistance is required. The fabric is twisted together to form a permanent air-tight joint. This type of duct is commonly used for short final connections to air terminal fittings.

duct, flexible connector A connector which joins a duct system to a fan or other rotating/moving equipment to obviate the transmission of fan vibration into the ducted system. It is commonly fitted between retaining flanges at the fan inlet and fan outlet. Asbestos used to be the common flexible material, but is no longer specified. Neoprene has taken its place.

duct, galvanized A duct constructed fron zinc-coated steel sheet.

duct, galvanized after made The requirement to have a completed duct system or section galvanized after it has been formed.

duct, grooved seam [Fig. 87] A method of jointing metal sheets to form an air duct by folding the surfaces to interlock with each other. The completed joint lies horizontally.

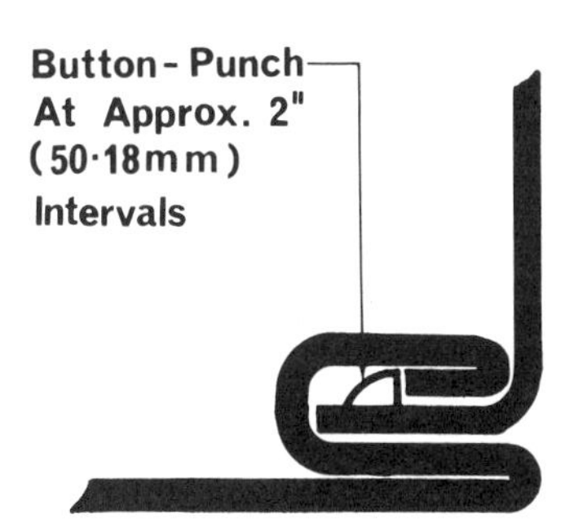

SNAP-LOCK JOINT

[Fig. 87] Two methods of joining lengths of ducts

duct, index See **index duct**.

ducting Arrangement or system of ducts.

ducting fabric Ducts are constructed of fabric components interlocked mechanically with a supporting metal spiral. They are commonly coated with butyl-coated cotton, but for a non-flammable finish the coating would be neoprene-coated glass.

ducting spiral Ducting consisting of a wire helix spring steel frame to provide flexibility. This is available in thermally insulated form with glass fibre insulation and a vinyl vapour barrier.

duct, liner splitter A partition inserted into a duct for the purpose of improving the sound attenuation.

duct lintel A lintel which enables the duct to pass through an opening without incurring the risk of masonry scratching the duct material and/or imposing its weight on to the duct wall.

duct, mid feathers Another term for duct splitters.

duct mounted This relates to an item which is mounted on a duct, such as a thermostat.

duct, Pittsburgh lock [Fig. 88] A method of jointing metal sheets to form an air duct by folding the surfaces to interlock with each other.

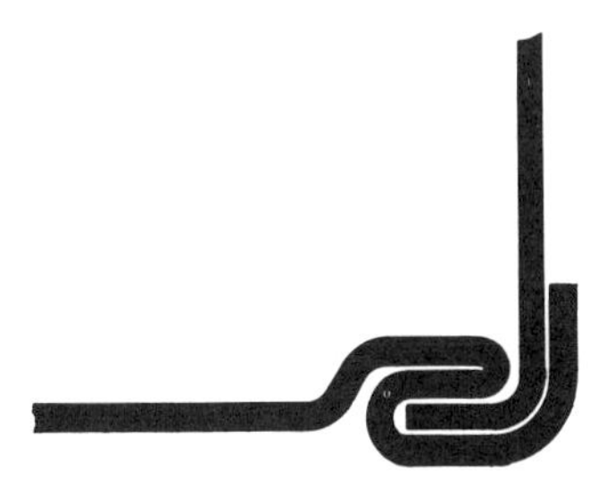

PITTSBURG LOCK

[Fig. 88] Methods of jointing metal sheets to form air ducts

duct sizing, by static regain See **static regain method of duct sizing**.

duct sizing, velocity reduction method See **velocity reduction method of duct sizing**.

duct, slip joint Two sections of duct which are slipped one inside the other. Airtightness is achieved by applying a jointing paste or a suitable self-adhesive tape to the mating surfaces and securing with self-tapping screws.

duct, snap lock joint A method of jointing metal sheets to form an air duct by folding the surfaces to interlock with each other and adding a button-punch at 50 mm intervals to secure the joint.

duct, spiral wound See **spiral wound duct**.

duct splitter A partition used to divide the internal cross-section of a duct to guide air flow. It is mainly used in air ducts having a high aspect ratio.

ducts, structurally integrated Airways formed within the building structure, e.g. inside hollow columns or beams.

duct, standing seam A method of jointing metal sheets to form an air duct by folding the surfaces in the vertical plane to interlock with each other. The completed joint stands upright.

ductstat A detector located inside a duct, to activate motorized valve(s), damper motor(s) and air controls to achieve control over temperature, pressure, humidity, etc.

duct, stiffeners [Fig. 89] Supports incorporated into large cross-sectional ducts to add structural stability under air flow conditions and to obviate drumming and vibration. They are commonly in the form of angle iron frames secured to the duct sides and may be supplemented with a flat iron framework placed inside the duct.

duct supports [Figs 90, 91] Supports which should ensure that the duct runs are kept horizontal and do not sag. Spacing of the supports must be adequate for this purpose. There is a great variety of supports depending on whether the ducts are supported from above or from walls. Arrangements at the supports must safeguard the ducts from damage due to the fixing of the supports to the ducts.

duct, swaged A method of stiffening a circular duct by forcing the metal into an upstanding ring.

duct systems, PVC PVC ducts are used where they convey corrosive fumes and/or pass through a corrosive atmosphere.

duct, transformation [Figs 92, 93, 94] A duct fitting inserted into ducting to change the dimensions from one section to another.

duct transition section An intermediate section of ducting which serves to change the dimensions from one section to the next.

duct, turning vanes Guides incorporated within sharp-angle duct fittings to direct the air flow with minimum friction and noise generation.

duct, vapour seal Impervious material incorporated with thermal insulation to prevent water vapour condensing on cold ducting.

duct, weathering skirt A transformation duct fitting, the upper part of which is of only slightly larger diameter than the duct being weathered and the lower part of which is of a much larger diameter, sufficient to cover an oversize sleeve through which the duct passes out of the building. It is complete with metal band, fastenings and joint sealing insert and is weatherproofed with suitable paint or compound.

duct weathering, through flat roof construction An upstand is formed in a roof where a duct passes through. An oversize metal sleeve is fitted over the duct and a metal weathering skirt is banded to the upper part of the duct, so that any water will drain off over the skirt on to the roof. The distance between the skirt and the sleeve must be adequate to prevent driving rain passing through the sleeve into the building.

duct weathering, through sloping roof A metal weathering plate is prepared to match the slope of the roof and an oversize sleeve is welded on to it to permit the duct to pass vertically through the plate. A weathering skirt is then banded on to the upper part of the duct to cover the sleeve and direct any water to drain down the roof slope.

dump condenser A means of dissipating excess steam (e.g. generated by a waste heat boiler) by condensing it in a fan-assisted heat exchanger.

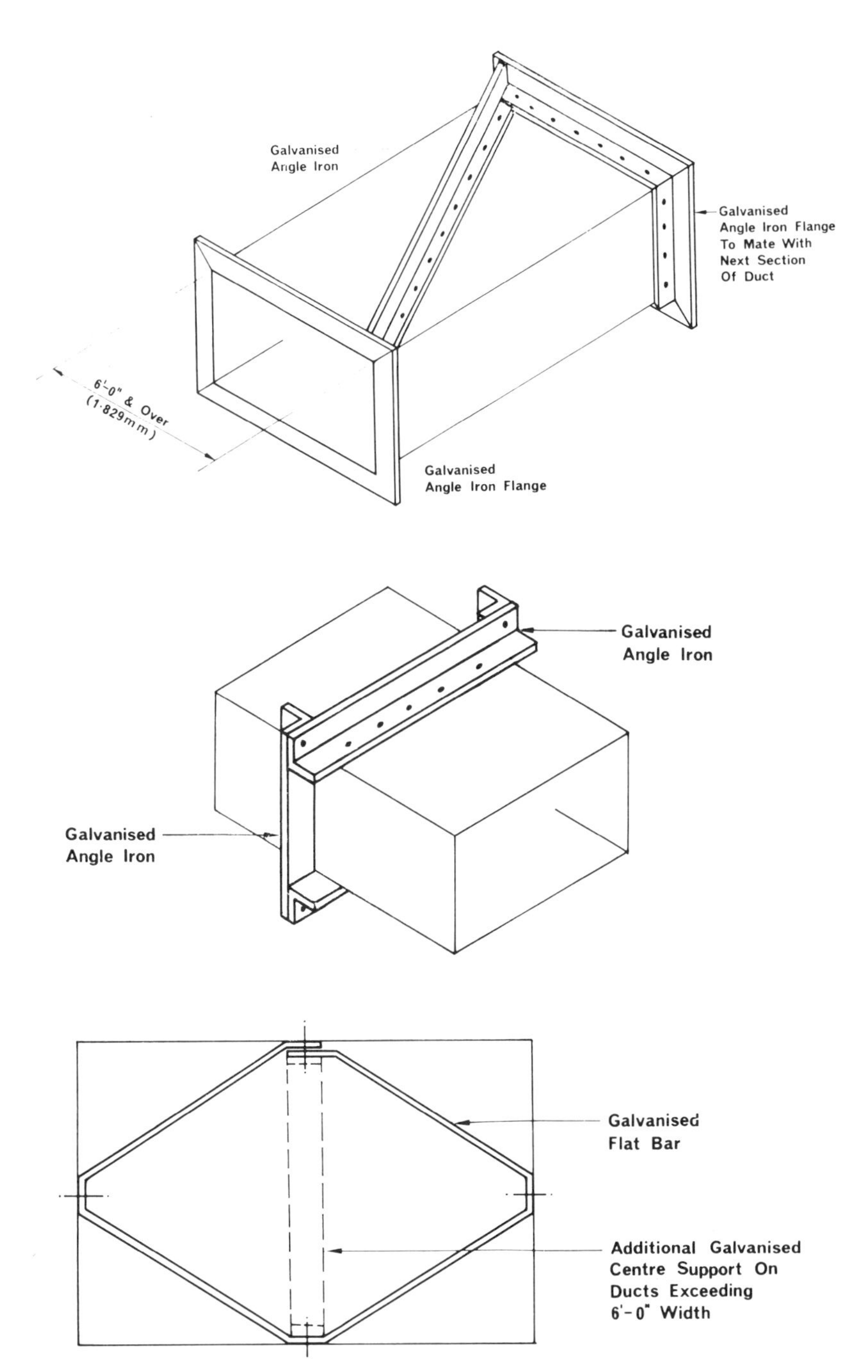

[Fig. 89] Methods of stiffening metal ducts

Heated air is usually dumped to atmosphere and condensate returned to the boiler.

dump heater See **heat dissipator**.

dump line A connecting pipe between the overflow connection of an elevated oil storage daily service tank and the main oil storage. It is a safety feature necessary to prevent overspill of oil at the service tank in the event of accidential over-filling.

dump valve A valve that actuates flow through the dump line from an elevated oil storage daily

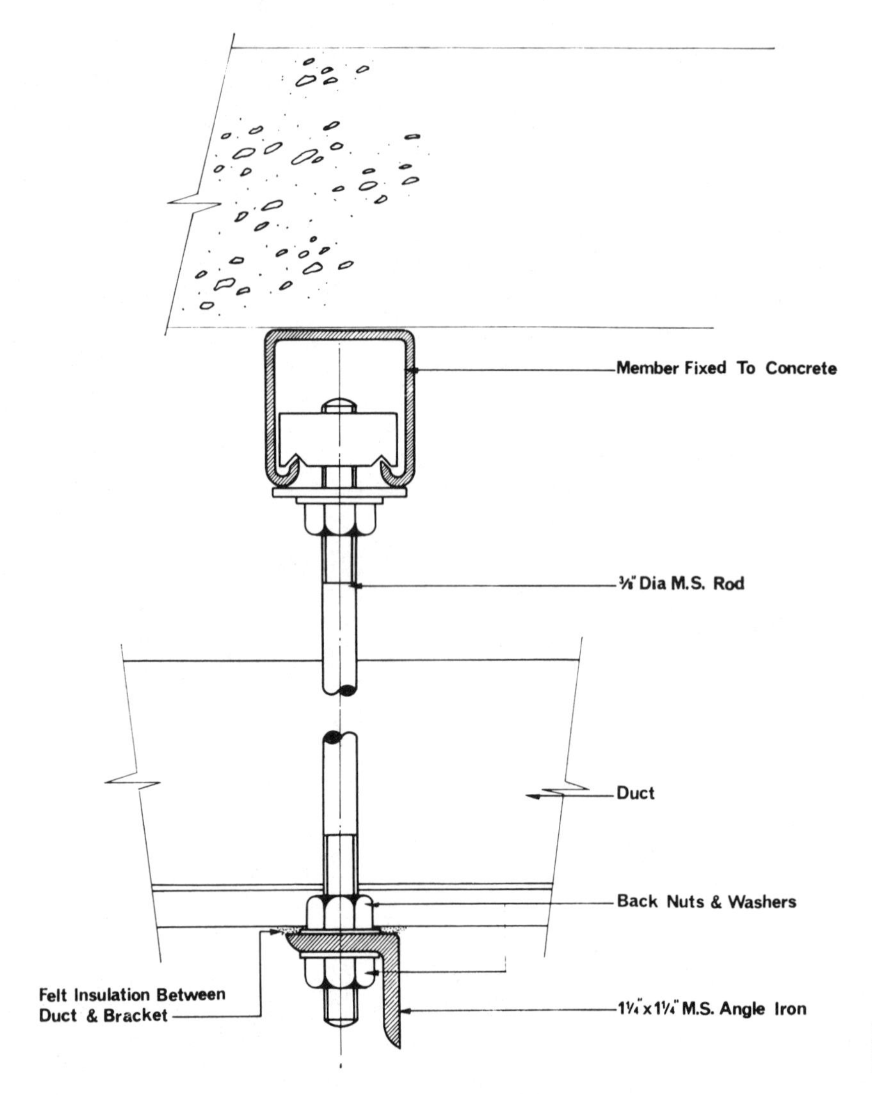

[Fig. 90] Support for horizontal metal duct

service tank in the event of an emergency. The dump valve is controlled by automatic level control or by manual push button.

duplex brass pipe fitting It should not be used in the UK for copper pipes on a potable water supply as it comprises a mixture of brass and zinc and may become subject to dezincification.

duplex pipeline strainer See **twin-compartment pipeline strainer**.

duplex pump set Two pumps arranged with one valved assembly.

duplex strainer Two strainer compartments separated by a changeover valve which permits one compartment to be serviced whilst the other is switched into use.

dust Solid particles smaller than grit particles which are formed by mechanical processes, such as crushing, grinding and blasting.

dust collector [Fig. 95] A mechanism that removes dust particles from an aerosol dispersion through displacement of the dust and subsequent removal of it. It may utilize gravity, or centrifugal or electrical motivating forces.

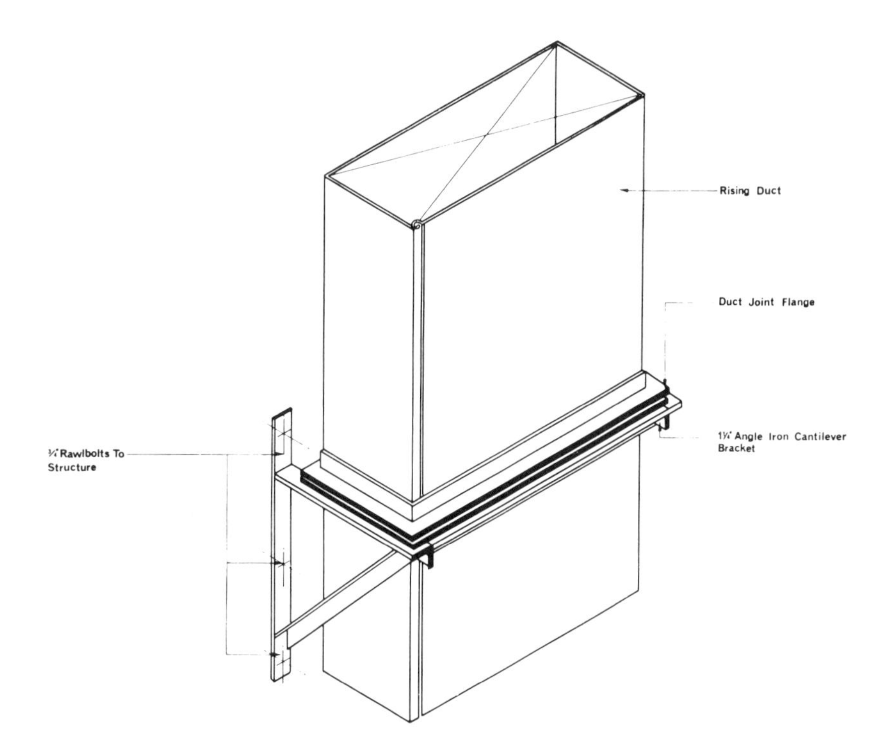

[Fig. 91] Support for vertical metal duct

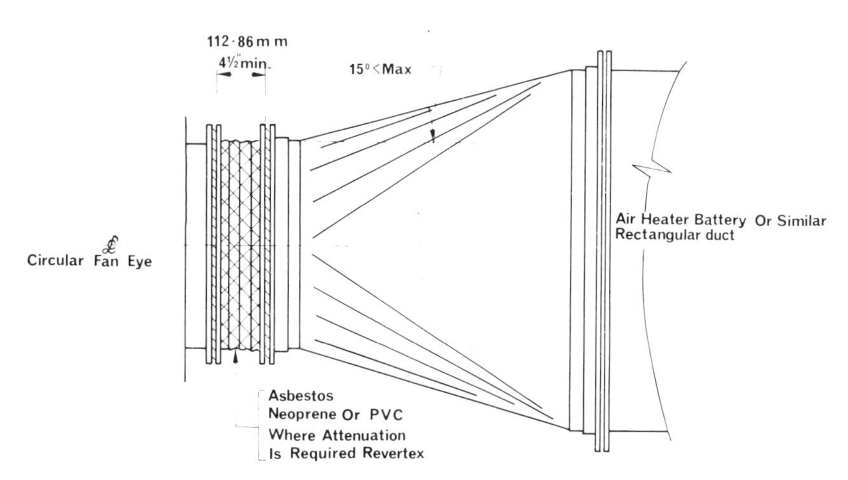

[Fig. 92] Transformation piece at the end of a fan

dust collector, centrifugal A dust collector that directs the dust particles to describe a number of revolutions within the dust collection zone. The resulting centrifugal forces are utilized for dust collection. A well-known example is the cyclone.

dust collector, combined gravity and centrifugal A dust collector that encourages centrifugal forces by single or multiple deflections situated within the collector, to assist the gravity forces in the dust collection.

dust collector, displacement zone A zone within the dust collector where the dust particles are dispersed out of the carrier medium in the direction of the removal interface.

dust collector, electrical (electro-precipitator) A dust collector that utilizes electrical forces to collect the dust from a carrier medium. Electric charges are introduced into the electrical field and exert a force on the charged particle in the direction of the potential gradient.

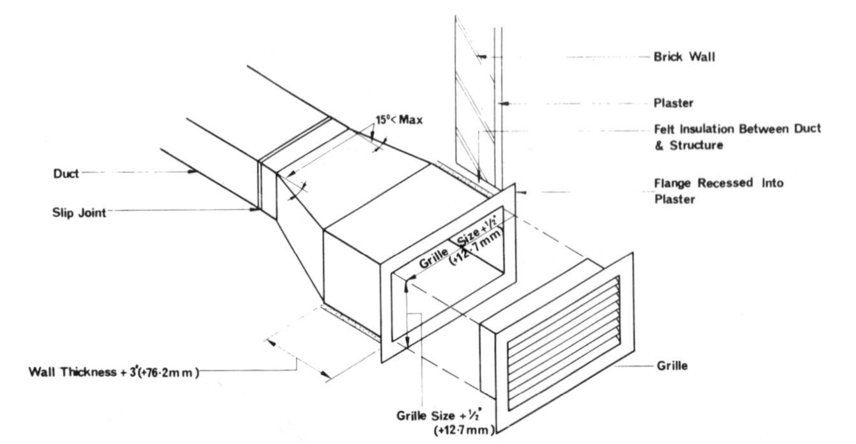

[Fig. 93] Final connection to air distribution grille

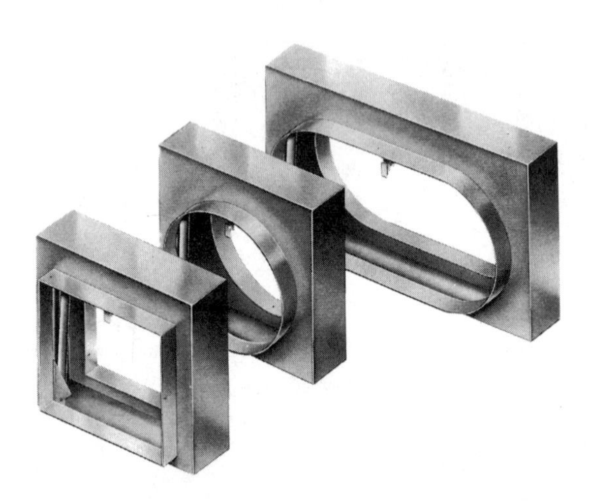

[Fig. 94] Transition fitments for insertion between air distribution ducts and terminal fittings (Waterloo–Ozonair Ltd.)

dust collector, filtration A dust collector that removes the dust particles through adhesion on to a suitable filtration material (fabric, viscous liquid or granulated material).

dust collector, gravity A dust collector that utilizes the forces of gravity to displace the dust particles at the settling velocity towards the removal interface.

dust collector, removal interface The boundary between the displacement and removal zones (assumed).

dust collector, removal zone The zone within the dust collector where the dust particles accumulate and may be considered removed from the carrier medium.

dust collector, scrubber A washing process that transfers the dust particles in the carrier medium from the fume to a liquid.

dust, explosion Explosion can occur in clouds (accumulation) of dust. The concentration below which explosions are unlikely to occur is of the order of 0.2–0.4 kg/m^3. Minimum temperatures below which ignition is unlikely are: 600°C for coal dust and 447°C for cellulose acetate.

dust, explosive Inflammable dusts, usually in the form of clouds of dust, will ignite and explode under

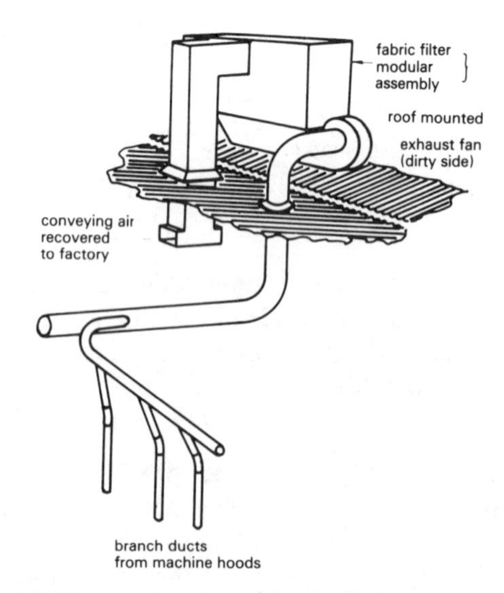

[Fig. 95] Waste extraction to fabric filter

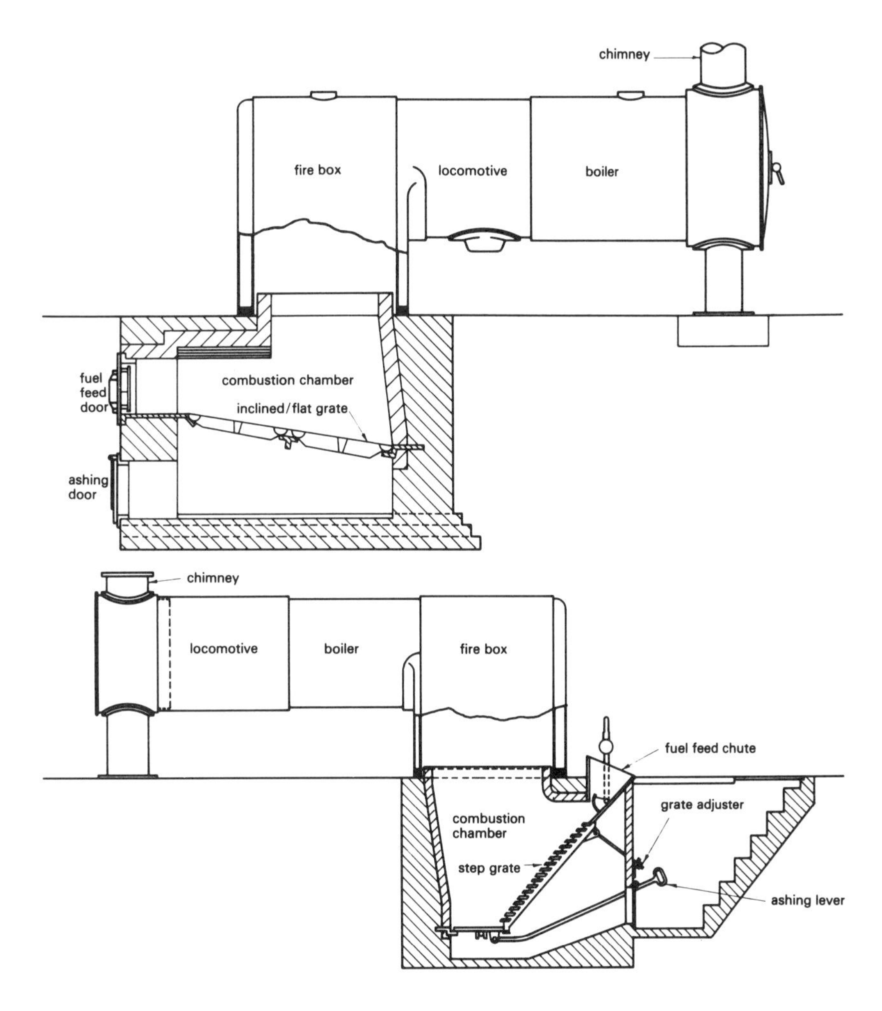

[Fig. 96] Dutch oven locomotive-type boiler underfired grate

certain conditions of concentration and temperature. Dusts of particular interest to the environmental engineer are wood waste, sander dust, coal dust and cellulose acetate.

dust-holding capacity The weight of dust which an air filter can retain at its rating as its resistance increases between the 'filter clean' and the 'filter dirty' conditions.

dust, sander See **sander dust**.

dust, test See **test dust**.

Dutch oven [Fig. 96] An oversized combustion chamber located beneath a waste heat boiler or in front of a furnace. It can burn hand-fed off-cuts, logs, paper, certain refuse, etc.

dwell time That period of time which a suspended fuel particle spends within the combustion chamber.

dynamo A generator of electricity from mechanical power.

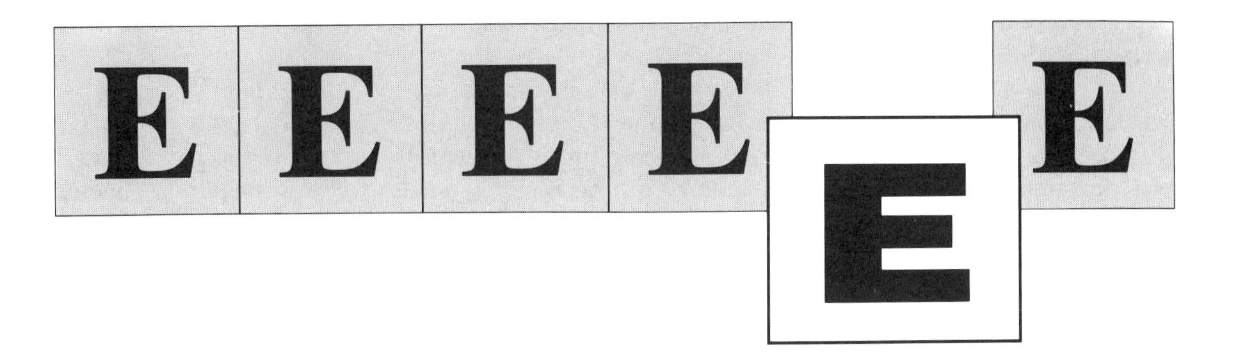

e.m.f. Electromotive force which causes the flow of an electric current when applied to a circuit.

EPA Abbreviation of the Environmental Protection Act 1990.

EPA fees Fees based on the principle that the polluter pays. Accordingly it is intended that the cost of operating the IPC (Integrated Pollution Control) system will be largely met by charging fees for authorization in accordance with published fee scales.

EPA prohibition notice A prohibition notice served by the EPA enforcing authority on the person carrying on a prescribed process if his continuing to carry it on, or to carry it on in a particular manner, involves an imminent risk of serious pollution of the environment. It may relate to any offending aspect of the process in question. The notice must specify the particular imminent risk of pollution, specify the steps that must be taken to remove it and also the period of time within which they must be taken. The notice renders the previous authorization ineffective, wholly or only to the specified extent.

earth connection A copper plate or copper rod buried in the ground. In rural and other isolated locations, electrical installations which carry a low load and where the subsoil conditions are favourable, the ultimate earth connection may be made in this way.

earth continuity A continuous path to the flow of electricity to earth which every electrical installation must offer for reasons of safety.

earthed circuit A circuit intentionally connected to earth at some point.

earth electrode A conductor which provides a means of connecting the electric network to earth.

earth fault loop impedance Impedance of the earth fault current loop (the current flowing between the phase of the faults circuit to earth loop) starting and terminating at the point of earth fault.

earthing For reasons of safety, an earth continuity conductor must be provided for every circuit of an electrical installation to ensure that earth leakage current can flow safely to earth. All metal-cased appliances, earth terminals of socket outlets, trunking, metal-sheathed cables, ducts and accessories must be connected to this earth conductor and, ultimately, to the metallic sheath of the incoming cable.

earthing conductor A protective conductor that connects the main earthing terminal of an electrical installation at an earth electrode or to other means of earthing.

earth leakage circuit breaker [Fig. 97] An electrical safety device. If earthing conditions are poor, it is possible for a phase-to-earth fault to draw insufficient current to rupture the protective fuse or operate a miniature circuit breaker. The faulty apparatus remains live and a potential safety hazard. However, even under such conditions, a small amount of current does flow and this is used to protect the installation by making connection between the earth point of the installation and the earth electrode through a current or voltage operated earth leakage circuit breaker, connected in series with the main fuse. Any current which flows through this coil causes the switch to trip, thereby safely disconnecting the main electric supply. IEE regulations state that, wherever a satisfactory value for earth impedance cannot be guaranteed at all times, an earth leakage circuit breaker must be installed.

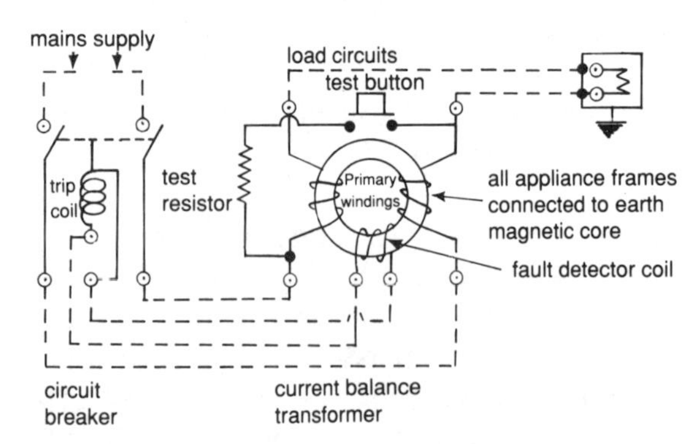

[Fig. 97] Circuit diagram: earth leakage circuit breaker

earth leakage current Electric current which flows to earth or to an extraneous conductive element within a circuit which is electrically in good order.

economizer A heat exchanger fitted into a boiler or furnace flue exit system to extract heat from the hot gases to heat hot water for general use or, in particular, to heat steam boiler feed water.

Economy 7 tariff A time-of-day tariff in which the supply of electricity, for whatever purpose it is used, is charged at a much reduced rate during a limited period of seven hours at night, generally within the period from midnight to 0800 (GMT). For charging purposes, the consumer is provided with a dual rate meter. Off-peak heaters and boilers utilize this charging period to minimize operating cost. Economy 7 was introduced in the UK in 1978.

eddy currents Electrical currents which are induced in the interior of conductors conveying alternating current due to variations in the magnetic flux surrounding the conductor.

efficiency The ratio of (energy) output to (energy) input. It is a measure of the effectiveness of any device.

effluent water Waste discharge into a sewer.

efflux Discharge; e.g. of flue gases from a chimney.

efflux velocity The speed at which gases leave a chimney or vent and are dispersed into the atmosphere. See also **chimney efflux velocity**.

ejector A device for lifting liquids, suitable for filling or emptying tanks, raising or transferring liquids and for priming centrifugal pumps. An ejector is operated by steam passing through an internal cone and it will lift water to a height of 10–20 m when operating at a steam pressure of 4–7 bar.

ejector heaters and circulators Devices for the heating and circulating of water in tanks and vats by the direct application of live or exhaust steam. The internal cone or nozzle is correctly proportioned for optimum utilization of the steam input without causing objectionable noises, such as would be caused by steam discharging under water from a plain open pipe. A variety of ejector nozzles are available and these may incorporate a discharge tube in the form of a venturi to promote efficiency and silent operation. For best results such heaters/circulators should be immersed low down in the water; according to a rule of thumb, the depth of immersion should not exceed fifteen times the pressure of the steam supply, expressed in bars.

elapsed-time clock A clock that indicates the passage of time from a set time, generally of up to one or two hours. It is of particular application in hospital operating theatres.

elbow action tap A quarter-turn tap fitted with a lever handle which can be operated by the forearm. It is suitable for use in hospitals by medical personnel.

electrical back-up protection In electrical installations which employ cartridge fuses for back-up protection, or where these are used for the protection of main circuits and of circuit-breakers for the protection of sub-circuits derived from them, the cartridge fuse should operate before the circuit-breaker when the value of the current flowing exceeds the rupturing or tripping capacity of the associated circuit-breaker. Where circuit-breakers are used for the protection of main circuits, and for the sub-circuits derived from them, discrimination should be obtained by selecting circuit-breakers of the appropriate rating.

electrical circuit diagram A diagram that indicates the various circuits served off a distribution board. It is an essential part of an electrical installation.

electrical distribution board A hinged box which houses fused electrical circuit terminals for connection to electrical equipment. Each circuit is labelled.

electrical, final connection A flexible cable connector between isolator and protected equipment.

electrical space heater, off-peak This comprises an assembly within an attractive cabinet of heat storage bricks or tiles and a circulating fan with an associated adjustable thermostat. The casing is thermally insulated to a high standard. This heater charges with heat during the hours of off-peak electric tariff under a programme control and dissipates most of the heat when the fan is actuated by a thermostat. The remainder of the heat loss is from the casing.

electrical block storage heater A heater that stores off-peak electricity in the form of heat retained within special heat storage bricks which are heated during the off-peak electric charging period and subsequently discharge heat.

electric motor A machine for converting electrical (input) energy into mechanical (output) energy. The following types are available.

Direct current: generally expensive relative to an a.c. motor, but used in areas served with d.c. electricity and for special applications where fine speed control and greater flexibility in the motor speed/torque characteristics are required.

Alternating current: supplied from an a.c. current distribution and may be single phase or polyphase. The motor speed relates to the frequency of the electric supply.

Single-phase: used generally for motors of up to 1 kW and/or where only a single-phase distribution is available. Most fractional kW motors are wound for single phase.

Polyphase: generally three phase. Most motors with output of over 1 kW are wound for three-phase supply. A polyphase a.c. supply permits greater output from a given current rating than a single-phase supply on the same voltage.

Squirrel-cage induction: extensively used on three-phase a.c. supply. Probably the simplest and most widely used of all the types of electric motors consisting of a stator wound for a three-phase supply and a rotor of squirrel-cage construction. The rotor winding comprises a series of bars fitted into the rotor slots and all the bars are connected at each end to a common conducting ring. The bars and the end rings form the 'squirrel-cage'. It has a lagging power factor due to the inductive component and is essentially a constant-speed machine, but some speed regulation can be made by voltage reduction. It is available as a pole-changing type with two speeds obtained by alteration of the connections to the stator windings.

Applications: for high-speed and/or arduous operating conditions. It is a fairly low-cost motor requiring only little maintenance and is associated with relatively cheap control gear.

Synchronous or autosynchronous induction: the field of such a motor is supplied with a.c. and the rotor with d.c., which is usually provided by a small generator (or exciter) driven off the motor shaft. The motor is not self-starting and must be run up to speed by an auxiliary motor. It has a leading power factor which can compensate for the lagging power factor of induction motors supplied off the same installation. There is radio interference from the exciter and this generally has to be suppressed.

Applications: very large motors (up to 5000 kW) for large steady power loads and where the leading power factor is desired.

Slip-ring: stator and rotor carry polyphase windings. The latter are brought out to slip rings to permit the connection of a separate resistance during start-up. Having reached the required speed, the rotor windings are short-circuited. The speed of the motor on any given regulator setting varies according to the imposed load. The starting current is 150–300% of full load current. To limit the starting current, the motor is usually connected on the lowest setting of the speed regulator. Starting torque at the lowest speed is in the order of 150–300% of full-load torque. With a well-maintained slip-ring motor there should be no radio interference. It is available in sizes from 5 to 1000 kW.

Applications: for larger machines starting against load and in cases where some limitation of the starting current is required.

electric motor enclosure A type of protective envelope provided to a motor to suit the particular application.

Screen protected (SP): internal motor parts are protected mechanically to obviate accidental contact. Ventilating openings in the motor frame and in the end-shields are protected by a wire screen or other perforated cover.

Application: relatively dry, clean and non-inflammable areas.

Drip proof (DP): Ventilation openings to the motor are protected to exclude vertically falling water or dirt.

Application: in areas of heavy condensation. Unsuitable for outdoors location.

Totally enclosed (TE): constructed to ensure that the enclosed air does not connect with the external air, though not necessarily an airtight arrangement.

Application: in contaminated areas.

Totally enclosed – fan-cooled (TEFC): motor cooling augmented by a fan driven off the motor shaft.

Totally enclosed separately air-cooled (TESAC): motor cooling augmented by a separately driven cooling fan.

Flameproof (FLP): constructed to withstand any explosion occurring within the motor enclosure and to prevent the spread of explosion or flame to the surroundings. Gas-tight bearings and provision of sealing glands to incoming cables are essential.

Application: locations where atmosphere may contain inflammable or explosive dust or gas.

Weatherproof (WF): construction permits outdoors location of the motor without need for additional weather protection. Note: associated electric wiring and switchgear must be similarly suitably constructed.

Application: outdoors.

Pipe or duct ventilated (PV, PVFD, PVID): has an enclosed case through which cooling air to the motor is continuously supplied via pipe(s) or duct(s). May be self-ventilating by natural air movement, forced-draught actuated or induced-draught actuated air flow.

Application: in locations where risk attends the emission of smoke from the motor windings into the building.

electric motor starter, direct-on-line [Figs 98, 99, 100] A starter that applies the line current to the electric motor on start-up. It is cheap and simple. The size of motor which may be arranged for this starting method depends on the permissible peak starting current. (Starting current is likely to be up to 650% of full-load current.) It varies with different electricity supply authorities and methods and is usually limited to motor capacities not exceeding 3 kW.

Star-delta: suitable for the larger motors. In the starting position, the stator windings are connected in star; in running position, in delta. The start winding connections must be taken separately to the starter. The method reduces the voltage across the motor winding on starting in star to about 60% of line voltage. During normal running in the delta method, each phase winding receives full line

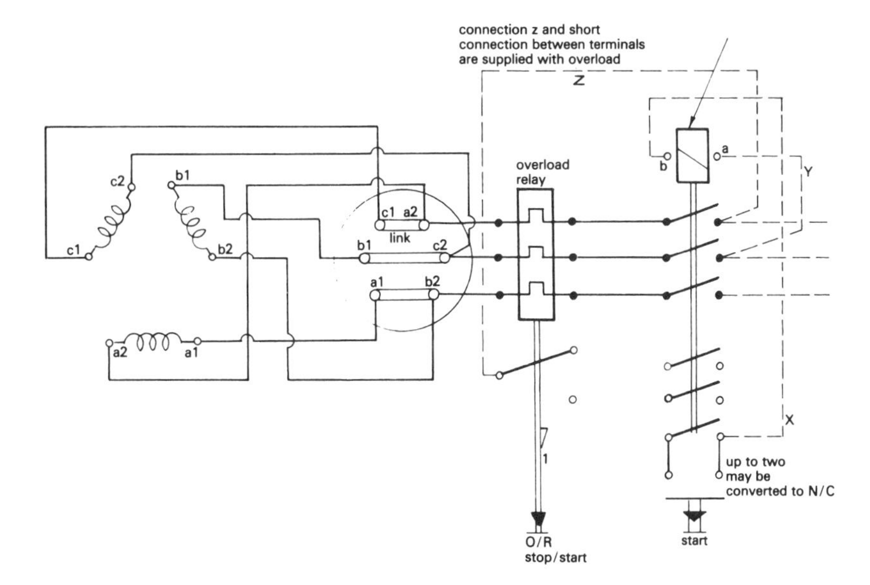

[Fig. 98] Direct-on-line starter circuit diagram

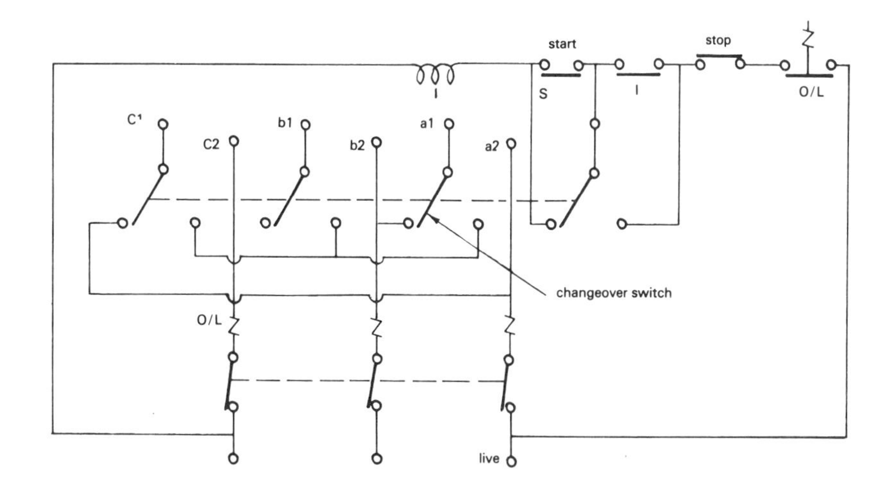

[Fig. 99] Star-delta starter circuit diagram

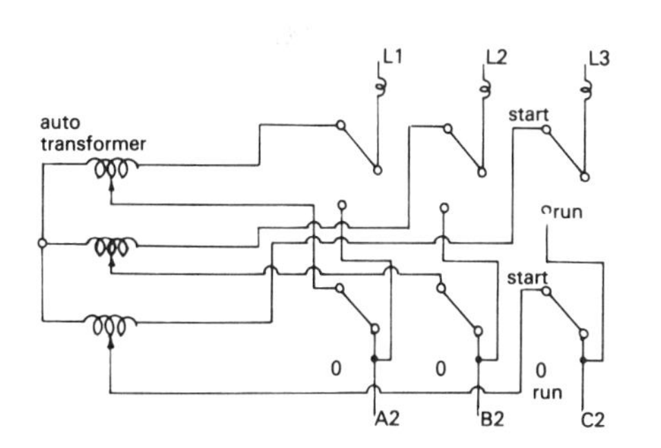

[Fig. 100] Circuit diagram of auto-transformer starter

voltage. A star-delta is thus effectively a change-over switch. It is suitable for starting motors of capacity not exceeding 15 kW.

Auto-transformer: a two-stage method of starting three-phase squirrel cage induction motors in which a reduced voltage is applied to the stator windings to give a reduced starting current. It consists essentially of a three-phase start-connected auto-transformer, together with a six-pole switch arranged as three double-pole change-over switches. Tappings are provided to give 40%, 60% or 75% of line voltage at the motor stator on starting. This provides a better starting method than star-delta where the motor must start under load or where slow action acceleration must be avoided. It is suitable for motor capacities up to 20 kW.

electric shock Dangerous pathophysiological effect caused by the passage of an electric current through a person or animal.

electric shock chart A wall chart which describes graphically how best to assist a person suffering from an electric shock. It should be displayed wherever electrical machinery is in use and as a basis for training staff in emergency routines.

electric surface heating Low voltage heating tapes or bandages applied to a pipe to maintain the fluidity of liquid being conveyed and/or as a means of frost protection. It is operated under thermostatic control.

electric water conditioner A metal screen to which is applied a low-voltage electric current which, it is claimed, breaks up the scale crystals in water flowing through the screen which are subsequently discharged with the outflow. The electric consumption is minimal but the effectiveness may depend on the water characteristics.

electrode A conductor which conveys an electric current into or out of a liquid or gas; usually one of a pair of an anode and a cathode.

electrode boilers Boilers for generating steam or hot water by electricity.

electrolysis The decomposition of a chemical substance by the passage of an electric current through it. It is the basis of the production of hydrogen from water by means of electrical energy.

electrolytic action Action that causes or intensifies corrosion in a system of water pipes, brought about by the presence of dissimilar metals. Particularly active are copper in close proximity to zinc (galvanizing), or aluminium in a circulating hot water system. In water mains severe pitting of iron pipes can be experienced in the vicinity of lead joints. Fitting together dissimilar pipe materials should be avoided.

electromagnet A core of iron or steel partly surrounded by a coil which, when an electric current is passed through the coil, behaves like a magnet attracting ferrous metal to itself.

electromagnetism The property of an electric current to induce a magnetic field around it. This is the basis on which electric generators and motors operate.

electrometic induction When an alternating current, or a direct current of varying strength passes through a conductor, then any other conductor in its close vicinity will have an electromotive force induced in it. This forms the basis of the operation of electric generators, transformers, induction motors, control systems, etc.

electromotive force The difference in electrical potential which causes an electric current to flow from a point of higher potential to one of lower potential. It is often designated e.m.f.

electropneumatic device The interface between electric and pneumatic control systems.

electrostatic air filter [Fig. 101] An air filter that comprises three major components within a common assembly: an ionizer, a dust collector and an electronic power pack. The ionizer comprises a series of earthed tubes and a fine metal wire is stretched between adjoining tubes and is charged to about 13 000 V d.c. A strong electrostatic field is established across the space between the ionizing wire and the earthed tube surface and, at the correct voltage, a corona discharge takes place from the wire causing ionization of the air molecules of the airstream passing through the filter and causing it to be greatly accelerated. Dust particles are impelled in the direction of the dust-collecting cell which consists of a number of flat, parallel, vertical plates. One set is earthed and the other is charged to about 6000 V. The plates are arranged alternately, causing the air and

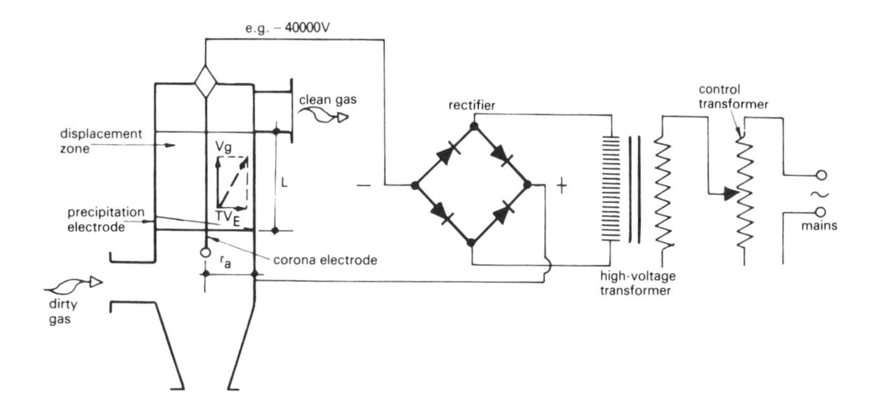

[Fig. 101] Electronic air filter: diagrammatic arrangement

dust to flow along narrow passages with an earthed plate on one side and a charged plate on the other.

The electrostatic field between the plates in the collector cell draws the dust particles to the earthed plate, to which they adhere until removed. Resistance to air flow at normal air system operating velocities seldom exceeds 0.25 mb. The average blackness test efficiency is about 90% and with a minimum particle size of 0.01 micron subject to the filter being maintained in good order and cleaned at the required intervals. Cleaning is carried out by washing the collector cells with warm water. It is suitable for filtering tobacco smoke, fog, etc. and is used in small units in places of entertainment and in large multi-cell assemblies for industrial uses. It must be coupled with a pre-filter to reduce the dust burden on the electrostatic filter.

eliminator plates A means of eliminating carried-over droplets of water; they are placed in the air stream from wet processes, such as cooling towers in such a manner that the wetted air impinges on them and allows the droplets of suspended water to settle out. Associated with systems using air washers or spray chambers. Construction must suit the condition of the water being used and the type of contamination being met. For example, in industrial and city areas, the eliminators will corrode quickly unless suitably coated or of inert construction, such as glass.

embedded panel A space heating panel which is embodied within the ceiling or floor construction.

emergency generator A diesel or gas-driven electricity generator installed in buildings to safeguard vital electric services in the event of a breakdown of the public electricity supply. In particular, it is required to operate exhaust systems to underground garages during a mains electricity supply failure, safeguard major computer systems, firemen's lifts, fire-fighting and protection installations, etc.

emergency lighting systems These may be either generator or battery systems but the former is seldom used as it is necessary to provide the required illuminance within 5 to 15 seconds of a power failure. This entails either a continuously running generator or a standby generator with a battery back-up.

emissivity That fraction of heat that is emitted from a hot body by radiation. Emissivity depends on the surface finish of that body, being unity for a perfectly black body and less than unity for others.

emittance The ratio of the radiant energy emitted (in the absence of incident radiation) from a given plane surface at a given temperature to that radiant energy which would be emitted by a perfectly black body at the same temperature.

employer A public authority, developer, private individual, etc. instructing the works and responsible for payment in respect of it.

emulsifier A device that automatically injects the correct amount of water into the oil stream supplied to heavy fuel oil burners to create a combustible emulsion. It incorporates pumps, valves and controls to accurately monitor the water flow at all stages of the operation. Fuel economy is optimized by adding 2–4% water to the maximum fuel flow.

endothermic Describes the reaction or process in which energy is absorbed.

endothermic reaction A chemical reaction which is accompanied by the absorption of heat.

energy audit The accounting of energy consumption.

energy centre A plant complex housing centralized boiler and refrigeration plants from which heating, cooling, water and electrical facilities are supplied to an extensive site or group of buildings.

energy, kinetic The energy a body has by virtue of its motion. More precisely, a moving object possesses kinetic energy (ke) by virtue of its mass (m) and velocity (v).

$$\text{ke} = \frac{mv^2}{2}$$

energy management and building control system A computer- or microprocessor-based system for the supervision and automatic operation of building services and systems. Programs are available for optimum start/stop, limitation of peak electrical demand, enthalpy optimization, efficiency monitoring, planned maintenance, actual energy consumption and cost, direct digital control of plant, access control, fire detection and alarms, tour patrol, flexitime administration, lighting control, intruder detection, elevator control and fault detection, etc. This system is suitable for large building complexes, chain stores or groups of buildings under common management. See also **building energy management systems**.

energy monitoring and targeting (M and T) A system or method which routinely measures energy consumption against laid down standards or targets for the detection of waste when this occurs, estimates its significance in cost terms, diagnoses the underlying cause(s) and then initiates or activates informed action.

energy, potential That energy possessed by a body or substance at rest by virtue of its height above a reference level or because of energy stored within it. Unburnt fuel can be considered to have potential energy by virtue of the heat energy locked up within the fuel.

energy recovery heat wheel See **heat wheel**.

energy, total The sum of the potential energy and the kinetic energy possessed by a body or substance.

enthalpy (*H*) The thermodynamic property of a substance given by the relationship $H = U + pV$ in which U is the internal energy, p is the pressure and V is the volume.

enthalpy chart A chart or graph representing the thermal properties of a substance with enthalpy as one of its coordinates.

entrain To collect and transport a substance by the flow of another fluid moving at higher velocity. For example, in induction air-conditioning room air is entrained into the supply air by the air being discharged at high velocity from the nozzles in an induction unit.

entrained air Currents of air which are induced to flow in certain directions under the influence of streams of air which move at high velocity. In a cross ventilation system the supply air into a space is injected via an air flow nozzle at higher velocity and traverses that space at an upper level towards an exhaust which is located also at the upper level; room air is entrained into the high-velocity stream and provides the ventilating effect in the space. In an induction air-conditioning unit the high-velocity supply air induces the air in the room to circulate over the cooling coils.

entrainment The result of entraining a substance by the flow of another fluid moving at high(er) velocity.

entropy (*S*) That quantity introduced in the first place to facilitate the calculations and to give clear expression to the results of thermodynamics. Changes of entropy can be calculated only for reversible processes and may then be defined as the ratio of the amount of heat taken up in the process (ΔQ) to the absolute temperature (T) at which the heat is absorbed, i.e. $\Delta Q/T$. Entropy changes for actual irreversible processes are calculated by assuming equivalent theoretical reversible changes. The entropy of a system is a measure of its degree of disorder.

The total entropy of any isolated system can never decrease in any change; it must either increase (as would be the case in an irreversible process) or remain constant (in a reversible process). The total entropy of the universe therefore is increasing, tending towards a maximum, corresponding to the complete disorder of the particles in it, on the assumption that the universe is itself an isolated system.

entropy chart A chart representing the thermal properties of a substance with entropy as one of its coordinates.

entry loss That loss of energy at the entry of a fluid system due to turbulence and friction around the entry position.

environmental assessment The technique and process by which information about the environmental effects of carrying out a proposed project is collected, both by the developer and from other sources. This is taken into account by the planning authorities in forming their judgment on whether the project should be allowed to proceed.

environmental engineer An engineer concerned with the totality of the building services and of associated aspects which bear upon the use and enjoyment of the environment inside and outside the building, being concerned also with matters such as noise, refuse, chimney effluents, water pollution, effects of industrial processes and disposals, etc.

environmental engineering A term embracing all aspects and disciplines relating to an engineered environment.

Environmental Protection Act 1990 A legislative framework aimed at the protection of the environment, incorporating laws for the control of pollution and liability for damage inflicted on the environment. The Act establishes the principle that 'the polluter pays'.

equal friction method A method of sizing a duct installation based on ensuring that each section of duct offers the same resistance per unit length. The air velocity varies accordingly. The method takes into account maximum permissible velocities.

equalizing screen A perforated metal plate inserted in a duct system of large dimensions after the supply fan to equalize flow and pressure over the cross-section of the duct.

equinox The moment at which the Sun apparently crosses the celestial equator. It is the point of intersection of the ecliptic and the celestial equator when the declination is zero.

equipotential bonding A safety connection between two or more conductors of electricity, but not necessarily part of the wiring system. This ensures that all metal within reach of a person is at the same potential and thereby obviates the risk of electric shock when touching metal components which are at a different electric potential.

equivalent circular duct diameter That diameter of a duct of a cross-sectional area which offers the same resistance to air flow as the square or rectangular duct section being considered.

equivalent duct length That length of straight duct which offers the same resistance to air flow as a specific duct fitting, such as a bend, off-take, junction, transformation piece, etc. It is a useful concept for duct sizing. Textbooks on ventilation generally include tables of equivalent lengths of duct fittings.

equivalent evaporation A term used to place steam output of boilers on the same basis by specifying the output as the steam generated by the boiler from and at 100°C.

equivalent fitting resistance Resistance in terms of the length of unobstructed straight pipe which offers the same resistance as the particular fitting.

equivalent pipe length That length of straight pipe which offers the same resistance to the flow of water as a specific pipe fitting, such as an elbow, bend, tee junction, reducer, etc. It is a useful concept for pipe sizing. Textbooks on pipe systems usually include tables of equivalent lengths of pipe fittings.

escape lighting Emergency lighting provided to ensure the safe and effective evacuation of a building in an emergency. Luminaires are located along the designated escape route.

Essex flange A proprietary flange arrangement used to insert a flange into a calorifier, tank, etc. with a built-in compression or screw pipe fitting. It is used for retrofit and maintenance purposes.

eutectic Relates to that mixture of substances which possesses the lowest freezing-point of all possible mixtures of the substances.

eutectic mixture A mixture of two or more substances which offers the lowest possible freezing-point.

eutectic salts Salts used with thermal storage heating and cooling applications which change state between solid and liquid at a temperature of 27°C.

eutectic salts air-conditioning system A form of thermal storage system based on a water storage tank which is filled with thousands of plastic hollow balls containing eutectic salts. These absorb latent heat from the water (or other fluid) circulating between the balls or, alternatively, release latent heat into the water through a change of state. The tank is connected to conventional air-to-water heat pumps and 'Versatemp' type ring mains which traverse the building being air-conditioned. In summer the eutectic salts are frozen overnight and allowed to melt during the day, thereby absorbing heat from the warm water in the ring main and cooling them. The reverse process is used to provide heating during the winter.

The system is suitable for use with low-cost off-peak electricity and can be designed to operate in these four modes:

(a) daytime cooling;
(b) night-time cooling charge;
(c) daytime heating;
(d) night-time heating charge.

During charging the tank and the charge heat pumps operate in series; when topping up is required during peak time, the charge heat pumps operate in parallel. Overall efficiency is enhanced by limiting both the tank and the heat pumps to their optimum sizes. As a result, neither can alone meet peak demands. Therefore a bypass is fitted across the eutectic tank to prevent its full discharge before the end of the working day.

The advantages of such a system are: the absence of a fossil fuel fired boiler plant with its associated chimney and pollution, smaller plant spaces, elimination of the need for a cooling tower (and the associated risk of *Legionella* problems), reduction in noise associated with the operation of the system (a eutectic tank is without moving parts) and economy in energy use which arises out of the ease with which the electricity input (mainly off-peak) can be controlled.

evaporation loss The loss of moisture from a liquid surface occasioned by the entrainment of water into a moving airstream in contact with that surface.

evaporative air-conditioner An air-conditioner that relies for its operation on the latent heat extracted from the liquid being cooled by some entrainment of the liquid into an airstream moving across its surface.

evaporative condenser A condenser that rejects heat to atmosphere by spraying water on to the condenser coils. In this way some heat is transferred to the water as well as to the air, increasing the capacity of the condenser. A pump, spray nozzles, associated piping and a collecting sump are required for the water recirculating installation. Fans force the air through the unit. Such condensers may be installed indoors as well as outdoors by connecting ductwork to discharge the exhaust air outside.

evaporative cooling A method of cooling in which air is evaporated into the passing airstream. It is at its most effective in conditions of low relative

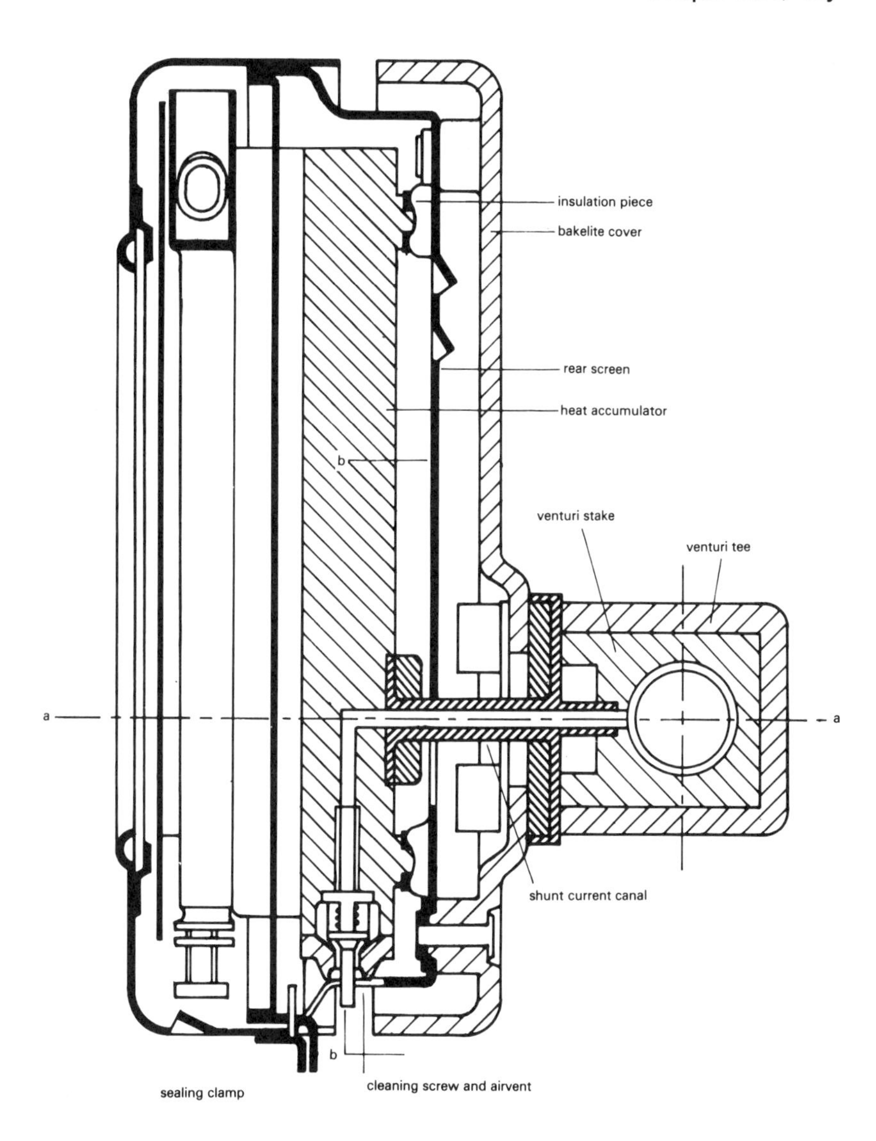

[Fig. 102] Clorius evaporation meter

humidity when the air can absorb more moisture from the sprays or water surface.

evaporation heat meter [Fig. 102] A meter that measures the heat consumption of heating or domestic hot water systems, based on the principle that the rate of evaporation of the liquid contained in the meter over a given time is proportional to the heat usage. Heat meters are attached to individual radiators and must be closely matched to their output. Such meters are relatively inexpensive and, if correctly installed, serviced and read, can provide a reasonable guide to the consumption of heat, but the meticulous identification of each such meter is essential to success. They are best suited to the proportioning of heat consumption in a multi-occupancy building or group of buildings. The best known make is the Clorius meter.

evaporator A heat exchanger associated with refrigeration and cooling equipment in which a liquid is evaporated to achieve the cooling effect by the process of abstraction of the latent heat of vaporization from the medium to be cooled (e.g. air, water, brine, earth).

evaporator, dry See **dry evaporator**.

evaporator, flooded See **flooded evaporator.**

excess air for combustion That quantity of air over and above the theoretical combustion air supply for a particular fuel to ensure the thorough mixing of the air with the fuel. Thorough mixing cannot be achieved by supplying only the theoretical minimum air quantity. The quantity of excess air requirement theoretically varies with variations of calorific value, bulk density, moisture and ash content of the fuel being burnt and with the particular method of incineration. For example, solid fuel bed-type furnaces require some 30–50% excess air quantity; cyclone combustion furnaces require as much as 100%. Careful control must be exercised over the exact quantity and distribution of combustion air (including excess air). Inadequate air supply causes smoke emission; excessive air supply reduces combustion efficiency.

exhaust air canopy [Figs 103, 104] A hood which covers the whole surface giving off fumes or vapour. For efficiency, it is fitted with internal baffles or intake slots.

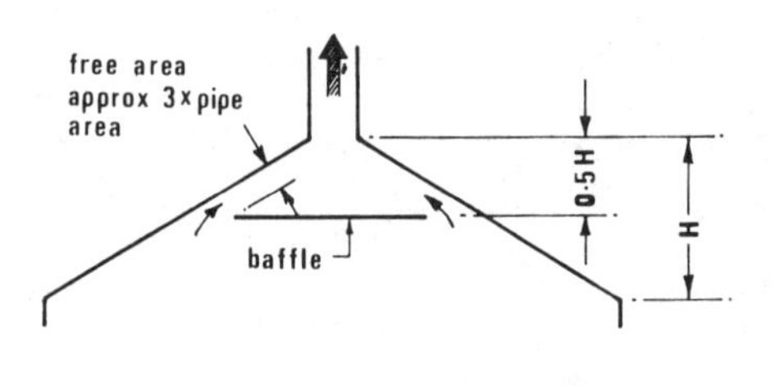

[Fig. 103] Large canopy hood with baffle

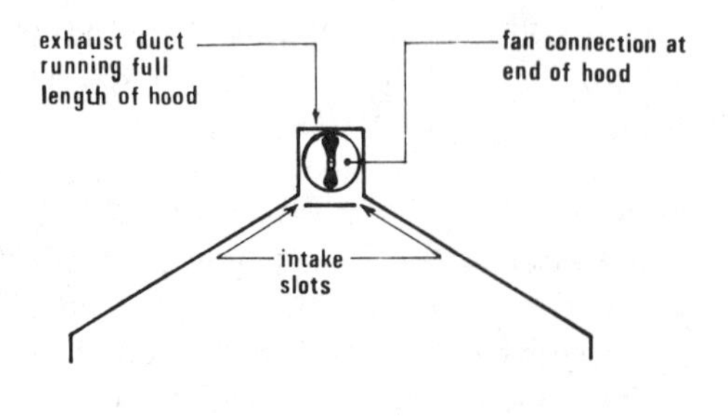

[Fig. 104] Large canopy hood with slotted suction pipe

exhaust air semi-canopy [Fig. 105] A construction angled to permit free access to a surface giving off fumes or vapour.

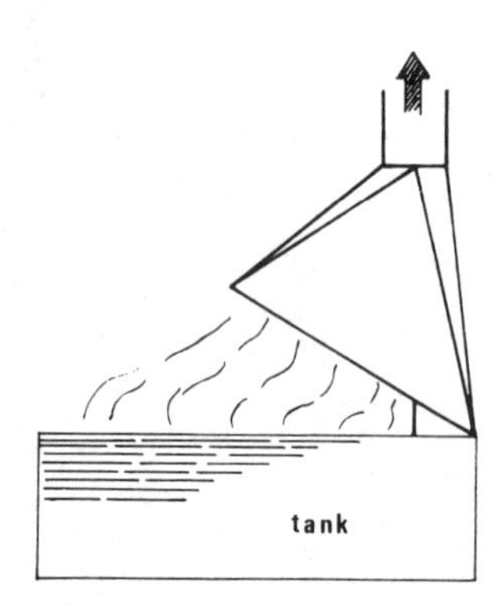

[Fig. 105] Semi-canopy arrangement

exhaust head A metal fitting installed at the terminal of a pipe or duct discharging vapours to protect the building fabric and persons from the adverse effects of the discharge.

exhaust hood A canopy placed above an appliance emitting vapour, dust or heat to collect and discharge them via a ducted installation in a suitable manner; e.g. into a cyclone or to atmosphere.

exhaust steam Low-pressure steam that is expelled from steam-using equipment such as a turbine or pump after it has done its work. The energy in exhaust steam may be recovered by use in back pressure steam generators or in waste heat recovery.

exhaust ventilation controller A device that measures cigar and cigarette smoke in bars, restaurants, etc. and switches on an extractor fan when a predetermined contamination level is reached, switching it off when the room is clear. This automatic operation avoids the waste of heat and fuel which usually occurs with a manually operated fan which may be left on when not needed.

exhauster A compressor employed to produce a vacuum.

existing drain, connection to This should be via a manhole or inspection chamber, but where this is impracticable, a junction should be inserted or a saddle may be fitted providing the receiving drain is at least one size larger than the branch drain and that the saddle connection is so positioned that the flow enters above the horizontal diameter.

exothermic Describes a reaction or process in which heat (energy) is given out.

expanded polystyrene insulation This insulation is supplied in various grades and densities in standard sizes 12 mm–500 mm thick. It is also supplied in linear runs.

expansion loop A factory- or site-made pipe loop which is inserted into a pipeline to compensate for thermal movement. It must be installed in conjunction with an appropriate system of guides and anchors.

expansion tank (cistern) A tank associated with open heating systems which provides space for the expansion of the water being heated. It comprises a tank (cistern), a ball valve with a copper or plastic float, an overflow connection to outside and a make-up water connection. The vent pipe usually terminates above the surface. A cover should be provided. The tank size should suit the water content of the system. The float control must be adjusted to permit an adequate volume to accommodate the expanding water. If the volume is inadequate or if the float is set too high, part of the expansion volume will overflow to waste and hard water will be admitted into the system. Vent pipe(s) must under no circumstances of operation dip below the water level in the tank. The balance of the system must be such that no water is discharged into the tank through over-pumping.

explosion The rapid instantaneous expansion of air caused by a sudden abnormally high release of energy.

external surface resistance A term used in U-value heat loss calculations which refers to the wind and weather exposure of the building fabric.

extended finned surface An extension of the area of a heat exchanger pipe surface which is attached to a tube conveying a fluid to increase the amount of heat transferred. The fins may be rectangular or circular, attached to the tube by a mechanical or soldered bond.

extended lubricator A small-bore pipe, one end of which is screwed into a lubricating terminal and the other end of which terminates with a removable closure cap in an accessible location at which the lubricant can be conveniently inserted.

extended surface A heat transfer surface which is extended by the addition of fins or by embossing the surface. See also **extended finned surface**.

extension of time A reasonable extension of time for the completion of the works allowed in respect of any detail in such completion which has been caused by any of the following factors:

(a) the execution of varied or additional works;
(b) weather conditions which make continuance of work impracticable;
(c) strikes or lock-outs;
(d) any other circumstance which is wholly beyond the control of the contractor.

extract plenum A plenum installed at the highest (hottest) level of a ventilated space which draws exhaust air back into the air handling unit and either recirculates or dumps it or a proportion of it to atmosphere.

extraneous conductive part A part liable to introduce a potential (commonly an earth potential) and not part of the electrical installation.

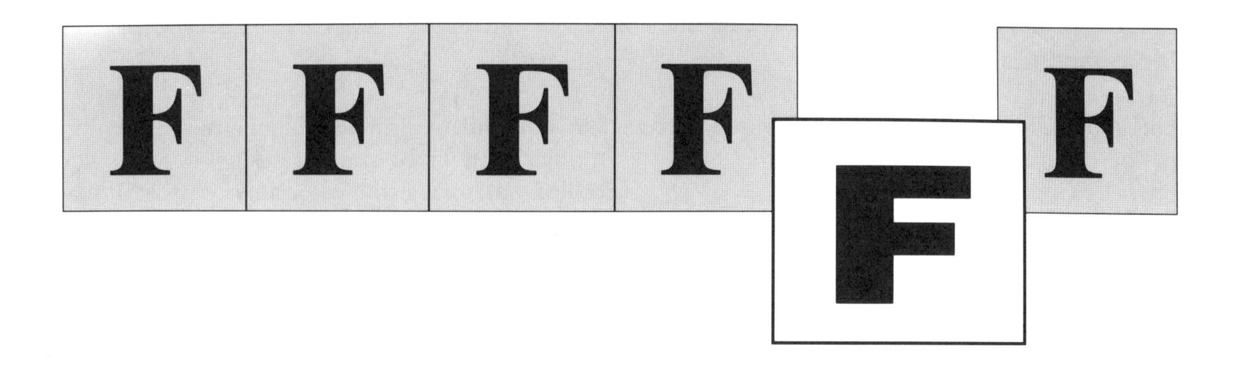

FIDIC contract A contract prepared by the Fedération Internationale des Ingénieurs – Conseils and the Fedération Internationale Européenne de la Construction. Originally intended to be used for civil engineering projects, it is also widely used for building work, including that sponsored by the World Bank.

fabric losses Heat losses from a building envelope.

face air velocity The velocity of an airstream expressed in terms of the velocity measured at the face (dimensions) of the heat exchanger or air filter.

face area That area formed by the dimensions of the face of the heat exchanger or air filter, i.e. excluding framing and flanges.

face velocity The velocity of air as measured across the face of air handling equipment, e.g. an air filter or heater.

factor of safety The margin over and above the theoretically required design parameters applied to allow for unforeseen circumstances such as snowfall, earthquakes, heat wave, extreme cold, etc.

fail-safe valve A valve that opens under remote control by hydraulic pressure in the operating cylinder. It closes automatically on loss of control pressure. The closing force is supplied by a spring and line pressure.

false ceiling A lightweight ceiling which is suspended from the main ceiling construction above for decorative reasons or to create space for light fittings and building services.

false floor Flooring supported off the main floor slab to create space for services at low level.

fan A device that conveys air in a controlled manner against the resistance to the flow of air offered at the fan outlet. Numerous types of fans are in use: see **axial flow fan**, **centrifugal fan**, **propellor fan**, and **bifurcated fan**.

fan-assisted warm air heating unit A space heating appliance which incorporates a fan to propel the air into the heated space or into a system of distribution air ducts.

fan bearing The support for a rotating fan shaft.

fan bearing, ball and roller bearing A fan shaft which rotates within a ring of hard steel balls or rollers housed within a cast-iron bearing. It requires grease lubrication. It can be packed with special grease (e.g. lithium based grease) to permit fan operation without further relubrication over a long period of time.

fan bearing, sleeve The fan shaft runs inside an oil reservoir which requires frequent replenishment or topping up. It is essential that the lubrication point is easily accessible.

fan bearing, storage pending installation It is essential that the fan shaft is turned over periodically. If it is allowed to rest in one position, the bearing can become sufficiently deformed to ruin it after a short period of running.

fan, bifurcated See **bifurcated fan.**

fan blade, pitch The angle at which the cross-section of the fan blade is set in relation to the direction of rotation.

fan casing The enclosure for a fan impeller and associated equipment.

fan, centrifugal in-line duct fan See **centrifugal in-line duct fan.**

fan characteristic [Figs 106, 107] The graphic representation as a curve of the relationship between the volume of air handled by a fan and the corresponding static fan pressure.

fan coil unit An air-conditioning or ventilation terminal unit which incorporates hot water and/or chilled water coils with associated controls. It can be arranged for floor or ceiling mounting.

fan, compact See **compact Fischbach fan.**

fan, cross-flow See **tangential fan.**

fan curve See **fan characteristic.**

fan, diluted gas boiler flue system A flue system that obviates a conventional chimney for all sizes of gas-fired boilers. It incorporates a dilution fan fitted into the flue duct off the boiler, arranged to draw air from outside the boiler room and to dilute the flue gases sufficiently to permit their safe discharge at low level to the outside. The system must incorporate as a safety feature a pressure differential or flow switch which will prevent firing of the gas burner until the flue has been purged of gases and an adequate air flow has been established.

fan drive The connection between the fan and the driving motor or engine.

fan drive, belt driven An arrangement of pulleys and belts which drives a fan shaft. The speed of the fan shaft depends on the ratio between the driving and the driven pulleys.

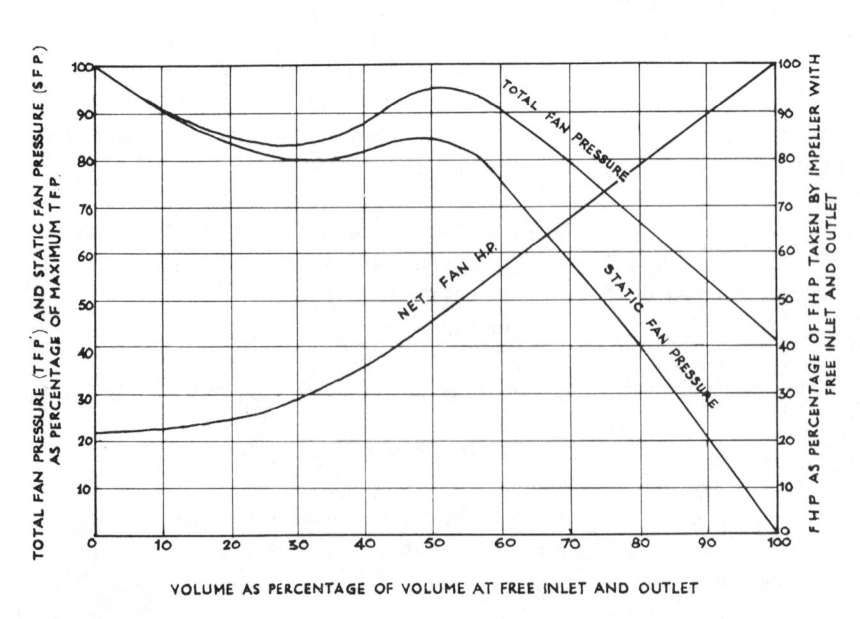

[Fig. 106] Characteristic curves for a forward-bladed impeller of a centrifugal fan

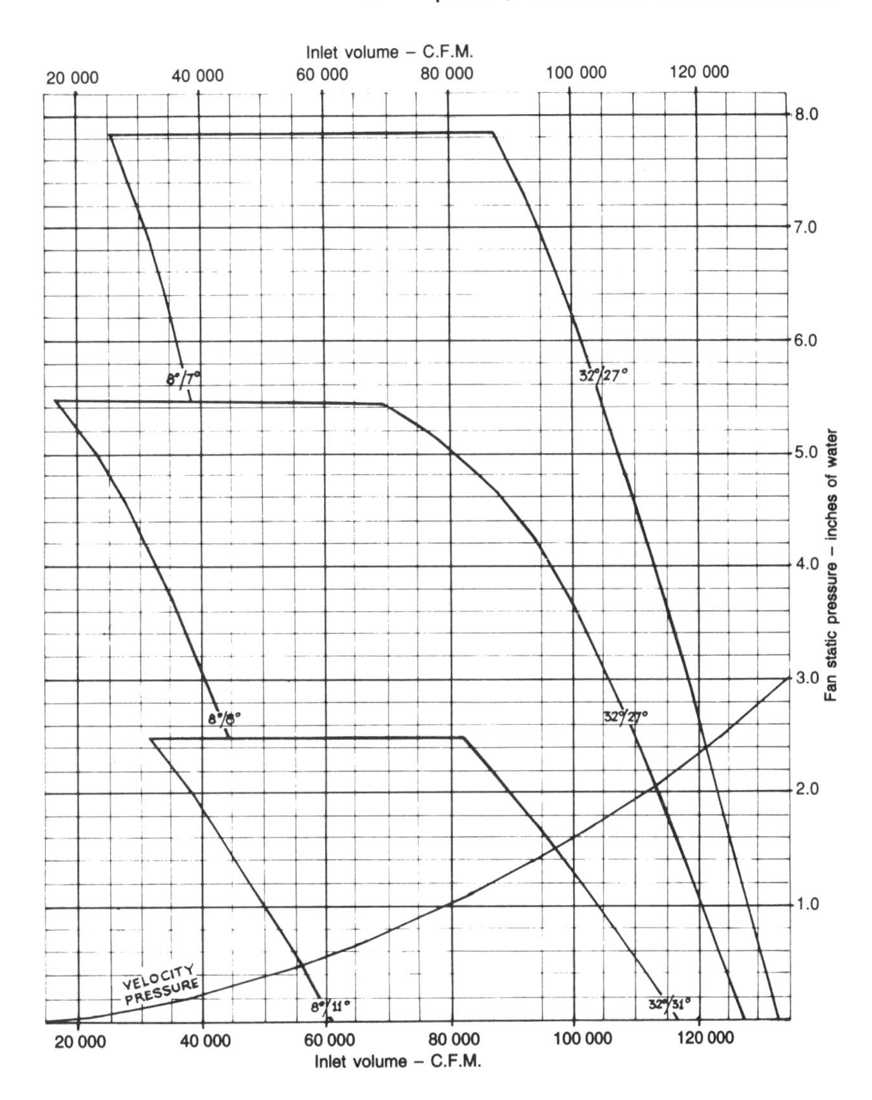

[Fig. 107] Characteristic curves for a typical axial flow fan

fan drive, direct A fan drive in which the fan shaft is coupled directly to the motor shaft. It rotates at motor speed.

fan drive, overhung A fan drive in which the driven pulley is fitted directly to the fan shaft without an intermediate support.

fan drive, pedestal A fan drive in which the fan shaft is supported on a pedestal with an integral bearing. The shaft is fitted with a driven pulley after it emerges from the pedestal bearing. This arrangement confers stability.

fan drive, vee rope A fan drive in which the driving belt is of vee-shaped section.

fan, efficiency The percentage ratio of output power to mechanical input in a fan.

fan impeller That part of a fan that imparts motive force to air. It is constructed of multiple vanes or blades and rotates within a casing having a spiral-cased contour. Air enters tangentially and leaves at right angles.

fan impeller, backward curved blades An impeller whose blade tips incline away from the direction of rotation. The blade angle is less than 90 degrees. The maximum power required is not much more in excess of that absorbed by the fan when operating at maximum efficiency. The driving motor will not overload.

fan impeller, forward curved blades An impeller whose blade tips incline towards the direction of rotation. The blade angle is greater than 90 degrees.

fan impeller, forward curved blades, performance A small increase in the volume of air handled results in a large corresponding increase in power taken which may overload the driving motor. This motor should be selected with a power margin of 25–30%.

fan impeller, radial blade An impeller whose blade tips (or the whole of the blade of a paddle blade fan) are radial. The blade angle is 90 degrees. It can be very stoutly constructed and is well suited to heavy duties, e.g. for air conveying solids, such as wood chips and sawdust.

fan laws

(a) The volume of air handled varies directly as the fan speed.
(b) The pressure developed by the fan varies as the square of the fan speed.
(c) The required power input varies as the cube of the fan speed.

Assuming constant fan speed but changing density.

(d) The volume flow of air does not change with a change in density.
(e) The pressure developed varies directly as the change in density.
(f) The power absorbed varies directly with the change in density.

fan, mixed flow See **mixed flow fan**.

fan motor protector A device fitted integrally with the motor windings that switches off the motor, thus protecting the windings against an excessive temperature rise due to a malfunction.

fan-shaped system A system of field drains laid to converge on to a single outlet at one point.

fan, static pressure The difference between the fan total pressure and the fan velocity pressure.

fan, tangential See **tangential fan**.

fan, total pressure The difference between the total pressure at the fan inlet and the fan outlet.

fan, velocity pressure That fan pressure which relates to the average velocity at the fan outlet.

fan, window A propeller-type fan inserted into a window pane. It can commonly be operated as an inlet or, via a reversing switch, as an exhaust fan.

farad A unit of electrical capacitance. It is that of a capacitor between the plates of which there appears a difference of potential of one volt when it is charged by one coulomb of electricity.

fast-track installation A system of carrying forward an installation with the utmost speed, taking short cuts to achieve speed.

feasibility report A report that presents the results, conclusions and recommendations of a feasibility study.

feasibility study A theoretical exercise to establish whether a particular project can be carried out within specified parameters such as time span, capital and operating costs, environmental and community concerns, location, etc.

featherweight check valve A valve installed within gravity circulations or circuits to prevent the unwanted natural circulation of hot water.

feed check valve A valve installed in a feed supply to a steam boiler to prevent back flow. See also **check valve**.

feed water economizer A heat exchanger installed within the flue gas system of a boiler to abstract the waste heat and utilize it for the preheating of feed water to a steam boiler.

feedback control/controller A controller which measures the value of a controlled variable in a control system, compares this with a standard representing the desired performance, and then manipulates the controlled system as necessary to achieve the maintenance of the required relationships.

feedback control system See **closed-loop control system.**

feeder A cable or other conductor which supplies electrical power to a distribution system.

female connection The terminal of a pipe fitting or boss which accepts the insertion of a pipe to form a joint.

fenestration The arrangement of glazing in a building.

ferrous metals Metals in which iron is the main constituent.

field drains A system of pipes employed in the control of ground water flow.

field test A performance check carried out on an operating installation of parameters such as pressure, temperature, air flow, velocity, air distribution (using smoke tests), etc.

fill thermal insulation See **thermal insulation, loose fill**. A method of insulation whereby thermal insulation material is injected into the cavity of the wall construction.

filled system thermometer A remote-reading thermometer. It usually has a circular chart or an indicating scale with the pen and pointer linked to a pressure element. The measuring element is connected by capillary tubing to a bulb located some distance away in the process. A fluid fill in the system creates an internal pressure which moves the pen or pointer in a definite relationship to the temperature change of the process medium. It is generally classified as vapour-filled, gas-filled or mercury-filled, depending on the fluid in the sealed measuring system.

film cooling tower A cooling tower that incorporates cellular fill to achieve closer surface contact between the air and the water streaming through the tower to enhance performance.

filter A device that removes impurities from air, gas or liquids.

filter drain A drain in which the pipe is surrounded with graded granular material or by a polypropylene or fabric filter. It is used where it is essential to prevent the migration of fine soil particles with ground water.

filter frame An assembly which supports an air filter.

fin An extension to a heating surface such as a heat transfer pipe. It may be a punched fin, a spirally wound fin or a crimped fin integral with the tube. Its purpose is to increase heat transfer.

fin drain A planar geocomposite structure which consists of a geotextile bonded to a polymer core to which a perforated plastic pipe is sometimes attached. This construction results in a lightweight fin drain and confers on it high strength. Such drains are designed to transmit ground water horizontally and vertically without clogging or silting. Their large surface area can remove water more quickly and effectively than any other highway drainage system. Reportedly, fin drains can be installed at half the cost of a conventional French drain and they are ecologically superior as it is not necessary to dig up acres of land to win gravel.

fin spacing The spacing between fins, expressed in terms of fins per unit length.

fin, spiral See **spiral fin**.

finned tube A heat transfer pipe which embodies fins to extend the heat transfer surface.

final certificate A certificate issued on the satisfactory completion of the works for final payment in respect of the contract.

final circuit A circuit connected directly to current-using equipment or to socket outlet(s) for the connection of such equipment.

fines Very small coal particles, not exceeding 3 mm, or wood-waste in the form of dust generated by sanders, sawdust and similar fine particles.

fire brick Brick manufactured of refractory material, commonly a mixture of silica (up to 78%) and alumina (up to a maximum of 38%). Fire brick is installed within combustion chambers to protect the equipment from contact with high temperatures and to provide a thermal mass to sustain combustion and retain heat. The exact specification of fire brick for a particular installation depends on the nature of the fuel being combusted and on the location of the brick within the combustion chamber.

fire brick bonding [Figs 108, 109] A method of joining together fire bricks which permits expansion movement under operating conditions.

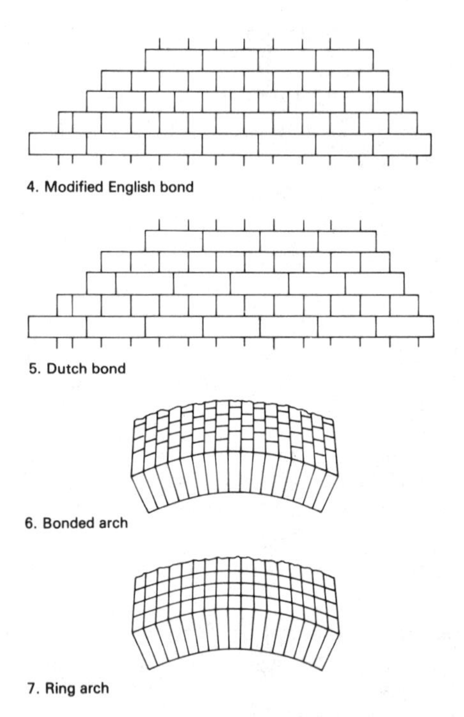

[Fig. 108] Standard brick bonding methods

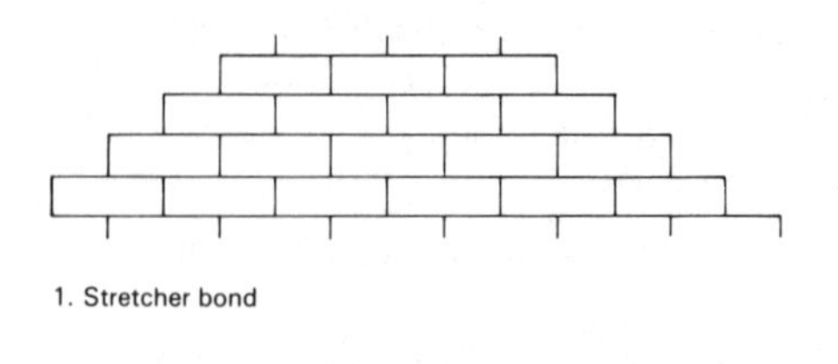

1. Stretcher bond

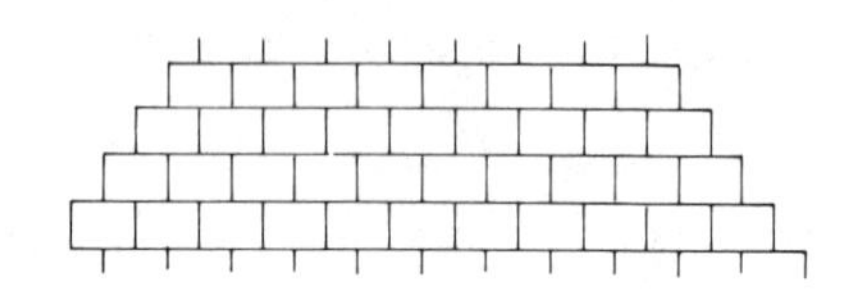

2. Header bond

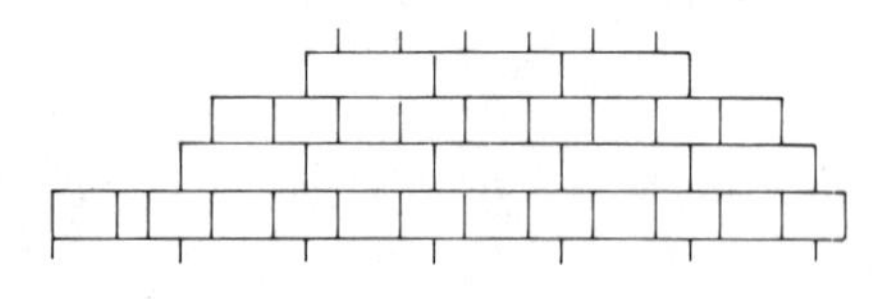

3. English bond

[Fig. 109] Standard brick bonding methods

fire brick door An access door into a furnace or combustion chamber consisting of a cast-iron or steel construction lined on the high temperature side with refractory material to protect the metal surface.

fire damper [Fig. 110] A substantially constructed metal damper within ductwork installed to isolate a section of duct in the event of fire or high temperature indication. It is held open by a tensioned linkage which incorporates a fusible link. Melting releases the damper to move from the open position to the tightly closed position thereby preventing the spread of fire along the airways. An access opening is required adjacent to the fusible link. An external indication is required of the status of the fire damper. It may be linked into a building management system.

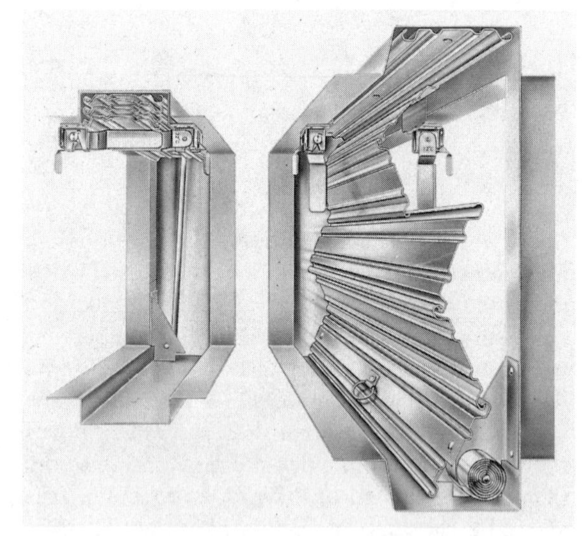

[Fig. 110] Fire damper assembly with fusible link (Waterloo–Ozonair Ltd.)

fire hydrant valve An assembly of pipe and valves which terminates with one or two valved connections for the use of the fire-fighting services. It may be of the pillar type with a floor support. The outlet valves are of fire brigade standard size and each one is fitted with a protective screwed gun-metal cap and chain.

fire offices committee (UK) A committee that sets standards and issues guidelines for fire protection, including the detailed design and specification of sprinkler installations.

fire precautions The measures taken and the means of fire protection put in place in a building or other fire risk to minimize the risk of fire to occupants, contents and structure from an outbreak of fire.

fire prevention The concept of preventing outbreaks of fire, of reducing the risk of fire spreading and avoiding danger from fire to persons and property.

fire protection Those design features, systems or equipment in a building, structure or other fire risk aimed at reducing the danger of fire to persons and property by detecting, containing and extinguishing a fire.

fire-rated pillow (Dow Corning) Firestop 300 is a fire-rated intumescent pillow that swells when heated to seal cables, pipes and ductwork that penetrate fire walls and partitions. Intumescent material, in the event of a fire, activates at approximately 260°C, completely sealing the penetration and thereby preventing the passage of fire, smoke or gases. It claims a four-hour fire rating for both integrity and insulation.

Fire-related pillows can be used to create a temporary or permanent fire seal around pipes, ducts and cables. They require no maintenance throughout their life, but allow existing cable arrangements to be altered when required. They will not emit halogenated byproducts under fire conditions.

fire stop An arrangement of sealing an opening (usually into a duct or around a pipe or duct) to prevent the passage of fire by placing approved non-combustible material into the opening or gap.

fire stopping Process of sealing off holes and gaps to prevent transmission of fire via same.

fire tube boiler A boiler that incorporates tubes through which the combustion gases pass within the water content of the boiler to transfer their heat to the water.

firebridge A refractory barrier within a boiler furnace placed there for the purpose of preventing coal from being thrown over the back of the grate into the furnace tube and to prevent combustion air from bypassing the grate at the back.

first cost The initial capital cost.

first law of thermodynamics The scientific law that states that heat and mechanical energy are mutually convertible.

fishtail burner A burner that discharges gas from a slot in the blunt end of the firing tube. Air mixes with the outer layers of the gas and the resulting combustion heats the gas which has not been exposed to air. It effects thermal deposition or cracking of the gas to carbon or hydrogen with yellow luminescence.

fixed foam installation A device that distributes fire extinguishing foam through a network of pipes to protect large installations by spreading a blanket of foam over the fire vulnerable equipment, e.g. in boiler houses or oil installations.

flame-failure device A device that protects combustion equipment (boilers, heaters, process plant) against damage which could arise if an automatic ignition system were to permit continuing fuel input into the combustion space in event of a satisfactory flame not having been established within a preset time. Assuming that ignition is delayed due to a fault and the fuel continues to pour into the combustion space, then a serious explosion would occur if an unduly large amount of the fresh fuel were permitted to accumulate and were then lit in an uncontrolled manner. Such devices may be of the photoelectric (magic eye) type which views the flame position and will cut off the fuel supply if the flame is not seen within a predetermined time or it may be a mechanical device which is fitted into the combustion (or burner) space and expands on sensing heat. If it does not detect heat within a specified time, it will close the fuel inlet valve.

flame form The shape of a flame. It is controlled by the air pressure within the furnace and by the feed of the material to induce combustion. This must be compatible with the shape and dimensions of the given combustion chamber to avoid impingement of the flame on heat transfer or refractory surfaces.

flame impingement A malfunction of combustion equipment whereby the flame generated in the process strikes the boiler or refractory surface.

flame speed The speed at which a flame moves into and through a combustion chamber. There must be a balance between the velocity of the gas at the burner and flame speed; excessive speed will cause backfiring and inadequate speed will cause extinguishing of the flame.

flame stability See **stable flame** and **unstable flame**.

flame, stable See **stable flame**.

flame, unstable See **unstable flame**.

flange, blank A flange in which bolt holes have not been drilled.

flange, blind A flange that seals the end of a pipe.

flange, companion See **companion flange**.

flange, Morgrip See **Morgrip flange**.

flanking transmission Transmission of sound between two rooms by an indirect path of sound travel.

flap and float valve A protective device installed in a drainage system to guard against the effect of back flooding caused by surcharge. Also known as a flap valve.

flash chamber A vessel in which hot water or other hot liquid generates (flashes into) vapour on reduction of pressure. It incorporates a space in which the vapour can accumulate before being piped off to a space heating or process application.

flash steam That steam generated from hot water when the pressure is reduced below that corresponding to the saturated steam previously in contact with the water. For example, low-pressure steam can be obtained from high-temperature condensate discharged from the high-pressure section of a process plant and can be used subsequently to service with steam the low-pressure section of that plant. When condensate is discharged into a vented receiver vessel, flash steam will be generated and the safe removal of this must be provided.

flash-over That stage in the development of a contained fire within an enclosed compartment at which fire spreads rapidly to give large merged flames throughout the space.

flaunching Weathering formed in mortar at the top of a chimney or base of a chimney pot.

flexible coupling A connection manufactured of a flexible material, such as rubber or neoprene, and installed for the purpose of obviating the transmission of vibration to the pipe to which it is connected or to permit easier dismantling for maintenance.

flexible ducting A duct manufactured of flexible material. A short length can be used to connect a branch duct to a ventilation or air-conditioning terminal for the purpose of obviating the transmission of vibration into the duct system, and to facilitate easier maintenance and dismantling of the terminal fitting or equipment. It is always provided at the duct connection to a centrifugal fan.

flexible pipe Pipe manufactured of flexible material. A short length may be used to make the final connection to rotating equipment, such as a pump or air compressor, to obviate the transmission of vibration into a pipe system, and to facilitate easier maintenance and dismantling of the connected equipment.

floating ball blanket A number of individual balls which are resistant to most acids, alkalis and solvents and have a heat resistance of up to 125°C which are placed on to the surface of a process tank with the object of reducing evaporation and smells with the resultant energy conservation.

floating control See **automatic control, floating control**.

floating floor A floor that offers high insulation against airborne noise and low transmission of impact noises, e.g. footsteps. It is constructed on top of the structural floor using resilient quilt and waterproof paper below the final screed, taking care to avoid any solid bridging path at the edges.

floating hollow balls Balls floated as a blanket on the surface of a hot or warm liquid contained within a tank to insulate it from loss of heat and/or to reduce the emission of steam or fumes and the absorption of oxygen.

float-operated steam trap See **steam trap, float-operated**.

floc A fine cloud of spongy particles that form in water to which a coagulant has been added. The particles are basically hydroxides commonly of aluminium or iron. They accelerate the settlement of suspended particles by adhering to them and neutralizing negative electrical charges which may be present.

flooded evaporation A process in which liquid refrigerant is fed to the expansion coil or evaporator of a refrigeration system through a float regulator which maintains the desired level of liquid refrigerant.

flooded evaporator See evaporator. In the flooded type of evaporator, a liquid pool of refrigerant is maintained by a float control device.

floor drainage gullies Traps normally connected to branch pipes of 75 mm or larger and therefore not subject to loss of water seal due to self-siphonage. Infrequent use can lead to the total loss of the seal through evaporation and therefore such traps should only be specified where there is sufficient usage of the system to ensure maintenance of the water seal.

floor heating A method of space heating that relies on the heating medium warming the floor.

floor heating, electric An arrangement of electric heating cables embedded within floor screed. It may be part of a ducted or withdrawable system which permits subsequent repairs or replacements or it may be embedded without facilities for future access to cables. It can be accommodated within 50 mm screed thickness (directly embedded) or 63 mm minimum thickness (withdrawable system). It commonly operates on a cheaper off-peak tariff and the programme is controlled by a time clock.

floor heating, floor construction To function effectively, all floor warming systems are thermally insulated from the sub-floor and building structure in order to direct the heat upwards into the occupied space and to avoid heat losses into the ground. Floor finish and carpets must be of the type that transmit heat effectively.

floor heating, headers Flow and return headers assemble the connections to individual floor coils. They incorporate air vents, drain cocks and control valves.

floor heating, heat pump Given the low temperature of the water circulating in floor heating systems, these systems lend themselves to operation in conjunction with heat pumps at a good coefficient of performance.

floor heating, metal pipes The arrangement of serpentine pipe coils (copper or mild steel) which are embedded within the floor screed above the structural slab. Pipe spacing depends on the required heat output; usually 120 mm or 240 mm. Water temperature must be controlled to protect floor materials to a maximum of 43°C and requires a thickness of floor screed of 50 mm–75 mm.

floor heating, plastic pipes Polypropylene pipes of 13 mm embedded within specially formulated cement or mortar floor screed. The pipes are located on spacing bars at pipe spacings to suit the required heat output. The water flow temperature is limited to 50°C and the required thickness of screed is 70 mm. The pipes are connected to headers above the floor.

floor heating, sleeved pipes Reducing the flow temperature to floor panel heaters involves automatic controls and possible risk of overheating the floor. Sleeved 'Panelec' pipes of 12 mm are encased in asbestos-cement sheaths and permit the use of water at temperatures up to 71°C within the floor panels without risk of damage to the floor materials. The pipe spacing is at 180 mm and 240 mm. The required thickness of floor screed is 50 mm to 75 mm. The floor panel pipes are connected to pipe headers above the floor.

floor panel heating See **floor heating**.

floor pipe ducting Ducting laid within floor screed to accommodate pipes and fittings within the thickness of the screed. Water Supply By-law 58 directs that underfloor pipes and fittings located within floor screeds must be laid within purpose-made ducts which may, if necessary, be exposed. Such ducts are commonly made of extruded plastic with a top flange and are available in 3 m lengths.

flow-regulating valve A valve that controls the rate of flow in a system.

flow sheet/diagram A diagram which indicates the various activities in a specific process.

flow switch, differential A switch that senses the pressure difference between two fixed points in an air or gas system. It actuates electrical apparatus, such as a boiler or a fan when the correct differential has been established.

flow switch, vane operated A flow switch that incorporates lightweight vanes which are propelled when air is passed through them. It is set to actuate electrical apparatus when a preset air velocity is sensed.

flue A conduit connected to the gas outlet of a combustion appliance to carry the waste gases travelling towards the chimney.

flue dilution The dilution of flue gases from gas burning appliances with fresh air to enable the products of combustion to be discharged at a low level without the provision of a conventional chimney. The suggested dilution ratio is 50:1; this is achieved using a dilution fan with safety features.

flue draught See **chimney draught**.

flue gas Waste combustion gas exiting from a combustion appliance.

flue gas isolator A substantial flanged steel casing which houses a hinged blade fitted with stainless steel sealing strip to ensure a tight damper closure. It is fully automatic with a motor drive for open/closed or modulating control and is equipped with an explosion relief facility. It is fitted into a flue system with or without a bypass.

flue gas loss The heat content of the flue gases which are discharged to atmosphere and which represent an energy loss.

flue gas recuperator [Fig. 111] A device that extracts waste heat from combustion processes. It is installed into an exhaust gas stream where nozzles spray water into the gas stream, the water absorbing the heat and utilizing it to indirectly heat the water for consumption or process.

flueless heater A heater that discharges its products of combustion into the space which it heats. Precautions must be taken to ensure that this space is adequately ventilated to prevent an undue build-up of combustion products which may also be accompanied by condensation.

flue liner See **chimney lining**.

flue stack heat recovery A heat recovery system that uses a heat exchanger installed within the flue stack of a boiler or furnace. Cold water is pumped through the heat exchanger and is heated by the hot flue gases. The flow rate is automatically controlled to maintain exit flue gas temperature above acid dewpoint. The hot water is used for a hot water supply or process heating.

fluidized bed Forms the heart of a fluidized combustion system. The fluidized bed is an expanded

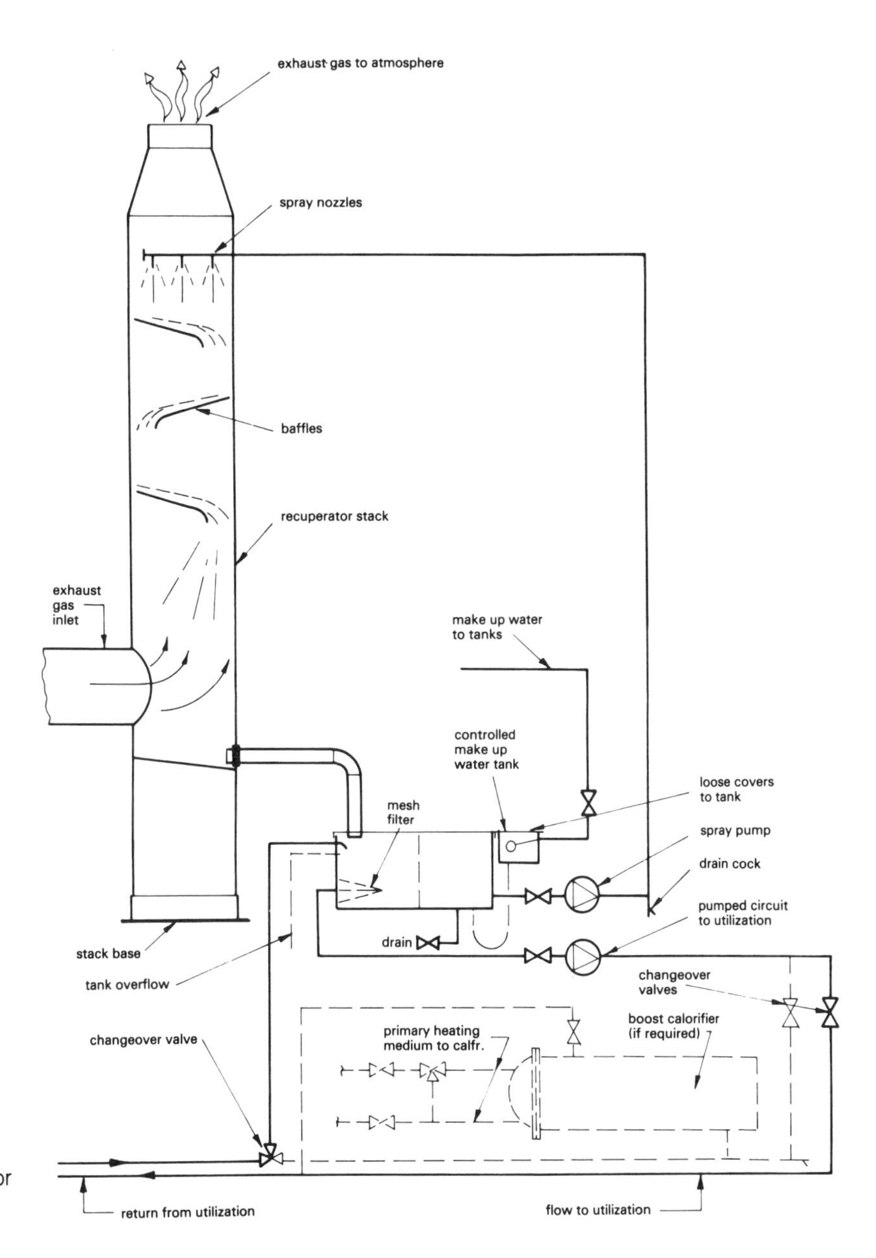

[Fig. 111] Basic recuperator system

fuel bed in which the solid fuel particles are suspended above the base of the bed by the drag forces caused by the gas stream which passes at some specific critical velocity through the voids between the fuel particles. The solids and the gas streams (also termed gas phase), intermix, acting somewhat in the manner of a boiling fluid. The solid fuel particles, when subjected to a current of air, may either be unaffected, if that velocity is below a certain value, or will become airborne if the velocity of the air exceeds a certain higher value. Between these two limiting velocities a bed of small particles will not become fully airborne but will have the effective particle density reduced. By adjustment of the air velocity, the solid particles can be conveyed almost as easily as if they were fluid. Any gas or vapour will achieve the same effect as air, but due allowance must be made for the relative velocity and density.

The fluidized bed is capable of considerable flexibility in being able to combust efficiently a wide range of fuels and wastes with high rates of heat transfer to the heat generating output system. The fluidized bed is relatively compact and operates at low temperatures; with the addition of limestone, it can desulphurize the fuel and thereby achieve very low emission of sulphur dioxide in the products of combustion which are discharged to the atmosphere.

fluidized bed combustion, shallow beds Bed depths of less than 456 mm can be considered as shallow beds. Deeper beds are referred to as deep bed systems.

fluidized bed combustion, support heat That heat required to raise the bed temperature from cold and to provide a trim heat facility during service. Various methods are used, including over-bed firing and an external form of direct fired air heater (oil/gas fired) for trim heating purposes.

fluidized combustion A method of combustion in which the fuel is burnt continuously in an atmospheric fluidized bed where the solids are suspended by the drag forces caused by the gas phase passing at some specific critical velocity through the voids between the particles. Solids and gas phases intermix in a manner similar to that of boiling liquid.

fluidized combustion, autothermic balance That heat release required to maintain the balance within the bed that stabilizes the temperature of the bed at the optimum level and equates the heat loss to the immersed heat transfer surfaces and to the gases exhausting from the bed.

fluidized combustion, bed depth Bed depth of less than 456 mm is considered a shallow bed. Deeper beds are referred to as deep bed systems.

fluidized combustion, bed temperature This is usually limited to a maximum of 950°C. At this temperature clinker formation by molten ash in the bed is avoided.

fluidized combustion of wastes [Fig. 112] Fluidized beds are suited to the incineration of waste materials, with or without heat recovery; e.g. hospital waste, oil refinery and coal preparation waste sludges, municipal refuse and sewage sludges. The usual requirement is the production of a dry inert solid for disposal and high throughput, rather than optimum combustion efficiency. Aqueous sludges are sprayed on to the bed surface in order to obtain some evaporation in the freeboard zone and to reduce the risk of fusing particles together.

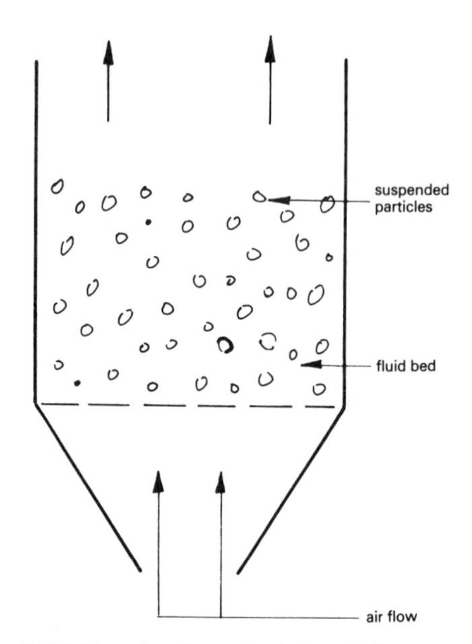

[Fig. 112] The basic principle of fluidized bed combustion

fluidized combustion, process The fuel, be it coal, oil, gas or refuse, is dispersed and combusted in a fluidized bed of inert particles, commonly sand. The temperature of the bed is maintained in the range of 750–950°C so that combustion of the fuel is substantially completed, but particle sintering is prevented. The gaseous combustion products leave the bed at the operating temperature, removing about half the heat which has been generated. The balance of the heat released is available for direct transmission to heat transfer surfaces which are immersed within the bed. In boiler applications, these comprise a bank of steam raising tubes; the low combustion temperature reduces the risk of corrosion and fouling of the heat transfer surfaces, allows a lower emission of nitrogen oxides and alkalis and permits the control of sulphur dioxide emission by the addition of limestone or dolomite to the fluidized bed.

fluidized combustor The complete assembly of equipment within one unit for the process of fluidized combustion.

fluidized combustor, bed material Sand or ash from the coal being combusted, crushed firebrick or other similar inert mineral material.

fluidized combustor, control Automatic control is essential. The temperature of the bed is constantly monitored and it is essential that the control of all trim features, including control of the load, be completely automated.

fluidized combustor, freeboard The dynamic surface zone established above the working bed, usually of about 305 mm effective depth.

fluidized combustor, rate of combustion The amount of fuel that can be burnt in a fluidized bed is determined by the air supply rate and hence by the fluidizing velocity and operating pressure. Velocities used are between 0.3 and 4 m/s measured at bed temperature, giving heat release rates of 0.2–3.00 MW/m^2 of bed area at atmospheric pressure. The efficiency of combustion depends on many operating factors, including the type of fuel, the level of excess air, bed temperature, fluidizing velocity, bed height and uniformity of fuel distribution. Volume (bed plus freeboard) heat release rates for coal are 2 MW/m^3.

fluidized combustor, separation of incombustibles [Fig. 113] An essential feature and requirement of every fluidized bed combustor plant is a facility for the continuous refinement of the fluidized bed inert material. Commonly a conveyor belt arrangement recycles the sand continuously and allows the separation of the incombustible components, such as ash, glass, or tins for separate disposal or recycling.

fluidized combustor, support fuel Consumption is minimized by arranging the level of the autothermic balance to correspond to the heat release requirement for the minimum turn-down condition of the incinerator or fluidized bed boiler. Experience indicates that the support fuel, after heat raising, does not usually exceed 1% of the total heat release required for the operation of the combustor at full load.

fluorescent lamp A tubular discharge lamp internally coated with a powder which fluoresces under the action of the electric discharge, producing a shadowless, white or specifically coloured light.

flush-fitting tank valve A valve which embodies a flange which closes flush with the inside surface of the vessel in order to eliminate the possibility of crevices where solids could collect and block the flow when the valve is opened.

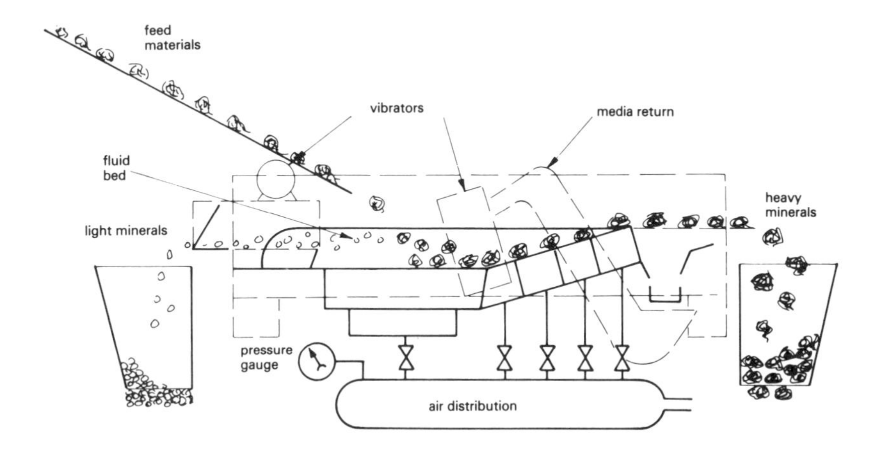

[Fig. 113] Fluidized bed operation: metals recovery

flushing cistern A water waste preventer fitted to a toilet bowl and arranged with a handle or push button to actuate the flushing mechanism.

flushing manhole A manhole at which water for flushing out the drainage system is introduced.

flushing valve A valve used where frequent flushing is required or where it is impossible to have a concealed or exposed cistern in the immediate vicinity of the pan. Flushing valves are not frequently used in the UK as the water utility companies generally require pans to be tank-fed to minimize wastage of water which might be caused by neglecting to maintain flushing valves routinely and effectively.

flush valve A valve installed to flush a toilet bowl. It is usually connected off the mains water supply.

fly ash Fine particle chimney escape from the suspended matter in the combustion chamber of a furnace which has not been captured by the grit arrester. It occurs more usually when the flue gases are cool.

flywheel A heavy wheel associated with rotating machinery (e.g. a steam engine) capable of spinning and of thereby storing energy. The attainable energy density is up to about 0.2 MJ/kg.

flywheel effect The storage of energy obtained by the use of a flywheel. The flywheel absorbs energy as the machinery slows down and gives out energy when it speeds up. In a wider context, a building of heavy construction is said to possess a flywheel effect. It will absorb into its structure solar energy and discharge it into the building as it cools down, thereby retarding the solar heating effect inside the building and the cooling down of the environment when the Sun effect has diminished. A similar effect relates to the space heating or cooling of the building by other means.

foam generation Equipment is generally carried on fire brigade vehicles comprising a small tank of foam compound with a flexible pipe connection for fitting to the foam inlet at the building. The flow of water through the pipe induces the foam compound into the water stream when aeration occurs and a jet of foam issues. The range of the jet depends on the available water pressure and lies between 9 m for 3.3 bar pressure and 30 m for 8.3 bar water pressure.

foaming The presence of tiny bubbles visible as a disturbance in the gauge glass of a steam boiler indicating the presence of contaminants in the boiler water such as concentrations of soap, oil, organic matter, suspended particles or sundry foreign substances. Foaming confirms an unsatisfactory water condition and that remedial action is required.

foam inlet A foam pourer placed at the distribution end of a fixed foam installation, located so that the foam issuing from it will form a foam blanket over the equipment and floor of the protected room. A foam inlet box at the face of a building provides a convenient entry point for the fire brigade's foam connection.

foam inlet box An externally located metal box which houses a foam inlet fitting for connection to the fire brigade's foam supply.

foam nozzles Nozzles attached to foam pipes, each located in a specified position above the burner equipment.

foam pipes An arrangement of pipes which connect between foam nozzles in a boiler room and a foam terminal box in a suitably accessible external location. In the event of a fire in the vicinity of or within the boiler room, the fire service pump fire-quenching foam through the pipes into the boiler room.

fog A concentration of mists in sufficient concentration to reduce the visual range. Natural fog contains particles larger than 10 microns. Artificial fog, as used in theatrical productions, is really water vapour generated in a fog-making machine.

foot valve The non-return valve inserted into the termination of a pump suction pipe inside an underground tank to permit the drawing of liquid from the tank.

forced draught condenser Part of a refrigeration system in which a fan-driven stream of air condenses the refrigerant gas passing through the condenser and absorbs the latent heat.

forced draught cooling tower A cooling tower that incorporates a fan (or fans) to move the airstream through it to enhance output capacity and performance.

forced vortex A vortex in which the fluid rotates at constant angular velocity. The difference in static pressure across the vortex is equal to the difference in tangential velocity pressure.

fossil fuel A deposit of fuel formed in the Earth or under the sea bed by the decay of organic matter which has been subjected to great pressure and temperature over a long period of time. The main fossil fuels are: coal, oil, gas and the various intermediate stages of fossil fuel formation, such as peat, shale, etc.

foul drain, connection to Connection to a foul drain from a discharge stack, sanitary appliance or branch drain is generally made to an intermediate inspection chamber or manhole; where it is made to a junction, this must be fully accessible for maintenance, clearing of blockages and for testing. The arrangement must be such that the discharge is swept in the direction of flow at the point of connection.

foul drain, ventilation of Commonly, adequate ventilation is achieved via ventilated discharge stacks, but where this is impracticable, a separate ventilating pipe should be provided near the head of the drain run or in such other position as may be required by local regulations. Where an intercepting trap is installed, ventilation of the drain is additionally required adjacent to that trap.

fouling factor A number that represents the thermal resistance of the water film on the heat exchange tubes. It applies to water-using equipment such as water-cooled condensers. Manufacturers rate water-cooled condenser and chiller capacity on the basis of the water fouling factor. A value of 0.0005 is considered clean water and ratings are often based on this value. Water treatment is intended to prevent formation of scale and the associated increase in the fouling factor.

four-pipe air-conditioning system A system that incorporates two separate flow and return circuits of hot water and of chilled water respectively, each with its own valves and automatic controls. The associated terminal units are a four-pipe fan coil or four-pipe induction units.

four-wire distribution Three-phase a.c. consisting of three 'phase' wires and one neutral wire.

frame room A designated room or area for the entry and main distribution centre of a telephone supply for a building.

free air flow Relates to compressed air practice and expresses the compressed air flow at atmospheric pressure rather than stating the flow at the actual compressed conditions for each application.

free cooling That amount of cooling which can be obtained from an air-conditioning system without operating the chillers.

free discharge valve A valve usually fitted at the end of a pipeline for the release of fluid under free discharge conditions. It may be of the needle, hollow jet or sleeve type.

free vortex A vortex in which the tangential velocity varies inversely as the radius and in which the total energy, as expressed by Bernoulli's theorem, remains constant.

French drain A drain constructed by excavating a trench and placing in it open-jointed pipes or perforated or porous material, the pipes then being surrounded with selected fill through which the water can percolate freely.

freons A range of chemicals used widely in refrigeration machines which comprise mainly fluorine with other halogen derivatives of hydrocarbons.

frequency The number of cycles (complete and reverse flow) of an alternating current in each second applied to electrical apparatus and measured in hertz (Hz).

frequency, definition The rate at which vibrations occur, expressed in hertz (Hz).

frequency, natural That frequency of a system or material at which it freely vibrates when a force is applied and removed, e.g. when it is kicked.

freezing-point That temperature of a substance at which it changes from the liquid to the solid state without any change in its temperature.

friction factor A function of the resistance to the flow of a fluid due to the roughness of the walls of the conduit (pipe), and the velocity and nature of flow (whether streamline or turbulent). It is calculated from the Darcy formula.

friction head The resistance to the flow of a fluid in a pipe.

friction loss Pumping pressure (head) used up in overcoming the resistance to the flow of a fluid in a pipe.

from and at 100°C The basis for comparing the evaporative capacities of steam boilers by assuming that the steam is generated at 'equivalent evaporation' from water at 100°C to steam at the same temperature at a pressure of one bar derived from the total heat in the steam of the particular application.

frost coil An air heater placed between the fresh air supply main shut-off damper and the pre-filter. It may be piped or electric and serves to pre-heat very cold air before it enters the main plant.

frost protection The prevention of freezing, usually in water, by various additives or by heating with a trace system keeping the fluid temperature above freezing-point (0°C or 32°F).

frost protection, automatic A controlled means of adding heat to a fluid under frost conditions to obviate frost damage; e.g. bypassing hot water to an exposed cooling tower.

fuel cell [Fig. 114] An electrochemical device in which the chemical energy of reaction between a fuel and an oxidant is converted directly into electrical energy. In a practical application, the chemical reaction between hydrogen (obtained from a hydrocarbon fuel) and oxygen of the air is employed. The fuel cell delivers its output of electrical energy so long as the fuel and the oxidant are supplied to it.

fuel efficiency monitor An automated portable flue gas analyser which simultaneously measures the temperature and oxygen content of the flue gases of a boiler or furnace to determine the combustion efficiency of the appliance.

fuel oil additive An additive designed to improve the combustion process and to keep boiler combustion paths clean and efficient. A liquid chemical is available for all grades of oil fuel and for all types of oil-fired plant. It can be added to oil in storage or injected into the fuel oil supply line. Specific additives are available to combat sludging, smutting, acid smutting, high and low temperature corrosion, etc.

full-modulation combustion control The adjustment of the fuel input to a furnace from zero input to 100% input in accordance with load variations. This is usually achieved by automatic control.

fume cupboard An item of laboratory furniture designed for the fume generating process. It commonly incorporates a safety-engineering enclosure and an extraction fan or a connection to a central extraction system; it affords the safe and efficient extraction of noxious fumes at source.

fumes Fine smoke of solid particles, formed by condensation of sizes usually below 1 micron.

fuse A safety device inserted into an electric circuit for protection against overloading. It comprises fusible wire or a cartridge installed within a ceramic fuse carrier. In the event of an overload, the wire fuses and breaks the circuit before the connected installation or individual appliance can be damaged.

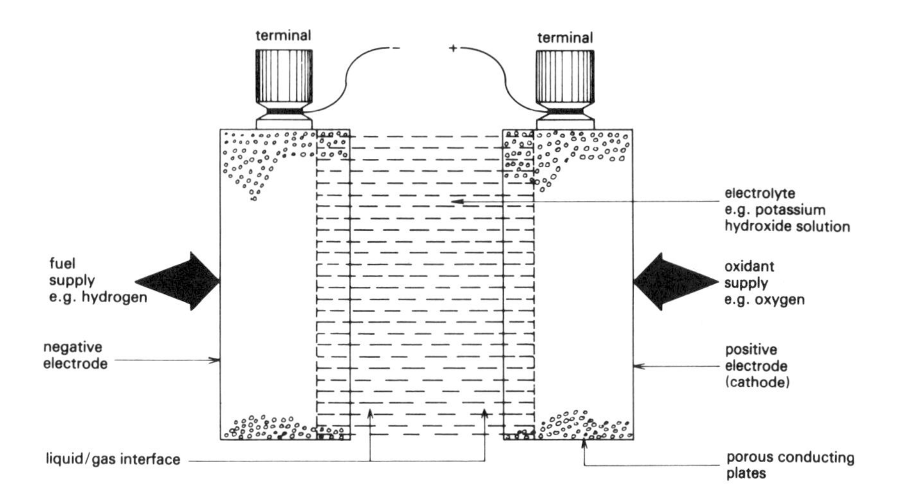

[Fig. 114] Principle of the fuel cell

fusegear A device that protects electrical installations or parts thereof from the effects of electrical overloads, relying for its functioning on the heat generated by an electrical overload.

fusible link A link made of a material with a specified low melting-point. It joins together two tensioned cables and in the event of the high ambient temperature of a fire in the vicinity of the link, the link melts and the cable parts, dropping a weighted valve lever or actuating an electric device. Its function may be to close down the fuel supply, ring an alarm, drop a fire damper, etc.

fusible link valve A spring-loaded or weight-operated drop valve fitted into the oil supply pipe to an oil burner as a safety device. It operates in conjunction with a fusible link (or links) fitted above the oil burner and connected to it by a system of tensioned wires. In the event of an excessive rise in the boiler room temperature (e.g. due to a fire) the link melts and slackens the wire linkage releasing the lever of the valve which then drops and closes the oil supply line to the burner(s).

fusible plug A device inserted into a solid fuel fired steam boiler as a safeguard against operation with insufficient water. When the fusible plug is exposed to excessive heat, its core melts, drops out and allows water and steam to be sprayed on to the fuel bed, smothering it.

fusing factor The fraction:

$$\frac{\text{Rated minimum fusing current}}{\text{Current rating}}$$

Fuse links (fuses) are rated according to four classes of fusing factor as follows:

Class	Fusing factor	Not exceeding
P	1.00	1.25
Q1	1.25	1.5
Q2	1.5	1.75
R	1.75	2.5

fusion joint Related to the smaller sizes of plastic pipes, it is a joint in which the surfaces which are to be in contact in a closely fitted spigot and socket are separately heated and whilst both are hot, the spigot is inserted causing socket and spigot to become fused together. Larger size pipes can be butt welded.

fusing temperature That temperature at which ash melts. Ashes that fuse in the range of 1000–1200°C are designated as low-fusing, in the range 1200–1400°C as medium-fusing and in the range above 1400°C as high-fusing. Ash which possesses a low fusion temperature tends to form clinker and this is likely to clog up the spaces between the bars of the fire grate and cause major operational difficulties to the combustion process and boiler maintenance. See also **ash fusion temperature**.

G

GRP Glass reinforced plastics.

galvanic series A list of metals in the order of their characteristic potential, commencing with the noble metals with positive potential and terminating with magnesium and magnesium alloys which have the greatest negative potential.

galvanized iron Sheet iron which has been coated with a layer of zinc to prevent corrosion by dipping the sheet into a bath of molten zinc.

galvanized pipe Steel pipe which has been coated with a layer of zinc internally and externally to prevent or slow down corrosion. It is used particularly for pipelines serving water draw-off services which are subject to frequent changes of water and therefore accelerated corrosion.

galvanized steel, welding Welding galvanized steel should be avoided as it destroys the galvanized coating adjacent to the weld and gives off objectionable fumes which cause discomfort to the welder.

galvanizing The application of a zinc coating to steel.

ganister A type of stone used with refractories. True Sheffield ganisters contain 97–98% silica. The term is also applied to other silica–clay mixtures which contain in excess of 70% silica.

gas An elastic fluid which cannot easily be liquefied at normal temperatures. (It is now known that all gases can be liquefied, though this involves very high pressure and/or very low temperatures.)

gas boiler flue diverter [Fig. 115] A device that prevents down draughts in a gas-fired appliance from interfering with efficient combustion. Such diverters prevent adverse flue action on the gas flame if down draught occurs in the flue (or connected chimney). It acts as a draught break and prevents wide variations in the amount of air being drawn through the boiler.

gas booster A compressor designed to handle gas. It is placed between the gas main and the appliance to impose the required gas pressure.

gas coke Fuel which is produced in the manufacturing process of town gas through the carbonization of the coal.

gas detection system A safety device for location in boiler rooms which house gas-fired boilers. A probe detects the leaking gas and stops further gas supply to the boilers by closing a solenoid gas valve.

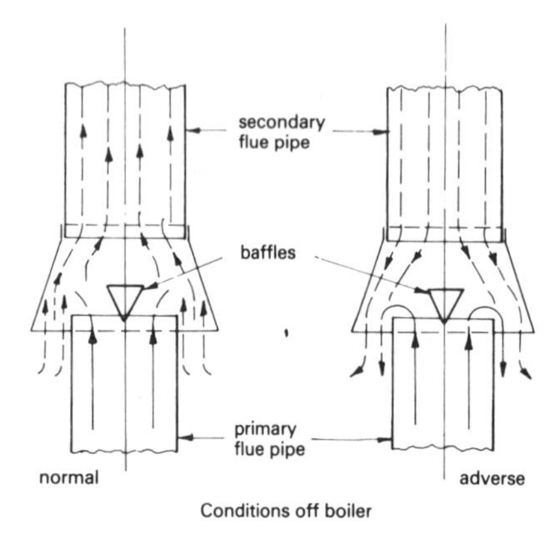

[Fig. 115] Flue gas diverter for use with gas-fired boilers

gas explosion The accidental ignition of an explosive mixture. It may be due to a defective flame failure device permitting unburnt gas into a combustion chamber where it accumulates and suddenly explodes when it is ignited or to a leakage in a gas pipe which is not vented from a confined space and is set alight by, for example, the lighting of a cigarette in that space.

gas-fired radiant heating system An integrated system of gas-fired overhead radiant heating comprising reflectors placed over the radiant pipes to direct the heat downwards into the area to be warmed. The combustion products are extracted by an induced-draught fan. It is best suited to industrial and commercial premises.

gas ignition controls Fully automated gas control incorporating electronics to control the ignition start sequence and embodying all the necessary safety features. It eliminates the need for a permanent pilot burner.

gasket A jointing ring of appropriate material placed between the two mating faces of a flanged pipe joint or between two sections of an assembly containing water or gas to achieve a fluid-tight joint when the flanges are bolted together.

gas laws Laws that define the relationships between the pressure, temperature and volume of a gas when two of these parameters change. See **Boyle's law** and **Charles's law**.

gas leak detector A device which detects the presence of a gas and sounds an alarm and/or closes a gas supply cut-off valve.

gas main burner An arrangement through which the operating gas supply is fed into the combustion space of a gas-fired appliance. It may be one single burner arm or be a grid of multiple burners. The gas is ejected through rows of nozzles which are fitted to each burner arm.

gas meter [Fig. 116] An appliance for measuring and recording the quantity of gas passing through it to the consumer.

gas migration A term usually referring to the seepage of gas from a landfill site.

gas monitor A device for locating and identifying the build-up of toxic fumes. It issues a warning by alarm or indication and it can be combined with a sophisticated survey camera to measure and indicate the exact location at which a build-up occurs. It examines the area for leakages or blockages that may cause the particular problem, transmitting the information to a screen console.

gas, natural See **natural gas**.

gas oil A petroleum distillate of the lighter range, having a viscosity of 35 newton seconds per m^2. It has a low sulphur content and does not require preheating in storage, pipeline or burner.

gas pilot burner See **pilot burner**.

gas plume A stream of gas exiting from a chimney terminal.

gas scrubber A device used to condition the combustion flue effluents, generally to cool the flue gases from an incinerator by the use of water sprays to lower the temperature sufficiently for the gases to be handled by an induced draught fan. It also eliminates fly ash. Each scrubber incorporates an induced draught fan and a fresh water facility to cool the gases, together with facilities for the removal of the contaminated water and solid deposits.

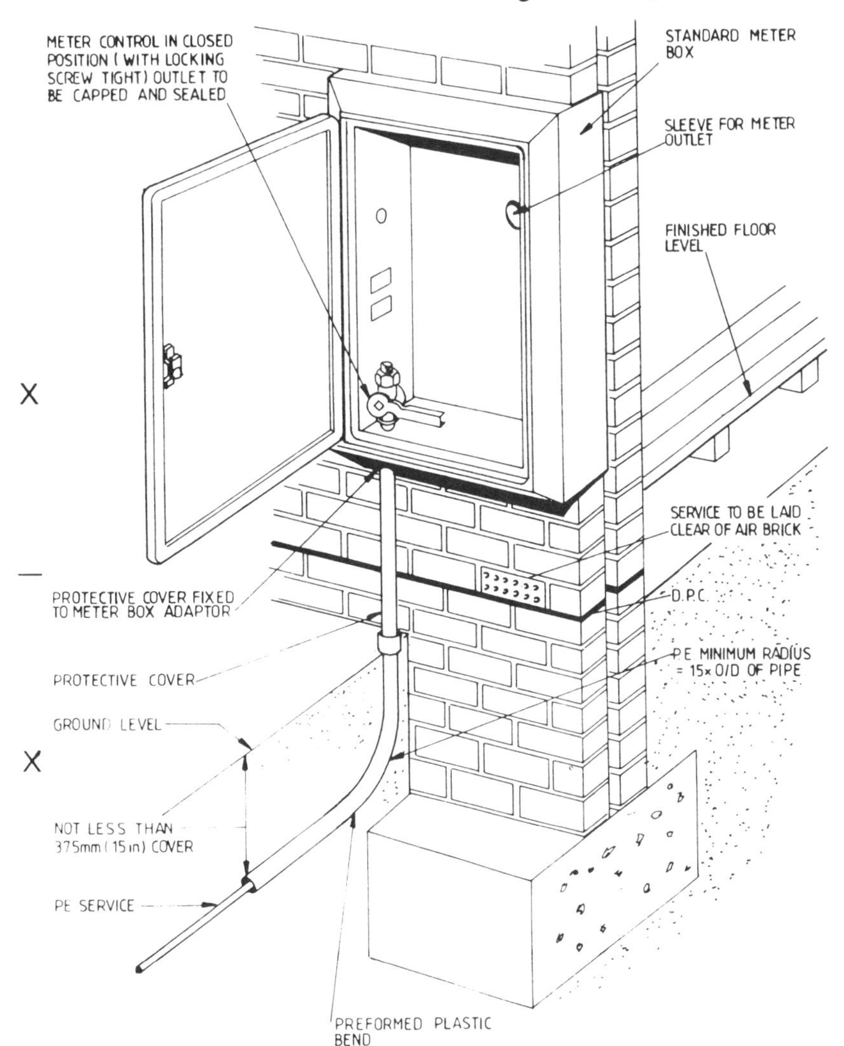

[Fig. 116] Typical meter box installation

gas, town See **town gas**.

gas train An assembly of control and safety devices installed in a gas-fired boiler or furnace.

gas turbine The prime mover for the generation of electricity which is fuelled by a gas supply, operating on the principle of the compression and expansion of a gas, usually air. Basically, it comprises a compressor, a combustion chamber for heating the compressed air and a turbine. The process is generally continuous and provides an even torque for power transmission.

gate valve A valve that controls the flow of fluid through the valve body by raising or lowering a gate attached to the valve spindle.

gate valve, double disc See **double disc gate valve**.

gate valve, inside screw A gate valve in which the actuating thread of the valve stem is contained within the valve. This may take one of the following three forms.

(a) An inside screw with a rising stem in which the handwheel is attached to the stem and rises with it when the valve is opened.
(b) An inside screw with a non-rising stem in which the handwheel is attached to a non-rising stem, the gate rising on the stem when the valve is opened.
(c) An inside screw with a rising spindle and rising stem in which the handwheel is attached to a

rising spindle and the stem rises within and with the spindle when the valve is opened.

gate valve, lever See **lever gate valve**.

gate valve, outside screw A gate valve in which the actuating thread of the valve stem is outside the exterior of the bonnet. This may take one of the following three forms.

(a) An outside screw with a stem rising with the handwheel which is attached to the stem and rises with it when the valve is opened.
(b) An outside screw with a stem rising through the handwheel which is attached to a yoke sleeve or bridge sleeve which revolves in the yoke or bridge and through which the stem rises when the valve is opened.
(c) An outside screw with a non-rising spindle, rising through the valve stem. The handwheel is attached to a non-rising spindle and the stem rises when the valve is opened.

gate valve, wedge See **wedge gate valve**.

gauge glass A steam boiler fitting which indicates the level of water in a boiler.

gauge protector A thick glass frame around a steam boiler water level indicator. This protects whilst also permitting visual inspection of the gauge glass.

gear pump or **geared pump** See **pump, geared**.

geothermal borehole A well drilled into the Earth's crust to tap a reservoir of hot water for subsequent use at the surface. This is a developing technique. There are existing geothermal working installations in Iceland, New Zealand and France, and there are at present test developments in the UK.

geothermal energy Heat extracted from underneath the surface of the Earth, in the form of steam (in volcanic regions) or as hot water pumped to the surface. It is tapped by means of boreholes or extracted from gushing geysers and may be exploited for district heating purposes.

geothermal fluid Hot water or steam obtained from subterranean strata, by natural geyser flow or from geothermal boreholes. The quality and temperature of the fluid are variable, depending on the locality. It may have a high mineral content requiring costly treatment before use in pipeline systems. There are recorded temperatures of fluid up to 200°C.

Applications: at higher temperatures for steam generation; at lowest temperatures for fish farming, greenhouse heating and similar uses. Also heat supply to district heating systems serving areas close to the geothermal stations.

gigajoule A unit which comprises 1000 MJ or 10^9 J.

gigawatt A unit which comprises 1000 MW or 10^9 W.

gill Alternative term for extended heat transfer surface fin.

gland A component of a valve or pump shaft which, with suitable gland packing, is screwed into the space between the valve body and the valve spindle. Functions to seal the valve stem or pump shaft against leakage of the fluid being handled.

gland cock A taper-seated cock in which the plug is retained in the body by means of a gland and appropriate gland packing.

gland valve A valve where the wheel head protrudes above the valve gland.

glare Excessively bright areas in the visual field that can separately or simultaneously impair visual performance or cause visual discomfort.

glare index An index that represents the degree of discomfort glare associated with a particular installation of luminaires. The degree of discomfort glare which can be tolerated decreases as the task difficulty increases.

glass fibre insulation Bonded insulation material comprising lightweight glass fibres. It is strong and free from shot and coarse fibres, and is easily handled, cut and installed. It is used for thermal insulation for a temperature limit of up to 230°C.

globe thermometer [Fig. 117] A mercury-in-glass thermometer designed for the subjective measurement of warmth. It comprises a thermometer placed within a blackened glass globe of 150 mm diameter and it measures the globe temperature.

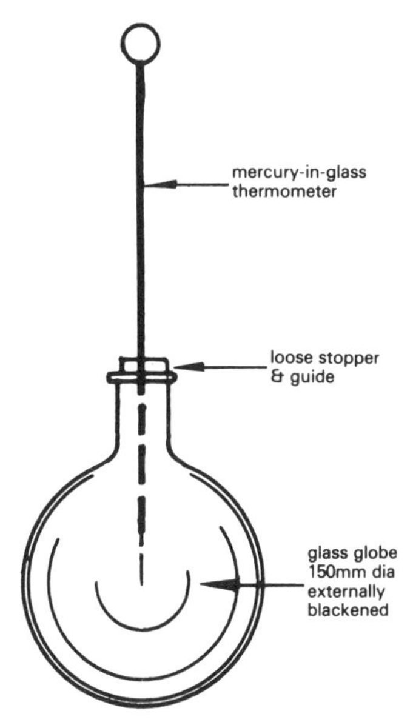

[Fig. 117] Globe thermometer

globe valve A valve that controls the flow of fluid (liquid or gas/vapour) by the action of a jumper and washer which acts on the valve seat. Fluid enters below the valve seat and exits above the seat.

gradient of discharge pipe The angle at which the pipe, drain or drainage channel slopes between fixed points in the system.

graph A diagram, commonly plotted between axes at right angles to each other, indicating the relationship of one or several variable quantities. The horizontal axis is usually referred to as the X axis and the vertical one as the Y axis.

grate A structure in a solid fuel fired combustion chamber which supports the fuel bed and allows the evenly distributed passage of sufficient primary air to pass through it. It separates the burning fuel from the resultant ash which falls through it into the ashpit. It may be horizontal or inclined, stationary or movable, or operated manually or automatically.

grate area The plane dimensions of a grate available for the burning of solid fuel.

grate bar One of a number of cross-members of a grate which rest on the frame of the grate. These are removable and offer a space between adjacent bars for the passage of primary air and the fall of ash into the ashpit. A grate bar is generally of cast-iron construction.

grated waste A waste fitting with an integral grating that is fitted flush with its inlet. It is incapable of accepting a waste plug.

grate, rocking See **rocking grate**.

grate, step See **step grate**.

grate, travelling See **travelling grate**.

grating A cover with slots or perforations to allow the passage of air or water through it; e.g. as used in surface water collection to pass the water into the drain.

gravimetric air filter efficiency See **air filter gravimetric efficiency**.

gravity circulation Circulation due to thermosyphonic action. At times this is undesirable, such as when the space heating is not required and only the hot water supply circuit is intended to operate. Unwanted wasteful gravity circulation may then arise into the space heating circuit.

Remedial action: the design of pipe circuits to inhibit thermosyphonic action. If this is not feasible or is unsuccessful, a featherweight non-return valve is fitted to cut off the circulation, located so that it will not interfere with the required space heating circulation.

gravity flow The movement of a liquid or other material along a channel or pipe by virtue of the gradient, as opposed to flow under pump pressure.

gravity heating system A system that operates by thermosyphonic action caused by the difference in density between the colder return water and the hot flow water. It employs no circulator (pump) to promote flow.

grease A semi-solidified lubricant which consists of emulsified petroleum oils and soluble hydrocarbon soaps. It is used widely in the lubrication of ball-bearings, motor shafts, etc. Grease is also the solidified residue from cooking and similar processes, requiring the provision of grease filters to protect connected systems such as exhaust ducts and drains.

grease filter [Fig. 118] An air filter installed within, or at the discharge of, cooker exhaust canopies to trap the grease from cooking processes and thereby to avoid fouling the exhaust duct system. The filter must be removable and be capable of being cleansed with ease using a degreasing liquid. See also **grease trap**.

[Fig. 118] Ventex cleanable grease filter panel (Waterloo–Ozonair Ltd.)

grease nipple A projection at a position requiring lubrication (e.g. a fan or motor bearing), with an opening for the entry of a grease lubricator which applies lubrication under pressure. A closing cap should be provided to obviate contamination.

grease traps in drainage systems These must be designed and located to promote the cooling, coagulation and retention of the grease within the trap. The temperature and velocity of the flow of the waste water must be such as to permit the grease to separate and collect on the surface of the water within the reservoir of the trap. The use of detergents will tend to impair or even prevent the process of separation as these emulsify the grease. The nature of the waste being discharged is relevant, as the reduced velocity of flow during the passage through the trap will allow solid matter carried in suspension to settle and collect, blocking the trap discharge. For reasons of hygiene, such traps must not be situated in food preparation and storage areas and provision must be made for the grease trap to be completely emptied and cleaned periodically to prevent the development of septic conditions within the trap. Easy access is therefore essential.

grid The high voltage transmission system operated by an electrical utility.

grid system Main field drains or drains which are laid close to the site boundaries and into which branches discharge from one side only. This is a useful method where the terrain has a general fall in one direction.

grille schedule [Fig. 119] A schedule that sets out in tabular form the full details of the various grilles required for a particular project, including the numbers of each item, the manufacturers' identification, dimensions, etc.

grilled cast-iron heating tubes These are commonly laid inside floor trenches and covered with open mesh gratings. They used to be popular for heating churches and greenhouses but are now seldom used.

[Fig. 119] Air distribution: grille with adjustable blades

grit Solid particles larger than 76 microns. Grit is collected by chimney grit arresters. Due to its abrasive nature grit must not be conveyed with the ash through a central system; it should be collected separately, bagged and dumped off site.

grit arrester A mechanical device for the arrest and separation of grit and dirt in flue gases. It is commonly in the form of a high efficiency cyclone.

grog Previously burnt fire-clay or fire bricks which are ground and mixed into the clay batch prior to moulding the fire bricks. Grog is used mainly with refractories to control the drying and firing behaviour of the bricks.

ground water Water that occurs naturally at or below the local water table.

ground water discharge termination The connection of a ground water drainage system to an outfall. It is commonly made through a catch pit which is designed to intercept and retain solids and suspended matter.

ground water drainage Localized drainage of ground water is at times necessary to increase the stability of the surrounding ground, avoid surface flooding, alleviate dampness in below-ground accommodation, protect the foundations and/or to prevent frost heave of the subsoil which might damage structural elements.

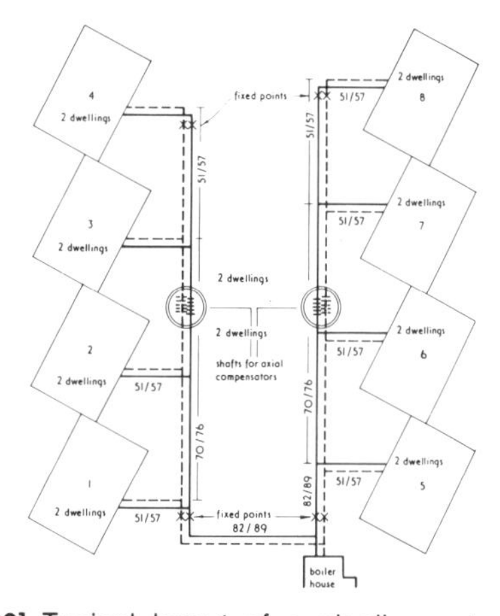

[Fig. 120] Typical layout of a pipeline network for a group heating system. Maximum heat supply 150 kW; supply temperature 110°C (230°F); return temperature 90°C (194°F). Figures on pipelines give internal and external diameters of pipes in millimetres

group heating [Fig. 120] The provision of heat services to an extensive site with a number of boiler plants in which each plant serves a separate group of buildings.

group heating station A boiler plant room which serves a group of buildings.

group heating system [Figs 121, 122] A heating system that serves a number of consumers off a central plant, usually via a consumer terminal which includes a heat metering facility. The heat distribution may be connected directly off the boiler headers or off heat exchangers which are served off

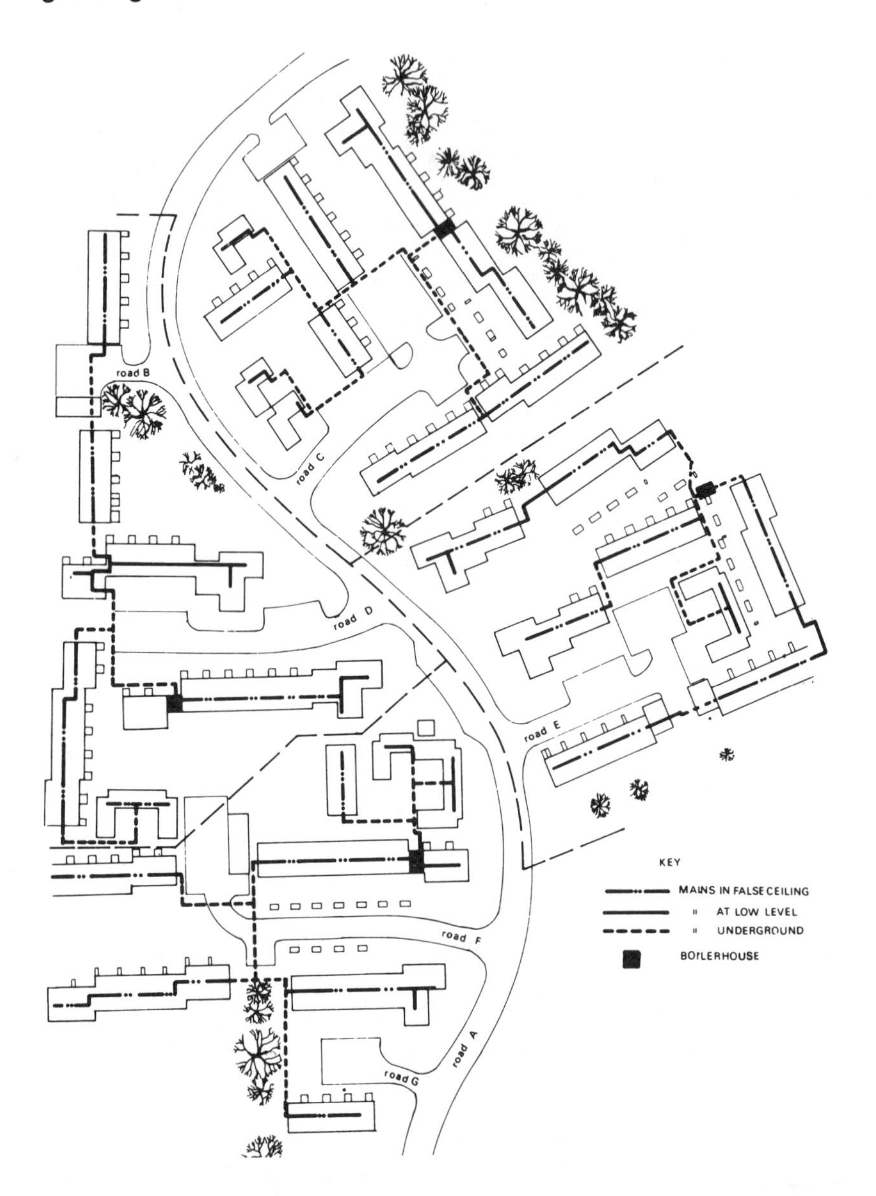

[Fig. 121] Layout of Totley Brook, Sheffield, group heating development

the boiler primary circuit. The latter arrangement permits the boilers to be operated at a higher temperature than the system serving the individual consumer.

grouting A material made of one part Portland cement and two parts clean sand that is used to seal spaces between foundation bolts and the concrete foundation. It must not be allowed to set too quickly and is liable to crumble under eventual vibration. Grouting should proceed without interruption from start to finish and the grouting must be well worked into the openings. When applied in a hot environment, such as an operating compressor room, the exposed portion should be dampened over a few days. The foundation bolts are pulled up tight after the grouting has firmly set.

guarantee A formal and legally binding undertaking by a supplier or installer of goods or services

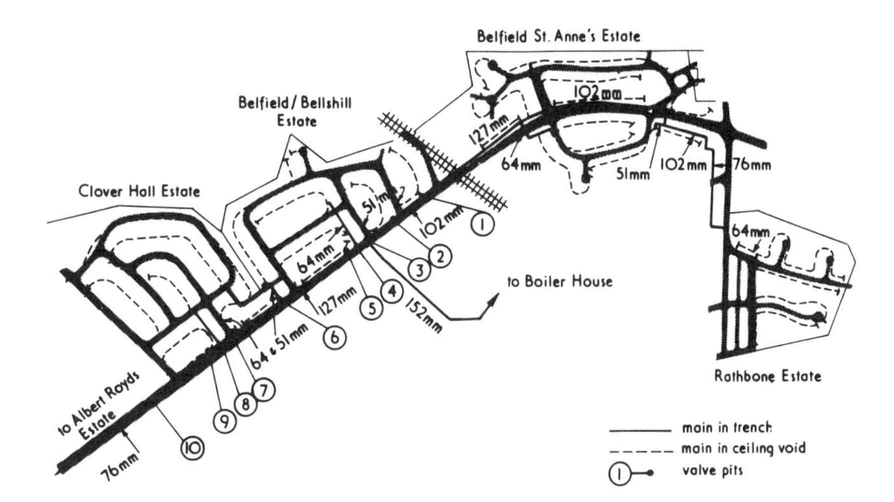

[Fig. 122] Layout of Rochdale group heating system

to the purchaser/user for maintaining quality, performance or service life.

guide vanes Vanes in axial flow fans, fitted downstream of the impeller. Their function is to remove the rotational energy component of the air thereby slowing down the air and converting some of the excess velocity pressure to more useful static pressure. They are also employed as pre-rotational vanes upstream of the impeller, where they function to rotate the air in a direction opposite to that of the impeller rotation. They are designed to ensure that the air leaves the fan in a truly axial direction.

gully A floor-located outlet through which water or other liquids can flow from a drained area into a drainage collecting system.

gully riser A fitting designed to extend the height of a gully or of a rainwater shoe. It may incorporate branch inlets.

gully, trapped A gully that incorporates a water seal which prevents the transmission of smells and the passage of vermin.

gully, untrapped An uncommon type of gully that does not incorporate a water seal.

gunmetal An alloy of copper and tin to which lead and nickel may also be added. It offers good resistance to wear and corrosion.

gutter A conduit for the collection of rain or storm water.

guttering A conduit system for the collection of water flowing off roofs. It is attached to the timber boards at the edge of the roof and incorporates outlets to rainwater down-pipes. It may be of cast-iron or plastic construction.

guy rope strainer A fitting incorporated with guy ropes to permit tensioning.

guy ropes Multi-strand wires supporting a structure, e.g. a free-standing chimney.

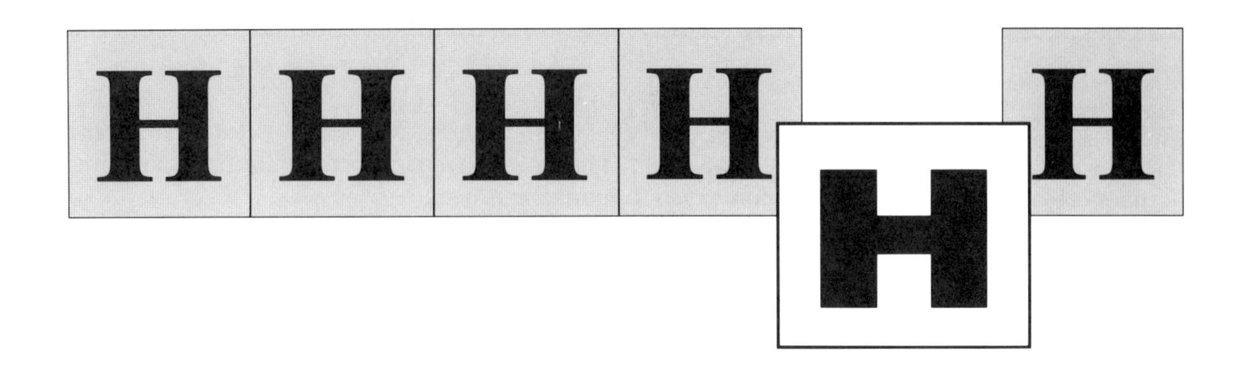

HMIP An abbreviation of Her Majesty's Inspectorate of Pollution. HMIP operates through a number of regionally placed pollution inspectors.

HMIP/NRA Cooperation by means of a Memorandum of Understanding between HMIP and the National Rivers Authority (NRA) which establishes the relationship between these pollution control bodies on the regulation of discharges to controlled waters.

HRC High-rupturing-capacity cartridge fuse in which the fuse element incorporates a centre section with a low melting-point.

halon A Gas used in fire-fighting in selected areas. It is stored under pressure in cylinders and is discharged into the space when the system is actuated by a fire alarm. Halon is one of a family of chlorofluorocarbons which, on discharge into the atmosphere, adversely affects the ozone layer of the Earth. As such, its use is nowadays severely restricted and it is likely to be phased out altogether. Generally, it is not permitted to discharge halon at low level; the discharge must be piped to a roof outlet terminal.

hammer mill [Fig. 123] A mill that reduces material passed through it to a specified size. It incorporates a hammer mechanism and is used in conjunction with waste recycling plants to prepare the waste material for combusion or separation.

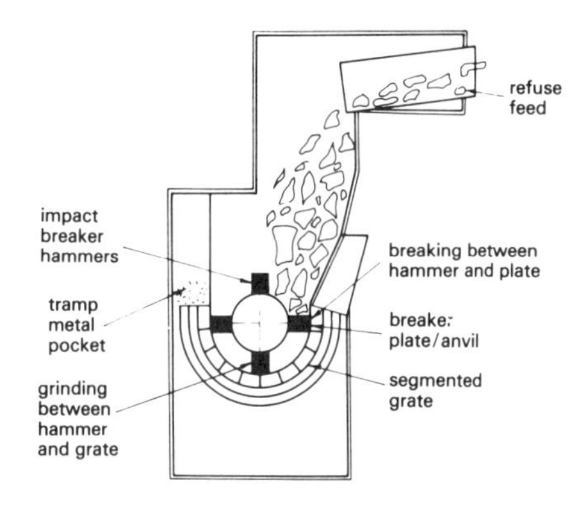

[Fig. 123] Arrangement of a typical hammer mill as used in mechanical sorting plants

hand dryer, automatic A device that provides for hand drying in washrooms, etc. It may be actuated by hands passing near the air outlet or by a push button.

Hand hole A small access cover in the shell or waterway of a boiler or other vessel, comprising cover, watertight seal and tightening device.

hard coal See **coal, hard**.

hardness of water The presence of salts of calcium and magnesium which have the property to destroy soap. Hardness is expressed in terms of temporary or permanent hardness. See **temporary hardness** and **permanent hardness**. A more precise evaluation refers to 'carbonate' and 'non-carbonate' hardness. Numerically, the carbonate hardness is usually identical with the alkalinity of the water. It is expressed as ' parts per 100 000' or as 'parts per million' (ppm).

harmonics (electrical connotation) Distortions of the fundamental waveform which can cause severe and inconvenient disturbance to consumers' electricity supply. The overriding advantage of alternating current is that the voltage can be readily changed and a sine waveform is delivered by the transformer. At present, power generated at 25 kV is transformed to 400 kV for bulk transmissions and is subsequently transformed down, in stages, to the mains voltage delivered to the consumer of electricity without serious loss of purity of waveform.

Thus, a good electrical supply is, among other criteria, one possessing a good sine waveform. There are a number of ways of corrupting the supply by generating harmonics, and modern factory (and some office) systems are increasingly likely to do so. Many factory and process computer systems as well as electronic controls are more sensitive to a corrupted supply and this enhances the importance of maintaining the good sine waveform of the supply to these consumers.

Severe distortion can be caused by the operation of rectifiers, thyristors and rapidly switching apparatus. Severe harmonic disturbances affect the frequency of the electric supply to the consumer and a major reduction, say, from 50 Hz to 35 Hz can upset the rotation of induction motors, the operation of sophisticated computers and the correct reading of wattmeters. Harmonic currents, once generated, can flow throughout the consumer network of electricity, upsetting the operation of electrical devices, often far from the cause of the harmonic disturbance. Their origin(s) may be very difficult to trace and eradicate.

harmonics of a wave motion Waves which are superimposed on a fundamental waveform, having a frequency which is a whole multiple of the fundamental frequency. For example, the second harmonic has a frequency twice that of the fundamental frequency, the third harmonic three times, and so on.

haunching The concrete surround to a drain pipe. It terminates at the top of the pipe.

head A unit of pressure.

header tank A small capacity tank installed above certain equipment to keep it flooded.

head loss Loss of pressure.

head pressure That pressure at which the refrigerant gas is discharged from a refrigeration compressor to the condenser circuit.

head pressure control One of a number of alternative methods of reducing the output capacity of a compressor as sensed by a pressurestat or thermostat.

Health and Safety at Work Act An Act promulgated in 1974, relating to the UK, and intended to provide a broad framework for health and safety measures. It supplements previous related legislation. The Act established a Health and Safety Executive and Commission and introduces Improvement and Prohibition Notices as alternative enforcement methods to prosecutions. Additionally it covers matters relating to the maintenance of the Employment Medical Advisory Services and the promulgation of Building Regulations.

The Act is very widely drawn in respect of the obligations it imposes on everyone concerned with the working environment; this tends towards difficulties as regards detailed interpretation. The Act is likely to be eventually tightened up by the issue of related Regulations and Approved Codes of Practice.

heat balance A balance sheet, expressed graphically or mathematically, of the overall heat input and output of a process or system. It establishes the magnitude of the various specific heat losses as an aid to minimizing them.

heat bank Air-to-air heat exchangers comprising essentially a series of heat pipes formed into a battery, creating a thermal path between two counter-flowing air streams. Such banks are manufactured to handle air at temperatures of up to about 200°C. They are utilized to recover waste heat from air or gases being exhausted to atmosphere from environmental systems and from industrial dryers.

heat bridge A conductor of heat which spans an insulated or cavity space, resulting in the undesired transmission of heat.

heat capacity The quantity of heat of a given body which that body takes in when its temperature is raised through one degree of temperature. See also **water equivalent**.

heat detector A device that is activated by a rise of space temperature to give an audible alarm and/or activate fire-fighting equipment.

heat dissipator A device that discharges excess heat from a process into the atmosphere. It commonly takes the form of a fan-assisted heat exchanger assembly and it is used particularly with incineration with heat recovery installations at times when the amount of generated heat exceeds the heat requirement.

heated ceiling A ceiling that provides radiant heating to the space below by means of embedded or suspended heating panels (coils).

heated towel rail A pipe assembly, usually chromium-plated and sometimes fitted with an integral radiator, for drying towels. It may be heated by circulating hot water or electrically. When heated by hot water, it should preferably be connected to an indirect (primary) circuit, as connection to a direct circuit will give rise to venting problems at the towel rail which will impede its function.

heat exchanger [Fig. 124] A means of transferring the heat from one fluid circulation to another; e.g. from a hot water fed battery to a stream of air passing over it, or from hot water coils within a storage calorifier to the stored body of water, heating it for domestic hot water use, etc.

heater battery, steam See **steam heater battery**.

heater battery, water See **water heater battery**.

[Fig. 124] Heat exchange coil of closed-circuit water cooler prior to installation

heater, boost A heater introduced into a central duct system to boost flagging air temperature at the end of the duct run or to increase temperature locally over that at which the central system is controlled.

heat gain The quantity of heat which is transferred to a space by a variety of sources, e.g. by solar energy passing through windows, or heat generated by machinery and the occupants of a space.

heat gain calculation Estimates and summarizes the calculations of all sources of heat gain in a particular building; e.g. from the building fabric, air changes, occupants, electric lighting, solar gain through windows and roofs or heat from office machinery. The calculation usually follows a codified method and appropriate computer software programs are available.

heating season The period of time, in weeks, during which a space heating system remains in use.

heating stack loss The loss of heat by convection and stratification to the upper (unused) level of a space.

heating surface The heat-emitting surface of a radiator, panel, etc.

heating system, gravity See **gravity heating system.**

heating system, ladder See **ladder heating system.**

heating system, microbore See **microbore heating system.**

heating system, one-pipe See **one-pipe heating system.**

heating system, small-bore See **small-bore heating system.**

heat lag The time that elapses between heat striking a surface and its penetration into the building. For example, solar gain on a concrete roof will take some time to penetrate, but a similar gain on a lightweight roof will penetrate almost immediately.

heat, latent That heat energy added during the evaporation of a liquid at constant temperature.

heat loss calculation Estimates and summarizes the calculations of all sources of heat loss of a particular building; e.g. from the building fabric, air changes, or the manufacturing process. The calculation usually follows a codified method and appropriate computer software programs are available.

heat meter [Figs 125, 126, 127] A meter that measures and records the consumption of heat by a consumer. It may be arranged to read out directly on an integral dial or register or it may transmit the data to a remote heat meter read-out station. The provision of a heat meter establishes a factual basis for charging the consumer for his consumption of heat.

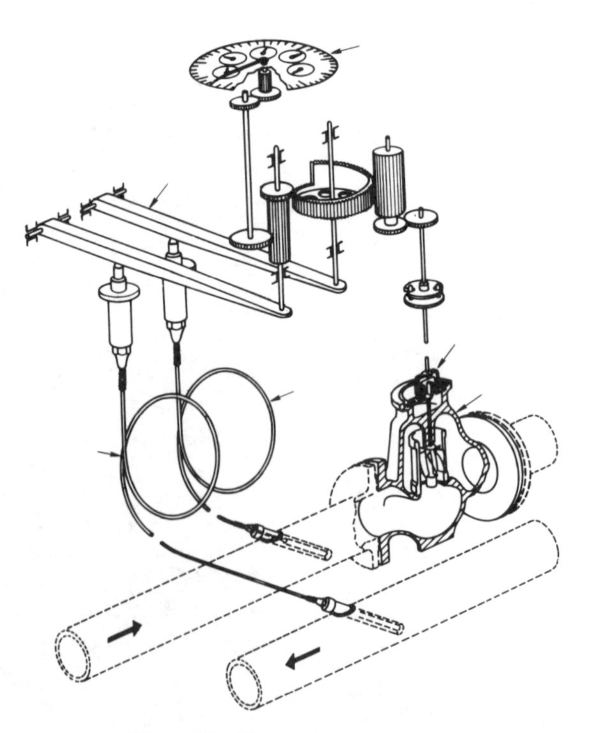

[Fig. 125] Aquametro heat meter

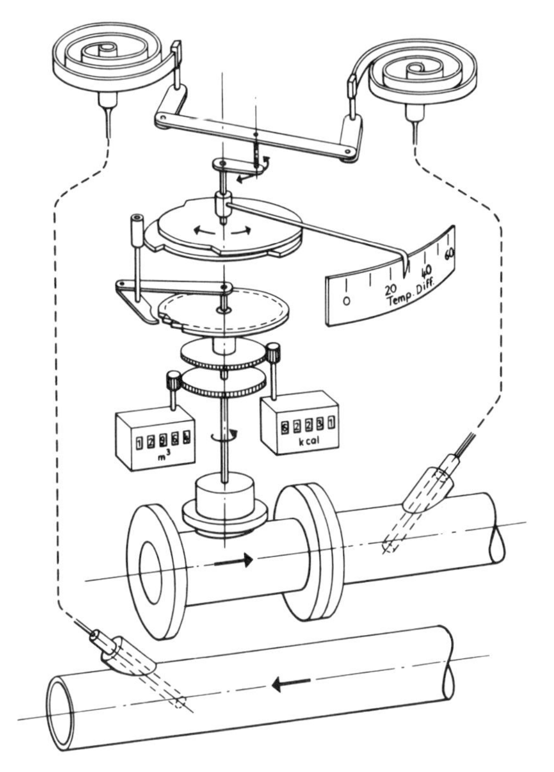

[Fig. 126] Pollux heat meter

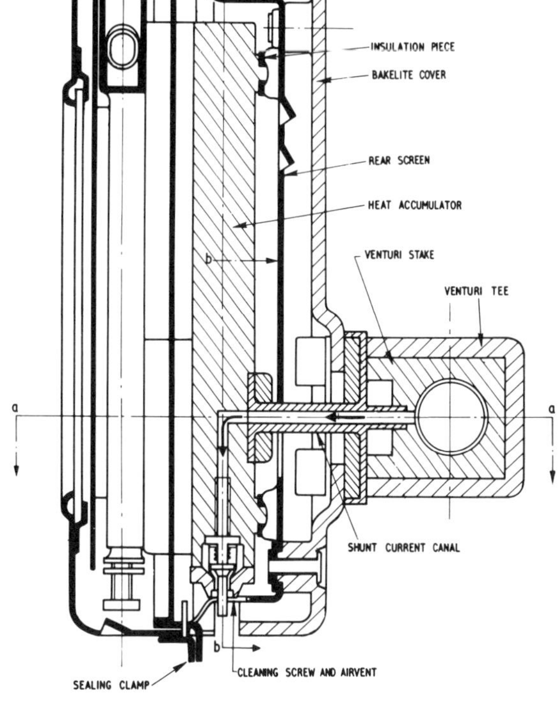

[Fig. 127] Clorius evaporation meter

heat meter, hot water: mechanical A meter that measures the quantity of hot water being supplied to the consumer together with the difference in temperature between the flow and return water flows. The two parameters are integrated and the heat consumption, in appropriate heat units, shows on a read-out panel. The degree of sophistication of the selected heat meter will relate to the metering needs and to the available finance.

heat meter, inferential See **inferential heat meter**.

heat meter, integrating An electronic temperature sensor and flow integrator unit which is utilized in conjunction with a water meter which has been compensated to measure the flow of hot water accurately. An integrating counter displays the consumption in kilowatt hours.

heat of fusion That quantity of heat which must be added or abstracted without change of temperature to effect a change of state from solid to liquid or vice versa.

heat pipe [Figs 128, 129] An finned pipe coil filled with a volatile liquid which possesses a low boiling point and wick. Heat is transferred via latent heat from one end to another by a cycle of evaporation and condensation, giving a vastly greater rate of heat transmission than can be achieved by conduction. It is effective for rapid heat transfer. Without moving parts, it can be built into a modular design with the lower part acting as evaporator and the upper part as condenser and it can be matched readily to specific air flow and temperature. It is used to recover heat from exhaust air and also to transfer cooling loads.

heat pipe system A heat recovery system utilizing heat pipes assembled in arrays to extract heat from exhaust gases and extracted air.

heat pump [Fig. 130] A pump that draws energy from a low-grade heat sink (water, air, sewage,

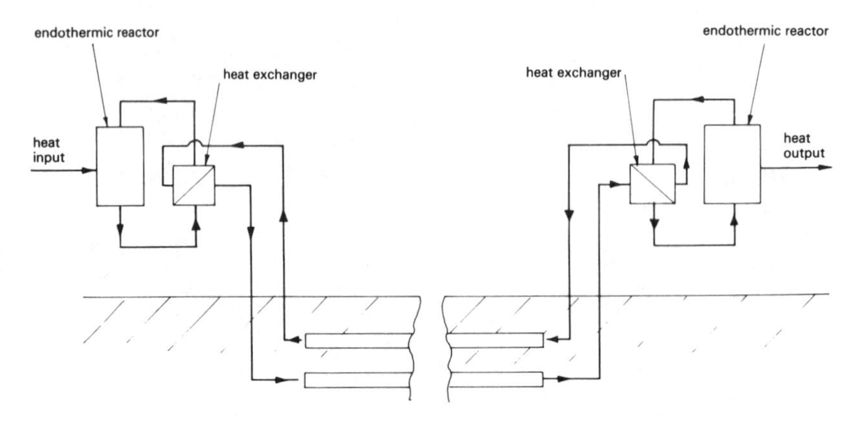

[Fig. 128] Flow sheet of a chemical heat pipe system

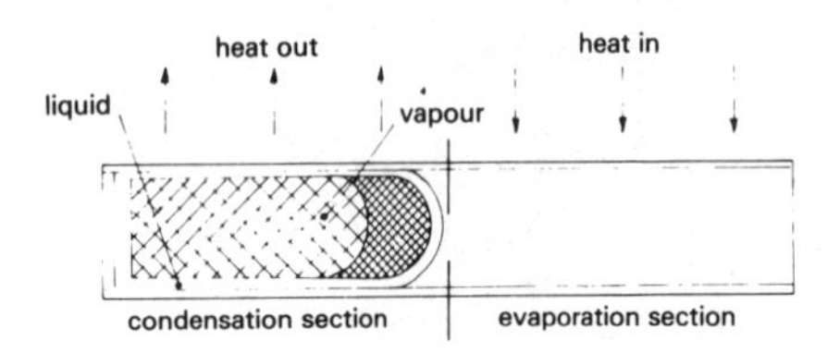

[Fig. 129] Evaporation heat pipe

earth, etc.) and increases the level of input energy to effect an increase in temperature of the working medium. The resultant energy can then be applied to heating or cooling purposes. Working fluid transfers heat by a cycle of evaporation and condensation. A heat pump may be driven by an electric motor, diesel engine or gas engine.

heat pump, coefficient of performance The multiple of the energy output related to the energy input at the heat pump driving mechanism is termed the coefficient of performance (COP) of that heat pump. It lies between 2.5 and 6, depending on the available heat sink, the heat pump arrangement and the operative temperature rise being achieved. A heat pump is at its most efficient (with the highest COP) when functioning to achieve a small temperature rise; e.g. the heating of swimming pools, underfloor heating coils and similar applications. A heat pump will generally function at its best when using an electric drive and competing with electric heating. Substituting commercial gas or biogas as the driving medium will improve heat pump viability.

heat pump dehumidifier A self-contained unit with an integral air moving system using a refrigeration cycle to remove moisture from air by cooling it below its dewpoint. Power supply and drainage connections must be provided.

heat pump, ground coil The heat source is the soil (ground). It is exploited by a pipe coil which is buried below ground. Approximately 30 m^2 of pipe surface is required for an output of 1000 kcal/h (1.16 kW). The temperature and temperature fluctuation of the coil is 5–15°C (23–59°F).

Limitations: suitable geological conditions are required for the pipe installation (i.e. rocky ground is unsuitable). It must be buried at a sufficient depth to safeguard against the coil freezing. When the coil serves a large installed load in continuous use, there is a risk of establishing an area of permafrost radiating outwards from the coil. The installation costs may be high.

heat pump, mode of operation A heat pump may operate in one of the following modes:

(a) air to air
(b) air to water
(c) liquid to air
(d) liquid to water
(e) earth to air
(f) earth to water.

In these modes the heat recovery exchanger is located respectively in the air, in liquid or in the earth.

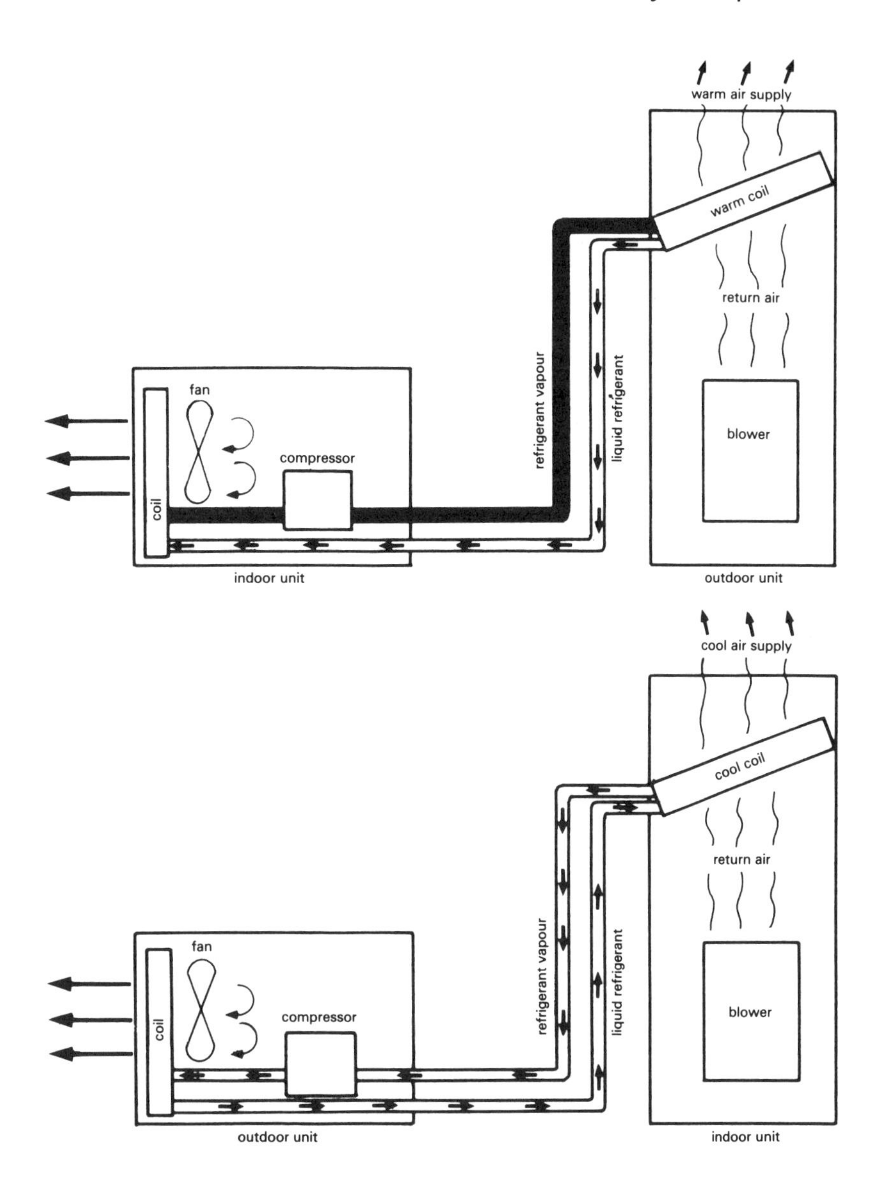

[Fig. 130] Illustration of heat pump operation

heat pump, operation Switched into the heating mode, the useful heat is extracted at the condenser. When operating for both heating and cooling outputs, cooling is obtained in addition at the evaporator and change-over valves are incorporated with the equipment. Heat pumps installed in colder climates must incorporate a means of defrosting or have their operation limited to those times when there is no risk of frost damage to the equipment.

heat reclaim The process of recovering heat which would otherwise be wasted.

heat reclaim coil A heat exchanger which reclaims heat. For example, it may be located within an exhaust duct or in circulating water to reclaim heat within a fresh air supply, prewarming it off waste heat.

heat reclaim system Any system which recovers heat that would otherwise have been discharged to atmosphere or into a drain.

heat recovery coil, process See **process heat recovery coil.**

heat recovery coil, run-around See **run-around heat recovery coil.**

heat recovery coil, wrap around See wrap-**around heat recovery coil.**

heat recovery wheel See **heat wheel.**

heat recuperator, plate type An air-to-air heat exchanger available for counterflow or crossflow operation, housed within a galvanized or stainless steel casing with removable covers for inspection and cleaning of the heat exchanger plates. It is used with ventilation and air-conditioning systems which handle large volumes of fresh air and for process plant.

heat release rate The quantity of heat liberated in a combustion chamber.

heat service contract An arrangement under which the heat or cooling requirements of a building are met by a contractor who is obliged to maintain an effective heat/cooling service under certain conditions based on short or long term obligations. A heat service contract commonly includes routine maintenance of plant on a replacement basis and usually includes warranties of minimum thermal efficiency to be maintained and a clause to compensate the contractor for the effects of inflation.

heat sink A facility into which surplus or waste heat may be rejected; e.g. atmospheric air, water or soil.

heat, total Relates to the heat content of steam or other vapour/gas; e.g. applied to steam of a particular pressure and temperature condition, it includes the sensible (heat in the water), latent (heat of steam formation) and the superheat (if any).

heat transfer fluid A specially formulated fluid with specific improved heat transfer properties and a high boiling-point which permits the achievement of high operating temperatures without accompanying high pressure, e.g. a range of tetrasilicates. It is suitable for high temperature hot water heating systems in which high temperatures can be achieved by atmospheric pressure.

heat trap A confined space subject to heat input via solar energy, hot pipes, warm chimney surfaces, etc., which is not ventilated to relieve the consequent temperature build-up.

heat wheel [Fig. 131] A rotary air-to-air heat exchanger which recovers sensible and latent heat. It is usually installed between the air supply and air exhaust of a ducted heating, ventilation or air-conditioning system with the object of recovering up to 90% of the total enthalpy (energy) from the exhaust stream before discharging it to atmosphere, transferring the recovered heat to raising the temperature of the air input stream.

heat wheel, applications Heat wheels are suitable for the recovery of heat from the air exhaust wherever large volumes of warm air must be discharged to atmosphere in proximity to the supply of unheated incoming air. Typical applications are indoor swimming pools, conference and exhibition halls, etc.

heat wheel, construction A typical heat wheel incorporates an energy exchange matrix with a permanent lightweight inert transfer medium. It operates with the moisture content of the air remaining in the vapour state, thereby keeping the matrix dry and free from the risk of bacteria and algae growth. The assembly is enclosed within a substantial steel casing which incorporates flanged connections to the supply and exhaust ducting. The equipment requires much space for plant location and needs to be placed within close proximity to the incoming and outgoing air streams.

heat wheel, operation and control The energy emission of the heat wheel can be continuously matched to the requirements of the connected air system by manipulating proportional controls to vary the speed of the wheel to achieve the required balance. The maximum rotational speed of the heat wheel is 10 r.p.m., which minimizes power demand and prolongs the life expectancy of the equipment.

height factor A compensatory allowance made in heat loss calculations for the extra loss occasioned

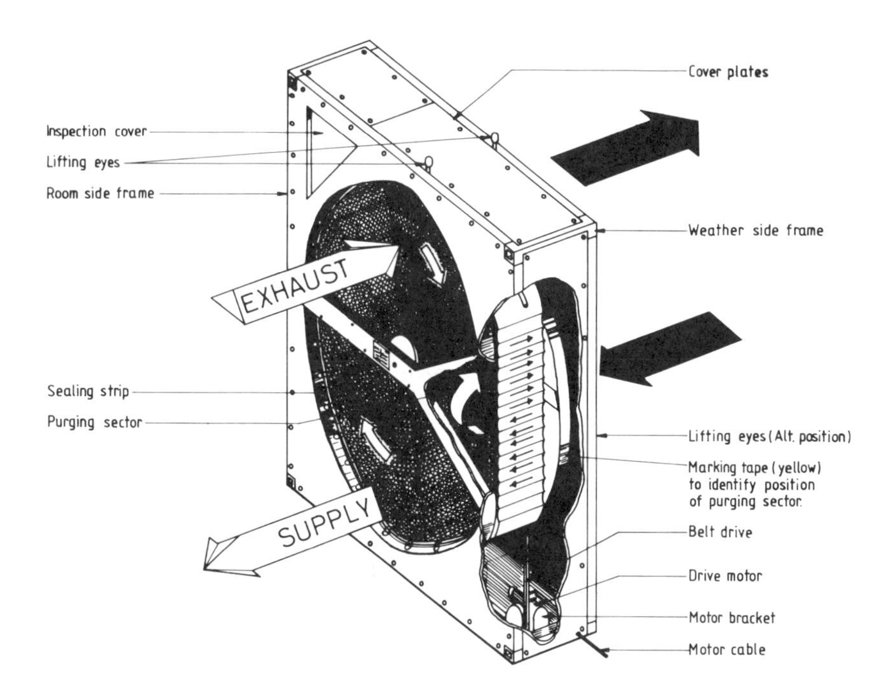

[Fig. 131] Heat recovery wheel

in tall spaces through stratification of the air temperature.

hermetic compressor A compressor manufactured with both the compressor and the motor sealed inside a casing to obviate the possibility of refrigerant loss from leaking around the shaft. The motor is cooled by refrigerant.

herring-bone system A main field drain into which smaller branches laid parallel to each other, at an angle to the main, discharge from both sides. A number of such herring-bone units can discharge into a collecting drain.

hertz (Hz) A unit of frequency equivalent to one cycle per second.

heterogeneous Describes substances which are not of uniform composition and exhibit different properties in different portions.

heuristic Describes the method of solving mathematical problems for which an algorithm does not exist. This involves the narrowing down of the field of search for a solution by inductive reasoning from the past experience of similar problems.

high alumina bricks Refractories used in applications involving high furnace temperatures. The alumina content may vary from 50% to 99%. The refractory behaves like a pure compound. High alumina refractory bricks undergo slow plastic deformation under conditions of pressure at high temperature.

high-pressure compressor cut-out A cut-out that protects a refrigeration circuit against overloads (e.g. failure of air flow).

high-pressure hot water system A system that functions at temperatures above the boiling-point of water by the imposition of an artificial pressure head. It is an alternative to a steam heating system. Pressurization is by a raised static head vessel or by an imposed nitrogen or air cushion.

high-pressure mercury lamp A lamp in which high-pressure mercury vapour operates in a quartz arc tube. The internal surface of the outer elliptical bulb is coated with a phosphor which converts the ultraviolet radiation from the discharge into light. It has a poor colour rendering and is used for street lighting (to a limited extent) and in dusty industrial premises, such as foundries.

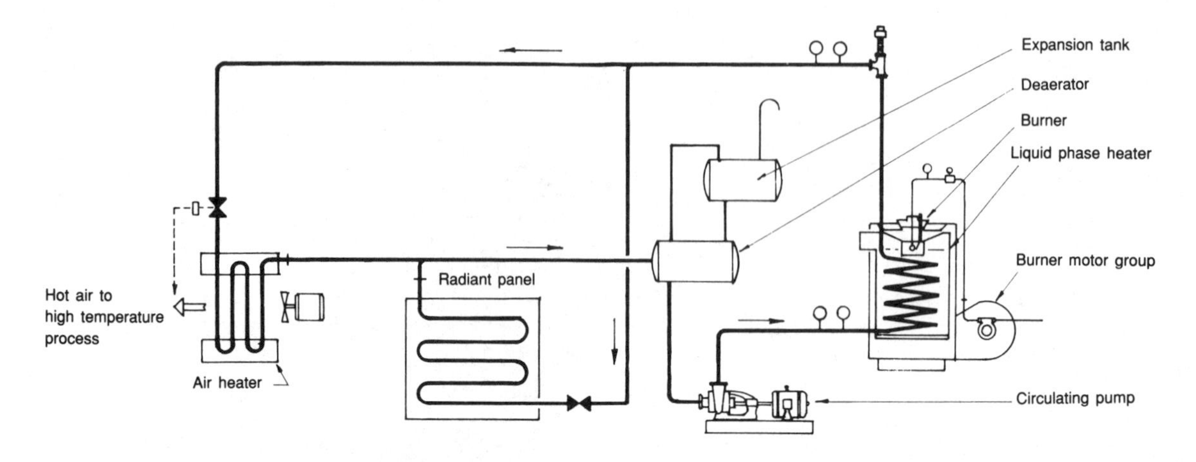

[Fig. 132] Diagrammatic arrangement of high-temperature heat transfer application

high-temperature liquid heating system [Fig. 132] A system that uses specially formulated heat transfer fluids to achieve heat exchange at high temperatures without accompanying high pressures (such as is the case with steam). It is suitable for process heating applications and also for factory space heating where this is appropriate.

high-temperature water system See **high-pressure hot water system.**

high-velocity air-conditioning system An induction air-conditioning system in which the air supply to the induction units is delivered at high velocity for discharge at the supply nozzles.

high voltage (hv) Voltage usually exceeding 650 V.

histogram A form of graphical representation which is used in statistics in which the frequency distributions are illustrated by rectangles. It presents data along *X* and *Y* axes using vertical bars. It may illustrate, for example, the relationship between measurements (along the *Y* axis) and the measured parameter, such as air velocity and turbulence intensity (along the *X* axis).

historic cost The cost of goods or services at the actual time of their purchase.

hogged material Material that has been reduced in size by passing it through a hogger. The material is commonly of chip size suitable for conveying via a wood-waste collection system.

hogger A device fitted with chipper knives used to reduce chunks of wood to the size of chips which can be conveyed via a wood-waste collection system into silo storage or into a furnace for combustion. Hoggers are commonly associated with wood-waste incinerators.

holiday A flaw, e.g. a hole, in the continuity of a protective pipe or duct coating.

holiday detector An electrical instrument which detects and locates a breakdown in the continuity of a protective coating.

hollow jet valve A valve that discharges small quantities of fluid at high pressure, the discharge being in the form of a hollow jet. It is similar to a needle valve, but suitable for operation at higher pressures.

homogeneous Describes substances which are of uniform composition throughout.

hooped ladder A steel access ladder provided with a series of hoops to prevent the user from toppling off the ladder backwards.

horsepower That power which raises 75 kg against the force of gravity through a height of 1 m. It is equivalent to 735.5 watts (J/s).

hose reel, automatic A hose reel that is fitted with a type of valve which is opened or closed by the revolving action of the reel when the hose is pulled out or rewound. It is only necessary to pull out the hose and open the shut-off nozzle to initiate water discharge.

hose reel, fixed type A hose reel that is mounted usually not less than 1.5 m above floor level. The hose can be rolled out only in one direction.

hose reel, hydraulic A hose reel that comprises a steel drum or reel on which is rolled a length of 20 mm or 25 mm reinforced non-kink rubber hose which terminates with a 5 mm lever-operated shut-off nozzle. The inlet is fitted with a stopcock and is connected to water supply of a suitable pressure. The length of hose is usually between 23 and 45 m. It provides a first aid means of fighting a localized fire.

hose reel, manual A hose reel on which a stopcock must be opened to activate water discharge from the appliance and closed to shut down the water flow.

hose reel, swinging-type See **swinging-type hose reel**.

hose union tap A controlled outflow device which has an integral connection for a garden or garage hose.

hospital radiator A column type of radiator with smooth rounded surfaces, particularly suited for use in areas requiring optimum cleanliness; e.g. in hospital wards. It is commonly of cast-iron. The heat output per unit length is less than that of an equivalent standard column radiator.

hot air disperser A thermostatically controlled fan which automatically switches on when the control senses a build-up of heat in the roof void at a predetermined temperature. The arrangement has an effective throw to drive the warm air downwards into the occupied area and it is intended for tall factory spaces with poor roof insulation.

hot water circulation, primary Water circulating within a closed heating circuit which does not come into contact with the hot water being drawn off for use. See also **closed system**.

hot water circulation, secondary See **secondary hot water circulation**.

hot water generator A boiler which heats hot water. The name is applied to a particular range of rapid heat-up hot water boilers which operate without intermediate storage of hot water.

hot water meter A temperature compensated meter that permits the monitoring and recording of hot water flows in domestic hot water and central heating installations.

hot water system, unvented See **unvented hot water system**.

hot well A tank which stores the feed water of a steam boiler. It may incorporate a blowdown heat recovery coil.

housekeeping, good The care of the plant in use. It should include preventive programmed maintenance.

humidifier fever An allergy caused by exposure to contaminated water from recirculating spray air washers. A potential risk is offered by spray-type humidifiers using recirculated water and by portable room-located aerosol spray humidifiers. The causes are likely to be contamination of the recirculated water due to inadequate cleaning of all parts of the humidifier(s), infrequent changes of water after the humidifier has been standing idle for a prolonged period, etc. Humidification by steam generating units is unlikely to promote such an allergy.

humidifier, mechanical See **mechanical humidifier**.

humidifier, spinning disc See **spinning disc humidifier**.

humidifier, steam A humidifier that evaporates water into the air or airstream by the action of a steam coil immersed inside an associated vessel. The steam is ejected via nozzles. This type of humidifier is preferred for central air-conditioning as it obviates the risk of generating legionnaire's disease which is more likely when using recirculating air washers. See also **steam humidifier**.

humidistat A moisture detector arranged to activate humidifier or dehumidifier controls.

humidity The presence of moisture in air.

humidity, absolute The weight of water vapour present in a unit volume of moist air expressed in g/m^3.

humidity, control The control of humidity in ventilation and air-conditioning systems is usually achieved by means of a hygroscopic detector activating a moisture generating device or damper system.

humidity control, dehumidification This is achieved by abstracting moisture from the air. Usually a moisture detector activates the cooling coil controls to reduce the air temperature below its dewpoint and shed moisture, or by passing the air over a bed of chemical adsorbents which dry the air.

humidity control, humidification This is achieved by adding controlled quantities of moisture into the air. Usually a moisture detector activates a humidification device, such as a steam generator, steam supply, air washer, spinning discs, etc.

humidity measurement Humidity is commonly measured by a wet and dry bulb whirling hygrometer in conjunction with hygrometric tables. More sophisticated instruments are also available.

humidity ratio This is defined as specific humidity.

humidity, relative Relative humidity is given by the ratio

$$\frac{\text{Actual vapour pressure of the air at a given dry bulb temperature}}{\text{Saturation vapour pressure of the air at the same dry bulb temperature.}}$$

This ratio is generally expressed as a percentage. Subjective perceptions regarding the 'feel' of ambient air depend largely on the relative humidity.

humidity, specific See **specific humidity**.

humidity, subjective perception Perceptions regarding the 'feel' of ambient air depend mainly on the relative humidity of the air.

hybrid control system A system that operates with a mix of pneumatic and electric/electronic controls.

hydrant A stand-pipe with a control valve connected to the water main. These are located at strategic points mainly for the use of the fire brigade for fire-fighting.

hydraulic gradient The loss of head in a liquid flowing through a pipe or channel expressed per unit length of the pipe or channel. In a channel it equals the slope of the free surface of the flowing liquid.

hydraulic hose reel A rubber hose wound about a drum connected via a valve to the mains water supply, ready to be run out to fight a local fire.

hydraulic test The application of a known water pressure to a vessel or installation storing or conveying a fluid to establish whether there are leaks at the applied pressure.

hydraulic test pump A portable device which incorporates a pump mechanism, gauge glass, operating lever and flexible pipe connections for attachment to pressure test points. For boiler testing, water treatment and dosing, power pumps are commonly used.

hydraulic valve A valve that controls either water, oil or a hydraulic system.

hydrocarbon Any compound of hydrogen and carbon. All fossil fuels are hydrocarbon compounds but commonly associated with some impurities.

Hydro-Electric One of two electricity supply companies in Scotland.

hydrogen sulphide A colourless gas with the characteristic odour of bad eggs. It arises in the decomposition of organic materials. It is corrosive to many metals and can sometimes be detected in corroding heating systems by applying a match and setting alight the gas which issues from the radiator air cock. It can be eliminated from such systems by the use of corrosion inhibitors and the avoidance of over-pumping.

hydrometer An instrument for measuring the density or specific gravity of liquids. It commonly comprises a weighted bulb attached to a slender graduated stem. The instrument floats vertically in the liquid being tested. A greater length of stem is exposed in liquids of high density than in those of lower density.

hydrostatic contents gauge A remote reading gauge for indicating the content of a liquid storage tank. The static pressure of the liquid acts on a diaphragm or balance chamber installed within the vessel and indicates the contents on the remotely installed gauge. It may be pump operated, hydrostatic or electrically actuated.

hygrometric tables A tabulation of the properties of air; usually for a particular dry bulb temperature, relative and absolute humidity, vapour pressure, enthalpy, volume and wet bulb temperature.

hygroscopic materials Materials that readily absorb or discard moisture from the surrounding atmosphere and therefore must be manipulated in an atmosphere of controlled relative humidity. Such materials include tobacco, rayons, textiles, photographics, flour, etc.

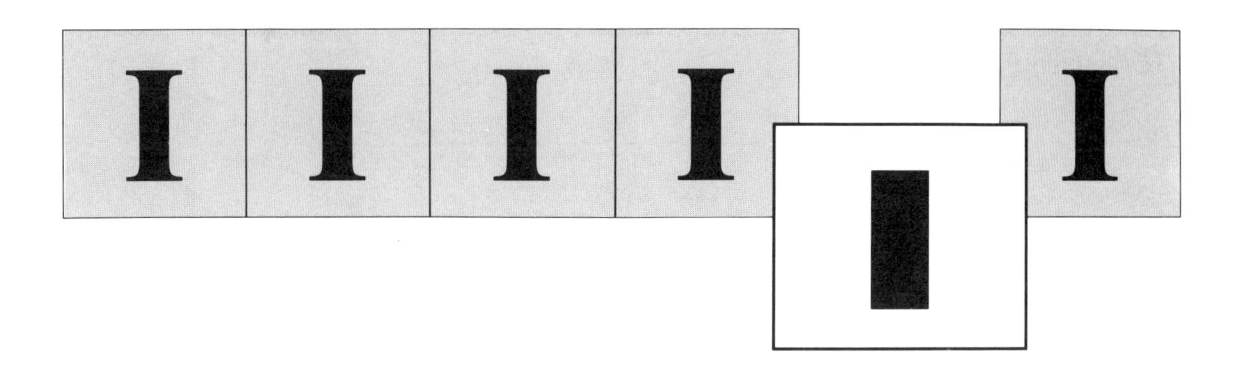

IEA The Institute of Environmental Assessment, established in 1990 as a non-profit-making organization for the purpose of establishing and maintaining high standards of environment assessment and to provide guidance and information for environmental consultancies and their clients. It audits the quality of environmental statements, maintains a register of environmental consultants, acts as an independent source of environmental advice and promotes best practice environmental techniques.

IEE The Institution of Electrical Engineers, a professional body for electrical engineers which promotes the technology of electrical engineering by means of publications, lectures and seminars, maintains a comprehensive library of books concerned with electrical engineering, and publishes and periodically revises the widely used and quoted Electrical Wiring Regulations for Buildings (the 'IEE Regulations'). The IEE possesses a Royal Charter and is authorized by the Engineering Council to confer the title of Chartered Engineer on suitably qualified members.

IEE Regulations A book of regulations issued, from time to time, by the Institution of Electrical Engineers which lays down guidelines and instructions for the safe execution of electrical installations. The current (1992) Regulations are the sixteenth edition. Specifications for electrical work commonly require compliance with the current IEE Regulations.

IEE Wiring Regulations Regulations that relate principally to the design, selection, erection, inspection and testing of electrical installations in and about buildings, published by the Institution of Electrical Engineers.

IMechE The Chartered Institution of Mechanical Engineers, a professional body for mechanical engineers, which promotes technology by means of publications, lectures and seminars, maintains a comprehensive library of books concerned with mechanical engineering and publishes books on this subject. It also publishes Conditions of Contract for the orderly documentation of contracts. The Institution possesses a Royal Charter and is authorized by the Engineering Council to confer the title of Chartered Engineer on suitably qualified members.

IPC Integrated pollution control.

ITOC station [Fig. 133] Intermediate take-off condensing electric power station suitable for combined heat and power schemes. Heat is obtained as a by-product of electric power generation. For example, a reduction of one GJ in electricity output

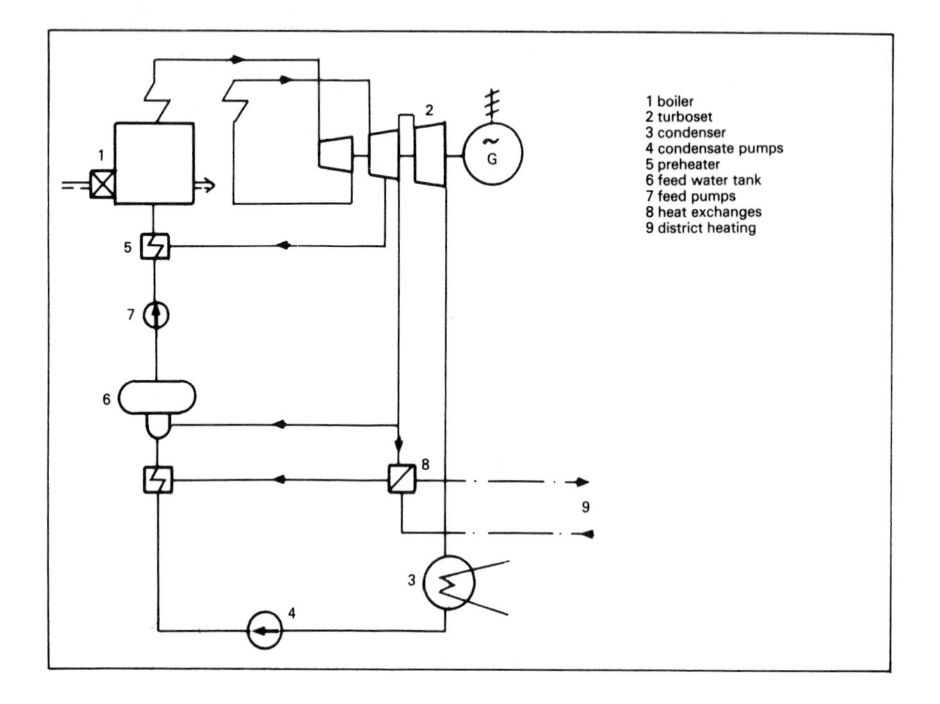

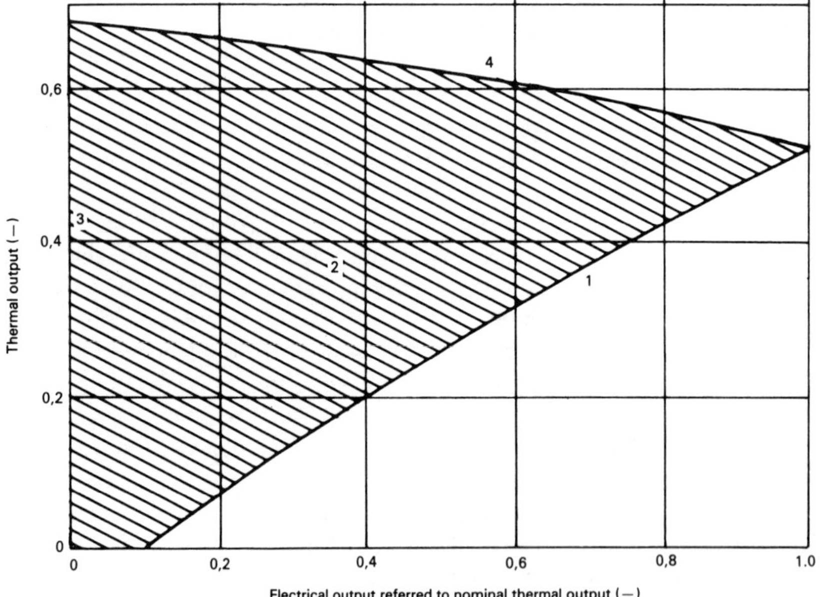

[Fig. 133] Sulzer fossil-fuelled ITOC plant

produces 5–6 GJ of heat (a heat generating efficiency of 500–600%).

ice banks Ice banks form the heart of a thermal storage cooling system. They comprise partitions which are suspended within the storage tank. The ice forms on these as refrigeration is applied. They are a means of storing a cooling effect by the phase change method. Ice is formed at times of cheap off-peak electricity and is melted at times of cooling demand. Since ice banks are required to provide a high output even when nearing depletion, combined with the relatively low flow temperature from the ice store, the ice water tanks are fitted with a means of compressed air injection. Agitation of the water content at partial storage conditions increases the release rate and resolves a potential problem with such systems.

ice harvesting The automatic removal by bursts of hot refrigerant gas of the sheet of ice formed on the evaporator plates of ice storage systems and the dumping of the dislodged ice into the ice storage tank below.

ice storage for air-conditioning See also **eutectic salts air-conditioning systems**. A 'full' storage system stores the entire on-peak cooling requirement during off-peak hours, which shifts all refrigeration energy requirements to off-peak hours (at a reduced electricity tariff) and materially reduces the required size/capacity of the refrigeration system, though at the expense of storage space for the ice store. A load-levelling 'partial' storage distributes the electrical requirements for cooling over a 24-hour period, reducing (but not eliminating) the on-peak demand.

identibands See **pipeline identification**.

idling The condition of low heat demand from a solid fuel fired boiler when the fuel bed must be maintained alight at the minimum level of output to prevent its extinction, pending renewal of heat demand. The term also applies to other situations and equipment which operate under 'no load' conditions.

ignition The commencement of combustion, effected by raising the temperature of a fuel to a point at which the rate of burning supplies the heat essential to sustain the process of fuel combustion.

ignition electrode An electric device within a fuel burner assembly which issues sparks to ignite the fuel.

ignition temperature That temperature at which combustion commences and below which burning is not possible. Some examples of ignition temperatures are: gas oil, 340°C; anthracite, 500°C; bituminous coal, 400–425°C; carbon monoxide, 650°C.

illuminance The quantity of illumination at a surface. Illuminance (E) at a point on a surface is defined as the luminous flux (F) incident upon a small area (A) of the element:

$$E = \frac{F}{A}$$

The basic unit of illuminance is the lumen per m^2 or lux.

immersion heater, conversion An assembly of heat transfer tubes which can be fitted into or to a liquid storage vessel to convert it to indirect heating via a primary heating medium.

immersion heater, electrical An assembly of electric heater elements suitably constructed for immersion in a specified liquid. It is wired for connection to the electricity supply and also commonly to a controlling stem of a thermostat. The latter may incorporate a fixed or variable temperature setting.

immersion heater, non-electric See **submerged combustion and calorifier bundle**.

immersion thermostat [Fig. 134] A temperature-detecting device that is immersed within a liquid or gas for the purpose of actuating a valve, damper or similar device to maintain the temperature of the fluid or gas at a set level.

impedance The total virtual resistance of an electric circuit or system to the flow of alternating

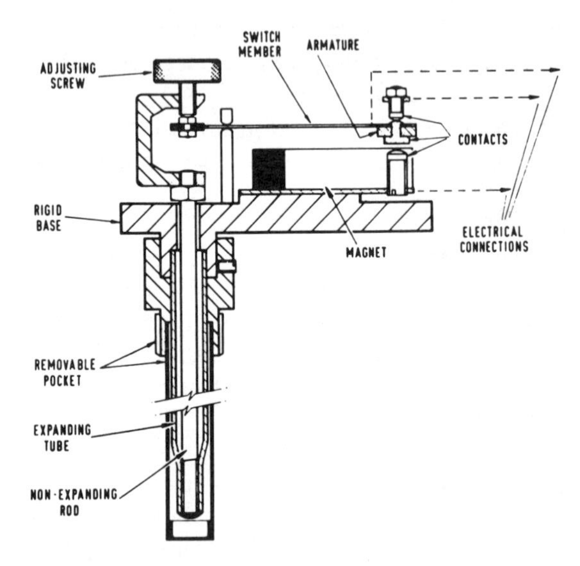

[Fig. 134] Electric immersion thermostat

current. It arises from the resistance and reactance of the conductor(s).

impermeable Describes a material or surface resistant to penetration by, for example, moisture.

impermeable area For the purpose of calculating the anticipated flow of rainwater into a surface water drain, it is that surface area which includes only the paved surfaces within the catchment zone. It is assumed that the rainfall from unpaved areas will disperse into the surrounding ground.

implosion The inward collapse of an evacuated vessel. This may happen to water storage vessels in which a vacuum is accidentally created, leading to damaging implosion.

incinerator [Fig. 135] A furnace designed to combust waste material. It must conform to pollution control standards, legislation and codes of practice and may function with or without recovery of heat.

incinerator, warm air heater A heat exchanger incorporated into the incinerator process for the indirect heating of air for environmental or industrial purposes.

incinerator, waste heat boiler A boiler incorporated into the incineration process and utilizing the hot flue gas discharge to generate steam or hot water. The difficulty in such a process is the balancing of the heat input via incineration with the available take-up of steam or hot water. At times

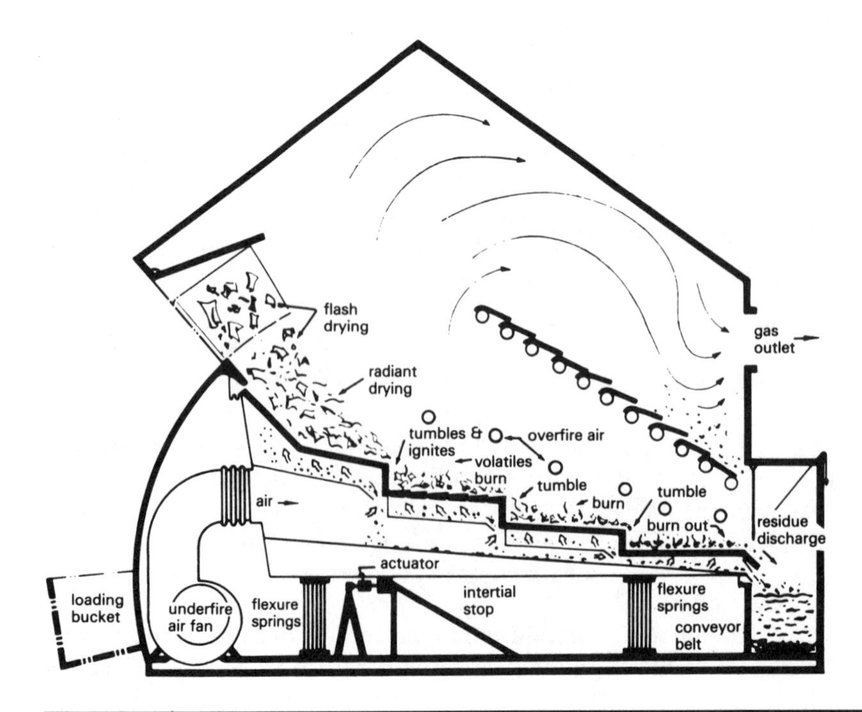

[Fig. 135] A typical fixed-grate refuse incinerator

of excess heat input, that excess must be dumped to waste or be stored as hot water in a heat accumulator. Because of the nature of the waste material being incinerated, the flue gas tends to be dirt-laden, calling for extra large bore boiler flue tubes.

inclined manometer See **manometer, inclined**.

incomplete combustion see **combustion, incomplete**.

index duct That run of ducting between the fan and the final terminal fitting within a ducted air system which offers the greatest resistance to air flow and which determines the resistance the fan must overcome (plus any resistance on the suction side of the fan).

index pipe circuit That pipe circuit within an installation which offers the greatest resistance to the flow of water.

indirect gas-fired heater A heater in which the heat exchanger is directly flued to the outside; i.e. exhaust gas does not enter the heated space.

indirect steam heating Heating that applies the heat to a liquid supplied via an immersed heating coil (serpentine or U-shaped) or radiator which is fitted with a steam trapping facility and thermostatic valve control.

Application: for permanent liquid heating such a system is preferable to direct steam injection; it provides improved steam utilization and does not dilute the solution.

See also **steam injection**.

inductance The property of a circuit by virtue of which a change of current in that circuit produces a change in the flux linkage.

induction air-conditioner A unit depending on the induction effect to entrain recirculating room air and pass it over a chilled water or a hot water heater battery. The primary air is supplied to nozzles by the central high-velocity fan plant.

induction air-conditioning A system that utilizes a centralized heating and chilling plant which provides hot water or chilled water to individual induction units which are located on the floor or within a suspended ceiling of the air-conditioned space. The primary air is supplied into the plenum box of each induction unit at high velocity. This discharge entrains (induces) secondary air circulation within the air-conditioned space. The air supply from the central plant is heated to a controlled programme and cooling is by chilled water pumped through the heat exchanger of the induction unit. The overall thermostatic space control is by room thermostat and heating programme. The high-velocity operation of conveying ducts and discharge from nozzles is susceptible to noise generation. Careful duct design and sound attenuation is required for ducts, fittings and induction units if complaints are to be avoided.

induction air-conditioning, changeover system Induction air-conditioning functions with a two-pipe circulation carrying either chilled water or hot water. At some stage during the year, the system is changed over from summer operation with chilled water to winter operation with hot water. Changeover can also be effected at other times.

induction effect Related to air-conditioning, the induced air flow by high-velocity discharge from air supply nozzles.

induction motor An alternating current motor in which the stator only is connected to the electric supply main in such a way as to generate a rotating field.

inductive circuit An electric circuit which exhibits self-inductance.

inert gases Gases which do not support combustion and which are non-reactive; e.g. nitrogen and carbon dioxide.

inert waste Waste that is not biodegradable.

inferential heat meter A meter that utilizes the secondary characteristic of heat flow from which it derives an estimate of total energy consumption. For example, the Minol meter measures the temperature differential between a radiator and the room air in which it is located, integrates that differential with time and then scales the result based on the specific particulars of the radiator to indicate the output in appropriate energy units.

infiltration See **air infiltration**.

infra-red radiation Electromagnetic radiation whose wavelength lies between microwaves and visible light.

infra-red scanner A hand-held electrical instrument which offers a number of applications to the heating and ventilating industry, including the detection of hot spots in electrical equipment which could indicate faulty connections, the identification of (thermal) insulation deficiencies, or the location of pipes and ducts buried in walls and floors.

inhibitor A chemically formulated substance or fluid which is added to the water of a heating or cooling system to check or minimize the effect of undesirable properties, e.g. scale or algae formation. It may also be added to fuel oil to inhibit undesirable properties which would adversely affect combustion or plant.

inorganic Refers to substances which are of mineral origin and do not belong to the large group of carbon compounds which are termed organic.

insertion loss The reduction of noise level achieved through the introduction of a noise control device.

in situ drain liners One of a number of available methods is Permaline which installs a new hard structural liner within an existing drain. A flexible felt liner is saturated in a resin and delivered to the site in a refrigerated vehicle, the sock-like felt liner is fed through the drain by filling it with water and the water in the lining sock is heated for about an hour, thereby curing the resin. When the curing is complete, the two ends are cut to fit the manholes.

insolation Solar energy received at a particular location on the Earth's surface.

inspection chamber or manhole, connection to [Figs 136, 137] The angle of entry of a branch pipe connection should not be greater than 90 degrees at the internal face of the inspection chamber or manhole. Wherever practicable, such an entry should be installed at half way up the main channel to provide a cascade entry and the connection should be so shaped that it permits the discharge to take place with minimum turbulence and without causing backing up into other connections.

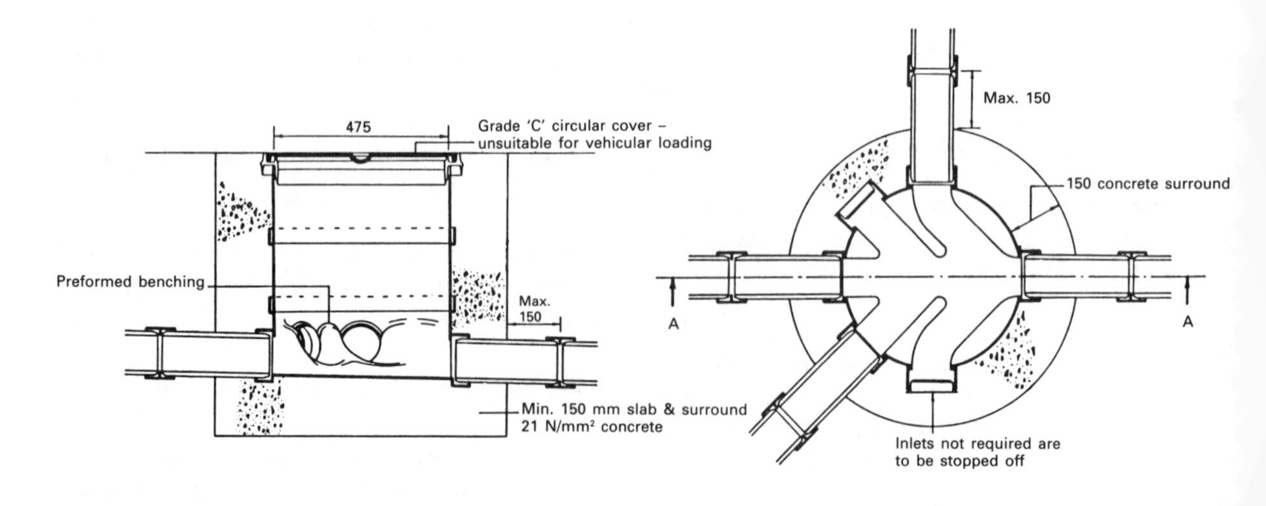

[Fig. 136] Polypropylene inspection chamber

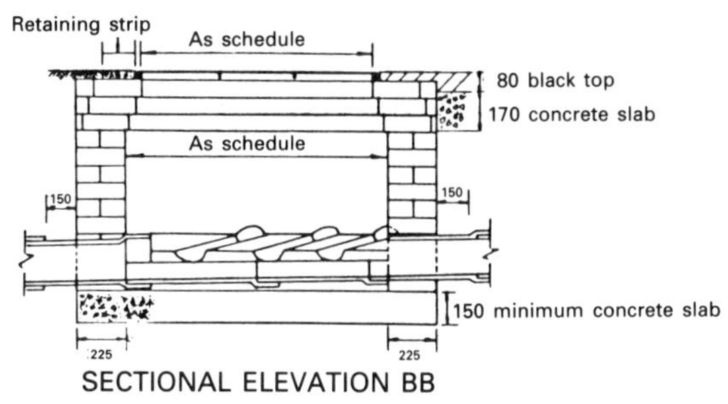

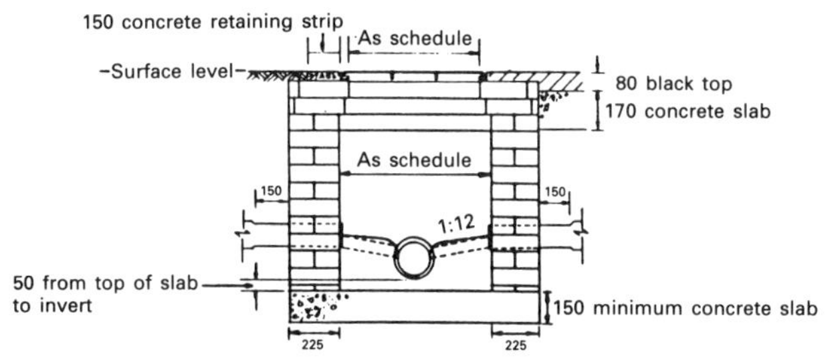

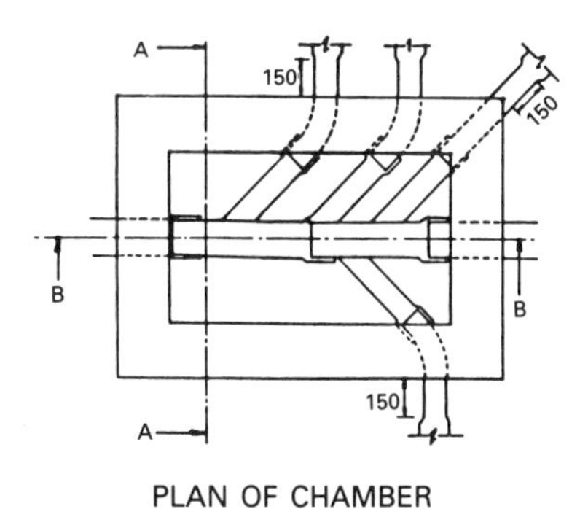

[Fig. 137] Inspection chamber

instantaneous water heater A heater that heats the domestic water as it passes through integral heating coils directly to the draw-off point. It may be electrically heated or gas-fired and the latter must include an adequate and safe flueing arrangement.

insulating firebrick A fireclay product of high porosity, formulated to withstand high temperatures with attendant good heat insulating value. It is used as an outer shield of a refractory system, such as an incinerator.

insulating materials, chemical resistance Materials which are acidic (pH below 7) should not be used in alkaline environments and materials which are alkaline (pH above 7) should not come into contact with acidic materials or metals such as aluminium. Many organic materials have poor solvent resistance. Some insulation materials (e.g. mineral wool) induce corrosion in steel; others may cause stress corrosion effects.

insulating materials, fire resistance Insulating materials must meet the following basic criteria for reasons of safety.

(a) Surfaces must be inorganic to confer fire resistance.
(b) Toxic fumes must not be given off by the material when subject to a rise in temperature or fire.
(c) The material must remain in place as protection when exposed to a fire for a certain specified minimum period.
(d) If flammable materials are used for insulation, these must be adequately protected by surface insulation or cladding using only approved fire resistant materials which do not generate toxic fumes when exposed to heat.

insulating materials, hygroscopic Such materials attract water vapour. The condensed water then forms either chemical compounds or associates with the material in some way on its surface. In either case, the thermal conductivity of the wetted material rises dramatically (with a corresponding fall in its insulation property) and there also arises the danger of corrosion upon the underlying metal.

Hygroscopy occurs when the water vapour pressure inside the insulation is lower than that of the surrounding air. If hygroscopic materials must be used, they must be fully protected against water vapour penetration by placing an impervious layer (vapour barrier) over the insulation, but in some cases, it may be best to follow the opposite course: ventilate the insulation layer with dry air to enable the water which has been absorbed by hygroscopy to be released.

insulation board A rigid fibre board with an open structure and density below 400 kg/m, suitable for insulation purposes.

insulation cladding Metal or other material placed over thermal insulation material to provide protection.

intake filter, air compressor A filter fitted at the air intake to a compressor to trap impurities in the air. It can be combined with a silencer.

integrated (electronic) circuits Semiconductors made of silicon and germanium which can amplify, store and control electrical signals. An integrated circuit may contain thousands of semiconductor circuits imprinted on a tiny silicon chip and may form the basis of microprocessors.

integrated pollution control A centralized pollution control of the most polluting industrial processes in respect of all their discharges to the environment. It aims to ensure that the method of waste disposal selected causes the least overall damage to the environment.

intelligent building A building monitored and controlled by a comprehensive building management system.

intelligent fire detection An addressable analogue fire detection and alarm system capable of monitoring a wide range of fire and smoke sensors as well as auxiliary equipment and providing comprehensive locational information. It may include a colour graphic mimic VDU and pocket data pagers giving an alarm signal and readout of the precise location of the fire.

interceptor A device for trapping foreign matter in a pipeline and storing it pending removal.

intercooler See **compressor intercooler**.

interface calorifier A calorifier with primary and secondary circuitry placed between the boiler system and the heating distribution, keeping the two circuits separate. It is of particular benefit for the protection of boilers from scale and dirt where modern boilers are installed to serve old heating distribution systems.

interim certificate A certificate issued during the course of the works for the purpose of certifying payment or for any other matter.

interline method of sewer or drain relining (by Rice–Hayward) This uses short lengths of unjointed uPVC pipes within plastic sleeves to effect drain repair. See **sleeved uPVc pipes for sewer and rain renovation**.

interruptible gas supply Gas supplied under agreement with the gas utility which permits the interruption of the supply at times when the network is otherwise overloaded.

Advantage: a cheaper gas tariff.

intumescent material Expanding material incorporated into plastic plumbing systems to prevent the spread of fire through plastic pipes. When the temperature resulting from a fire reaches about 150°C, the intumescent material expands and thereby closes the uPVC pipe and effectively seals the hole (the pipe bore) and stops the passage of smoke and flame.

intumescent sleeve [Figs 138, 139] A metal container or collar into which has been placed a liner of intumescent material. It is used to prevent the spread of fire within a plastic plumbing installation and comprises two half-collars which are jointed by a full-length hinge which offers structural integrity and ease of fitting of the sleeve around the pipe. Robust multiple toggle clamps ensure permanent security after installation.

The sleeve is fitted around the plastic pipe where it passes through a fire barrier. It is equally suitable for installation in walls, floors and ceilings either built in or surface mounted. When the sleeve has closed following a fire, it is imperative that the seal maintains its integrity for a period of time which is at least equal to the fire rating of the barrier which it penetrates. Such ratings are usually, with some specific exceptions, in the order of four hours.

inversion See **temperature inversion**.

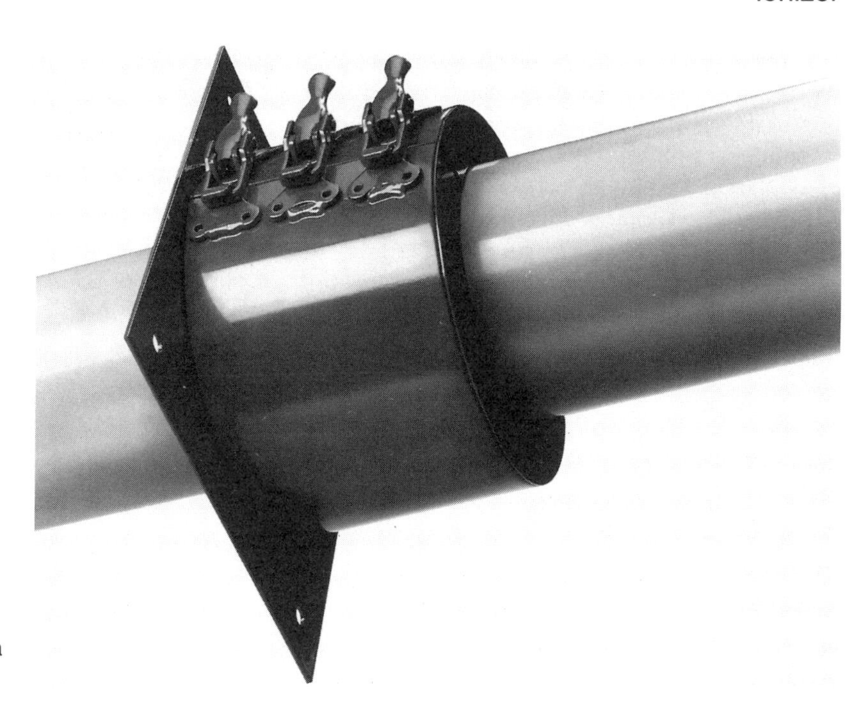

[Fig. 138] Intumescent 'Terrain Firebrake' sleeve fitted around a section of a uPVC pipe

[Fig. 139] Intumescent 'Terrain Firebrake' sleeve in open position

invert The lowest point of the internal surface of a pipe or channel at any cross-section or the lower portion of the internal surface of a drain, sewer or channel.

inverted bucket steam trap See **steam trap, inverted bucket type**.

inverter An inverted rectifier for converting direct electric current into alternating current. It operates with batteries and offers an alternative to a conventional generator for certain applications, with particular advantages for computer systems.

ionization The production of a concentration of negatively charged ions within a space. It is claimed to encourage a healthier environment and has been monitored to such effect in clinical trials.

ionizer Equipment that ionizes air. It may be portable or fixed in position within an ionized space.

iris damper A circular damper with 'camera'-type action. It is an efficient means of balancing air flow in ducts by reducing the orifice configuration. It maintains a uniform flow over the cross-section of the duct and avoids an increase in high-frequency noise which is common when dampering. The adjustment of air flow by a single extended lever by rotation through 90 degrees regulates the air volume from full free area (100%) to tight shut-off.

isobar One of a series of lines of designated areas of constant pressure.

isolator (electric) A simple switch for opening or closing a circuit under conditions of no load or of negligible current.

isothermal process A process conducted at constant temperature.

J J J J J J

JCT 1963 Joint Contract Tribunals Standard Form of Building Contract; 1963 edition.

JCT 1980 Joint Contract Tribunals Standard Form of Building Contract: 1980 edition (in current use).

Jacuzzi (trade name) A type of bath that generates a whirlpool. It is used as a normal bath and the water is drained after each bath.

jet nozzle air diffuser [Fig. 140] An air diffuser that discharges the supply air through a nozzle at high velocity. It is particularly suitable in situations requiring a long air throw.

joinder junction A junction pipe fitting in which the branch connection is made with a closed end. This can be easily cut off if the branch is required for active use.

joint ring Circular jointing material comprising asbestos, neoprene or flexible corrugated metal placed between flange faces to secure a tight joint.

joule A unit of energy, including work and quantity of heat. It is defined as the work done when the point of application of a force of one newton is displaced through a distance of one metre in the direction of that force. One joule equals one watt second.

[Fig. 140] Jet nozzle pattern air diffuser (Trox Brothers Ltd.)

Joule–Thomson effect Also known as the Joule–Thomson cooling effect. The small increase in the specific heat of air with a reduction in pressure will permit a slight temperature reduction (0.25°C or each atmosphere drop in pressure) even if no work is done.

jubilee clip A flexible metal clip in two hinged halves, secured by tightening a screw into a matching threaded hole. It is used with flexible hose-pipes.

jumper A fitting inside a water supply outflow tap to which a rubber or fibre control washer is attached.

junction box A device that connects the electrical distribution to electrical appliances or switches.

junction, connection to The connection between a branch pipe and another drain made to a junction must be so arranged that the tributary discharges into the receiving drain in the direction of flow and, if practicable, at half-pipe level or above.

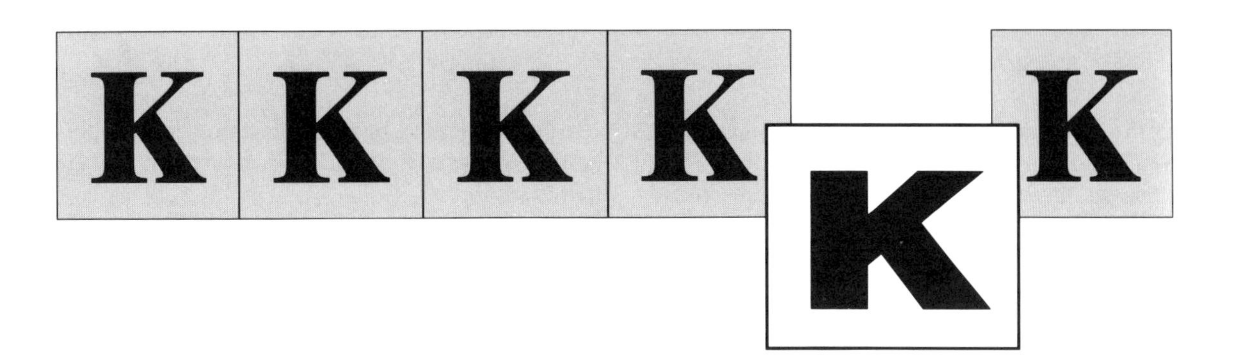

Kelvin (K) The name given to the scale of absolute temperature. The zero point of the scale is the absolute zero of temperature. The divisions within the scale correspond to degrees Celsius. Conversion from Kelvin to Celsius is achieved by subtracting the number 273.

kelvin temperature scale A kelvin (K) is the SI unit of thermodynamic temperature. It is the basis of the absolute temperature scale. The temperature of melting ice is 273.15 K (zero on the Centigrade scale) and that of boiling water at standard atmospheric pressure is 373.15 K (one hundred degrees on the Centigrade scale). It is named after the scientist, Lord Kelvin.

kerosene See **paraffin oil**.

kicker joint A sheet of waterproofing material placed during building construction between the concrete floor of a manhole or inspection chamber (before application of the screed) and the shuttering of the adjoining concrete wall. The vertical dimension of the waterproofing is commonly 76 mm.

kilowatt A commonly used unit of electric power equivalent to 1000 watts.

kilowatt hour (kWh) A commonly used unit of electrical energy equivalent to 3.6 MJ (megajoules) which equals 3600 kJ (kilojoules).

kinetic energy See **energy, kinetic**.

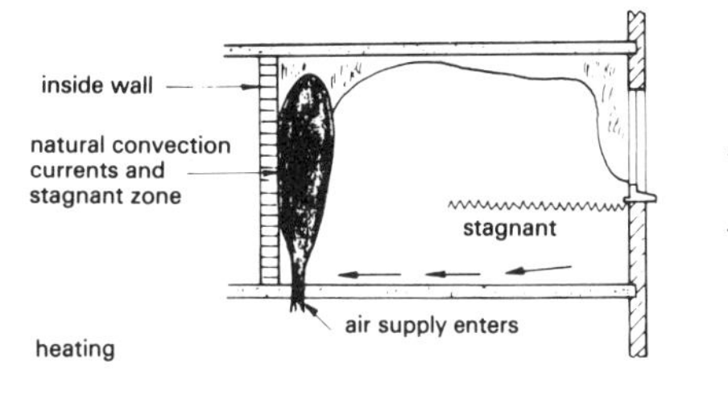

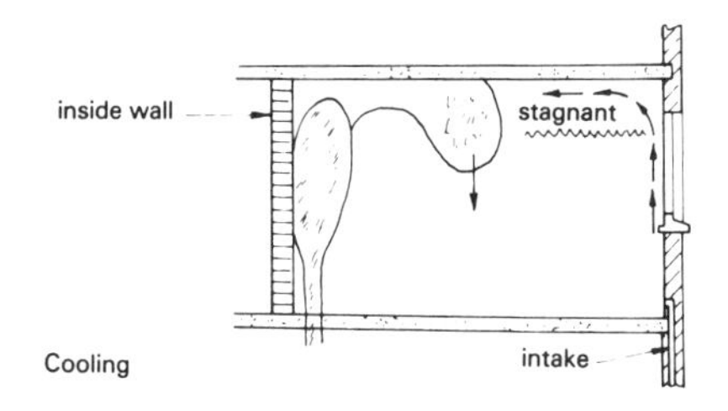

[Fig. 141] Illustration of the Koanda effect

kinetic ram See **cleaning and descaling of sanitary installations**.

Kirchhoff's radiation law A scientific law that states that the radiating capacity of a body is proportional to the absorbing capacity of that body at any given temperature and wavelength.

kitemark See **BSI kitemark**.

knuckle bend A short-radius bend in a sanitation system.

Koanda effect [Fig. 141] This refers to the presence of a stagnant layer of air which occurs with the air movement from mechanical ventilation/air-conditioning supply outlets. When cooling, the natural convection currents form a stagnant zone between the stagnant layer and the ceiling; when heating, the stagnant zone is formed between the stagnant layer and the floor.

Krantz system A system of microclimate (personal) air distribution. See **microclimate air distribution**.

Kuterlite 'GM' pipe fitting A proprietary make of zinc-free fitting for copper pipes which cannot be affected by dezincification.

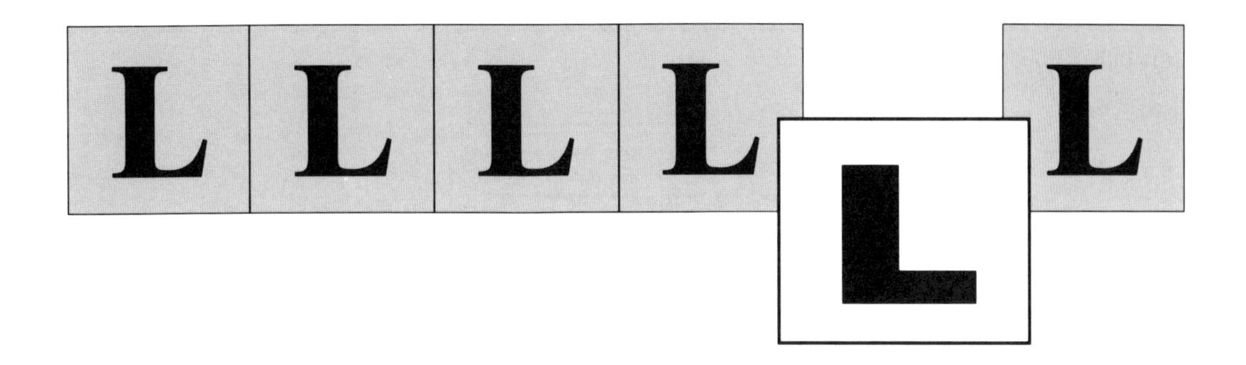

LED Light emitting diode. An LED provides low-power illumination and is used widely to indicate the status of controls, instruments, push buttons, etc. It offers a choice of indication colour.

LNG Liquefied natural gas. LNG is mostly methane maintained as a liquid under pressure for bulk storage or transportation.

LPG Liquefied petroleum gas. LPG is stored in a container under high pressure in liquid form.

Ladder heating system A form of one-pipe heating system in which the main flow and return pipes are located at opposite ends of the building and the single-pipe branch pipes are run horizontally between them. The heaters are connected to the horizontal single-pipe run. See also **single-pipe heating system**.

Lag To apply thermal insulation to a pipe or surface.

lagging Describes insulation. See **pipe insulation**.

laminar flow The flow of fluid in straight lines.

La Mont boiler [Fig. 142] A water tube boiler with controlled forced circulation through the water tubes. The heating surface comprises a number of small-bore tube elements which operate in parallel, the ends of which are expanded into the distribution header(s) of the boiler. The design offers a high rate of heat transfer. This type of boiler is suitable for large high-pressure hot water and steam installations. In view of the construction, requires close monitoring and control of the boiler feed water as any scale formation in the tubes is likely to lead to localized tube overheating and rapid ruptures.

lamp cut-off circuit Certain high-frequency electronic ballasts contain such devices to prevent failed lamps making repeated restart attempts and causing annoyance by their consequent flickering.

landfill gas [Fig. 143] Gas generated within rubbish tips and landfill sites. Its constituents are methane and inert gases in about equal proportions. It tends to be a dirty gas with calorific value about half that of natural gas. It is utilized for heating and/or power generation via a number of wells drilled into the landfill from which the gas is pumped into holding tanks for onward transmission to the consumers. Existing installations utilizing landfill gas commercially tend to have a payback period of two to five years.

landfill gas, danger Seepage of gas from unvented landfill sites which are close to built-up areas has

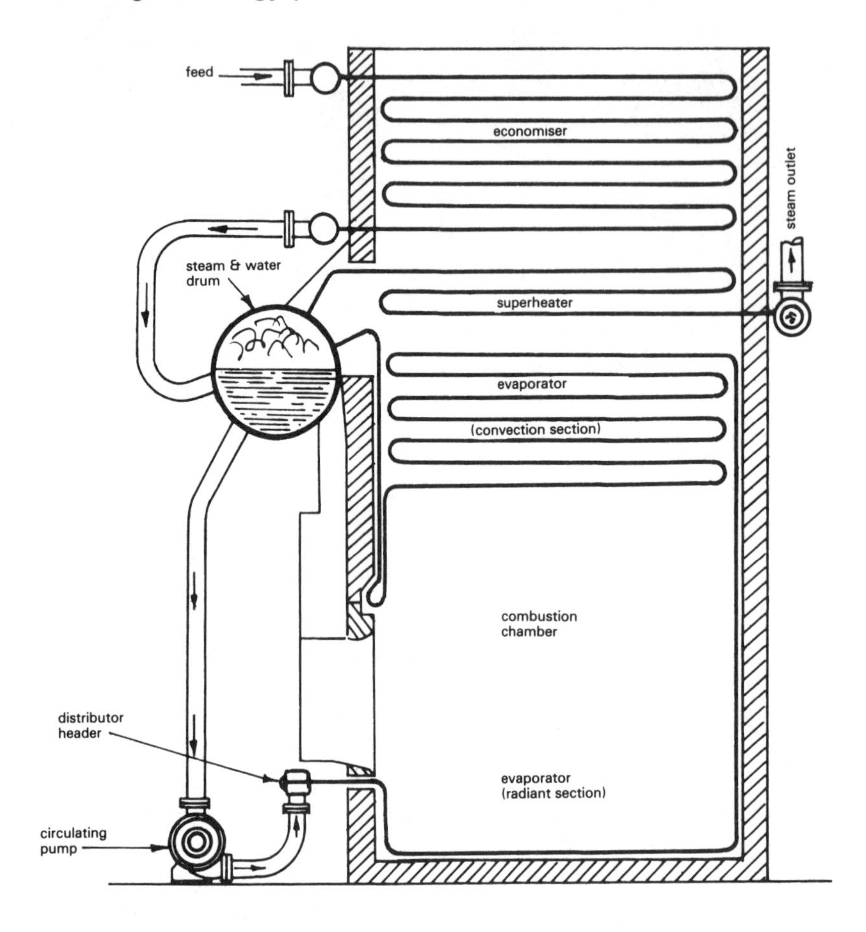

[Fig. 142] Forced-circulation water-tube boiler: diagrammatic arrangement

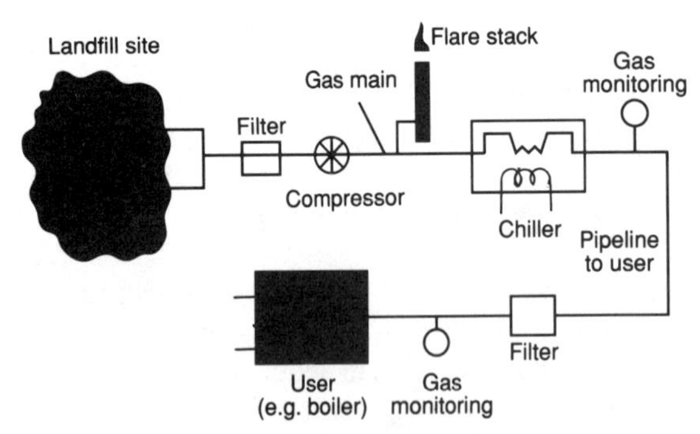

[Fig. 143] Progress from landfill gas well to end user

caused damaging explosions where gas has accumulated in pockets in, or close to, dwellings and been ignited.

landfill gas, energy potential Theoretically, each tonne of waste in a municipal refuse landfill site can provide about 400 cubic metres of gas. Conservative estimates allow for a high level of gas production over a period of 10 to 15 years. Biogas will continue to be produced over the life of the landfill for perhaps 50 to 100 years. Over a period of 10 years, each tonne of domestic solid waste can generate more than a hundred times its own volume in biogas, equivalent to some 3000 kW of energy.

landfill gas, uses [Fig. 144] Landfill gas can be used directly in boilers, furnaces and kilns; for the generation of electricity for use in the locality or for sale to electricity generators; or for the upgrading to higher equality fuel, such as natural gas. The calorific value of landfill gas is about half that of natural gas. This assumes that customers for the use of landfill gas exist close to the landfill site.

landfill gas well A well located at a suitable point within a landfill site for the extraction of landfill gas. It connects to gas collecting pipe(s).

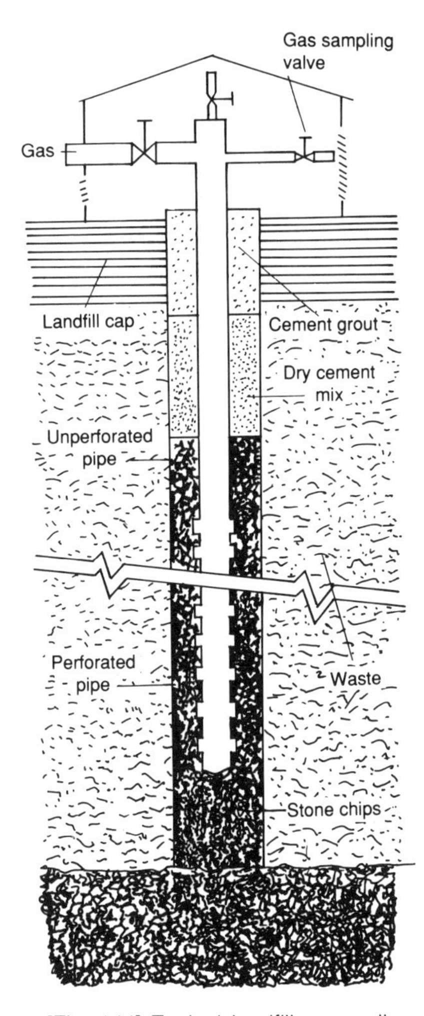

[Fig. 144] Typical landfill gas well

landing valve A valve that isolates fire-fighting dry or wet risers at each floor level of a building. It is used for the connection of canvas hoses to combat a fire.

landscaped offices Open-plan offices which are divided into different work and rest areas by the positioning of plants, fish tanks, etc. Permanent partitions between work or office areas are not provided. The object of such a layout is to promote good communications and flexible work practices.

latent defect Defective workmanship or materials which become evident at some stage after the final completion of the contract, possibly after some years. The law concerning latent defects is complex.

latent heat See **heat, latent**.

latent heat of fusion The quantity of heat required to melt a unit mass of a substance at constant pressure without change of its temperature.

latent heat of vaporization The quantity of heat required to convert a unit mass of a liquid into the gaseous state at constant pressure and whilst its temperature remains unaffected.

laundry tumblers Drying machines which take their air from the room and discharge the exhausts outside the building through ducting at high level. The outlet vents incorporate protection against the effects of wind and rain without restricting the air flow.

leaching Washing out a soluble constituent; e.g. from a landfill site.

lead boiler That boiler or boiler module within a sequence-controlled multiple boiler system which is scheduled to accept the initial load imposed on the boilers. As this load grows and exceeds the output capacity of the lead boiler (or module), the other boiler units are progressively switched into operation to meet the increasing load. As the output declines, the sequence operates in reverse.

lead-free solders These must be used in plumbing applications for potable water systems. They incorporate alloys of tin with additions of antimony, copper or silver.

lead pump That pump within a sequence-controlled group of pumps which is scheduled to meet the load on start-up of the system. When the required output exceeds the capacity of the lead pump, the other pumps in the group are automatically switched into operation to meet the increasing load. As the output declines, the sequence operates in reverse.

leak detector A valuable maintenance tool that detects compressed air and steam leaks, bearing wear, steam trap malfunction, corona discharge, etc. by acting on the ultrasonic frequencies generated by the passage of a gas through an orifice. The signals are converted into an audible hiss and/or display on a meter.

legionnaires' ACOP The Health and Safety Commission's (HSC) Approved Code of Practice on the prevention and control of legionnaires' disease and of other illnesses caused by the *Legionella* bacteria is now in effect. This code provides a basic framework for preventing further outbreaks of the disease. It includes advice on the requirements of the Health and Safety at Work etc., Act of 1974 and on the Control of Hazardous Substances to Health Regulations 1988. See also **risk assessment**.

legionnaires' disease A disease caused by contamination of water in the presence of *Legionella pneumophila* bacteria. Contamination has been traced to the warm spray emanating from spray cooling towers serving air-conditioning installations and to pockets of stagnant water within hot water distributions, particularly at showers. Remedial and preventative measures include: the substitution of blast coolers for cooling towers (long term); closely monitored chemical treatment and cleanliness of the cooling tower ponds; pipe design to ensure circulation of hot water to all draw-offs without creating stagnant pockets. Also, the cooling tower discharges must be sited to avoid the possibility of the spray entering via the air intakes of the air-conditioning system.

level control, automatic A means of switching a transfer pump on and off by reference to the required level in the tank.

level control, electrode operated The rising and falling liquid level moves between the location of two stationary electrodes. The upper electrode stops the transfer pump; uncovering the lower electrode restarts the pump.

level control, float operated A float moving freely on the surface of a liquid moves a lever between two electrical contacts which switch the transfer pump on and off.

level control valve A valve that maintains a predetermined level of liquid.

level invert taper A taper provided for the purpose of connecting together pipes which have their invert in the same straight line.

lever gate valve A valve intended for quick operation in which the gate is moved between the body seats by a stem or cam or is link-actuated by a hand lever. It may incorporate a sliding stem whose axis is at right angles to the body ends. The lever lifts with the stem when the valve is opened. Alternatively, it may incorporate a rotary stem in which the lever is attached to a stem which is connected to the gate through a cam or link whose axis is parallel to the body ends. The lever turns approximately a one-quarter turn when the valve is opened.

life-cycle cost The total costs incurred (or estimated to be incurred) by an installation, assembly or plant item throughout its active operating/service life.

lift check valve A valve in which the check mechanism incorporates a disc, piston or ball which lifts along an axis in line with that of the body seat. It is described in accordance with the type of embodied check mechanism. See also **disc check**, **piston check** and **ball check**.

lifting beam A tee section steel girder firmly secured above an item of heavy equipment to permit it to be lifted for inspection, repair, withdrawal or replacement. It is marked with a maximum safe lifting load.

light meter A direct-reading hand-held instrument which measures illumination. It incorporates a photoelectric cell and an indicator with a suitably calibrated scale.

light output ratio The ratio of light emitted from a luminaire to that emitted from the lamp(s) it houses.

lightning conductor A conductor of electricity which is connected from spikes at the high point(s) of a structure or building to earth. The spikes attract electric charges due to transient factors and conduct them safely to earth via the lighting conductor(s).

lignite See **coal, brown**.

limiting glare index The upper limit of the glare index for different situations, selected through experience and practice, to represent a standard of visual comfort acceptable to most people. Limiting glare indices are specified in the range from 13 to 28 in steps of 3. They indicate the point at which, for a particular activity, the lighting provides an acceptable standard of visual comfort.

linear Arranged in line, or having only one dimension. It may describe a system having output directly proportional to the input.

linear diffuser An air supply outlet which has greater width than height. The outlet tends to be in form of a slot or slots.

linear expansion The expansion of a body in one direction under the influence of heat, being expressed in terms of millimetres of expansion per metre. Different materials possess different coefficients of linear expansion. The concept is important when considering pipe and duct installations, it being essential to make adequate provision for the thermal movement of pipes and ducts to avoid placing undue strain on them and on their supports.

linear motor An induction motor which has the stator and the rotor in linear configuration instead of cylindrical, and parallel instead of coaxial.

litmus indicator A paper coated with litmus, a soluble substance of vegetable origin. The paper is turned red by immersion in acids and blue in alkalis.

liquid electric motor starter [Fig. 145] A liquid electrolyte resistance starter which offers an efficient means of bringing electric motors of large output on line and permits smooth continuous acceleration to take place. Various sizes of such starter systems are manufactured: up to 800 kW for squirrel cage and up to 1600 kW for slip-ring motors. For

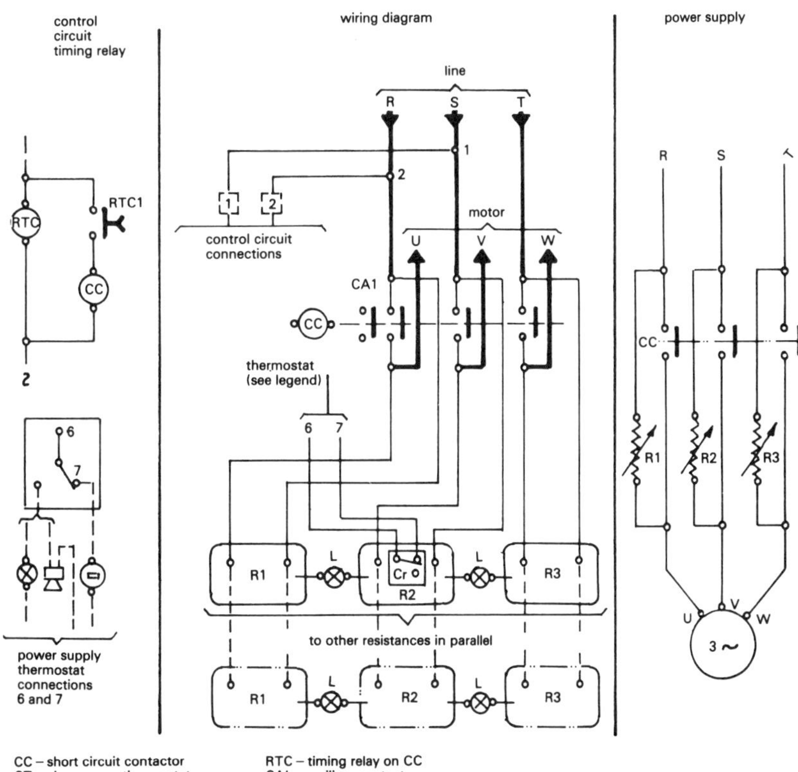

[Fig. 145] Liquid electric motor starter circuit diagrams

CC – short circuit contactor
CT – changeover thermostat
RTC1 – timed contact on RTC
RTC – timing relay on CC
CAI – auxiliary contact
R1, R2, R3 – thermo variable resistances

Note: in the case of separate coil supply connections 1 & 2 are taken to a terminal block

motor sizes above 20 kW they are considerably cheaper than auto-transformer starters.

liquid resin grouting An effective method of sealing cracks in drainage systems by means of a resin which is preblended with a catalyst. The resin is received at site in powder form requiring only water to produce a low viscosity grout solution. To apply, the lower end of the drain is sealed off and the liquid is poured into the higher end to a sufficient height to force the solution into every crack or porous joint in the drainage run. The liquid resin will penetrate through cracks in the drain into the surrounding material, thereby grouting it as well as the drain. This method is suitable for the treatment of bends and branch pipes which would be otherwise inaccessible for repair.

liquidated damages Damages paid by the contractor in the event of delay to the completion of the works. They are costed in accordance with the particulars of the penalty clause to the contract.

liquefied natural gas See **LNG**.

live (electrical) Describes a circuit or conductor of electricity actively connected to a source of electro-motive force.

live steam Steam discharged from a steam line or steam boiler.

load factor The ratio, expressed as a percentage, of the average load supplied in a given time to the highest load which could have been supplied during that period. For example, in relation to an electricity supply system, it represents the ratio the average load supplied throughout a calendar year to the highest load on the system during that period.

load shedding The process of disconnecting from the supply items of electrical equipment to avoid overloading the supply or to prevent maximum demand charges.

loading unit Used in the design concept of sanitary services. A specific number of such units are allocated to each water-using appliance and these express the relative water requirements as multiples of the loading unit, and the rating of an appliance is determined by the duration and rate of water flow required at the appliance and on the interval between usage.

lockshield valve A valve that incorporates a protective metal shield (instead of a valve wheel) usually screwed on to the valve body. It is used for the permanent adjustment of water flow to radiators, etc. by recording the number of turns the valve must be closed to maintain the required adjustment.

log A record of service, maintenance and call-out visits entered in a designated log book.

long-throw sprinkler A sprinkler designed to throw fire-fighting water horizontally. It is particularly suitable for fire protection of atria where such sprinklers would be located at ground level. See also **atria, active fire protection**.

loudness The intensity of sound waves combined with the receptivity of the ear. It depends on the response of the ear to sound which is not equal at all frequencies of sound. Loudness is measured in phons.

looping The behaviour of an unstable chimney plume, due to inadequate efflux velocity or local environmental factors.

louvre, acoustic see **acoustic louvre**.

louvred grille [Figs 146, 147, 148] An air supply terminal fitting equipped with adjustable horizontal, directional or fixed blades.

louvred shutter A security grating with integral air inlet slots and louvred blades to prevent ingress of rain or snow.

louvred solar screen system See **solar screens system**.

louvred ventilator A grating fitted into an external wall for the purpose of ventilation and provided with protective fixed louvres to prevent ingress of rain or snow.

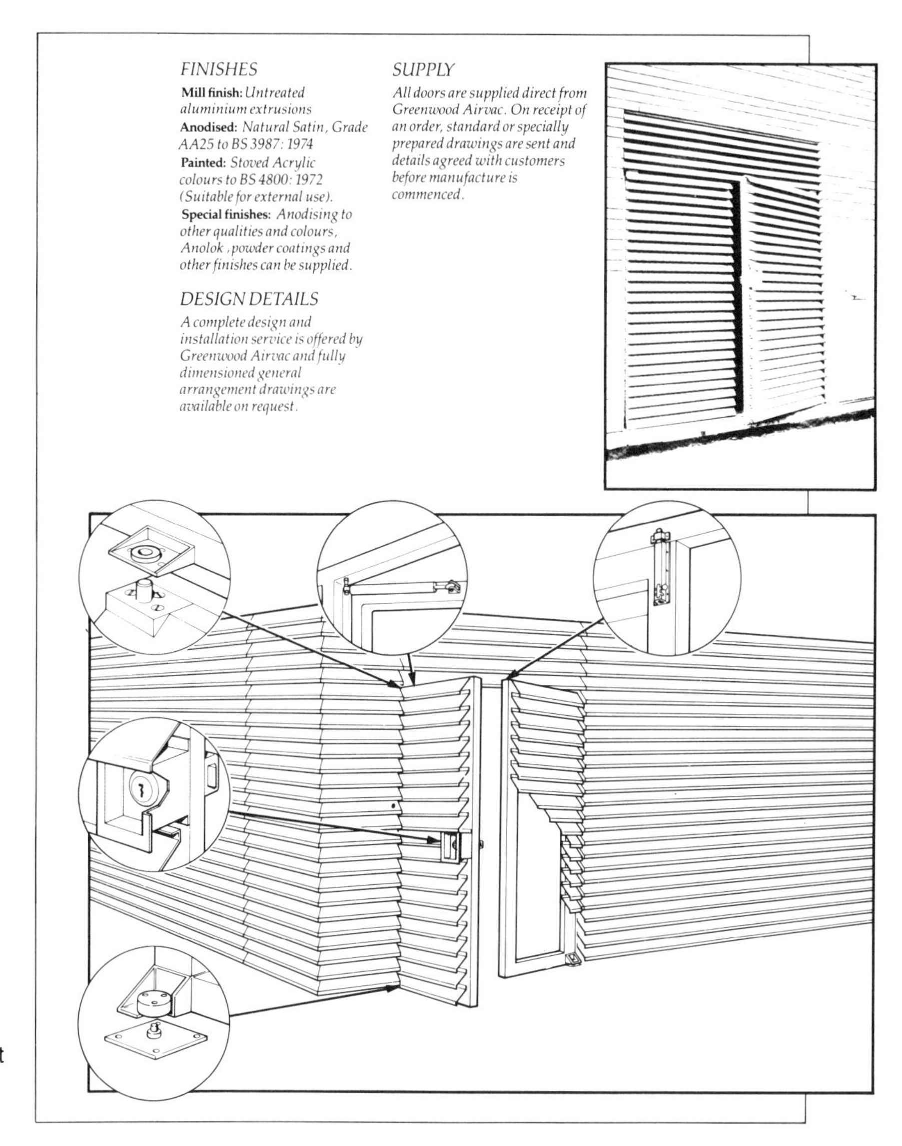

[Fig. 146] Louvred door and wall construction in untreated aluminium extrusions suitable for plant rooms and boiler houses (Greenwood Airvac Ltd.)

low invert trap A vertical inlet in which the level of the outlet invert is low relative to the inlet.

low-pressure air-conditioning system A system that operates with relatively low fan pressure, resulting in low velocities of air flow and relatively large air ducts. The advantages are silent operation and absence of vibration, whilst the disadvantages are the requirement for large duct spaces and generally higher first costs.

low-pressure air switch A switch that actuates controlled appliance(s) through differential pressure when the monitored air pressure has reached the low set-point. It typically controls in ranges of between 13 Pa and 500 or 3000 Pa. It embodies an adjustable set-point, a non-conductive changeover switch and a conduit connection.

low-pressure hot water space heating Space heating that utilizes the circulation of hot water at a temperature below the boiling-point of water.

low-pressure safety cut-out A device that guards against the danger which may arise under conditions of excessively low pressure. It is used to stop a refrigeration compressor when the refrigerant

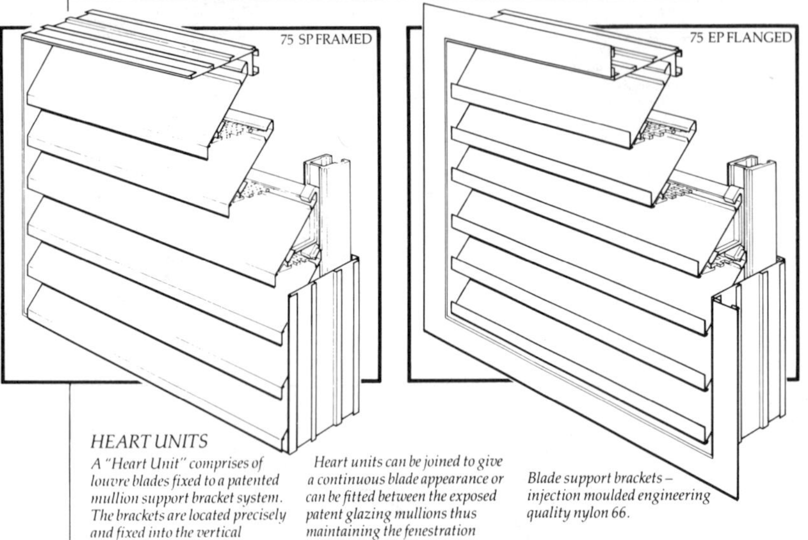

[Fig. 147] Louvre 'Heart Unit' comprising louvre blades fixed to a patented mullion support bracket system: brackets are located precisely and fixed into the vertical mullions by automatic jig assembly (Greenwood Airvac Ltd.)

suction pressure is below a safe limit, usually intended as a temperature safety device. The pressure setting of the cut-out corresponds to a temperature at which frosting might occur.

low-pressure sodium lamp A discharge lamp which includes a small quantity of sodium and adds sodium vapour to the discharge. It is characterized by monochromatic yellow light which lies close to the maximum sensitivity of the eye (555 nm) and therefore the lamp is very efficient. It exhibits very poor colour rendering and is therefore mainly used in applications where this is unimportant, such as for floodlighting and street lighting.

low-side float valve A flow control device used with flooded chillers. If too much liquid refrigerant accumulates because the flow is in adequate, the

PLUSAIRE 75
WEATHERLIP ADJUSTABLE LOUVRES

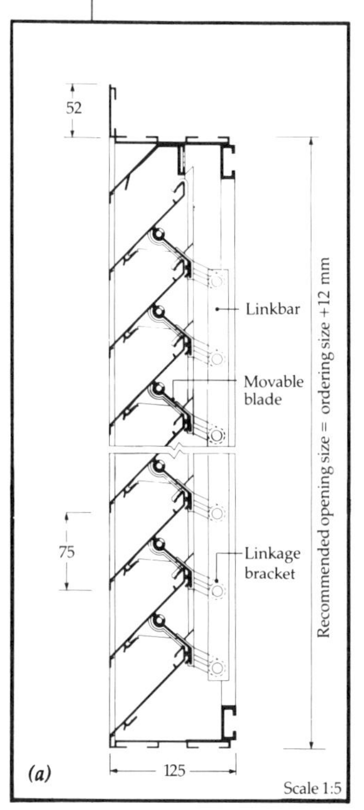

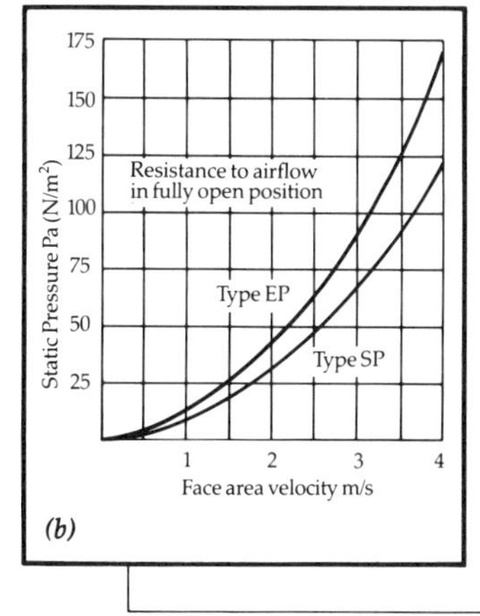

INTRODUCTION

The Plusaire 75 Adjustable Louvre is available in SP, EP and HP weathering profiles, and provides a controllable means of ventilation whilst maintaining the continuous linear appearance of the Weatherlip fixed blade system. Both systems can be used in one assembly without affecting the aesthetic appearance even when the adjustable louvre is closed.

Unlike conventional types of adjustable blade louvres, the Plusaire Weatherlip design incorporates an external fixed blade with a concealed adjustable damper ***(a)****.*

DESCRIPTION

The fixed blades, with the adjustable dampers complete with brushed nylon seals, are snap-locked onto patented brackets mounted on to mullions to which flanged, framed or frameless peripheral trims are attached.

The adjustable damper blades are interlinked to suit a variety of controls.

USES

Adjustable louvres can be fixed into wall openings, window frames, or as terminals for ventilation systems to clad and weather the openings.

Typical uses are for providing controlled air movement, operating in conjunction with powered and natural ventilation schemes in factories, plant rooms, power stations and similar projects.

Adjustable louvres are also ideally suited to provide controllable air inlets linked to smoke venting systems in warehouses, supermarkets, shopping complexes and equivalent types of developments where the louvres normally remain closed until an emergency arises.

MANUFACTURE

Louvre panels are purpose made from extruded aluminium sections to BS 1474: 1972, HE9TE.

Blade support brackets – injection moulded engineering quality nylon 66.

CONTROLS

A comprehensive range of controls are available to operate the Plusaire Adjustable Louvres to meet the needs of powered and natural ventilation requirements or to work in conjunction with automatic sensoring equipment for fire and smoke.

TYPES OF CONTROLS

DIRECT MANUAL
Hand operated control. A push-pull lever with locking nut is attached to the link bar.

ELECTRIC MOTOR
There are two different types of motor dependent on the mode of operation required, either for daily ventilation or for fire and smoke control.

ELECTRO-MAGNETIC
Normally used in smoke venting applications.

The electro-magnet is directly coupled to the end of the link bar which operates counter-weighted louvre blades. Once the electric power is interrupted in any way, the electro-magnet will de-energise, releasing the link bar.

PNEUMATIC CYLINDER
The pneumatic control system is the most versatile and flexible method of controlling large louvre assemblies.
The simplicity of the control and ease of installation makes it the cheapest form of remote control of ventilators where compressed air is available.

FUSIBLE LINKS
Fusible links automatically open the louvre at a pre-set temperature, to allow smoke and hot gases to escape to the atmosphere.

[Fig. 148] Adjustable louvre in weathering profile providing controllable means of ventilation whilst maintaining continuous linear appearance (Greenwood Airvac Ltd.)

float rises and a connecting linkage opens the valve, allowing greater flow of refrigerant.

low voltage (lv) Not exceeding 250 V between any two conductors or between one conductor and earth.

low water steam alarm An alarm fitted to a steam boiler and actuated when the water level falls below the set minimum water line. Steam then passes upwards into the tube of the device, melts a fusible disc and sounds a whistle alarm. The standard device incorporates an isolating valve or cock to permit replacement of the fusible disc without having to shut down the boiler.

lubricant A material used for lubrication, e.g. oil, grease, graphite, etc.

lubricated plug cock A lever-operated valve with provision for lubrication of the gland to maintain free movement.

lubrication The process of applying a film of oil or other substance with similar properties to the contact face between two moving or rotating metal components with the object of greatly reducing friction between the surfaces to achieve longer wear and/or quieter operation.

lubricator An appliance for applying lubrication or the terminal through which lubricant is inserted.

luminaire A fitting which houses the light source. It is a widely used term for an electric light fitting.

luminaire, air handling See **air handling luminaire**.

luminance The quantitative expression of brightness.

luminous efficacy The ratio of the luminous flux from a lamp to the power consumed, measured in lumens per watt.

luminous flux The rates of flow of luminous energy, measured in lumens. One lumen of luminous flux at a wavelength of 555 nanometres corresponds to a radiated power of 1/680th of a watt. At a wavelength of 400 nanometres, one lumen corresponds to a radiated power of 3.5 watts. The relationship between lumens, watts and the wavelength at which the power is radiated is fundamental to the design of luminaires.

luminous intensity The strength of a light source in a given direction, measured in candelas.

lux A unit of illumination. One lux is equal to one lumen per square metre, i.e. the illumination produced by a light from a source of one international candle falling directly on a surface at a distance of one metre from the source.

M

MCB Miniature circuit breaker. It replaces a conventional fuse arrangement. After it has interrupted the circuit, the fault is diagnosed and rectified and the MCB is then simply reset.

MICC See **mineral insulated cable**.

MSW Municipal solid waste.

macerator An electrical apparatus which grinds waste products in the presence of water and reduces them to a liquid stream before discharging the waste into a drain. For example, a sanitary towel macerator is used to avoid the direct discharge of sanitary towels to drain where they would be likely to cause blockages.

magazine boiler A boiler that incorporates an integral coal hopper which usually holds fuel for about 24 hours' continuous operation at full load. The coal flows into the furnace by gravity and the rate of burning is controlled by a thermostat which adjusts a combustion air inlet damper. It is vulnerable to dangerous overheating through careless opening of the ashpit door, when the thermostat cannot control the boiler temperature.

Magna Line system a method of lining existing drain-pipes by inserting the same size or larger new pipes. See **pipe bursting**.

magnetic field The effect of moving electric charges. These charges may be in a circuit carrying an electric current or in a permanent magnet where the charges reside on spinning electrons.

magnetic field strength The strength of a magnetic field measured in amperes per metre.

magnetic force The attractive or repulsive force which is exerted by a magnetic field on a magnetic pole or an electric charge.

magnetic separator [Figs 149, 150] A device that separates ferrous items by magnetic attraction. It protects materials/fuel/waste-handling plant from suffering damage through the presence of unwanted ferrous items, such as loose nuts and bolts.

magnetic valve A valve equipped with a solenoid (magnetic coil) which is activated by a detecting device to open or close the valve by drawing the stem of the valve through the magnetic field.

magnetic water-conditioning [Fig. 151] It is claimed that passage of water through a magnetic field breaks up the scale crystals which are subsequently discharged with the outflow of water. Electric connections are not required. Such conditioning is particularly popular with the once-through district heating systems which were widely installed in the former Soviet Union.

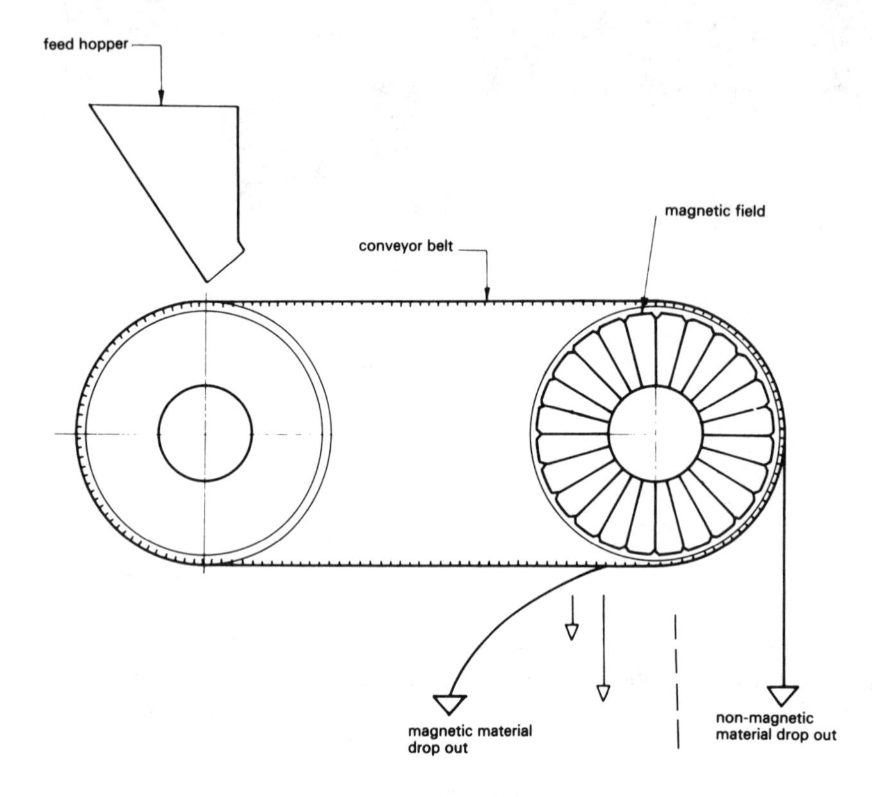

[Fig. 149] Magnetic separator: conveyor belt arrangement

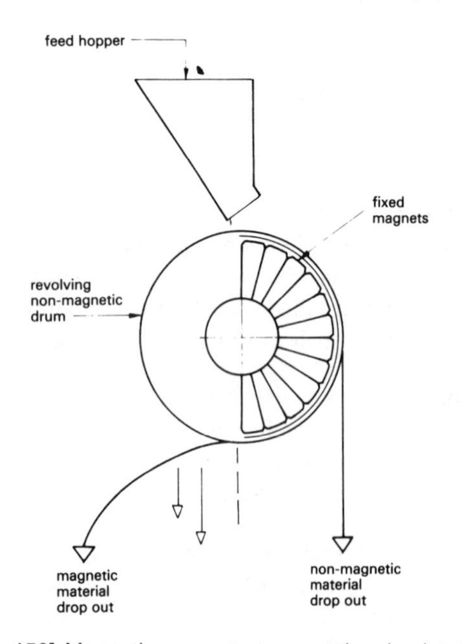

[Fig. 150] Magnetic separator: magnetic wheel type

magnetism The branch of physics concerned with the properties of magnets and magnetic fields.

magnetohydrodynamics, heat recovery [Fig. 152] A method of heat recovery whereby liquid metal is circulated through a heat source (e.g. solar collectors, sewage, industrial wastes) in conjunction with a vapour circuit. The vapour is a highly volatile refrigerant which assists the liquid metal to flow past a magnet at which an electric field is generated and electricity is produced. The latent heat of the vapour can be recovered at the condenser heat exchanger. The system is being evaluated and appears to have great potential for heat recovery and electricity generation at low cost.

main contractor The lead contractor who is responsible for the total package of contracted works.

mains water supply Water drawn directly from the water utility's water main.

male connection A pipe end which enters into a mating fitting.

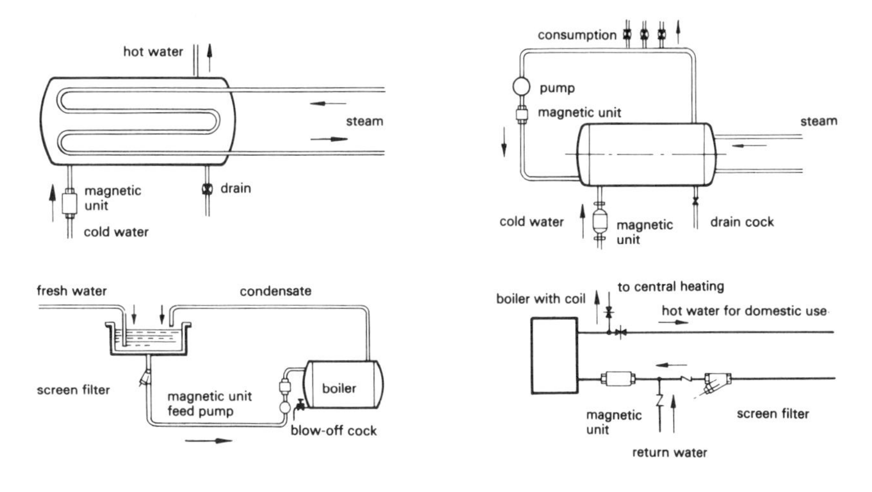

[Fig. 151] Diagrams showing magnetic water conditioning process

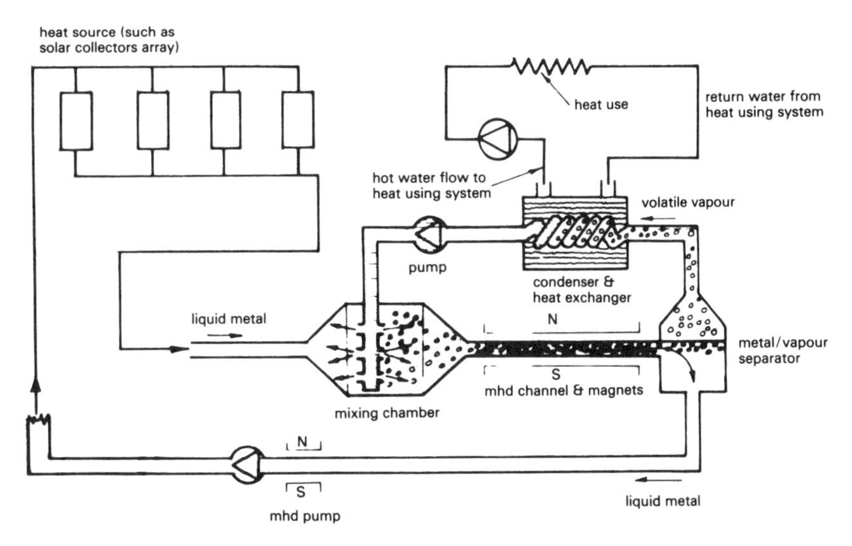

[Fig. 152] Exploitation of low-grade heat source (solar, waste, etc.) using magnetohydrodynamic principles

malleability The ability of a metal to be hammered out into thin sheets.

manganese bronze A copper–zinc alloy containing up to 4% manganese, also known as manganese brass.

manhole [Figs 153, 154] A substantially built access chamber located at those sections of a drainage system where branch drains enter a main drain and/or where full access is required to that section for inspection or clearance of blockages. The bottom of the manhole incorporates open drainage channels; branch ducts must connect into the manhole swept in the direction of flow. Deep manholes must be fitted with access step irons. Manholes may be constructed *in situ* of brick or may be assembled from precast rings, etc.

manhole, channels and benching An open channel of half-round section should extend the whole length of the manhole. A vertical benching should be formed from the top edge of the main channel to a height not less than that of the soffit of the outlet. It should be rounded off to a radius of about 25 mm and then slope upwards at a gradient of about 1:12 to meet the wall of the chamber. Side branches should discharge in the direction of flow, preferably at half-pipe level. Vertical and side benching must be so shaped that the flow is contained and directed without permitting

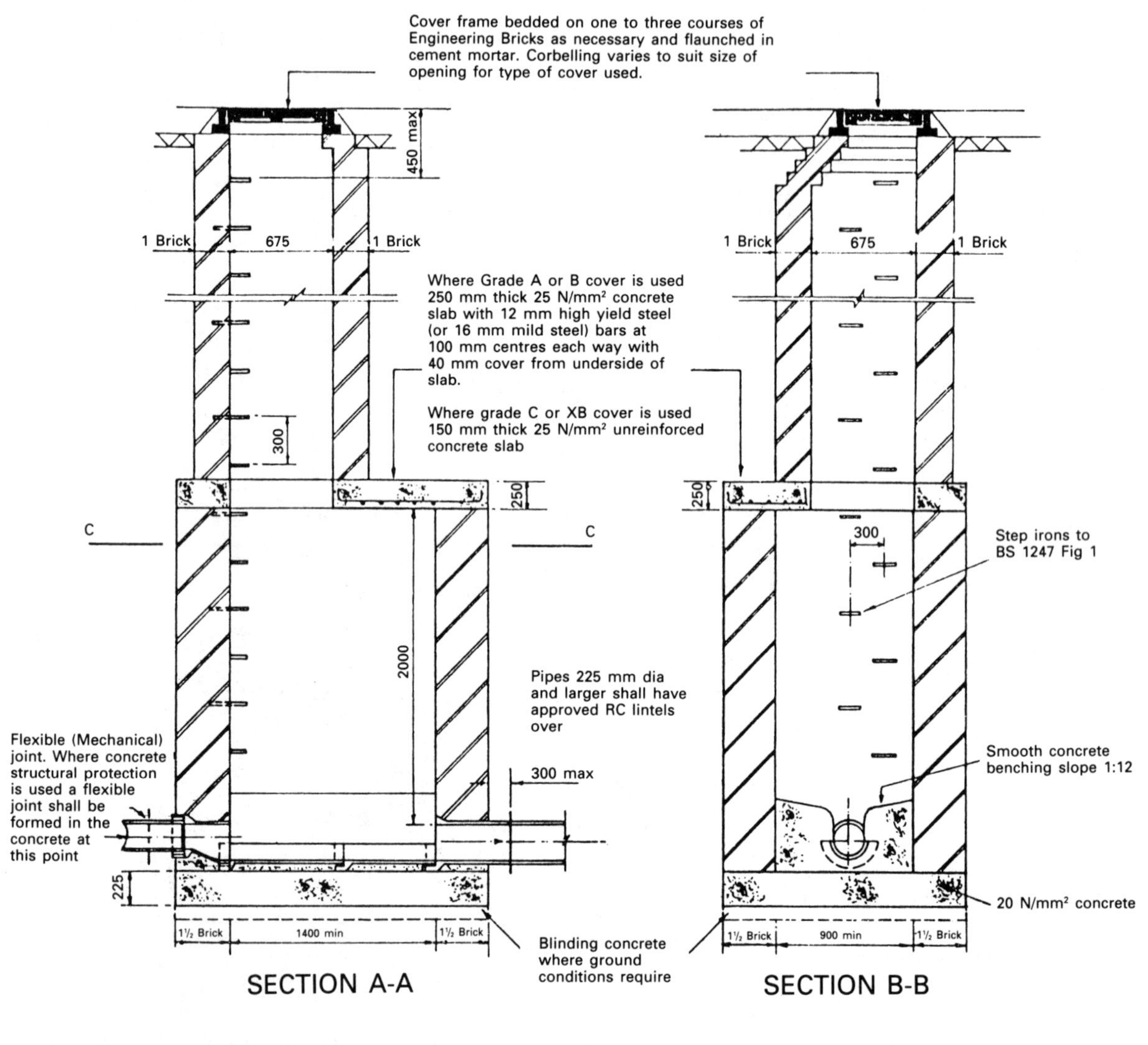

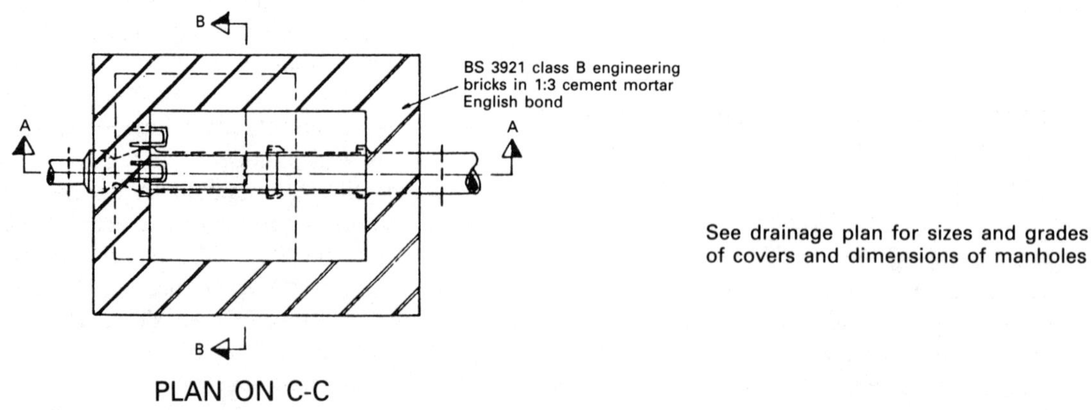

See drainage plan for sizes and grades of covers and dimensions of manholes

[Fig. 153] Brick manhole (open channel)

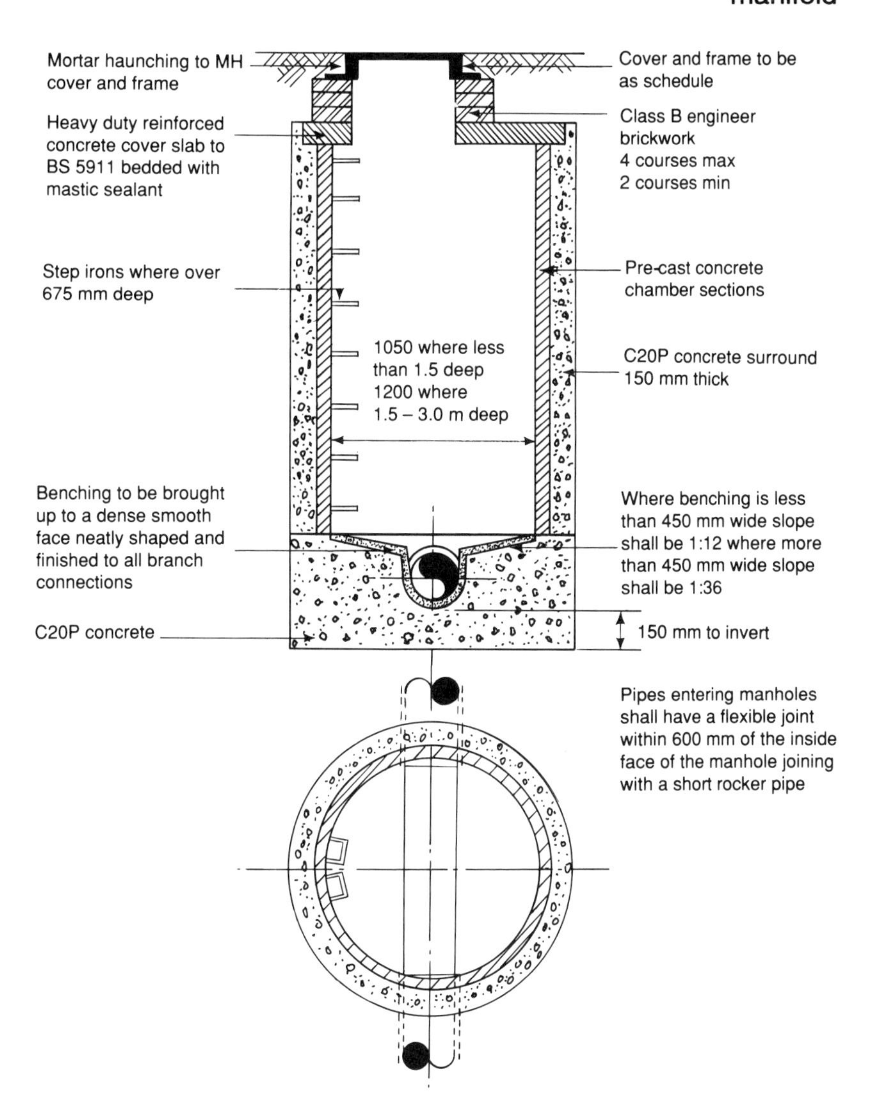

[Fig. 154] Precast concrete manhole

fouling within the chamber and to facilitate rodding of the branch drains.

manhole cover A substantial metal cover fitted over a manhole incorporating lifting rings. The quality of the cover will depend on location; whether light duty or heavy traffic bearing. The cover is commonly complete with frame which is embedded at the mouth of the manhole and is tightly sealed to prevent the escape of odours and the ingress of surface water into the drain.

manhole, flushing See **flushing manhole**.

manhole, plastic Manholes and inspection chambers are available moulded in thermoplastic and thermosetting materials such as polypropylene, GRP or uPVC, supplied either as integral bases or as complete chamber units.

manhole, vitrified clay The integral base of such inspection chambers offers a variety of inlet and outlet connections. They have flexible joints, incorporating elastomeric sealing rings, to connect raising pieces to the base and permit variations of depth.

manifold A purpose-made distribution centre for branch pipes.

manipulated variable That quantity or condition which is varied by the automatic controller so as to affect the value of the controlled variable.

manometer, inclined A manometer in which one leg of the manometer U-tube is inclined permitting greater sensitivity of pressure reading.

manual reset A push-button control which is pressed by hand to repeat an operation; e.g. filling a daily service oil storage tank. It is also the means of adjusting the set-point of a controller by hand to effect changes in the control sequence, the controlled variable, etc. The set-point of a controller is at some intermediate value of the controlled variable within the proportional band, usually at or near the middle. With load changes, however, it is generally necessary to change the amount of the valve opening in order to provide the normal flow rate of the control agent. This is accomplished by manual reset adjustment in the controller.

marsh gas See **methane**.

masking of sound The process by which the threshold of hearing of one sound is raised due to the presence of another sound.

master clock [Fig. 155] The principal timekeeper in a central clock system which emits impulses at predetermined intervals for the operation of other timekeepers.

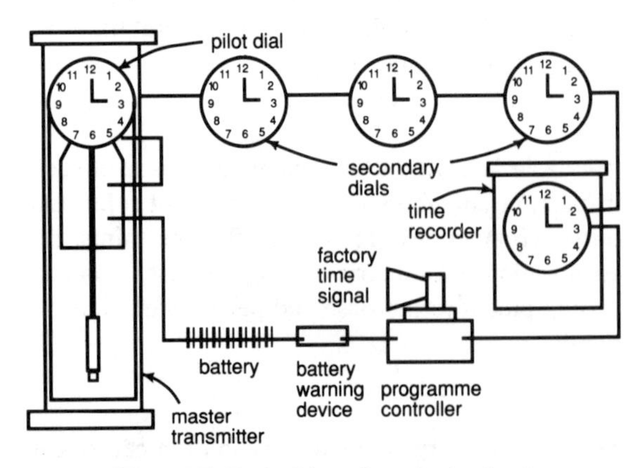

[Fig. 155] Typical impulse clock circuit

master controller A device which actuates a number of slave devices; e.g. one master clock will control numbers of slave clocks which follow its directions.

maximum demand The maximum current or power that has flowed through a circuit in a given period.

maximum demand charge An appropriate limit of power consumption in commercial or industrial premises set by the utility company. Excess consumption commonly entails increased quarterly or annual charges for the excess consumption, irrespective of the period of time over which this demand persists. It is therefore essential to monitor and control the consumption and, if necessary, shed load, so as not to exceed the set maximum demand.

maximum demand (electric) The maximum current or power that has flowed through an electric circuit in a given period. It is used in connection with tariffs for electricity consumption which are based on a specific maximum demand and impose a cost penalty when that demand has been exceeded.

maximum demand meter A meter that measures and records the maximum demand taken from the supply in a given specified period.

maximum demand monitor A device that monitors the consumption of electricity in commercial or industrial premises and warns visually or audibly when the electricity use approaches the maximum demand set by the utility company so that an immediate load-shedding programme may be activated to avoid increased maximum demand charges.

maximum simultaneous demand That total outflow from draw-off fittings or output from electric points which would occur if all these demand terminals were in use together – an unlikely occurrence. Systems are therefore designed on the basis of an assessed probable peak demand. See also **simultaneous demand**.

means of escape The route along which occupants will leave a building in the event of a fire. The route must be protected by design so that smoke and fire cannot be drawn towards it.

mechanical humidifier A humidifier that works by atomization. A sealed motor drives a disc and pump tube which draws water from a reservoir which is maintained at constant level by a float valve, up to the top surface of a horizontal disc which rotates at high speed. Water is cast by centrifugal action against a breaker comb which divides the water into fine particles, i.e. achieves atomization. Air enters under the disc and picks up this atomized water and carries it into the humidified space or into a ducted airstream where the moisture is absorbed in the surrounding air.

The use of such humidifiers is not recommended in areas supplied with hard water, as there will be a tendency for salts to be precipitated on to duct walls or surfaces within the humidified space in the form of white deposits.

mechanical stoker Equipment attached to solid fuel fired boilers and furnaces by means of which the fuel is automatically fed to the combustion chamber, as opposed to manual stoking. Numerous different types of mechanical stoker are in use, classified by the manner in which the fuel is handled (e.g. underfeed stokers, spreader stokers, rotating grates, etc.).

mechanical stoker, chain grate See **chain grate stoker**.

mechanical stoker, coking See **coking stoker**.

mechanical stoker, spreader type See **spreader mechanical stoker**.

mechanical stoker, sprinkler type See **sprinkler mechanical stoker**.

mechanical ventilation [Fig. 156] Ventilation that relies on a fan plant for its operation.

mediation A process that is voluntary, non-binding and without prejudice in which a specially trained third party intervenes in a dispute and attempts, through negotiation techniques, to bring the parties to the dispute into a settlement agreement.

mega A prefix denoting one million.

megawatt A unit equal to one million watts.

megohm One million ohms.

medium voltage (mv) Voltage from 250 V to 650 V between any two conductors or between one conductor and earth.

medical gases A collective term for piped gas installations provided for medical purposes, encompassing medical quality oxygen, nitrous oxide, nitrous oxide/oxygen mixture, carbon dioxide, medical compressed air and medical vacuum.

medical gases, terminal identification Terminal units should be identified by labelling with the appropriate recognized name, abbreviation or symbol and by following colour coding.

(a) Oxygen: white.
(b) Nitrous oxide: French blue, British Standard colour code 166.
(c) Nitrous oxide/oxygen: French blue/white quartering.
(d) Carbon dioxide: light grey, British Standard colour code 630.
(e) Medical compressed air: white/black quartering.
(f) Medical vacuum: primrose, British Standard colour code 354.

Identification must be of a permanent type, not painted on.

medical gases, terminal units Features required in a standard terminal are the following.

(a) The unit should accept, retain and release the probe by means of a quick-release mechanism designed for single-handed operation.
(b) An independent valve should be incorporated in the assembly to permit maintenance on the outlet without having to isolate the pipeline.
(c) The unit should incorporate a check valve which is self-sealing when the probe is removed.

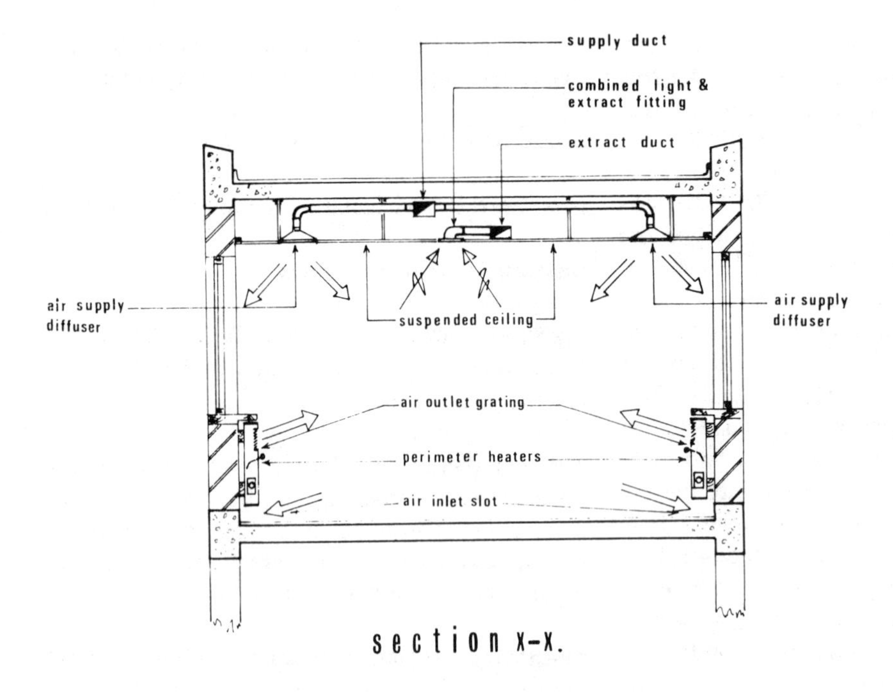

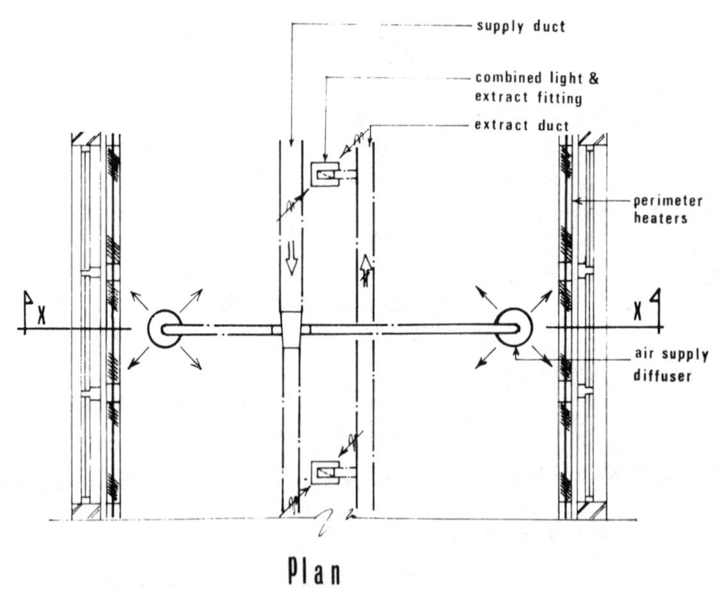

[Fig. 156] Combined mechanical ventilation warm air heating/warm wall system

(d) The terminal/unit probe connection should be of the non-swivel type, so that secondary equipment, such as a flow meter, is not tilted by the probe hose.

(e) A secondary locking mechanism should be incorporated in order to prevent damage to auxiliary equipment by the accidental ejection of the probe.

(f) The connecting probe, socket and check valve assembly for each medical gas should be designed to prevent interchangeability with equipment used for other medical gas supply systems.

medical vacuum As used in hospitals, a piped system with two outlets available at each hospital bed position. Each outlet incorporates a vacuum restrictor and shut-off valve. Vacuum gauges should be fitted to the vacuum bottles. Each medical vacuum station should be capable of permitting a free air flow of not less than 40 litres per minute when the vacuum in the pipe line is 500 mm mercury below standard atmospheric pressure of 760 mm mercury.

membrane, ceramic See **ceramic membrane**.

membrane expansion vessel A vessel used for the pressurization of water in a heating system whereby a cushion of air or nitrogen maintains pressure above a membrane.

meniscus When two liquids of different density, or when a liquid and a gas, are in contact, the surface of contact forms a curve which is called the meniscus.

metabolic heat Heat energy released from human or animal bodies arising from on-going chemical processes. This contributes towards heat gains in an occupied space.

metabolism Chemical changes within living cells by which energy is provided for vital functions.

metal halide lamp A lamp that utilizes a high-pressure discharge in mercury vapour with metallic additives operating within a quartz tube. It provides a good quality white light and is used wherever such quality of lighting is required.

meter pit A brick-built or concrete chamber which gives access to the water meter at the boundary of a consumer's estate.

methane Odourless inflammable gas which forms an explosive mixture with air. Chemical symbol CH_4. The first carbon of the paraffin series. It is formed from decaying organic matter, such as from deposits of refuse in landfills, and in coal mines. Occurs in natural gas and in coal-gas. Also known as marsh gas, firedamp and landfill gas. Heat values in the range of 16 300 to 20 000 kJ/kg.

methanol (CH_3OH) A methyl alcohol. It is a clean liquid fuel with a calorific value of about 20 MJ/kg.

method of measurement A method of measurement should be carried out in accordance with a standard method laid down or accepted within the building industry and commonly defined in the preamble to the bill of quantities.

methylene blue air filter efficiency test A test that uses the staining effect of the dust as a criterion of filtration efficiency.

microchip A form of microprocessor.

microclimate Climatic conditions which affect a particular site due to peculiar local conditions; e.g. wind effects arising from surrounding tall buildings.

microclimate (personal) air distribution A method of supplying primary air at relatively high velocity through a special outlet (e.g. an angled desk outlet) adjacent to the occupant of a space, inducing secondary air flow around that person to maintain localized comfort conditions. Primary air volumes are usually between 7 and 8.5 litres per second per person. The induction ratio is 1.5 to 1 with a minimum air supply volume of 10 litres per second per person.

micromanometer A device that measures low differential pressures in dry inert gases (usually air) on a digital scale. It is hand-held and powered by a dry cell or rechargeable battery.

micron A unit of length equal to 1×10^{-3} mm or 1×10^{-4} cm. The symbol for micron is μm. The micron is relevant in particular to the specification of filtration systems, e.g. air filters.

microprocessor An integrated circuit that utilizes the benefits of microelectronics to achieve sophisticated control of electrically motivated applications. Microprocessors are utilized in building services engineering, in energy management systems, climate controls, programming, lift supervision, etc.

mill scale A deposit left on a pipe or other metal surface from the manufacturing process. It must be removed by shot blasting or a similar process before the surface can be satisfactorily treated, painted, etc.

[Fig. 157] Mixed flow axial fan: variable speed and fixed guide vanes and pitch (Woods of Colchester Ltd.)

mimic diagram A diagram related to computerized controls, describing visually the extent and status of controlled or monitored installations.

mineral insulated cable (MICC) A cable that has its conductors encased inside a metal sheath insulated with a mineral powder. It is an alternative wiring system to conduit systems etc. and is particularly suitable for outdoor use, in hazardous locations and in areas of heavy condensation. The cable consists of single-strand conductors of either copper or aluminium enclosed within a thin tube of the same metal, packed with finely powdered magnesium oxide under pressure. The outer sheath is usually employed as the earth continuity conductor. The smaller size of cable incorporates up to seven conductors, intermediate sizes one to four and the largest size one single conductor only.

mist Liquid particles, generally smaller than 10 microns. When present in sufficient concentration to reduce the visual range, it is termed fog.

mixed flow fan [Fig. 157] An axial flow-type fan which also has a radial component of flow (generally small) in addition to the main axial flow component. The pressure development of a mixed flow fan is comparable to that of those types of axial flow fans which are capable of developing higher pressures.

mixed reflection A mixture of diffuse and specular reflection.

mixing valve A device which mixes together different streams of water to achieve a set mixed water temperature. For example, it mixes high-temperature flow water with the lower temperature return water of a heating system.

moat or cut-off system A system of field drains which are laid to protect a building. They are laid on one or more sides of the building to divert the flow of ground water.

modified single stack system See **sanitary plumbing, modified single stack system.**

modular boilers A range of boilers for central heating or hot water supply, each unit comprising a number of identical heat exchanger modules which vent into a common flue outlet. Typically, each module has an output capacity of 50 kW. The units are available with up to 12 modules. The arrangement offers flexible boiler performance as the modules are taken on and off the line as required by the controls.

modulate To gradually adjust a controlled device to achieve or maintain a set condition.

modulating control Control that gradually adjusts a controlled device between set limits, e.g. operates an oil burner between high flame and low flame.

mole drain A drain formed by means of a mole plough which comprises a beam to which is attached a steel blade, on the lower end of which is fixed a bullet-shaped steel tool which forms the drain. The plough is pulled by a tractor or by winch-operated cables to form a cavity in the subsoil for the purpose of field drainage. It is best suited to draining clay subsoils. The life of a mole drain system is unlikely to exceed ten years when it must be replaced.

Mollier chart A diagram plotting total heat or enthalpy against entropy.

monitoring system A means of maintaining a continuous or intermittent check or observation on certain performance parameters of an operating system, such as electric power consumption, oil flow, environmental conditions, integrity of underground pipes, etc.

Morgrip flange (Pilgrim Morside Ltd.) A flange that relies for its operation on a unique gripping system and on a double-seal arrangement which together allow the coupling to withstand very high external loading and seal high internal pressures. A series of spring-loaded ball-bearings is positioned circumferentially around the inside of the Morgrip body. Both primary and secondary seals are used, the primary seal being located between the pipe ends and the secondary seal arrangement inside the unit. Tensioning of the bolts causes the ball bearings to swage into the pipe, pulling the pipe ends together and compressing the primary and secondary seals.

The flange obviates welding and requires only a minimum of pipeline preparation. It provides a permanent joint, accommodates pipe ovality and some degree of misalignment, and is easily dismantled and removed. It is also reusable. The design ensures that there is no restriction to fluid flow. The use of the double-seal design permits the testing of joint integrity without having to pressurize the whole pipeline. The connection can be designed to be stronger than the pipe in axial tension, bending and torsion. Pipeline corrosion inhibitor injection ports are incorporated as standard. Morgrip flanges are available in the form of low-pressure or high-pressure top side couplings in nominal pipe sizes of up to 300 mm. They are ideal for a range of media at temperatures of between –200°C and +500°C.

motor bearing The support for a rotating motor shaft.

motor bed plate A metal base on which a motor is assembled and secured to a foundation.

motor, disc rotor See **disc rotor motor.**

motorized valve A valve that is coupled to an electric motor, enabling it to be actuated under automatic control.

mould A growth of fungus on moist organic matter.

mould formation To develop and grow, moulds require water to germinate. This is achieved if the surface relative humidities exceed 80%.

mud hole An inspection opening built into the water side of a boiler. It is provided with a watertight seal and cover.

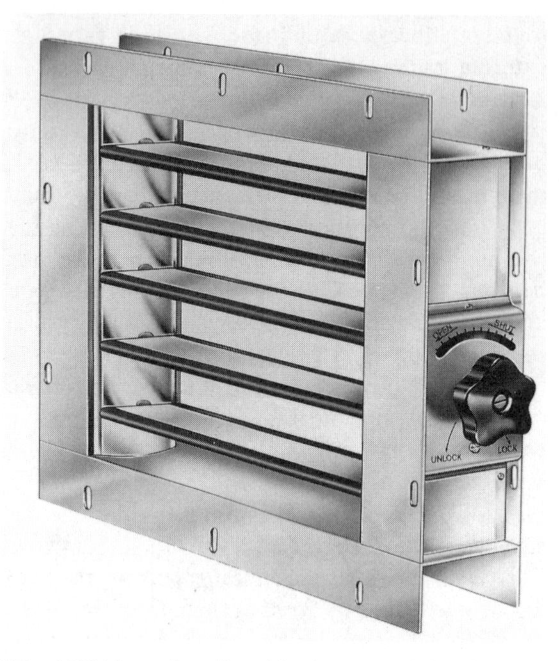

[Fig. 158] Manually adjustable damper suitable for tight shut-off arranged for attachment to flanged ductwork: incorporates positioning and locking controls (Waterloo-Ozonair Ltd.)

[Fig. 159] Manually adjustable damper for attachment to flanges on ductwork incorporating position and locking control (Waterloo-Ozonair Ltd.)

multi-addressable fire detection and alarm system A system that locates the source of fire and, by means of automatically illuminating fire exit signs, provides selected safe routes to evacuate the building. It is particularly suitable for shopping malls.

multi-compartment chimney See **chimney, multi-compartment.**

multi-flue chimney A chimney that incorporates two or more separated compartments within one shell, each connected to only one boiler or furnace. This permits accurate chimney management and is the recommended arrangement for optimum flue draught management.

multi-leaf damper [Figs 158, 159] A device for the control of air or gas flow which comprises a number of horizontal blades mounted on a common spindle and which move together to reduce or increase air flow through a duct. The shape of the blades may be of aerofoil section to give smooth air flow and the tip of each blade may be felted for tight shut-off. The arrangement may be motorized or manually manipulated.

multi-pass boiler A boiler that incorporates a number of smoke or fire tubes through which the flue gases traverse the boiler whilst transferring heat to the boiler water before exiting to the chimney.

multi-port valve A valve that embodies three or four ports which are used to change over the fluid flows in separate pipelines.

multi-stage axial flow fan See **axial fan multi-stage, contra-rotating.**

multi-stage compressor A compressor that incorporates multiple cylinders; permits a number of steps of capacity to be achieved, affording greater flexibility of operation.

multi-zone units Comprehensive heating/ventilation/air-conditioning assemblies which are fed with a heating and/or cooling medium from a central plant and which are arranged for connection to a number of separate zones in a building. Each zone has an independent environmental control or is separately connected as an outstation to a building management system. Some available units serve up to twenty zones.

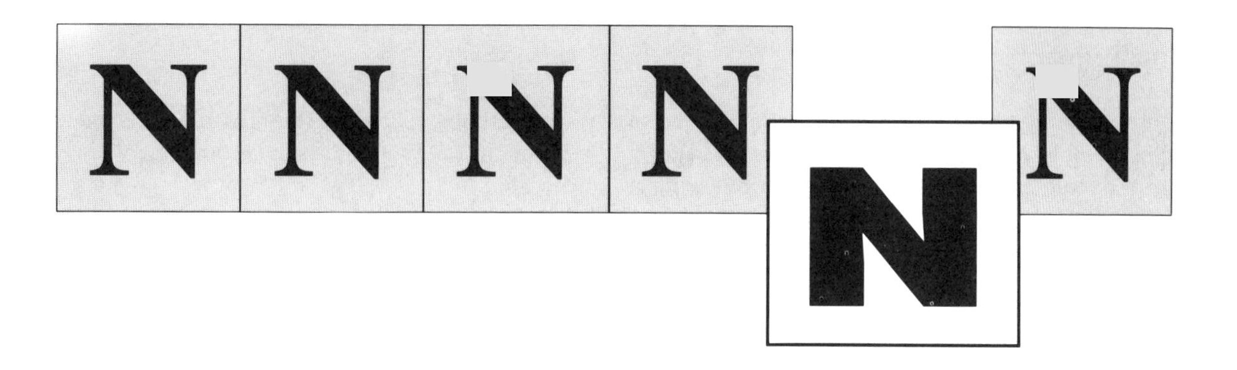

NCB National Coal Board.

NCSIIB The national certification scheme for in-service inspection bodies. It was launched in 1989 to provide assurance of a high level of competence in the in-service examination of safety-sensitive plant and machinery. It specifies the experience, qualifications, training and supervision of the engineers carrying out on-site assessment of plant and equipment to ensure consistent standards, enhanced safety and value for money.

NICEIC The National Inspection Council for Electrical Contractors. This body monitors the technical performance of members relative to conforming to the current IEE Regulations.

NICEIC certificate A certificate issued by member firms of NICEIC to certify that a particular electrical installation completed by that member firm fully meets all the requirements of the IEE Regulations.

NRA National Rivers Authority.

National Coal Board The organization that controls and operates the British state-owned (nationalized) coal mining industry. Changed name to British Coal.

National Power One of two primary electric power generating companies in England and Wales.

National Rivers Authority The organization responsible for the quality control and avoidance of pollution of waterways and in certain cases for sewerage disposal.

National Water Council A national body which concerned with the national interest in water; formulates guidelines as regards its supply and use. It issues by-laws and approves equipment related to the water industry.

natural convection Unforced air movement in a space which is caused by the difference in temperature of the lower and upper layers, causing the lighter, warmer air to rise. See also **stratification**.

natural frequency See **frequency, natural**.

natural gas Hydrocarbon gas derived from natural underground or below-sea sources, often in association with oil deposits. Natural gas generally consists mainly of methane, together with varying smaller proportions of ethane and some inert gases, such as carbon dioxide, nitrogen and helium. The gross energy value is usually in the order of 39 MJ/m^3. Natural gas is suitable for fuelling prime movers (e.g. gas turbines) and is very widely used for industrial and domestic heating applications.

needle valve A form of screw-down stop valve, generally restricted to small sizes. Body ends are in line or at right angles to each other or may be of the oblique type. There is a control disc in the form of a needle point. It is used to drip-feed or inject very small quantities of fluid.

neon (Ne) A colourless, odourless and invisible inert gas which occurs in the atmosphere. Discharge of electric current through neon at low pressure generates an intense orange-red glow – the effect used in neon signs.

neoprene A fire-resistant plastic sheet material used widely for flexible connections and cladding in mechanical services engineering.

net present value analysis A technique for analysing the economic features of systems which have a long life expectancy. It involves the discounting of future economic benefits to their present value and using appropriately selected interest rates. The assumed interest rate is critical to the outcome of the analysis.

neutral (electricity) Possessing neither negative nor positive net electric charge.

neutral point That point in a pumped heating system where there is neither negative nor positive pump pressure on the system.

Newton The SI unit of force. When applied to a mass of 1 kg, 1 newton gives it an acceleration of 1 m/s^2.

Newton's law of cooling A scientific law that states that the quantity of heat which passes per unit time from a liquid which is cooling is proportional to the difference in temperature which exists between the liquid and its surroundings.

night set-back A control programme feature which acts to reduce the temperature in a building overnight and during other periods when the building is unoccupied as a means of frost protection or to prevent excessive cooling of the building at such times.

nipple See **coupling nipple**.

nickel plating The deposition of a thin layer of metallic nickel by an electrolytic process.

nitrogen generator A generator that produces nitrogen in-house and eliminates the need for storage of individual cylinders. It operates by compressed air at a minimum pressure of 7 bar passing through a carbon sieve. Separation occurs by pressure swing adsorption where two beds of carbon material are exposed to the compressed air in a cyclic process. The generator can be on stream at a set purity within 20 minutes of start-up, and the resultant nitrogen is usually stored in steel cylinders at a pressure of 5 bar.

noise Any unwanted sound.

noise breakout The transmission of noise from an attenuated system to the outside.

noise criteria curves (NC) USA-based set of curves which relate to the sensitivity of the human ear. They give a single figure for a broad band of sound. Used for specification of indoor design noise criteria.

noise rating curves (NR) European-based set of curves related to the sensitivity of the human ear. They give a single figure rating for a broad band of sound frequencies and are used for interior and exterior design criteria levels. They offer a greater decibel range than the NC curves.

noise frequency, effect The frequency of a noise determines its audibility, character and nuisance. For example, a centrifugal fan generates most of its noise at low frequency, giving rise to a rumble, whilst an axial flow fan generates a higher frequency noise giving rise to a sound approaching whining. The frequency of the noise to be attenuated largely determines the method and material used for the sound attenuation.

noise pollution Distress caused by exposure to excessive noise. This is an important issue of growing concern.

nominated subcontractor A subcontractor selected by/for an employer who instructs the main

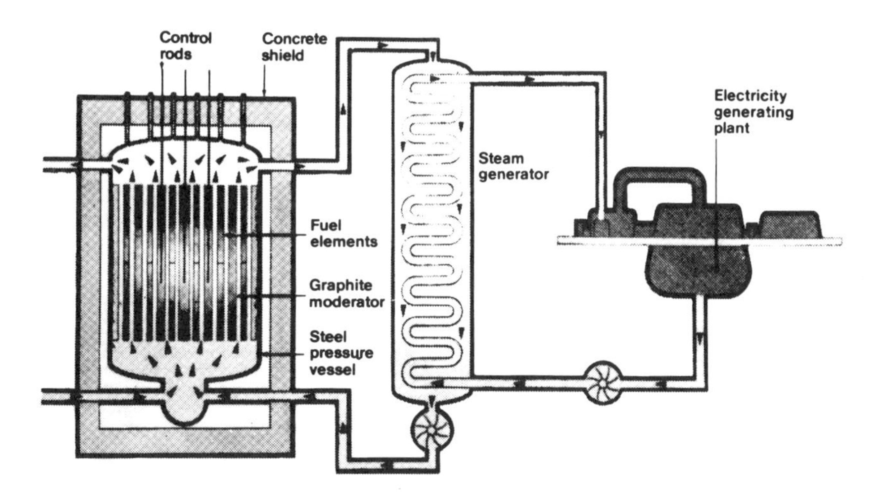

[Fig. 160] The gas-cooled (Magnox) nuclear reactor

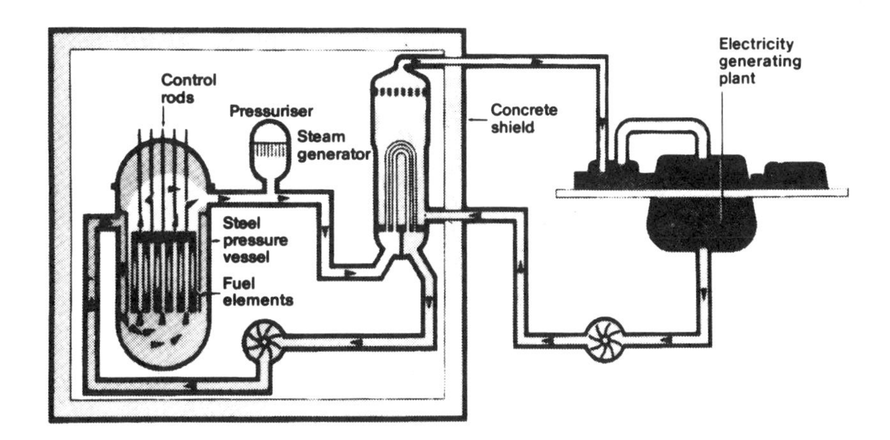

[Fig. 161] A pressurized water nuclear reactor (PWR)

contractor to place a subcontract with the selected firm, though the latter has the right under most types of contract to challenge such a selection. Whilst the main contractor is responsible for integrating the subcontract into his overall contract works, planning, payment certifications and final accounts will be separate from those of the main contract and the main contractor's responsibilities are less onerous than for domestic subcontractors.

nomogram A diagram or chart which co-relates a number of different parameters. For example, a nomogram of the flow of steam might co-relate steam velocity, the volume of steam, the internal diameter of the conveying pipe and the weight of steam conveyed.

non-ferrous metal Any metal other than iron or steel.

normalizing Applying heat treatment to steel to relieve internal stresses through heating above a critical temperature and then allowing the steel to cool in air.

notation The representation of numbers or other quantities (or entities) by a system of symbols.

notice to Local Authorities Formal notifications of relevant matters in accordance with Acts of Parliament, Local Authority by-laws, police regulations, etc. commonly the contractor's responsibility.

nuclear power [Figs 160, 161] Energy generated by the destruction of matter in a nuclear reactor. At the present time this employs the principles of nuclear fusion (using uranium fuel), but in the future nuclear fission may be employed (using hydrogen).

OFFER Office of Electricity Regulation.

oblique valve A valve that generally has a spherical body in which the body ends are in line with each other and in which the axis of the valve stem is oblique to that of the body ends. It is also described as a Y-valve. It is suitable for good performance when controlling slurries, super-saturated solutions or corrosive and abrasive fluids. Most valves of this type can tolerate severe corrosion and wear and can be restored to their original performance by the simple fitting of a new seat and disc.

octave A range of frequencies of which the upper limiting frequency equals twice that of the lower limiting frequency.

octave band A convenient division of the frequency scale identified by its centre frequency; typically 63, 125, 250, 500, 1000, 2000, 4000 and 8000 Hz.

off-peak electric heating Heating that operates by charging with electricity at off-peak (low tariff) times and discharging the stored heat as required at other times.

offset The steady-state difference between the control point and the value of the controlled variable corresponding with the set-point. It is an inherent characteristic of positioning controller action.

oil burner quarl A specially shaped refractory-lined opening through which a burner fires the fuel into the boiler or furnace.

oil-free air compressor A compressor that incorporates a piston cylinder which is not lubricated from the compressor sump. The cylinder is sealed with special rings made of Teflon or a similar hard-wearing material. Its construction ensures that there is no contact between the lubricating oil and the air being compressed and it is a valuable feature when supplying compressed air to highly sensitive instruments or process equipment. Such compressors must be manufactured to fine tolerances and this makes them more costly than conventional, lubricated compressors. Since very efficient filters have become available for compressed air systems, the use of oil-free compressors has been greatly reduced.

oil meter An appliance for measuring and recording the quantity of oil passing through it to the consumer.

oil preheater See **preheater, oil**.

oil separator A device that separates oil and oil vapour from the discharge of a refrigeration compressor.

oil shale A fossil fuel resource in the form of sedimentary rock which contains a wax-like hydrocarbon. Vast quantities of these deposits are found in many countries. It is expensive to extract by present techniques relative to conventional oil extraction, but it will become a valuable source of energy in the future.

oil transfer pump See **pump, oil transfer**.

one-pipe heating system A system that relies on the circulation of hot water for space heating through a system of single pipe loops. The flow and return connections to individual heaters are taken off the same pipe which may be a horizontal or a vertical run. Circulation of hot water through each heater is due to the difference in pressure between the points in the main pipe at which the flow and return connections enter. Tongue tee fittings may be used at the mains connections to direct the water flow in and out of the heater. The circulation usually relies on a circulating pump, but such a system can also function by thermosyphonic action alone, although this involves the use of larger bore pipes.

on-peak current That current supplied/consumed during periods when general usage is at or near peak levels, i.e. at times when the heaviest loads are imposed on the electricity power stations.

on-peak period That period of time during a week when industrial and commercial demand for electricity is high. Other times are termed the off-peak periods, during which electricity can be supplied to consumers at reduced rates. Typical on-peak periods are from 06.00 to 20.00 on Monday to Friday and Saturday from 06.00 to 13.00. Sundays are generally off-peak periods.

open circuit An electric circuit in which the continuity of current is temporarily interrupted.

open compressor A compressor that has an exposed shaft to which the electric motor or other driver is attached externally. Most such compressors use mechanical seals, rather than packing seals, to reduce refrigerant leakage around the shaft.

open-cycle gas turbine [Fig. 162] A gas turbine that takes air from the atmosphere and compresses and heats it at constant pressure by the combustion of fuel. The resultant exhaust gases are discharged at very high temperature to atmosphere or they can be utilized to produce additional thermal energy.

open vent pipe A safety pipe which vents a boiler or heat generator system to atmosphere and prevents excessive build-up of pressure due to accidental boiling in the system. It prevents the accumulation of air at the high point of the system.

operating theatre, air flow pattern This controls the temperature and humidity of the air supply to theatre, dilutes the airborne bacteria and directs the air in such a manner that the clean areas are not contaminated by air from less clean areas.

operating theatre, displacement air flow Air flow directed in a downwards vertical direction from ceiling-located discharge fittings and support nozzles. It is commonly used for the air distribution in hospital operating theatres where the air is displaced downwards towards the operating table. Bacteria set free during the operation are rapidly displaced and inductions to the directional supply air flow are prevented.

opposed-blade damper A multi-leaf-type damper in which adjacent blades rotate in opposite directions to provide tighter control over the air flow.

optical pyrometer An instrument for the measurement of high temperatures, the hot object being observed through a telescope which incorporates a standardized electric lamp. It measures electric current with an ammeter which is calibrated in units of temperature. It is suitable for the measurement of temperatures of up to about 3500°C.

optimum start control A control that senses the inside and outside air temperatures and controls

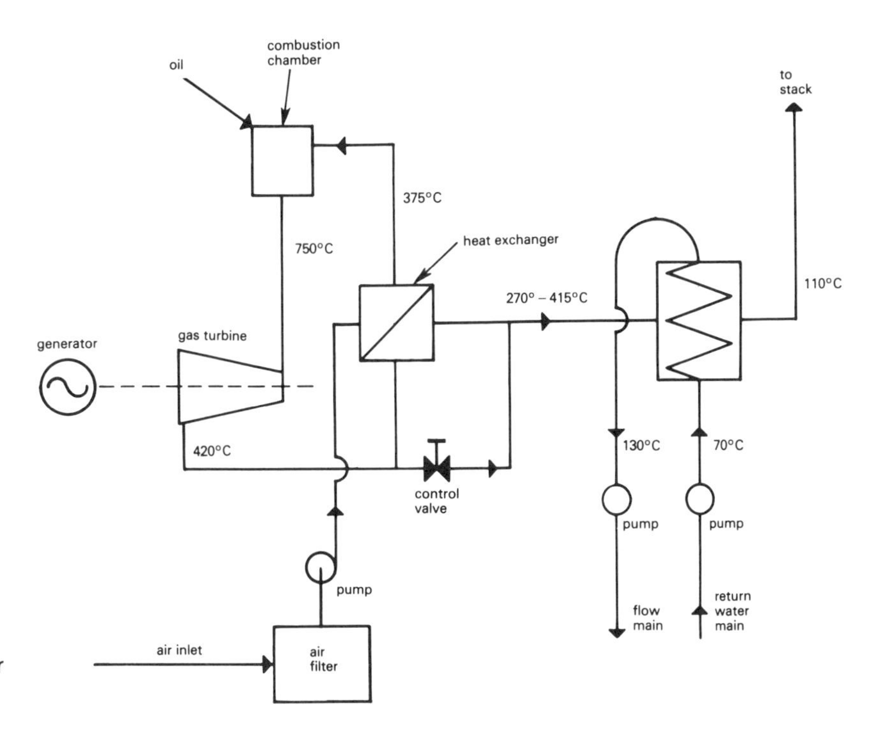

[Fig. 162] An open-cycle gas turbine circuit used for district heating supply

space heating to ensure that the building reaches a specified temperature at the start of the occupancy period. It is enclosed inside a plug-in model and utilizes electronic microcircuitry. It is suitable for the space heating control of commercial and industrial premises requiring improved control combined with optimum energy efficiency. Also known as optimizer.

organic Refers to substances which belong to the large group of carbon compounds. These are combined with hydrogen, oxygen, nitrogen and/or other elements.

organic chemistry Chemistry concerned with the organic compounds; that is, the chemistry of substances produced by living organisms as distinct from those of mineral origin.

organic compounds Chemical compounds which contain carbon combined with hydrogen. In addition they are often combined with other elements such as oxygen and nitrogen.

organic material All chemicals which are based on carbon.

orientation The direction in which a structure faces (e.g. north, south, east or west). This is of importance in heating and air-conditioning, as orientation impinges on the magnitude and timing of solar heat gains, glare and wind effects.

orifice A hole cut into a nozzle or plate which is inserted into a pipe or duct as part of a flow measurement or commissioning device.

orifice plate A nozzle or plate into which a circular hole has been drilled which is inserted into a pipe or duct to restrict the flow of fluid through it. Measuring points are established upstream and downstream of the orifice constriction to indicate the pressure loss through it and this permits the calculation of the velocity of flow through the pipe or duct. An orifice plate can also be a purpose-tailored disc of gunmetal, brass or other appropriate material which has a concentric hole with tapered edges. It is inserted between flanges or other means of support to create the venturi effect for the purposes of measuring the flow of fluid through a pipe or to facilitate commissioning. See also **venturi meter**.

orimulsion Liquid fuel comprising a mixture of water and bitumen with black tar suitable for fuelling high output boilers. It is imported from Venezuela.

O-ring A sealing ring of circular cross–section used in plumbing and drainage installations, inserted into the collar of a pipe to seal push-fit joints. It is commonly manufactured of rubber or polypropylene to the required specification. The joint is smeared with grease before insertion of the O-ring.

Orsat apparatus An instrument for the determination of carbon dioxide, oxygen and carbon monoxide proportions in a given flue gas. It comprises essentially three glass vessels which are partly filled with absorbent liquids (pyrogallic acid absorbs the oxygen, caustic potash absorbs the carbon dioxide and a solution of ammoniacal cupreous chloride absorbs the carbon monoxide) through which measured samples of the flue gases are passed. Correct use of the Orsat apparatus requires training and skill and glass components must be handled with care.

oscillation Repeated movements about a fixed point or axis.

osmosis [Fig. 163] A diffusion process across a membrane placed between two different fluids. It occurs because of a tendency in nature to equalize concentrations. For example, when a semipermeable membrane (permeable to water but not to salt) is placed between sea water and pure water, both being under the same pressure, diffusion of the fresh water into the sea water will then occur because of osmosis.

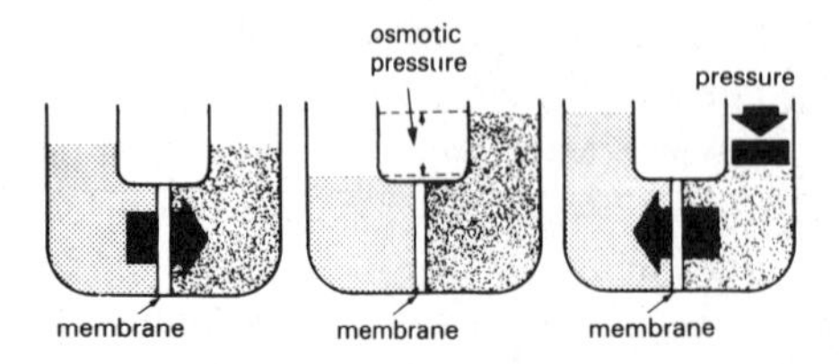

[Fig. 163] The principle of reverse osmosis

Otto cycle Introduced by Otto in 1876 as an ideal cycle of operation of gas and petrol engines against which other practical cycles may be critically compared. It comprises the following series of operations on a gas:

(a) adiabatic compression;
(b) heating at constant volume;
(c) adiabatic expansion;
(d) rejection of heat at constant volume.

outage That proportion of time by which plant availability differs from the total capacity of the installation through it being taken out of service due to breakdowns, inspections, maintenance, etc.

outfall to water course The invert level of the outfall should be about 150 mm above the normal water level in situations where periodic backflooding is likely and the outflow termination cannot be placed at a higher level. A non-return valve should be installed in the outfall pipe. The termination must be placed and shaped to avoid localized soil erosion and be protected against accidental damage or vandalism.

outstation A term for each individual control centre within a comprehensive building management system.

overcurrent That current that exceeds the current-carrying capacity of a conductor or the rated value of an electrical appliance.

overcurrent detection A method that establishes that the value of electric current in a circuit exceeds a predetermined value for a minimum specified time.

overfeed combustion A process in which ignition takes place from the bottom of the charge of fuel, the ignition phase travelling upwards in the same direction as the flow of combustion air. It tends to cause smoke and any caking properties of the coal will develop during such a method of combustion.

overfeed stoker A mechanical stoker which operates on the principle of overfeed combustion.

overflow pipe A pipe that serves to discharge excess water from a sanitary appliance, sewer or chamber. It commonly terminates with a chamfered end and bird-flap to direct the flow downwards in a location in which the water flow generates a nuisance.

overflow, standing See **standing overflow**.

oxyacetylene welding A widely used industrial process which employs a mixture of acetylene gas and oxygen in a two-port torch to weld or cut metals at high temperature. The high-temperature flame is adjustable to suit the required welding or cutting operation. See also **acetylene**.

oxidation The combination of oxygen with another substance or the removal of hydrogen from it. Any reaction in which an atom loses electrons, e.g. change of a ferrous ion.

oxygen analyser A device that provides a precise continuous measurement of boiler flue gas oxygen concentration and temperature, thereby supplying the plant engineer with all the information he requires to manually adjust the boiler/burner for operation at peak efficiency.

oxygen sampling apparatus Custom-built, permanently installed or versatile trolley-mounted system for the continuous monitoring of the oxygen content of flue gases. It commonly provides a digital readout and is equipped with power relays to actuate an external contact or alarm. It can be fitted with a chart recording facility. It is used to monitor the combustion performance of boilers and furnaces.

P

PABX Private automatic branch telephone exchange.

PF Pulverized fuel. This is used with large coal fired power station boilers.

PFA Pulverized fuel ash which results from the combustion of pulverized fuel. It is used as a constituent of certain building materials.

pH test kit A kit used for the periodic monitoring of the acidity/alkalinity of water. It usually comprises a small comparator. A few drops of the testing solution are introduced into the water sample and the ensuing colour is compared with standard slides mounted on a disc. Most liquids used for pH measurement are effective only over a narrow range, so that the suitability of the test liquid must be established for each particular application. Once the required range is known, one or two test liquids are likely to cover the range of operations.

pH value A measure of the degree of acidity or alkalinity of water related to the concentration of hydrogen ions. In water of absolute purity at 21°C the concentrations of hydrogen and hydroxyl ions are equal and each may be expressed as 10^{7} gram ions per litre. The pH scale uses the logarithm to base 10 of this value and changes the sign. Thus the hydrogen ion content of pure water at 21°C is stated as pH 7, the neutral condition. Water having a deficit of hydrogen ions (pH > 7) is alkaline and water having an excess of hydrogen ions (pH < 7) is acidic.

PI insurance Professional indemnity insurance. Professional practices use PI insurance cover to protect themselves against claims of negligent performance of their duties. Many employers of professional teams will only employ firms which are adequately PI insured.

PMBX Private automatic manual branch telephone exchange.

PVC strip doors and windows Use of PVC strips reduces heat loss through doorways whilst still allowing free access and visibility.

packaged air-conditioner [Fig. 164] A factory-made assembly of air-conditioning components installed within one casing. It is brought on to site in one piece and requires only electrical and duct connections (if required) to become operational.

packaged boiler A factory-assembled matched arrangement of boiler, combustion equipment and controls.

packaged boiler room [Fig. 165] An off-site-assembled boiler room which houses all the

[Fig. 164] Packaged air-conditioner: centrifugal fans on supply and condenser air side

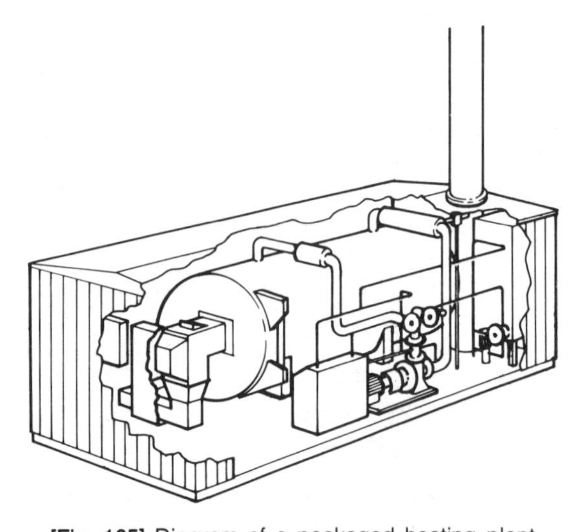

[Fig. 165] Diagram of a packaged heating plant

essential boiler, combustion, control, feed water, water treatment and flue pipe components and lifting hooks. It is usually located on the roof of the building and is placed into position by crane.

package deal A contractual arrangement between the employer and the main contractor under which the contractor is responsible for all aspects of design and construction.

packaged door heater A warm air door heater which is supplied as a pre-assembled unit ready for mounting on either a ceiling void or bulkhead. It comprises duplicate centrifugal fans, heating elements, controls and safety devices, outlet duct and grilles. Discharge ducts and grilles are available for a range of different door widths. The heating medium may be hot water, gas or electricity.

packed cock Any type of cock in which the packing material is inserted to effect a seal between the plug face and the body seat.

packless-gland flow control valve A valve suitable for high-pressure (temperature) hot water services to combat the problem posed by leakage at the valve glands using conventional gland packaging. The packed gland hermetically seals the fluid passage of the valve from atmosphere. When designed with an inclined valve spindle, a mushroom-shaped valve body and easy contours of inlet and outlet passages, minimum pressure drop across the valve is ensured.

panel heater, electric An oil-filled steel panel heater which incorporates a thermostatically controlled electric heater.

panel radiator A space heating heat emitter comprising more or less flat surfaces. It is commonly made of steel, but it is also available in cast-iron (particularly as the inset heater of a hot water towel rail) and aluminium. It can be purchased in an assembly of one, two or more panels, one placed in front of the other, offering enhanced heat output per unit length.

panel radiator, rolled top A panel radiator that has a smooth rounded upper top surface. It affords easy cleaning and a reduced risk of accidental damage to persons or their clothing.

panel radiator, seamed top A panel radiator constructed with a seam (joint) at the top surface, resulting in a sharp edge.

parabolic reflector A concave reflector, the section of which is a parabola. It produces a parallel beam of electromagnetic radiation when a light source is placed at its focus.

paraffin oil A mixture of hydrocarbons obtained in the distillation of petroleum. It has a boiling range of 150–300°C. It is also known as kerosene.

parallel circuit or connection Electric conductors or circuits are said to be in parallel when the respective terminals are of the same polarity and are connected together in such a manner that the current divides between them.

parallel slide valve A valve that controls the flow of fluid through the valve body by vertical movement of a slide moving between parallel guides.

parasitic circulation The unwanted circulation in a pipe circuit related to a domestic hot water system which is temporarily intended to be inactive. This is usually caused by the effect of thermosyphonic movement and is obviated by fitting a suitable non-return valve. The term is also applied to unwanted circulation in a vent pipe which may be caused by incorrect pipe circuitry or incorrect positioning of the vent pipe terminal.

pargeting Rendering on the faces of a chimney to form a smooth surface.

pascal The SI unit of pressure. It is that pressure produced by a force of one newton uniformly distributed over an area of one square metre.

partnering A long-term commitment between two or more organizations which are jointly engaged on a project, the purpose being to achieve specific business objectives by maximizing the effectiveness of each participant's resources. This requires a spirit of trust and co-operation rather than one of confrontation and hostility which has been usual in the construction industry. The concept derives from the USA. Refer to the NEDO report 'Partnering: Contracting without Conflict', available from NEDO Books, London SW1.

passively heated solar building A building in which solar input is stored and used internally without the aid of mechanical equipment.

pass-out turbine A steam turbine which generates electricity and from which steam is also bled for process or heating uses. It differs from a back-pressure turbine in that the remaining steam is passed through the low-pressure zone of the turbine to the condenser.

patch panel A conducting fitting which serves as a means of making and changing the various telephone connections in the main telephone (frame) room of a building.

pathogenic Refers to a substance which causes disease. A pathogen is an organism which causes disease.

payback condition (simple) An economic stipulation which requires that the achieved life of an energy conservation system shall exceed its initial cost, divided by the annual value of the fuel saved by installing the system ignoring all interest charges and inflation.

payment certificate A certificate prepared by the relevant professional, based on application for payment by the contractor, and submitted to the employer for payment.

pea gravel Small size gravel used for bedding on, or around, buried drain-pipe.

peat A fossil fuel formed comparatively recently (relative to coal). It has a low calorific value which mitigates against the transport of the fuel over long distances and it is usually consumed close to source. There are major users in former Soviet Russia and in Ireland.

Pemberthy ejector A well-known make of ejector.

penalty clause A clause in a contract stating that monies may be deducted from payments to the contractor and/or subcontractor in the event of shortfall in his contracted performance.

Penstock A single-faced type of valve which consists of an open frame and door and is used only in terminal positions. It is usually installed in tanks or channels as a means of controlling fluid flow into a pipe.

performance bond A money-backed guarantee given by the contractor to the employer that he will meet his obligations under the contract.

performance certification scheme A scheme that confirms that an accredited third party either measures, monitors or witnesses the performance testing of a product or installation.

Permaline A method of renovating a drainage system by means of a resin saturated felt liner. See **in situ drain liners**.

permanent hardness Hardness that is retained in the water after heating and boiling. It therefore does not cause scale deposition.

permeability A body is said to be permeable to a particular substance if it allows the passage of that substance through itself. For example a material that is permeable to moisture is not suitable for weatherproofing.

petrol interceptor [Figs 166, 167, 168] A means of preventing petrol entering a drain. It is a offence under the Public Health Act to discharge petroleum spirit into a public sewer; hence, some form of interceptor is required between the outlet from a garage, car park, etc. to any drain and thereby avoid the accumulation of explosive gases in the drainage system. The interceptor will take the form of a multiple chamber arranged to separate the petrol and to intercept mud, oil and other foreign matter washed into the drains.

A more dangerous case could arise in areas where tankers park to discharge petrol to storage tanks

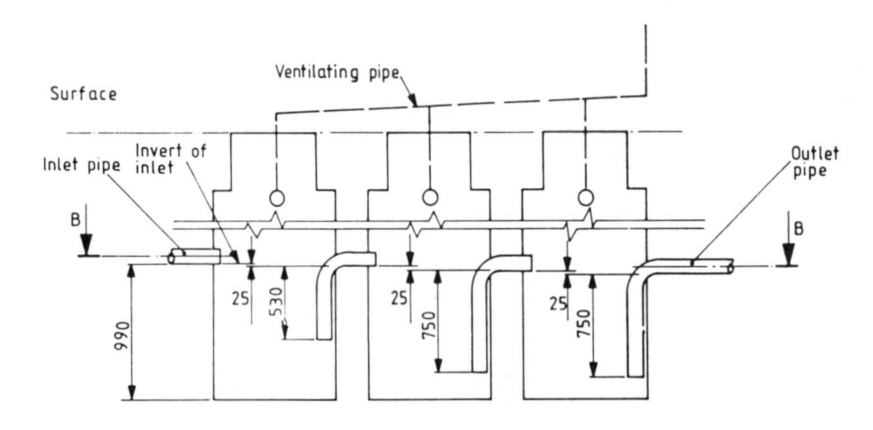

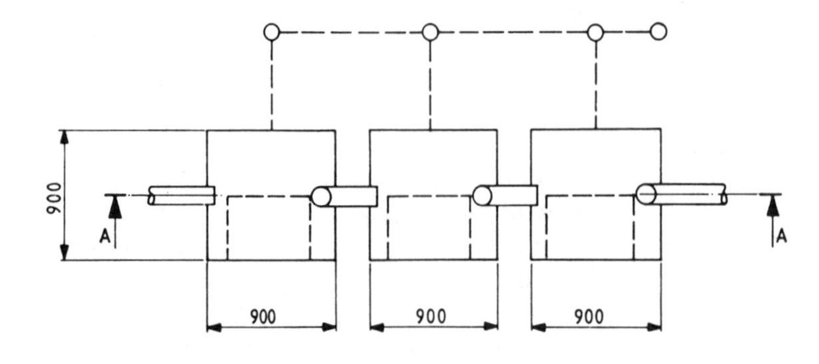

[Fig. 166] Traditional three-chamber petrol interceptor

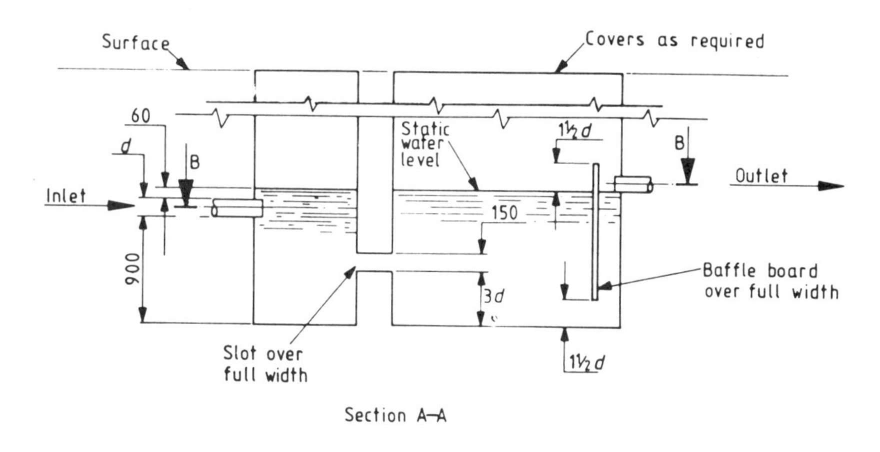

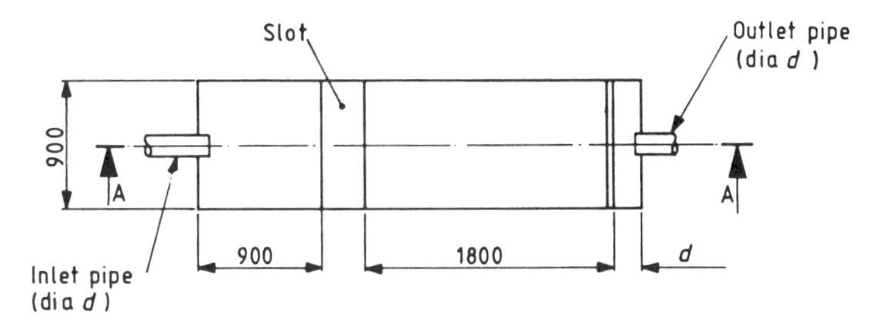

[Fig. 167] Alternative two-chamber petrol interceptor

or to have their tanks supplied with petrol. The area must be drained to a petrol inceptor so constructed that it will effectively prevent spilled petrol from entering the drain. The interceptor should serve this area alone. It requires to be adequately ventilated using a ventilating pipe which should be as short as practicable and be terminated not less than 2.5 m above the paving level nor less than 1 m above the head of an openable window or other opening into a building within a horizontal distance of 3 m. Such chambers usually have a number of interconnected separate chambers.

phase change heat storage Heat storage that utilizes the latent heat changes which occur when a solid changes into a liquid and vice versa. These phase changes take place at constant temperature and for certain materials the process of melting and freezing can be repeated over an unlimited number of cycles without change to their physical or chemical properties. Such materials have therefore the potential for being used to store substantial amounts of energy where only a small difference in temperature is permissible. By comparison, a conventional sensible storage system would occupy a volume several times greater than that occupied by a phase change storage system. Water/ice is the most commonly available material for phase change storage purposes and is commonly used in storage systems for refrigeration.

phase change heat storage, floor heating Specially formulated compositions with more appropriate fusion temperatures than water are available for heat storage for use in conjunction with heating systems and these are particularly well suited to floor warming to a high degree of comfort. In one particular system (TH30 system) capsule strips containing the phase change material are directly embedded within the floor construction below the screed and above a layer of insulation and are connected to a heating cable or pipe distribution. Because the latent heat is released at almost constant temperature such a system will maintain the floor surface temperature close to the ideal desired operating condition.

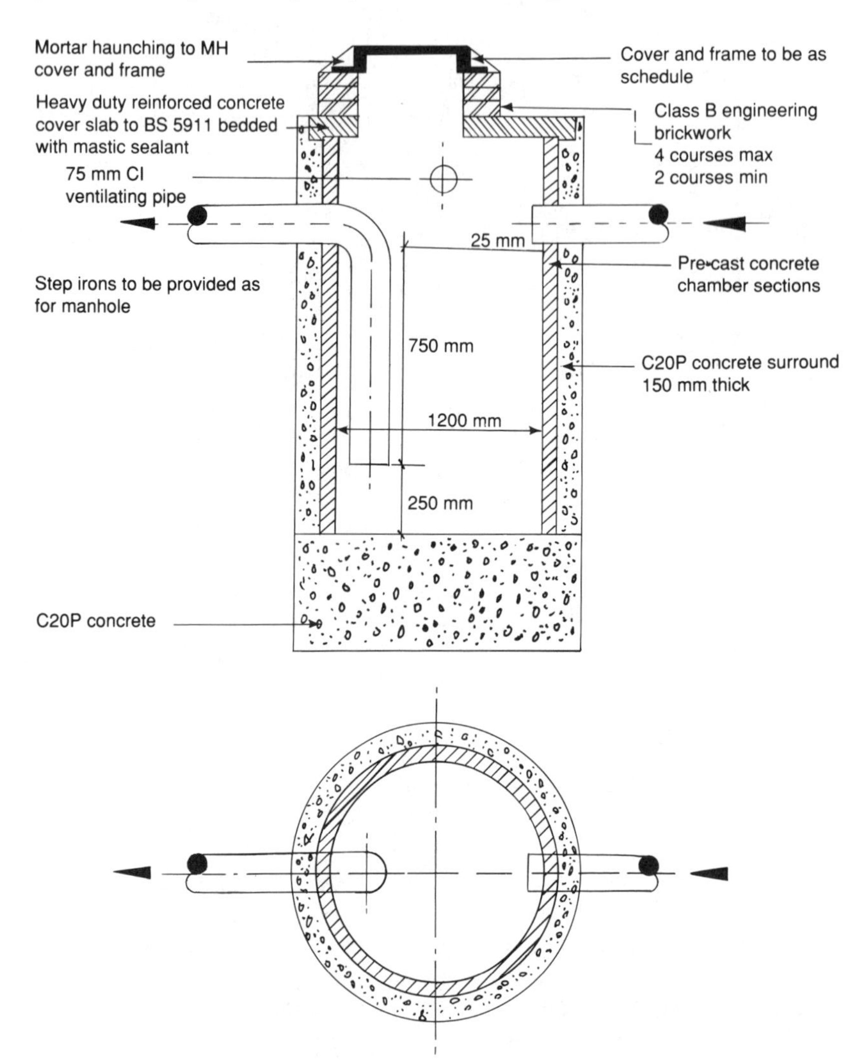

[Fig. 168] Single-chamber petrol interceptor

phase change storage system, advantages Such a system permits the storage to be heated or cooled at times of low demand, such as overnight and at weekends when the building heating/cooling requirements are low or non-existent. The stored energy is released at other times. Advantage is taken of low overnight energy tariffs and the storage capacity permits a corresponding reduction in refrigeration or heating plant capacity and capital cost.

phon A unit of loudness. It is used when establishing the sensitivity of the human ear.

photocontrol unit The automatic switching of lighting, activated by diminishing or increasing levels of natural light, independently of time-switches, etc.

photosphere The visible, intensely luminous portion of the Sun which is at an estimated temperature of 6000 K.

photovoltaic cell [Fig. 169] A cell that converts sunlight directly into electricity. It is currently used in isolated locations where the high capital cost can be justified. This is a developing technology

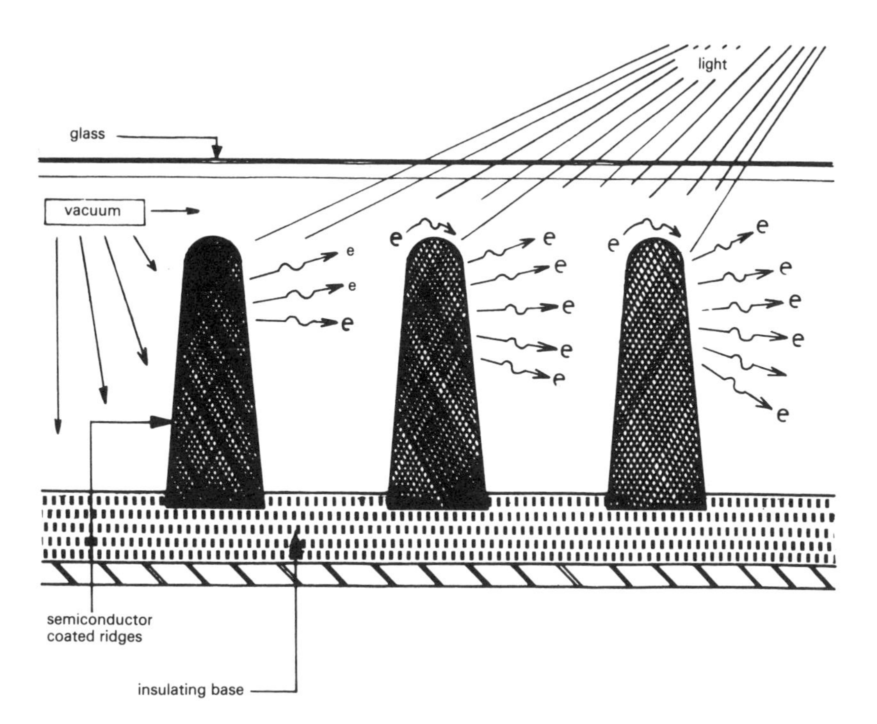

[Fig. 169] The principle of micro-corduroy (thin film) photovoltaic cell for electricity generation

which, it is anticipated, will play a major role in the future energy scenario when the cost of producing such cells will have been reduced to compete with conventional electricity generation sources.

pig trap An access arrangement to a transmission system for the introduction or retrieval of a piped pig travelling device.

pigging A technique commonly employed for separating products in, or cleaning, energy pipelines.

pigment A coloured substance which, when applied to a surface, colours it. In general, pigments are insoluble materials which may subsequently be removed by mechanical means as opposed to dyes which penetrate the fibres of tissues below the surface.

pillar tap A hot or cold water outflow control valve designed for a basin or bath. The pillar of the tap is pushed through the integral tap holes of the bath or basin and connects to the pipe union of the pipe below. A washer and a backnut are applied to the long thread of the pillar to seal the opening.

pilot burner A separate burner section within a gas (or other fuel) firing system which maintains a constant pilot flame from which the main burners (jets) are lit in response to heat demand. A pilot burner pipe connection is usually fitted with a separate isolating valve or cock. Since this burner remains in use continuously, it must be carefully adjusted (with instruments) to maintain a flame of minimum output related only to the requirement of the flame remaining stable and being adequate for lighting the main jets. The most modern boilers tend to dispense with pilot burners and employ electric or electronic main jets ignition methods. See also **pilot flame**.

pilot flame A flame maintained in continuous operation from which the main gas jets are lit on heat demand. It relates mainly to gas-fired boilers, heaters and process equipment.

Caution: the flame must be stable to withstand the disruptive effect of draughts.

pilot light See **pilot flame**.

pilot-operated reducing valve This valve is operated by means of a piston which is actuated by the pressure from a pilot valve. It is used in applications which demand large flows of fluid and a high degree of critical accuracy. Such a reducing valve can cope with large variations in flow (from 10 to 100%), and also ensures a dead-tight shutoff under conditions of 'no flow'.

pilot-operated safety valve A valve that incorporates a 'full-bore' property, ensuring maximum discharge capacity, with the additional feature of relay operation of the safety valve by the initial release of an associated pilot or control valve.

pinch valve A valve in which the flowing medium is isolated or controlled by the compression of a shoe on a flexible sleeve, usually of rubber, and which also isolates the operating mechanism from the line fluid. It is suitable for handling corrosive fluids which would otherwise damage the valve mechanism.

pipe A conduit for conveying liquids or gases. It is commonly circular and manufactured of mild steel, stainless steel, copper, plastic, glass or to special requirements.

pipe anchor A fixed point (a structure or special steel or concrete post) from which a compensated pipeline will expand or contract, there being no movement at the anchor point.

pipe bore The internal diameter of a pipe.

pipe bursting A method of lining an existing drainpipe by inserting a new pipe of the same size or larger by means of a pneumatic mole which is placed into the existing pipe in order to burst it open to form a hole to suit the diameter of the pipe being pulled along which may be larger than the existing pipe. Magna Line (by Rice–Hayward) is one such more easily controllable system which involves the use of a hydraulic buster together with either uPVC clips or expanded metal to form ducts into which conventional pipes or special liners can be inserted.

pipe chase A continuous space cut into a wall or structure for the passage of a concealed pipe.

pipe clip A purpose-made two-part support for pipes of smaller bore. The lower section is secured to the structure and the pipe is laid on to it. The upper part is then screw attached. Clips are particularly suitable for attachment to baseboards and skirtings.

pipe coating An internal off-site paint coating applied to protect the pipe material from the aggressive action of the fluids or gases being conveyed. It may also be an external (commonly off-site) paint coating for protection of the pipe installation from aggressive soil or ground conditions.

pipe coil heater A pipe used as a useful heating surface. It may be a single or double pipe, etc. loop. The coil is connected to smaller flow and return pipe connections via eccentric reducers to obviate airlocking.

pipe connector A short length of pipe, say 200 mm long, with a conventional screw thread at one end and a long-running thread with a pipe socket and back nut at other end. It is used to connect and disconnect sections of pipe.

piped pig, gas transmission system A surface-controlled travelling device, termed a 'pig', which has been developed by British Gas and is widely used for the internal inspection of gas transmission pipes.

pipe duct An enclosure which houses one or more pipes.

pipe fitting A fitting that interconnects with a pipe to direct and/or control the flow of fluids or gases within a piped system.

pipe fitting, bellows compensator This is constructed of stainless steel or chrome-molybdenum with a specified number of convolutions by one of a number of patented processes. It is installed in line with the pipe together with anchors and guides to compensate for thermal movement. It is also used to smooth out vibrations. See also **bellows**.

pipe fitting, boss An integral or welded female connection to permit direct pipe connection to a tank or pipeline.

pipe fitting, bush This connects a male pipe end into a female connection at equipment. It is short and compact. See also **bush**.

pipe fitting, capillary See **capillary pipe fitting**.

pipe fitting, compression See **compression pipe fitting**.

pipe fitting, concentric reducing socket A socket that connects one section of pipe to another of unequal bore with a common centre line.

pipe fitting, eccentric reducer This connects one section of pipe to another of unequal bore. The pipe entries are offset to prevent the formation of an air pocket at the fitting.

pipe fitting, elbow A short-radius right angle pipe connection.

pipe fitting, end cap A cap that seals the open end of a pipe.

pipe fitting, expansion loop A factory-made pipe loop which confers flexibility on a pipeline and permits thermal movement. The dimensions of the loop are designed to meet anticipated movement. It is used in conjunction with anchors and guides.

pipe fitting, female A fitting arranged for the insertion of another fitting.

pipe fitting, flange A fitting that has two correctly machined matched faced mild steel components which are each screwed on to one pipe end or matched up with an integral equipment flange and are then drawn together by nuts and bolts. A tight fit is ensured by the insertion of a flexible joint ring. The dimensions of flanges for different pressures are laid down strictly in British and other Standard Specifications.

pipe fitting, floor plate A steel plate with a female boss in the upper part and holes for screw fixings in the base. It is used in conjunction with a ring clip and support rod and the latter is screwed into the boss. The assembly is then secured with screws into the floor.

pipe fitting, guide Compensation for thermal pipe movement requires expansion loops or bellows joints. These must be placed between firmly secured anchors to direct the compensation. Guides are placed between the anchor points to move the pipes without stress and buckling.

pipe fitting, long-sweep bend An extra long radius bend, used within a pipe circulation to reduce pipe friction and ease flow.

pipe fitting, male A fitting arranged to be inserted or screwed into another fitting.

pipe fitting, reducing bush A fitting that connects a pipe end into a female connection of unequal bore at equipment.

pipe fitting, ring clip A clip used for supporting pipes from above or below. It comprises two cast-iron halves which fit over the pipe circumference. The upper half-ring incorporates at a highest point a female boss into which a suspension rod is screwed. The two halves are assembled with two screw fixings.

pipe fitting, socket A fitting that connects one length of pipe to another.

pipe fitting, strainer [Fig. 170] A fitting installed within a line of fluid flow to trap impurities, such as scale and sludge. It incorporates a removable metal basket.

pipe fitting, sweep tee A tee fitting with a long radius sweep into a junction.

pipe fitting, tank connector A water tank attachment which, through a purpose-made hole, connects the inside of a tank with the pipe off-take. It comprises a long thread and two back nuts

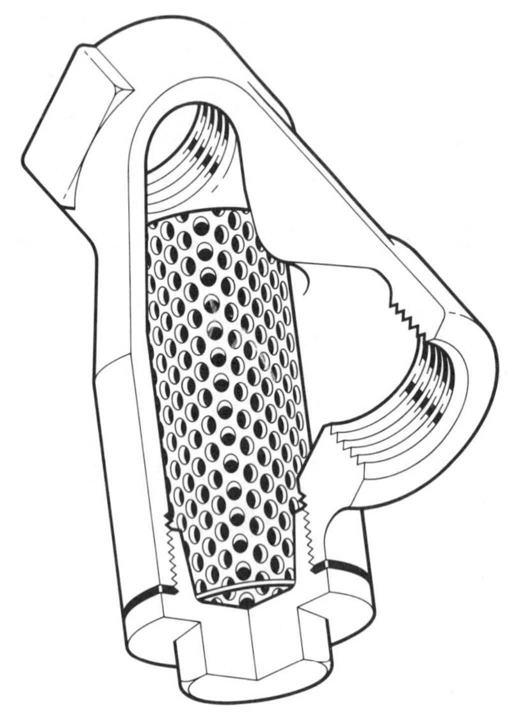

[Fig. 170] Pipe strainer (Spirax–Sarco)

to secure the joint at the inside and outside of the tank, respectively.

pipe fitting, tee A fitting inserted into a pipeline to permit off-take to a branch connection.

pipe fitting, tongue tee A tee fitting which incorporates an internal projection (tongue) to direct flow into a junction.

pipe fitting, union See **union**.

pipe float A pipe header fitted below a range of sanitary fittings to which all the fittings in the range are connected.

pipe freezing A process of pipe isolation by applying freezing jackets to each side of a section to be frozen with the object of producing an ice plug which will isolate the section (like a closed valve). Freezing jackets operate by the injection of carbon dioxide. The freezing operation avoids the need to drain down a complete water-carrying system when work is required on only one section of it.

pipe guard Protection for pipework from damage through someone standing on it. It may take the form of timber, metal step-over pipe or strong metal surround.

pipe, index circuit See **index pipe circuit**.

pipe-in-pipe A term generally applied only to an underground pipe assembly which consists of thermally insulated service pipe or pipes encased in a pressure-tight casing of a suitable material. It may incorporate an air gap between the thermal insulation and the outer protective casing.

pipe insulation Thermal wrapping or spray-applied lagging of a pipe system carrying hot or chilled fluid to conserve energy and/or as a protective covering and/or to obviate frost damage.

pipe interrupter A non-mechanical device through which water passes and into which air can enter via an unobstructed annular aperture, or apertures. When a vacuum condition occurs on the inlet side of the device, a corresponding vacuum is generated on the outlet side, thereby preventing back siphonage. It must be correctly located to avoid any restriction downstream of the interrupter that could impose back pressure on the device.

pipe interrupter, up to 40 mm bore An interrupter fitted in pipes which are not subject to mains pressure located downstream of a flow control valve. Water passes through a section of pipe into which air may be drawn through holes or slots. In the event of subatmospheric pressure conditions arising upstream of the device, these induce air flow into the pipe to prevent back flow or back siphonage.

pipe jointing material Material used to reinforce the tightness of joints. It may be special jointing compound, hemp, solvent cement or jointing tape.

pipeline A pipe joining service points some considerable distance apart. Smaller branch pipes may be connected to the pipeline.

pipeline identification Identification applied to multi-service pipe systems to facilitate the speedy identification of the service provided by a particular pipeline. Colours are specified by BS 4800 as follows.

Water services: basic pipeline (insulation) colour: green (12D 45).
Drinking water: auxiliary blue.
Cooling (primary) water: white.
Boiler feed water: white between crimson.
Condensate: green between crimson.
Chilled water: green between white.
Low-temperature central heating: crimson between auxiliary blue.
High-temperature central heating: auxiliary blue between crimson.
Cold down service: auxiliary blue between white.
Hot water supply: crimson between white.
Hydraulic power: salmon pink.
Sea or river water (untreated): green.
Fire-fighting service: red.
Gas: basic colour: yellow ochre (08 C 35).
Manufactured (town) gas: green.
Natural gas: yellow.
Oils: basic colour: brown (06 C 39).
Diesel fuel: white.
Boiler and furnace fuel: brown.
Lubricating oil: green.
Hydraulic power: salmon pink.
Transformer oil: crimson.
Sundry services: compressed air: light blue (20 E 51).
Vacuum: white on light blue.
Steam: silver grey (10 A 03).
Drainage: black.
Electrical conduits and ducts: orange (06 E 51).
Acids and alkalis: violet (22 C 37).

Note: B.S. 4800 provides a numbered code for each of the above basic colours. The identification may be applied by painting coloured strips on to the surface of the service pipes or, more commonly, by the application of adhesive colour strips to the surface of the pipes.

pipeline strainer See **pipe fitting, strainer**.

pipeline strainer, twin compartment See **twin-compartment pipeline strainer**.

pipeline, thermal compensation The arrangement of compensators, anchors and guides to permit unstressed thermal movement of the pipeline.

pipe, pre-insulated Insulation that is applied off-site.

pipe riser A vertical distribution pipe.

pipe sizing A codified method of selecting pipe sizes for a distribution network, based on required flow rates and permissible pressures or temperature drops. Appropriate computer software programs are available for this.

pipe supports, intervals Pipes must be supported at minimum intervals to prevent sagging with consequent airlocking and drainage defects. The intervals depend on pipe sizes. Appropriate recommendations are published by the CIBSE.

pipe thread A screwed pipe termination.

pipe trace heating, electric See **electric surface heating**.

pipe tracing, steam A method of maintaining a specified temperature of a fluid, e.g. fuel oil, whilst conveying same through a pipe. It comprises a small-bore steam pipe which is fitted to be in contact with the carrier pipe throughout its length and which incorporates a thermostatic control valve, steam trap set and efficient thermal insulation covering, water-proofed where the pipe travels outside buildings.

pipe union connector A two-part connector in which the two faces link together and are secured by a nut which runs from one end over the other. It is used to assemble and disconnect sections of pipework without having to disconnect the remainder.

pipe, vapour seal Impervious material incorporated with the thermal insulation of pipes conveying gases or fluids at low temperature to prevent condensation of atmospheric moisture on the pipe wall. Its main use is for chilled water pipes in air-conditioning and refrigeration systems.

pipework An assembly of pipes and fittings into a piped installation.

pipe wrapping Water-resistant impregnated bandage site-applied to pipeline to protect it from the effects of aggressive soil, ground or atmosphere.

piping See **pipework**.

piping complex A multiple assembly or network of pipes associated with major engineering installations, such as factories, district heating or refineries.

piping, ladder system Pipework where the flow pipe rises at one end of the premises being heated and the return pipe drops at the other. The heaters are connected to a single pipe run between the two risers.

piping, one-pipe system A system where the flow and return connection to each heater is taken off the same distribution pipe.

piping system A system that may consist of pipelines and/or pipe complexes.

piping, two-pipe system A system where each heater is connected to separate flow and return pipes.

piston check valve An enhanced disc check valve which incorporates a dashpot, consisting of a piston and cylinder which cushions the valve movement.

pitch fibre pipes, jointing Such pipes have machined tapered joint surfaces and are jointed by a driven fit. The pipes depend on their inherent flexibility to accommodate ground movement. Where large-scale ground movement is anticipated, the rolling ring type of joint should be employed. The joint depends for its watertightness on the accurate machining of the joint faces.

pitot tube Portable apparatus commonly used for field measurement of air velocities and pressures. It measures the velocity head from which the velocity in the immediate vicinity of the instrument can be calculated. In use, the tube is inserted into the duct or pipe facing the direction of flow.

pitting Wasting of the inner surface of a pipe or item of equipment due to corrosion or erosion.

plastic pipe, earth bonding Since plastic pipes do not conduct electricity, they must not be used for the purpose of earth bonding.

plastic pipe, jointing by compression fitting All compression fittings are supplied fully assembled with rings and compression nuts. The following steps are required to make a joint.

(a) The pipe end is cut square and the end is deburred.
(b) A 45 degree external chamfer is formed on the pipe end with a flat file or chamfering tool and wiped clean.
(c) The pipe end is marked to the required socket depth (see manufacturer's tables).
(d) Lubricant is applied to the pipe end as well as to the 'O' ring.
(e) The pipe is inserted into the fitting past the slight resistance of the 'O' ring up to the stop inside the socket.
(f) The ring nut on the fitting is tightened as far as possible by hand and then finally tightened using a strap wrench. A pipe wrench or steel grips must not be used on either pipe or fitting as these are likely to damage the plastic.

This type of fitting can be repeatedly dismantled and reassembled without impairing its sealing efficiency provided that the component parts remain undamaged.

plastic pipe, jointing of threaded fittings These are screw-jointed with the insertion of PTFE tape to give a tight seal. The fittings must not be assembled dry. Hemp and/or linseed oil based jointing compound must not be used as these are likely to have a detrimental effect on the polyethylene pipework.

plastic pipe jointing, solvent cement The plain pipe end is coated with solvent cement prior to being push-fitted to the plastic pipe fitting.

plastic pipe, jointing with solvent cement The pipe end and the fitting must be cleaned of dirt and grease. The end is painted with solvent cement paste and is immediately pushed into the fitting to make a sealed and permanent joint which cannot be dismantled.

plastic plumbing system A system constructed of plastic pipes and fittings.

plastics A term referring to materials which are stable when in normal use, but which at some stage of their manufacture are plastic. They can be shaped or moulded by heat or by imposed pressure or by a combination of heat and pressure.

Plastronga A proprietary make of high-density polyethylene pipe.

plate heat exchanger A heat exchanger that typically comprises a pack of metal plates in corrosion-resistant material enclosed in a frame. Liquids flow in thin streams between the plates while peripheral gaskets control the flow. Troughs pressed in the plates can give extreme turbulance and together with a large surface area produce high heat transfer. Typical applications: condensers, evaporators, interface heat exchangers, process systems, etc.

plate-type condenser See **plate heat exchanger**.

plate-type evaporator See **plate heat exchanger**.

plate valve A valve that incorporates a slicing effect.

plenum A space in which ventilation supply or extract air can accumulate.

plenum air extract chamber The designated space into which exhaust air is directed from a number of extract terminals and from where it is recirculated to the main plant or is exhausted to atmosphere.

plenum air supply chamber The designated space into which air is supplied and from where it is distributed to ventilation or air-conditioning terminals.

plug cock A taper-seated cock in which the plug is retained in the body by means of a washer, screw and nut at the smaller end of the plug.

plug valve A form of shut-off device which incorporates a plug, either parallel, taper or spherical in shape, which can be turned to move its port or ports relative to the body seat ports to control the flow of fluids. It incorporates design features which minimize the friction between the plug face and the body seat during the turning of the plug and/or seal them against leakage.

plug valve patterns Plug valves are available in four patterns having the following port shapes and areas.

(a) Round opening pattern: has full-bore round ports in both body and plug.
(b) Regular pattern: has substantially full-area seat ports of rectangular or similar shape.
(c) Venturi pattern: has reduced seat ports of round, rectangular or similar shape in conjunction with a body throat approaching a venturi.
(d) Short pattern: has substantially full-area or reduced-area seat ports of rectangular or similar shape with restricted face-to-face dimensions.

plumbing, final connection This is commonly the pipe between the stopcock controlling a sanitary fitting and the hot or cold water tap.

plumbo solvency The ability of some waters to dissolve lead.

plume A visible discharge to atmosphere of chimney gases or cooling tower drift.

plunger See **cleaning of sanitary installations**.

pneumatic Relates to operations conducted with compressed air.

pneumatic controls Devices which are actuated by compressed air.

pneumatic control system [Fig. 171] A system that incorporates compressor(s), air receivers, an air drier, compressed air pipes and pneumatic valves, dampers, etc. Its applications include automated processes and building services installations.

pneumatic control valve A valve which embodies a diaphragm chamber and a piped connection to a compressed air supply which acts on the diaphragm to modulate the operation of the valve.

pneumatic damper A damper in a duct which incorporates a diaphragm chamber and a piped connection to a compressed air supply which acts on the diaphragm to manipulate the damper setting.

pneumatic ejector A type of sewage lifting apparatus. It relies on a supply of compressed air to lift the sewage from a low to a higher level.

polluted air controller A controller that monitors cigar and cigarette smoke in the atmosphere and switches on an extractor fan when a predetermined contamination level is reached, switching it off when the room is clear. The controller can be switched to an alarm condition to monitor carbon monoxide, methane and hydrogen.

polyethylene pipe, quick joint fittings A proprietary all-plastic design of compression joint suitable for use with all classes of low-density and high-density polyethylene pipe to a maximum pressure of 12 bar. They have high impact resistance, are corrosion resistant and resistant to ultraviolet light. They are easily dismantled and

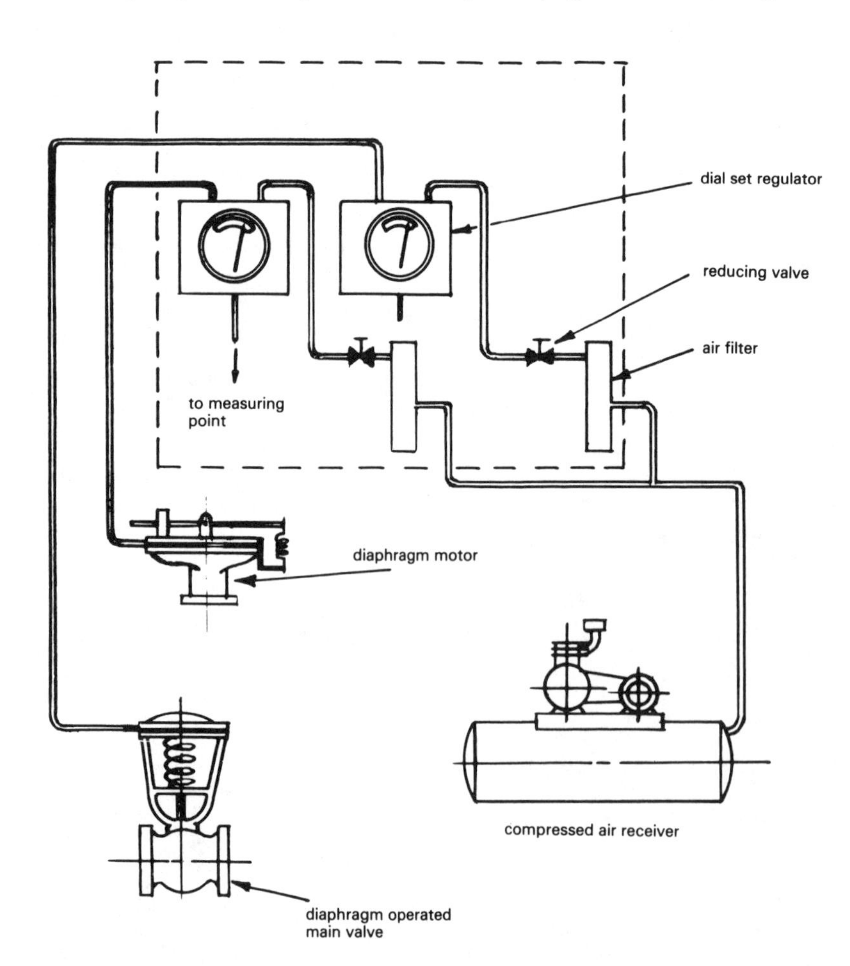

[Fig. 171] Pneumatic control system: diagrammatic arrangement

re-assembled and therefore are particularly useful for temporary pipelines.

polyethylene (plastic) pipe This pipe is widely used for water distribution. It is durable, non-toxic, non-contaminating, unaffected by corrosive soils, aggressive waters and chemicals as well as being lightweight, flexible and easily handled. It is supplied to site in coils.

polyphase Describes an alternating current system where the circuit is divided into several branches or phases.

polypropylene ball blanket A blanket for covering tanks, baths, etc. to minimize fumes or to retain heat.

polyurethane foam Rigid polyurethane thermal insulating foam sprayed *in situ* internally or externally on roofs or walls.

polyvinyl chloride pipe (PVC) Unplasticized PVC pipe, injection moulded or fabricated. It is supplied in a rigid length and offers an advantage for pressure mains, especially with pumped systems. It is used widely in domestic and commercial plumbing installations and is jointed with solvent cement type plastic fittings.

Ponding (drainage) Term describes the condition where a pool of water is held within a section of a drain run; the result of the drainage pipe at this location being lower than the sections adjacent to it. Can be caused by ground settlement or disturbance of the drainage during construction works.

pop safety valve A direct-loaded spring safety valve.

popular light sources, efficiencies [Fig. 172] Fluorescent tubes are currently the most popular light sources for offices and factories, and with some domestic applications, as they are relatively energy efficient, but the luminaires are more expensive than tungsten and mercury light sources. A wide choice of fluorescent tubes is available, some more suitable to a specific purpose than others. The conversion of electrical energy to light differs for alternative luminaires; from about 80 lumens per watt to only 30 lumens per watt input.

A development of the conventional fluorescent lamp is a miniature version which has the fluorescent ballast and the starter also miniaturized and fitted into the base of a bulb which may have a screw or bayonet form of attachment. A particular application of this is the 'Philips SL' bulb, which uses phosphors from the Philips Colour 80 series fluorescent lamps, derived from colour television technology. It offers a high light quality combined with enhanced light output. It is claimed, following in-use tests, that the SL lamp will reduce the electricity consumption, compared with tungsten lighting, to one quarter and will

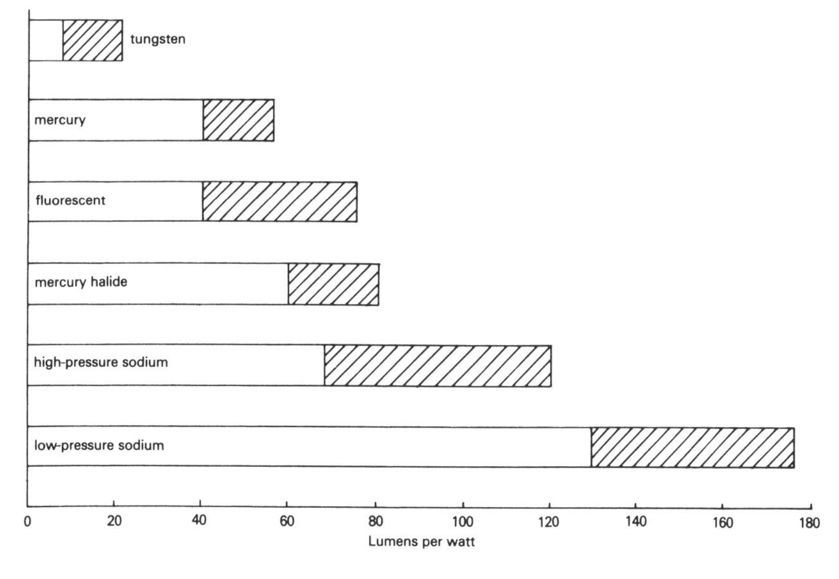

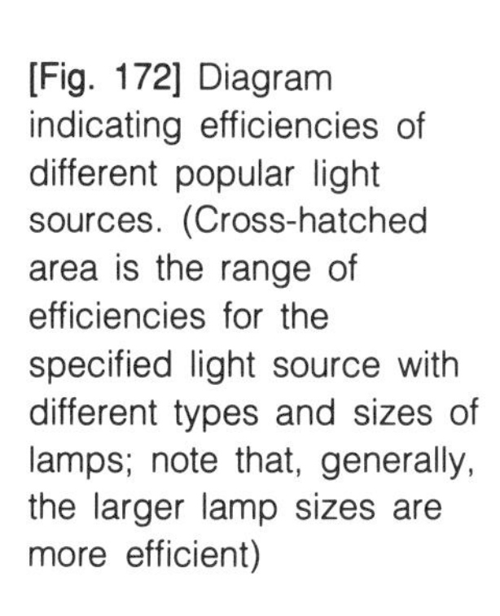
[Fig. 172] Diagram indicating efficiencies of different popular light sources. (Cross-hatched area is the range of efficiencies for the specified light source with different types and sizes of lamps; note that, generally, the larger lamp sizes are more efficient)

increase the life of the bulb five times to about 5000 hours.

One note of caution: before deciding to replace all tungsten bulbs within existing luminaires with SL bulbs, one must check that the different bulb dimensions can be accommodated.

Mercury halide and sodium lamps are also now available.

porthole An inspection opening through which the interior of a furnace, etc. is viewed.

positive displacement compressor A compressor that functions by reducing the volume of gas in a confined space, thereby raising its pressure.

potable water supply Water that is suitable for drinking.

potential Electrical pressure, measured in volts.

potential difference The difference in the electrical states existing between two points.

potential energy See **energy, potential**.

potentiometer An instrument which compares the electromotive forces and potential differences in an electric circuit by the null method, i.e. does not depend on measuring deflection. It is incorporated in detecting and control equipment.

pour point The lowest temperature at which a liquid will flow at the temperature of 2.8°C above that at which a sample of oil just ceases to flow when cooled under standardized test conditions.

powered roof ventilator An assembly of a fan and an electric motor within a weatherproof casing which is located on the roof of a building and commonly serves to extract air to atmosphere. The casing is hinged or is otherwise easily accessible.

power factor The ratio of power output (watts) to the total equivalent volt–amperes input. It relates to alternating current only, and is used in calculating the non-inductive electrical loading of an electric network. The greater the inductive load (e.g. in induction motors) the lower the power factor. The majority of industrial loads have a power factor of less than unity; in most cases where induction motors predominate, the power factor will be in the order of 0.7 lagging. A low power factor is undesirable as it reduces the effective current-carrying capacity of a distribution network. Most supply companies penalize the consumer with a low power factor under the terms of the supply tariff agreement.

power factor correction equipment Equipment to convert a lagging power factor to, say, an overall factor of 0.95. The necessary leading current may be supplied by employing phase advancers, synchronous motors (running over-excited) or capacitors.

PowerGen One of the two primary electricity generating companies in England and Wales.

power recovery turbine A turbine that recovers power from high-pressure processes. The turbine can be coupled to a motor-driven pump with the motor providing the balance of power required between the pump power demand and that recovered by the turbine.

power shower A shower that incorporates an electric pump to boost shower outlet pressure. It is particularly suitable for apartments with low headroom which cannot obtain adequate pressure at a shower from a header tank. The shower is powered directly off the mains water supply and does not require electrical input. It incorporates a non-return valve.

practical completion That stage in a project when the building is handed over to the employer for beneficial use, subject to the rectification of a schedule of remaining outstanding works which is commonly attached to the practical completion certificate.

pre-charged insulated refrigerant line A line employed for the speedy connection of air conditioners. These lines are available in lengths of up to 15 metres.

pre-filter A coarse air filter installed ahead of the main air filter of a ventilation plant to arrest the larger particles and thereby reduce the dust burden on the main, more efficient, air filter.

preformed new sewer replacement pipe A continuous slip lining using long lengths of polyethylene insertion pipes to line sewers from manhole to manhole by means of size-for-size or oversize pipe bursting, to pull through a larger diameter pipe. The Snap-Lock system (by Stewarts and Lloyd Plastics) uses short lengths of polyethylene pipe for relining sewers from within the manholes; these pipes are assembled by just snapping together and are then pushed through the drain. Demco sectional slip liners use integral threaded jointed sewer lining pipes of modified polypropylene in lengths of between 500 mm and 6 m long. These are assembled within a manhole by being screwed together and are then pushed through the drain as a slip liner. Vitrified clayware and reinforced concrete jacking pipes are used for lining large diameter sewers.

pre-heater A heater inserted into a system to warm the incoming fluid before it enters the main heater.

pre-heater, air A pre-heater placed into a duct system close to the outside air intake to avoid damage to the ductwork by very cold saturated air.

pre-heater, oil A pre-heater placed into the oil supply pipe of a heavy fuel oil plant close to the off-take from the oil storage to permit free flow of oil to the burners.

pre-heating A process of heating a fluid before it undertakes its main function, for the purpose of permitting easier flow or handling or to achieve better combustion. For example, heavy fuel must be pre-heated at the discharge from the oil storage before it can be economically conveyed to the oil burner.

pre-insulated pipe [Fig. 173] Pipe that has its specified thermal insulation applied off site under supervised factory conditions. The pipe is increasingly used on the buried pipes of district heating installations. Systems are available which incorporate valve pits, anchors, inspection manholes and leak detection provision.

preliminaries Those items in the specification which refer to the basic maintenance of the contractor's establishment at site, such as site huts, canteen, site telephone, progress photographs, toilet facilities, etc.

premises distribution system A system that permits telephone and data systems to share a common bus which utilizes a variable-twist cable suited to voice and data transmission. Universal information outlet (UIO) connectors can be used to connect either telephone or computer terminals. The system may use optical fibre cables which offer major improvements in transmission speed as well as immunity to electrical interference, coupled with the ability to be used over long distances. It will support voice, text, video and graphics traffic.

prescribed processes and substances The IPC system empowers the Secretary of State to designate (or prescribe) those industrial processes which, because of their potential for significant releases of harmful substances, will require to be assessed in the light of their impact on the environment taken as a whole, before they can be authorized to operate. The relevant processes are designated in the Environmental Protection (prescribed processes and substances) Regulations 1991. These came into force in England and Wales on 1 April 1991 and took effect in Scotland on 1 April 1992. Prescribed processes required authorization from H. M. Inspectorate of Pollution.

present worth A sum, *X*, expressed in monetary terms, to be paid in the future will be worth less than its present monetary value. Invested now, it would earn interest. To establish the true value of *X* in today's terms, one calculates it on the basis of present worth plus interest using the compound interest formula in reverse.

pressure, absolute See **absolute pressure**.

pressure balanced valve A valve that controls low pressures, particularly in gas control to burners. The controlled pressure is measured under the valve diaphragm, above which is applied a preset balancing pressure.

pressure control valve A valve that controls pressure in a fluid system. When not requiring an

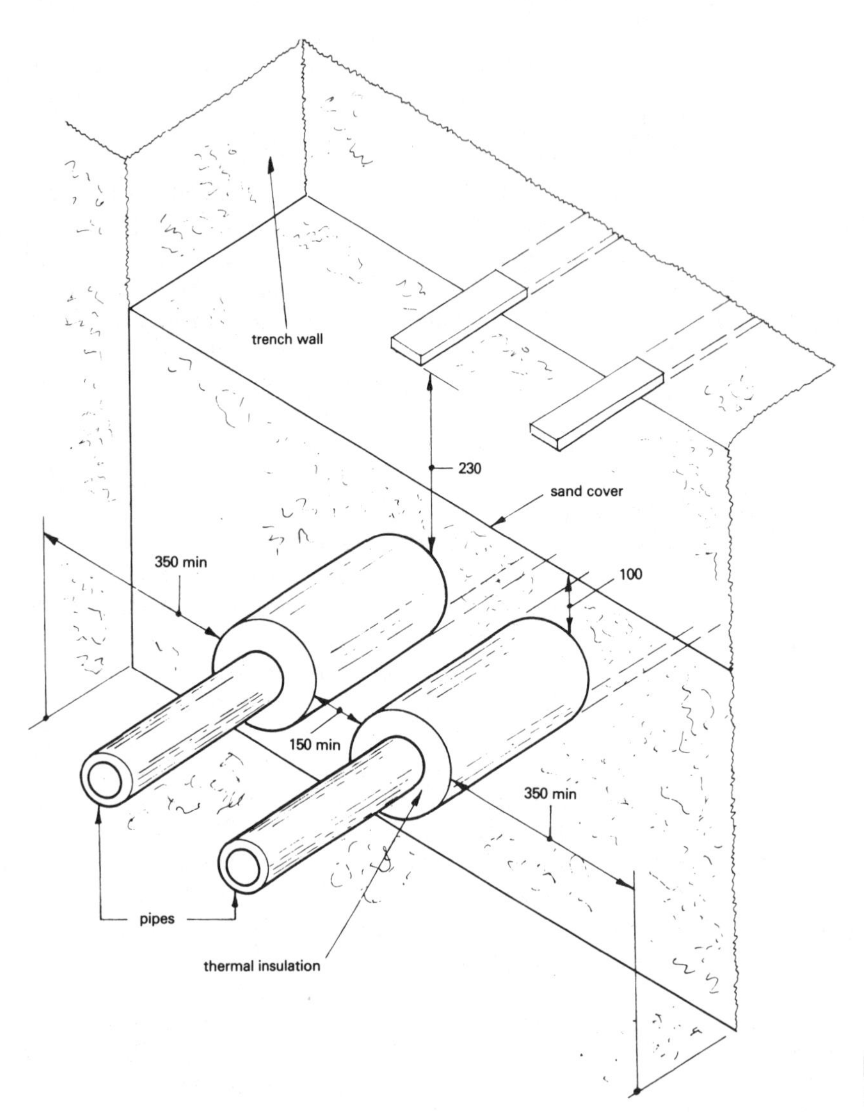

[Fig. 173] Pre-insulated pipes in position

external power source it is termed 'self-operated'. When relying for its actuation on external power, it is a specific form of an 'automatic process control valve'. It is available in various types and modes to suit particular applications.

pressure cut-out A device which switches off an installation or item of equipment when its safe set pressure is exceeded.

pressure differentials measuring equipment for commissioning of heating and chilled water systems A portable test set comprising single-column type manometer connected to measure the pressure difference across an orifice balancing valve, complete with bypass, isolating valves and lengths of connecting plastic tubing which terminate with probe units.

Most heating and air conditioning applications operate with velocities which result in a low energy loss, so that the pressure loss across the orifice balancing valves will require differential measurements of up to 450 mm water head loss. Higher pipeline velocities require measuring devices reading up to 3000 mm water head loss and use mercury-filled test sets. It is easiest to read head loss signals up to 400 mm water on a fluorescent-filled manometer (specific gravity 1.88), rather than on a mercury-filled set. The test set must be used in the upright position.

pressure filter A filter contained within a steel pressure vessel which can operate under pressure if hydraulic conditions in the system require it to do so.

pressure jet burner A combustion appliance which discharges a jet of oil from a suitably sized nozzle into the combustion space under high pressure before igniting the oil. An oil or gas burner which incorporates a fan and supplies the combustion air under pressure. It does not rely on the chimney draught for the induction of combustion air, but requires a suitable chimney or exhaust arrangement for the evacuation of the combustion products.

pressure of water supply, definition of high and low pressure Low pressure is up to 13.7 m; 1.35 bar. High pressure is up to 30.5 m; 2.98 bar.

pressure reducing valve [Fig. 174] A valve that enables point-of-use steam (lower pressure) to be provided as variously required from a single boiler operating at its optimum generating pressure. Under dead-load conditions the valve should give an absolute tight shut-off.

pressure regulator A spring-loaded device that delivers a constant, reduced differential or relief pressure in any application which requires the control of water, steam, oil or other fluid without using an external power source.

pressure regulator, compressed air A regulator that reduces the input compressed air pressure to the required operating level for efficient operation of the connected equipment.

pressure relief damper A damper that maintains specified ventilation rates and specified pressure.

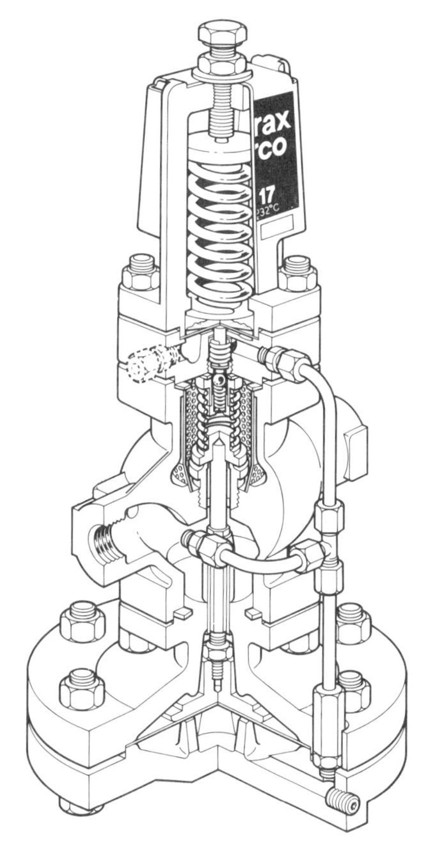

[Fig. 174] Pilot-operated pressure-reducing valve (Spirax–Sarco)

pressure retaining valve A self-operated valve that maintains predetermined pressure upstream.

pressure set A device that stores a specified volume of water under a cushion of air to permit a variable demand for water to be met economically. It typically comprises two or more centrifugal pumps, a suitably piped pressure tank, pressure switches, non-return and sluice valves and a control panel, all mounted on a common base plate.

pressure stat A switching device which is actuated by a change in pressure.

pressure switch A switch that senses pressure differentials. It offers an advantage over a flow switch, and is without moving parts. It activates controlled equipment when the specified differential has been achieved, lost or exceeded.

pressure vessel A container in which the fluid is maintained at a pressure above atmospheric; e.g. the receiver of an air compressor, a steam accumulator, a storage cylinder or reservoir associated with steam or pneumatic processes. It must be equipped with appropriate safety devices.

pressurization The process of applying an external pressure to a fluid. The purpose may be to achieve more economical distribution of the fluid, to permit a boiler system to operate at higher temperature or to obviate the need for a feed and expansion system.

pressurization unit A unit purpose-made to pressurize (usually by nitrogen) water systems (hot or cold) to provide an artificial head for the purpose of preventing boiling in a high-temperature (high-pressure) hot water system or to supplement the static pressure in a cold water distribution.

preventative maintenance Maintenance that is planned and carried out to maintain plant in good operating order to anticipate and prevent breakdowns.

primary air Air applied to a bed-type furnace, where it is introduced below the bed of solid fuel to provide the main or primary air for combustion. In liquid or gas fuel burning, the primary air is introduced into the combustion chamber with the fuel, often in a pre-mixed fashion. In a cyclone or vortex furnace, the primary air is often the only source of combustion air and is also used to pneumatically convey the solid fuel material into the combustion chamber.

primary air filter A filter that serves to arrest coarse and heavier particles in a ventilation air stream.

primary circuit A circuit directly connected to the main energy source, e.g. a boiler or chiller.

primary feedback A signal related to the controlled variable which is compared with the reference input to obtain the actuating signal.

primary generating companies The privatization of the electricity supply has resulted in the establishment of two primary generating companies in England and Wales: Powerqen and National Power. In Scotland two new supply companies have come into being: Scottish Power and Hydro-Electric.

primary hot water circulation See **hot water circulation, primary**.

prime contractor The contractor who carries the major responsibility for the execution of a project.

prime cost A lump sum included against a specific item of equipment within the specification on which the contractor is entitled to a specific percentage payment.

priming The starting of a pump by the action of atmospheric pressure on the surface of the pumped liquid causing a flow of liquid. It is essential to ensure that a pump can be primed and that the priming is sustained. To this end, the top of the pump casing or the open end of the suction pipe should always be submerged by 1.5 times the diameter of the open end of the suction pipe or by 150 mm, whichever is the greater. Where a suction pipe is at a lower level than the pump casing, an air venting device must be fitted to release the air as the liquid level in the sump rises. The maximum suction lift is in practice usually limited to between 2 m and 4 m.

The term is also related to steam boiler plant when the discharge of steam contains excessive quantities of water in suspension because of violent ebullition in the boiler. It is induced by surging and aided by a high water level. Vigorous movement of steam bubbles to the surface and their explosion causes surging and artificially raises the water level. The ensuing spray tends to be carried over with the steam flow from the boiler. Priming can be avoided by limiting the concentration of dissolved solids to the recommended levels related to the particular boiler type and steam discharge pressure.

process drain A drain that conveys waste discharges from industrial or laboratory processes or trade wastes only.

process heat recovery coil See also **run-around heat recovery coil**. Pipe(s) conveying water or steam which is used to transfer heat from one process or medium with the object of recovering heat which would otherwise be wasted. When using such coil arrangements in conventional building ventilation systems, the exhaust ducts and coils can be protected adequately by air filters, but with process plants, the condition or composition of the discharge air generally requires other forms of protection.

For use in process plants, the coil arrangements must be designed specifically to either avoid the risk of clogging the coils and/or to enable them to be easily cleaned. The nature of the contaminant that will be in contact with the coil must be known and understood; this may be granular or fibrous, soluble or insoluble and one must establish whether the exhaust air will be cooled below the dewpoint and thereby cause a flow of condensation. Bare tubes minimize the risk of clogging, but offer a very poor cost/performance ratio. Finned tubes are less of a problem as the fin spacing is increased. Horizontal coils are more easily cleaned where gravity assists the hosing-down process. Multiple-row finned coils are difficult to clean unless they are divided into sections with intermediate access for cleaning when, ideally, each section would be two rows deep which leaves about 35 mm depth into the coil for cleaning. Where intermediate cleaning access is provided, the fin spacing may be somewhat closer.

The coil construction will face a corrosion risk in most applications, so that suitable materials must be used to prolong the life of the coils. In cases where severe corrosion is likely, it may be more economical to accept a lower working life (with arrangements for ease of coil replacement) rather than use very expensive alternative materials which may give a longer life.

process of sterilization In this process the water storage tank/cistern is cleared of all debris, the whole system, including the water tank, is filled with water and is thoroughly flushed out via the drain cock at the bottom of the system. The system is then refilled with water and, as the water tank is filling, a sterilizing chemical containing chlorine is added to promote thorough intermixing of the water and the chemical, the dose being such as to result in 50 parts of chlorine to one million parts of water. Proprietary brands of sterilizing chemicals should be added in the proportions instructed by the manufacturers.

After filling the system, the incoming mains water supply is shut off and each tap on the distribution pipes opened successively, commencing with that nearest to the water storage. When the water which issues from a tap smells of chlorine, that tap is closed. This is repeated until all the taps have been checked and turned off. The water storage is then once more refilled to the water-line with water to which the correct dose of the sterilizing chemical has been added. The system is then allowed to remain charged with the treated water for a period of not less than 3 hours, after which a test is made by smell for the presence of residual chlorine. If there is no such smell, the sterilization has not been satisfactory and the process must be repeated.

Before any water is drawn for domestic purposes, the whole system must be emptied of water and be thoroughly flushed out before being refilled for use.

programmer, heating and hot water A more or less sophisticated time clock, arranged to cater for a variety of time and heating/hot water functions: e.g. 'hot water only', 'heating only' or 'heating and hot water', with several switching operations in a 24-hour period. It activates electric switches or valves to achieve the desired programme.

programming control A sequence of operations which is based on specific timing.

progress photographs A set of photographs taken during the progress of the works which visually records important aspects of them.

project management A layer of management inserted between the employer and the design team which acts as a briefing and client control body.

project manager The head of a project management team.

prolongation and disruption claim A claim that arises when the contractor properly and directly incurs any expense beyond that otherwise provided for or reasonably contemplated by the contract in complying with any of the supervising officer's instructions, so long as these were not rendered necessary as a result of any default by the contractor.

proof loaded A method of establishing the safe maximum loading of a lifting beam.

propane A hydrocarbon obtained from crude oil. See also **LPG**.

propellor fan The simplest form of fan. It commonly comprises a sheet metal impeller with relatively large clearance inside an orifice. The fan blades – usually four or six – are bolted to a cast-iron or aluminium impeller hub. The performance depends on the shape of the blades; a broad-bladed fan will handle more air more quietly than a narrow-bladed fan of the same diameter running at the same speed. Air enters the fan from all directions and is discharged mainly axially. When the air meets resistance, it tends to flow backwards through the impeller.

proportional band The range of values of the controlled variable which corresponds to the full operating range of the final control element in proportional-position control. It is usually expressed as a percentage of the full-scale range of the controller. The adjusting dial is calibrated as a percentage.

proportional control Control in which the final control element assumes a definite position for each value of the controlled variable.

protective multiple earthing (PME) A procedure that enables the electricity utility company to provide the consumer with a reliable earthing terminal connected to the neutral conductor at the supply intake. For protection against the presence of earth leakage currents, electrical connections are made between the earthing terminal and all metal structures, metal pipes, metal banisters and all other metalwork installed within the confines of the electrical installation which might come into contact with persons.

provisional sum A lump sum included within the specification for some aspect of the proposed works which had not been finalized when the tender documents were prepared. Such sums of money will be resolved and instructed during the course of the contract.

psychrometric chart [Figs 175, 176, 177] A graphic presentation of hygrometric data on the properties of air for a given pressure.

PTFE pipe jointing tape Made of polytetrafluorethylene on narrow strips wound on a roll. Employed widely for sealing the screwed joints of pipes conveying fluids; wound over screwed portion during construction to strengthen joint against leakage.

P trap A trap that connects to a sanitary stack horizontally.

public or private foul sewer, connection to Arrangements must be agreed with the authority which is responsible for the public sewer. The branch drain must be laid to give an adequate hydraulic gradient and to ensure that the minimum length of drain is subjected to back flow in the event of the sewer surcharging.

puddle flange [Fig. 178] A flange welded to a pipe at its point of penetration into a water-retaining structure, e.g. a brick wall, to minimize or obviate the seepage of water along the exterior of the pipe into or through that structure.

pulse combustion A totally sealed system developed by Lennox, with combustion air taken from outside the building by means of 50 mm PVC pipe. The products of combustion are so low in temperature, after having passed through a series of heat exchangers (including a condensing coil) that they can be discharged through a similar small-bore pipe with a small-bore condensate drain. Pulse units generate heat in a manner unlike other warm air heaters; they incorporate neither pilot light nor atmospheric burner but predetermined quantities of gas an air ignite in a combustion chamber 60–70 times per second. The system provides economical gas-fired warm air heating, suitable for stores, warehouses, etc.

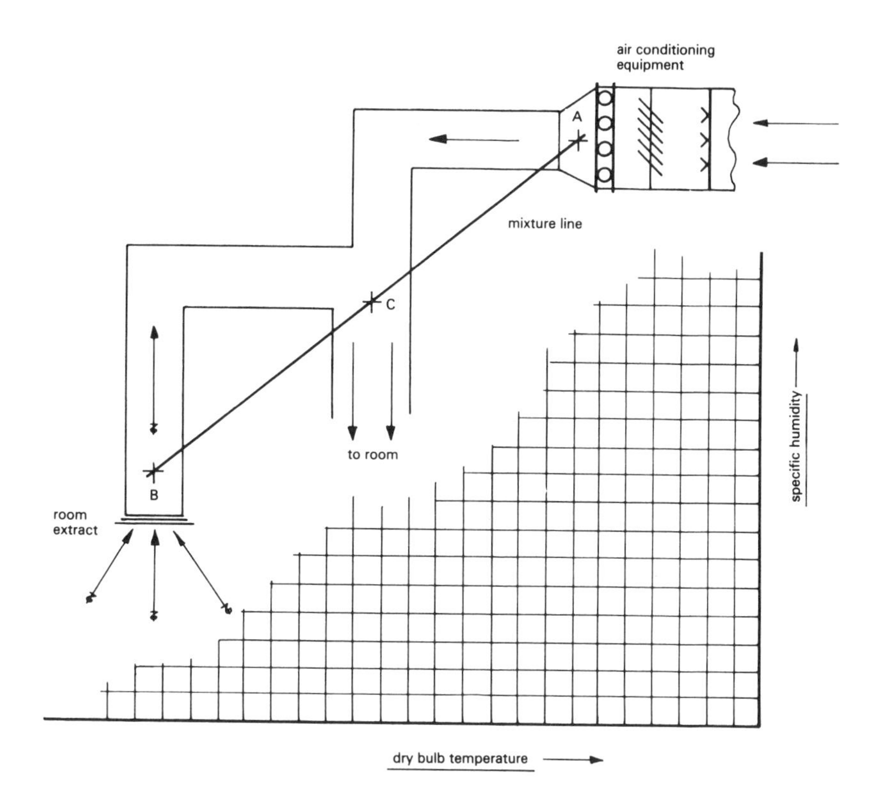

[Fig. 175] Application of a psychrometric chart: construction of a mixture line

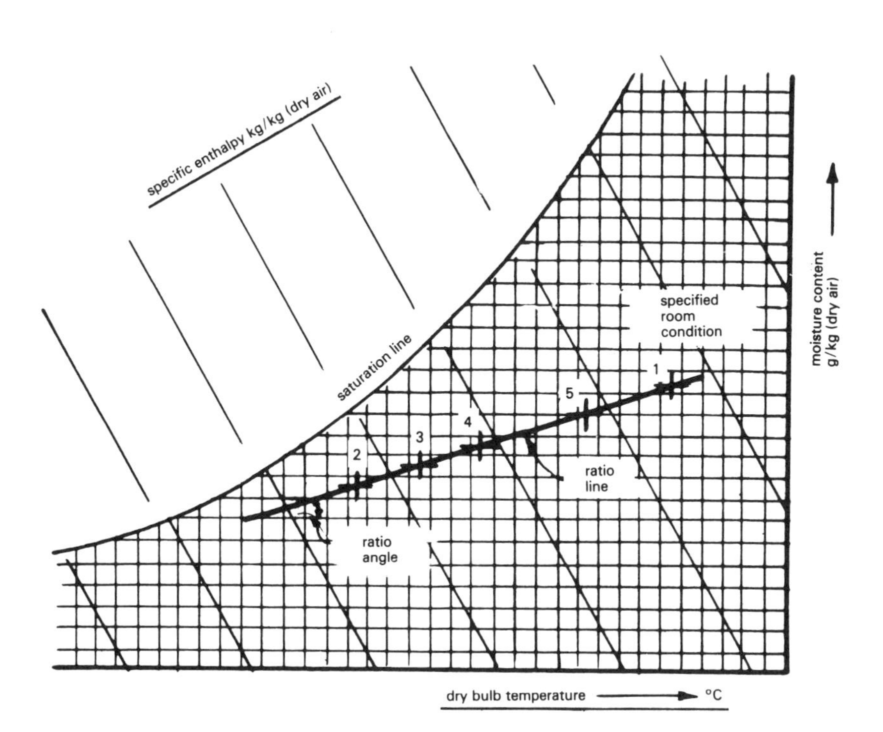

[Fig. 176] Application of a psychrometric chart: construction of a ratio line

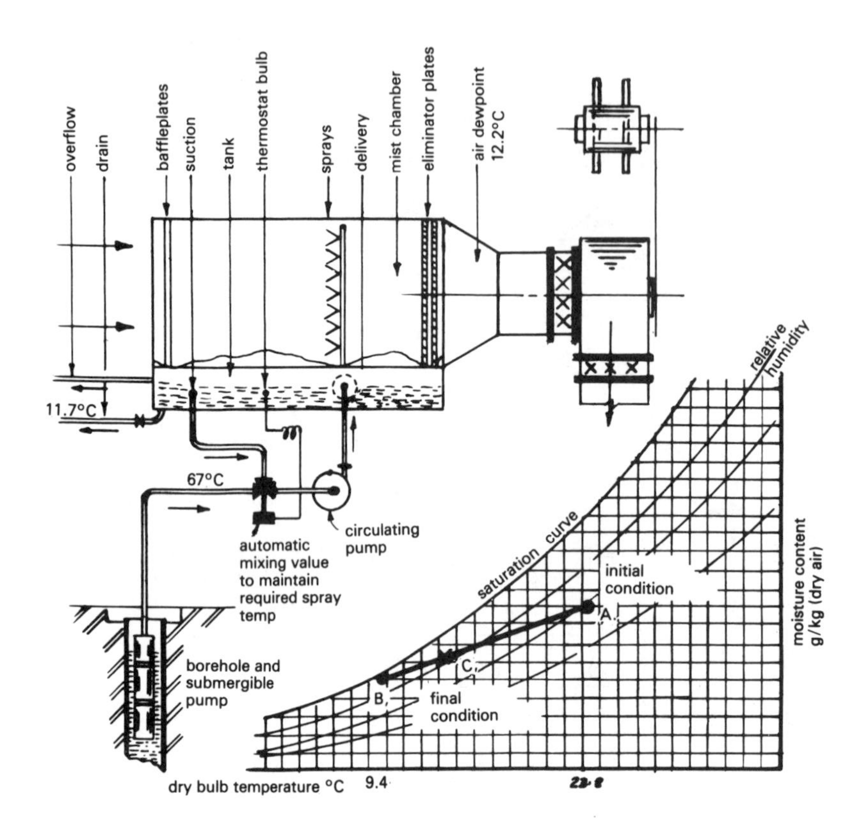

[Fig. 177] Application of a psychrometric chart to air-conditioning

pulverized fuel Finely ground fuel, usually coal, prepared for a specific combustion process. The most suitable type of fuel for pulverizing is bitumenous coal with over 20% of volatile matter and of less than medium coking power. The ash content may be up to 20%.

pump A motorized device which draws in a fluid at its suction end and delivers it at an elevated pressure to a pumped system.

pump, booster See **booster pump**.

pump characteristic curve A curve that expresses the relationship between the capacity and head of a particular pump, each pump possessing its own unique characteristic curve which is published by the manufacturer and which allows the designer to predict and specify pump performance.

pump, condensate return See **condensate return pump**.

pump delivery The delivery from a pump under positive pressure.

pump, direct-in-line A pump placed into a pipeline as an extension of it.

pumped storage An energy conservation method which uses low-cost off-peak energy to pump water to a high level and allows it to fall back to its original low level to generate electricity during peak time. It is applicable particularly to hydroelectric and tidal electricity generation schemes.

pump, end suction A pump mounted on a bedplate with connections at right angles to each other.

pump, final connection This is carried out with a flexible pressure pipe to avoid the transmission of pump vibration.

pump, geared A pump designed to handle very viscous liquids, such as heavy fuel oil or process liquids, incorporating two gears in place of the

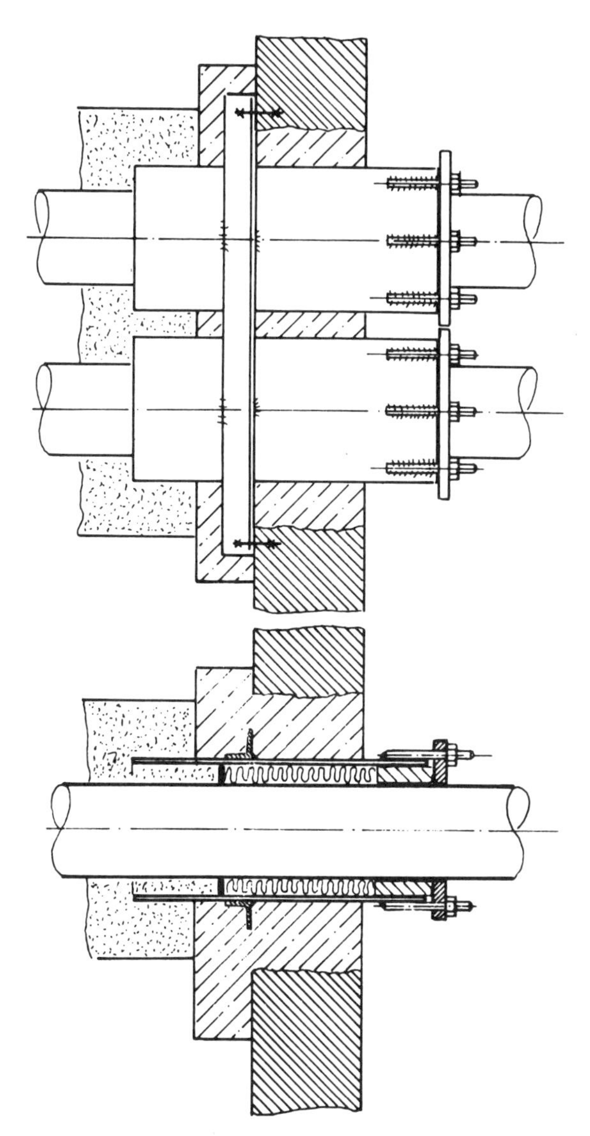

[Fig. 178] The entrance of pipelines into a building combined with fixed-point structure

conventional impeller. It can develop a high discharge pressure.

pump governor A device that maintains constant discharge pressures on steam-driven turbines or reciprocating pumps. It may be diaphragm actuated; a large diaphragm with a wide selection of casing and body sizes provides reliable pressure control with a minimum drop of steam pressure, in conjunction with double or single ported globe valve bodies with a quick-opening valve plug and screwed-in seat rings.

pump, holding-down bolts Bolts that secure the bedplate of a pump to a concrete base.

pump, insulated base A special vibration-attenuated base. It may be of concrete incorporating anti-vibration material in sandwich form.

pump, lead See **lead pump**.

pump, magnetic drive A self-priming pump for handling corrosive liquids, constructed of polypropylene material with a magnetic drive. A shaft seal is not required.

pump, oil transfer A device to transfer fuel oil from the main storage tank to a daily service tank. It operates through automatic level control and a manual start switch.

pump, reciprocating A pump in which the action relies on a piston driven shaft, vertical or horizontal. It is used as a steam-driven boiler feed pump.

pump, screw See **screw pump**.

pump seal A seal that prevents seepage of water along the shaft to the outside.

pump, self-priming See **self-priming pump.**

pump shaft The shaft that connects the pump with the driving motor.

pump, steam heating vacuum See **steam heating vacuum pump**.

pump, submersible A pump enveloped within a capsule suitable for immersion and operation under water. It is used as a borehole pump and is capable of developing high discharge pressures.

pump suction Pump operation under negative pressure, e.g. drawing liquid from an underground tank.

punkah fan A room-located slow-rotating suspended propellor fan with large blades.

pure tone A single frequency signal.

pure water See **de-ionized water**.

purge The process of cleansing equipment from resident fumes, impurities, etc.

purge period The time interval set for clearing fumes, etc., from a system before the combustion equipment fires.

push joint A method of joining pipes by pushing one component into another and securing tightness by soldering, compression or cementing.

Pushlock (Terrain Ltd.) plumbing system [Fig. 179] A proprietary system of uPVC plastic plumbing fittings and the method of jointing them.

putty Material composed of powdered chalk mixed with linseed oil used as a sealant by glaziers and plumbers.

pyrolysis [Fig. 180] A process which is a modified form of incineration, involving the physical and chemical decomposition of organic matter under the action of heat within an environment which is deficient in oxygen. In this, organic matter is readily converted into gases, liquids and inert char. Such products constitute about half of the initial volume of the materials introduced to the process; these can be converted into useable energy output or be directed either to sustain the process itself or to generate electric power. It is particularly suitable for the destruction of refuse. When employed for the recycling of refuse, the pyrolysis process requires the heating of the refuse in a retort in the absence of oxygen at temperatures ranging between 500 and 1000°C, the actual required temperature being a function of the specific processing method and of the required end products. The refuse can be converted into convenient liquid or gaseous forms to provide a liquid or gaseous fuel which can be either stored or used immediately in an associated process, such as power generation or heat recovery. It thus offers the practicable option of heat storage, which option is not available in the conventional incineration system. The calorific value of the resultant gas depends somewhat on the final temperature at which the operation is carried out; at 1650°C, it is in the order of 19.5 MJ/m. The heat value of the residual char is low, making it virtually useless as a fuel.

The basic pyrolysis equipment comprises a refractory-lined combustion chamber into which the waste is loaded and to which only a restricted flow of air is admitted. The waste is initially heated by

[Fig. 179] Selection of MuPVC plumbing fittings from the Terrain Ltd. range

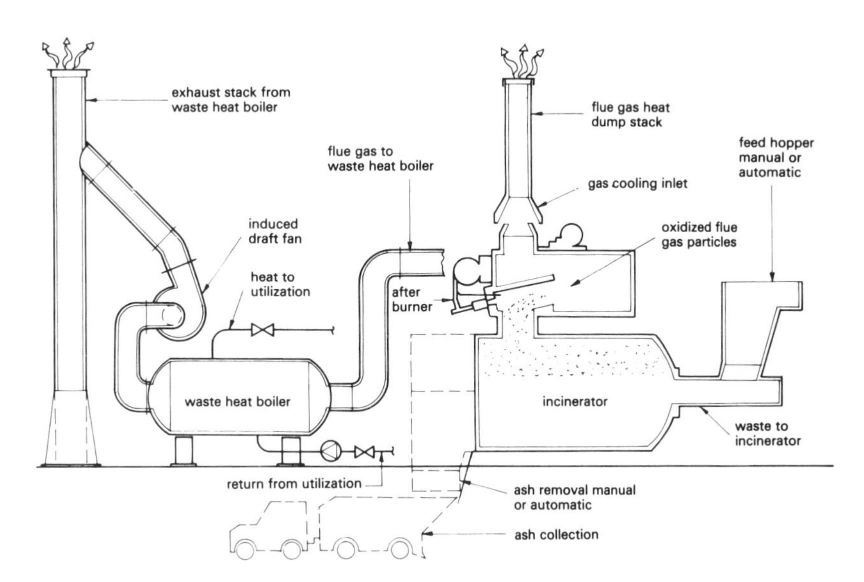

[Fig. 180] Diagrammatic arrangement of Consumat pyrolisis unit: single module

small-capacity oil or gas burners and undergoes the pyrolysis process. Normally, the reaction becomes autothermic (self-sustaining as to temperature) when the operating temperature of 800°C has been achieved. This method of primary burning offers the advantage that the waste decomposes under quiescent conditions and, consequently, the carry-over of particulate matter (which would otherwise contribute towards the emissions from the chimney to atmosphere) is minimized. The partially combusted products flow directly into an after-burning chamber which is mounted immediately above the main combustion chamber. The gases are then admixed with additional air and elevated to a temperature of 1000 to 1200°C to ensure the successful burn-out of the smoke. Before discharge from the pyrolysis chamber, the gases are cooled to approximately 800°C by entrained ambient air. The air supplied for combustion is supplied to the lower and upper pyrolysis chambers by independent centrifugal fans. The rate of air supply to both chambers is adjusted during commissioning to achieve the correct combustion conditions for each specific application. The whole process is automatically controlled, programmed and monitored by an array of sophisticated instruments and controllers to provide efficient and safe operation throughout the process.

Pyrolysis is capable of converting the organic portions of domestic refuse, rubber tyres, all types of plastics, sewage sludge, etc., efficiently and with minimum harmful chimney emissions.

pyrometer A device that measures temperatures above 800°C for industrial or laboratory use.

Energy-saving potential: typically allows detection of hot spots before plant shut-down stage has been reached. It can eliminate guesswork in measuring temperatures of moving targets (e.g. porcelain) which might be damaged by contact.

Q trap A trap that, connects to a drain or stack at an angle. It is an intermediate arrangement between P and S trap types.

quality assurance All those planned and systematic actions necessary to engender confidence that a product or service being offered will satisfy specific requirements for quality.

quality control Operational methods and activities used to fulfil requirements for quality.

quality management, standards International Standards No. ISO 9004 gives guidance to all organizations for quality management purposes. ISO 9001, 9002 and 9003 are used for external quality assurance purposes in contractual situations.

quality system The organizational framework, defined responsibilities, established procedures, processes and resources for implementing quality management.

quantity surveyor A person, also termed a building cost accountant, who prepares a measured bill of quantities, which includes the specification for the proposed works and the measured quantities of the various materials and fitments required for the contract, prepares and updates cost plans, recommends payment for executed works and settles the final account.

quarl A refractory throat placed around the port of a burner to direct air into the flame. The heat radiation off the hot refractory assists the maintenance of stable combustion conditions.

quartz clock A clock regulated by a quartz crystal which vibrates with a definite constant frequency under the effect of an alternating electric field tuned to the resonance frequency of the crystal. Its great accuracy suits it to applications in precise work.

quartz, halogen lamp A tungsten filament lamp in which the filament is enclosed within a small quartz envelope containing the vapour of a halogen (bromide or iodine). Such a filament can operate at a much higher temperature than in a conventional lamp with much greater brilliance and longer service life.

quartz linear lamp A lamp that emits heat at the same frequency as the Sun. It is comfortable even in close proximity and is available in capacities from 1 kW upwards. It is well suited to spot heating. It is a white light heat source with coloured filters available to suit application.

quick-action valve A valve used in an emergency to rapidly perform a safety function.

quick-release coupling This permits easy uncoupling of items of equipment.

R

RDF [Figs 181, 182] Refuse-derived fuel, particularly as supplied from waste recycling processes. See also **waste-derived fuel**.

radial circuit A circuit in which the circuit cables radiate from a distribution board.

radiant cooling system A system that incorporates passive cooling devices, such as pipe coils or panels, commonly located within suspended ceilings.

radiant floor panel A steel panel with flat fronts and waterways or serpentine coils at the back, set into floor screed with the flat surface uppermost and lined up with the top of the screed. It can operate at the temperature of the general heating distribution and is used mainly for spot heating near workstations or at door openings to counteract cold draughts.

radiant heating panel A heating panel that comprises serpentine pipe loops secured to one steel plate or placed between two such plates. It may be suspended vertically in the area to be heated with both surfaces radiating heat, or horizontally, in which case the upper surface is thermally insulated to direct heat downwards. It is particularly suited for use with high-temperature hot water heating systems.

radiant heating system A heating system that incorporates panels or pipe coil heat emitters which provide their heat output by radiation into the heated space.The system may be fuelled by electricity, gas or low-temperature/high-temperature primary hot water. Heat emitters may be embedded within the structure, suspended from it, located at low level or hung within the heated space.

radiant strip heating Heating that comprises an overhead system of hot water pipes which are bonded in some way to a reflective metal surface. The upper part of the joint arrangement is thermally insulated to direct heat primarily downwards. It may incorporate single or multiple pipes.

radiant tube heating system A gas- or oil-fired radiant heating system which utilizes a direct or indirect air heater connected to a range or ranges of air ducts insulated on top through which air is continuously recirculated by a centrifugal fan which is mounted on the heater unit. The system is well suited to the heating of tall spaces such as aircraft hangars, warehouses, repair shops, etc.

radiation The transfer of heat from the heat source to other bodies remote from it without raising the temperature of the intervening space.

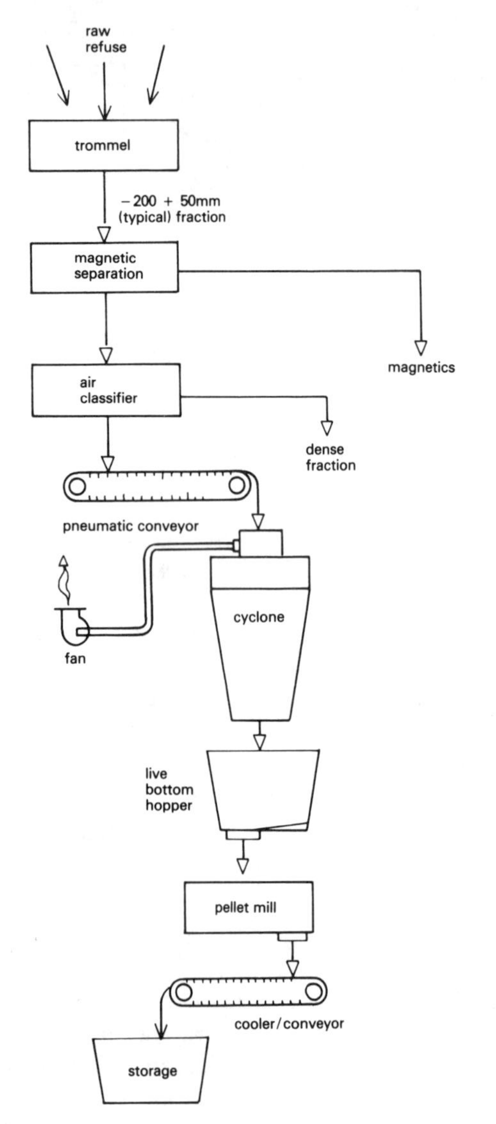

[Fig. 181] Simplified flowsheet for the production of refuse-derived fuel pellets

radiator A heat emitter installed as part of a space heating system, available in various types, models, materials and configurations.

The term is also applied to a heat dissipator forming part of an electricity generator installation. Within building services it usually applies to the heat dissipator of an emergency diesel generator plant.

radiator bracket A purpose-made steel support arranged to suit the type of radiator being supported. Brackets for wall-hung column radiators are shaped to give even support and are secured firmly to the wall behind the radiator. The minimum of required brackets is two; long radiators require additional brackets for firm support.

radiator, column See **column radiator**.

radiator, convector See **convector radiator**.

radiator foil Aluminium foil which is self-adhesive and protected by a peel-away backing paper for quick application. It is supplied in easily handled rolls and is placed on the wall behind a radiator with a reflective surface facing into the room. It improves the effectiveness of the radiator and reduces direct heat transmission from the radiator through the wall to the outside.

radiator, hospital See **hospital radiator**.

radiator, panel See **panel radiator**.

radiator reflective foil See **radiator foil**.

radiator schedule A schedule that sets out in tabular form the full details of the heaters required for a particular project, including numbers of each item, manufacturers' identification, type, dimensions, manner of required connections, etc.

radiator stay A steel rod with a running thread, one end of which is attached to the wall behind the radiator (either built into the wall or screwed on to the wall via a wall plate) to provide top fixing for a column-type radiator. The stay is passed between two adjacent column sections and the front is secured by means of a backnut run on to the stay. The surplus length of the rod is cut off. Long radiators will require more than one radiator stay for firm support.

radiator valve [Fig. 183] A valve fitted at the end of a radiator for control or regulation of the flow through the radiator and/or for isolating it. Commonly, a wheel valve is fitted at the flow connection and a lockshield regulating valve at the return connection.

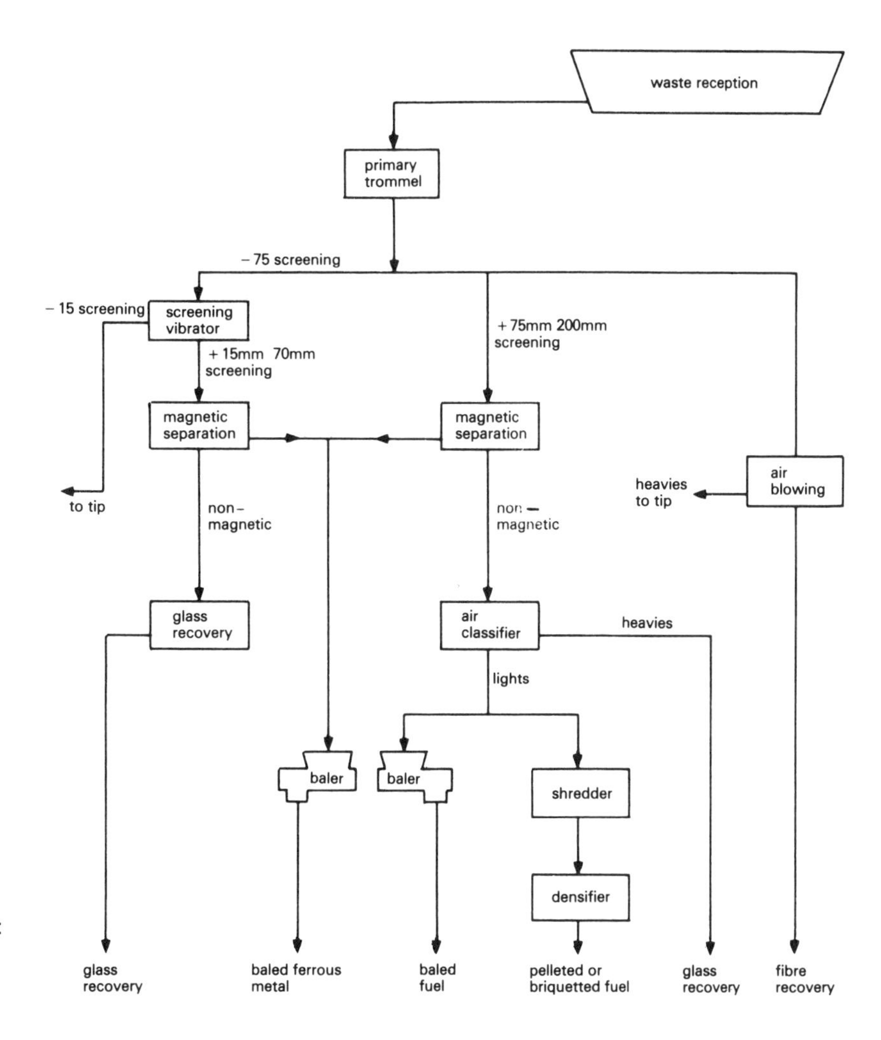

[Fig. 182] Doncaster plant: process flow diagram for RDF

radon An element of atomic number $Z = 86$; a product of uranium. It is found as a gas in certain geological areas, particular those comprising granite. Radon gas tends to leak into buildings constructed in such areas and is deleterious to health. Its presence within buildings can be neutralized by adequate and purpose-designed ventilation.

radon gas A naturally occurring gas which percolates through the ground and is particularly evident in rocky and granite terrain. The gas is said to be poisonous and possibly a cause of cancer to persons living or working in a radon gas environment. Ventilation of the affected premises is a cost-effective remedy.

rag bolt A bolt used for securing an item of equipment to concrete or masonry. The end of the bolt is opened and spread out to secure optimum support.

ramp A short length of drainage pipe or channel which is laid at a much greater gradient than the adjoining pipe(s) or channel(s).

Rankine cycle The thermodynamic cycle on which the principle of the steam turbine operates. The steps of the cycle are as follows.

(a) An increase in pressure as water is pumped into the boiler.
(b) Constant heat input (latent heat), raising steam.
(c) Isentropic expansion of the steam, generating mechanical power.
(d) Condensation of the steam as it gives up its remaining heat at atmospheric pressure.

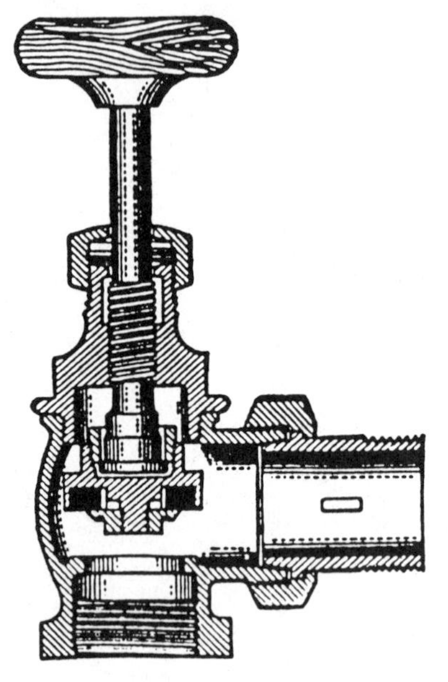

[Fig. 183] Typical radiator valve with tail for connection to radiator

random start A method of starting up a group of electrical motors, such as in a number of air-conditioners fitted with compressors, whereby they do not all take full starting current at the same time. See also **random start relay**.

random start relay A relay typically incorporated with the Versatemp heat reclaiming air-conditioning system employing a number of such units on one programme-controlled circuit. Since each relay incorporates a hermetic motor compressor which operates with a starting current usually in the order of five times its running current, it is essential/desirable to ensure that all the units on the circuit do not start simultaneously and blow the protective fuse. A completely automatic factory-tested random start relay is therefore normally fitted and this ensures that only one unit starts up at a time, although the power supply is made available to all units simultaneously. However, at normal voltage, all the units will start within a total period of approximately 90 seconds.

rapid gravity filter A filter that removes the finer particles carried over from the settling of a filtration system. It works at high rates and can be washed easily and quickly. It is now widely adopted where filtration is necessary.

rapid steam generator A generator typically capable of producing saturated steam at pressures of up to 10 atmospheres in about three minutes following switch-on. It is available in stationary or mobile form, usually fired by 35 sec gas oil or gas. Standard fitments include a steam safety valve, lack-of-water unit, excess steam pressure switch and steam temperature cut out.

rates of rainfall For any urban or rural area in the UK these can be obtained on application from the Meteorological Office in Bracknell, Berkshire, and such data will enable soundly based rainwater system calculations to be made.

rational number A number which can be expressed as the ratio of two whole numbers.

raw water Water in its untreated state as it enters the treatment plant.

reactance The virtual resistance of an electric circuit to an alternating current, related to the inductive and capacitive components of the circuit.

reactive attenuator An attenuator in which the noise reduction is brought about typically by changes in cross-section, chambers and baffle sections, e.g. a car exhaust silencer.

reagent A chemical substance used to produce a chemical reaction.

real-time working The method of operating a computer as part of a larger system in which information from the computer output is available at the time it is required by the rest of the system.

reciprocating compressor A compressor that is normally single acting and may employ up to 16 cylinders, frequently arranged in V or VW formation. On larger machines, capacity reduction is achieved by cylinder unloading. Capacities for refrigeration compressors usually range up to 565 kW, depending on the nature of the refrigerant being used, at a speed at 50 Hz of approximately 24 revs/sec. At capacities above 38 kW, direct drive is preferred. Electric driving motors may be of the open or hermetically sealed type.

reciprocating pump See **pump, reciprocating**.

recirculated air That proportion of the air supply into a space which is returned to the air handling plant via the extraction system for mixing with the incoming fresh air (if any).

recirculation The process of recycling a proportion of a fluid flowing in a system back to the supply plant to be mixed with the supply of the fluid into system. It is used extensively with ventilation and air-conditioning systems to conserve heat and cooling energy. The process may be automatically controlled or be set by dampers which are permanently positioned to achieve a certain predetermined proportion of recirculated or fresh air in the air being supplied into the space. It is the usual practice to employ full recirculation to warm up or pre-cool a space before occupation in, say, an exhibition hall, theatre, etc. and to switch to part or all fresh air during periods of occupation. It is suspected that recirculation of air within an occupied buildings can contribute to contamination by the spreading of germs. Excessive recirculation of air leads to feeling of stuffiness.

record drawings A set of drawings which accurately records the installed works, commonly the responsibility of the contractor.

rectifier A device that converts alternating current into a unidirectional current. It has no moving parts.

recuperator See **flue gas recuperator**.

recycling yard [Fig. 184] An at-source waste recycling facility for a limited variety of waste constituents, such as glass, paper, cardboard, textiles, batteries, tyres, light bulbs, fluorescent tubes, etc.

reducing agent A substance which removes oxygen from, or adds hydrogen to, another substance.

reducing valve See **pressure reducing valve**.

reducing valve, direct acting See **direct-acting reducing valve**.

reducing valve, pilot operated See **pilot-operated reducing valve**.

reed switch A switch that comprises two magnetic tongues with electroplated contacts, sealed into a glass tube with protective gases.

reference input The reference signal in an automatic controller. The output signal of the reference

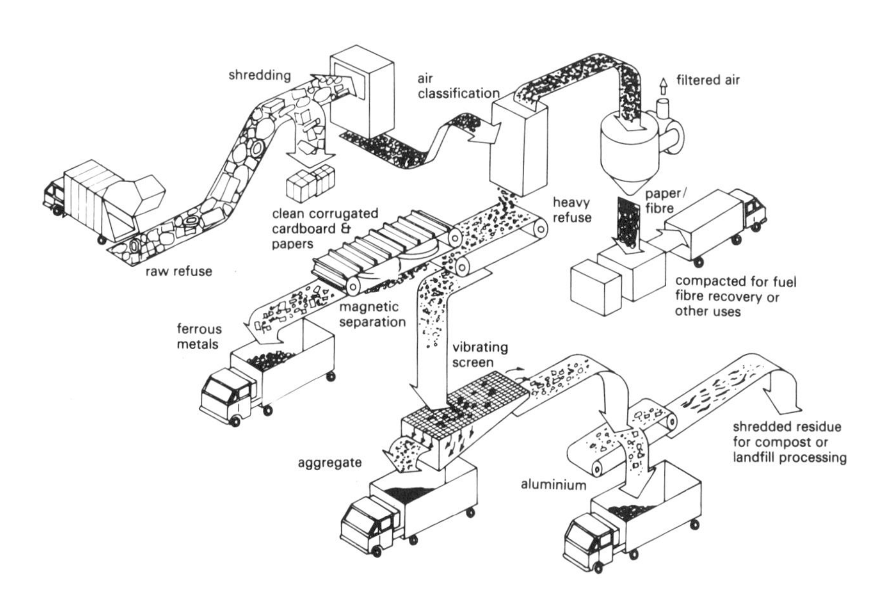

[Fig. 184] Resource recovery (recycling) at Monroe County, USA

input elements is determined by their response to the set point input signal. It has the same units as primary feedback.

reflectance The extent to which a surface is capable of reflecting incident radiation. It is the ratio of the intensity of the reflected radiation to the intensity of the incident radiation.

reflection When light strikes a surface that will not transmit light (an opaque surface), some of the light is absorbed and some is reflected. The ratio of luminous flux reflected to the luminous flux received is termed the reflectance.

reflection, diffuse See **diffuse reflection**.

reflection, mixed See **mixed reflection**.

reflection, specular See **specular reflection**.

reflectivity See **emissivity**.

reflector Any surface which reflects radiation.

reflector, parabolic See **parabolic reflector**.

reflux The back flow of a fluid.

reflux condenser A condenser that receives vapour given off by a process, condenses it and returns the condensate to the process medium.

reflux valve See **check valve**.

refraction The bending of light as it passes from one transparent medium to another of different density.

refractories, monolithic Mixtures of plastic refractories prepared under controlled conditions and used widely for repairs of furnace refractories or for complete refractory linings. They can be formulated for specific applications; and offer long service life under arduous furnace conditions.

refractories, plastic See **refractories, monolithic**.

refractoriness The degree of resistance to softening under the action of heat offered by a refractory material.

refractory Relates to any material which is difficult to fuse, melt or soften. The primary characteristic of a refractory is the ability of the material to withstand the action of heat without deforming or softening. Different materials vary in their ability to resist softening and the degree of the resistance offered by each material is called its 'refractoriness'.

refractory baffle A structure associated with a combustion chamber which functions to change the direction of the flow of air, fuel or combustion products.

refrigeration system condenser This rejects from a refrigeration cycle the heat energy gained through evaporation in the evaporator part of the system and in the compressor. The most convenient heat sinks into which this energy may be rejected are atmospheric air or circulating water. The heat in the condenser water may be utilized to provide space heating in equipment such as reversible cycle heat pumps.

refrigerant An operating/working fluid used in a refrigeration machine. It is subject to cycles of evaporation and condensation.

refrigerant, leaks Statistically, the average commercial chiller loses its complete charge three times in ten years. As such loss occurs, there is a drastic reduction in efficiency, as well as likely compressor breakdown. Correctly designed, installed and maintained, chiller installations which suffer no leaks can look to 30 and more years of trouble-free operation without compressor replacement.

Major contributory factors towards leaks are as follows.

(a) The chiller vibrates, shaking connections; the remedy is to place the chiller on anti-vibration mountings or frame.
(b) Tight pipe connections put stress on the pipework; the remedy is to use long sweep connections and long flexibles.
(c) Thermal movement due to changes in tem-

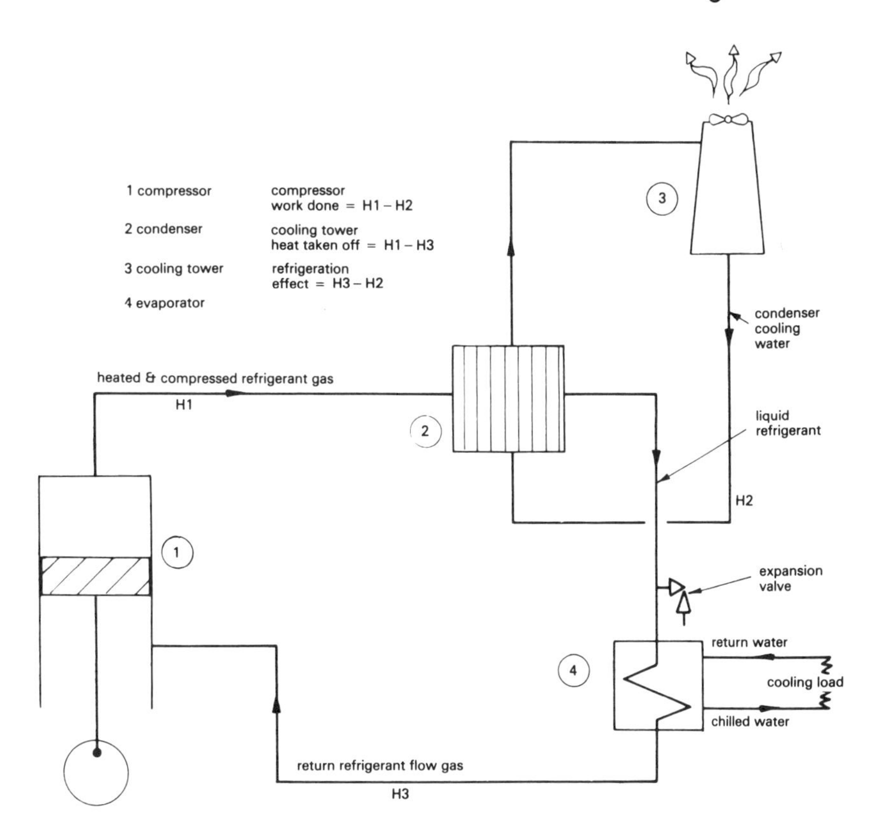

[Fig. 185] Schematic operation of the compression refrigerator

perature of the refrigerant; the remedy is to compensate for pipe movement.

(d) Inadequate routine maintenance: the remedy is to check for leaks on all joints at frequent intervals and repair any such leaks without delay. Also check for uncontrolled vibration and pipe stresses.

refrigeration The process of cooling a fluid or substance by a specific method.

refrigeration compressor [Fig. 185] See **chiller**.

refrigeration, steam-jet system See **steam-jet refrigeration**.

refuse incinerator [Fig. 186] An incinerator especially adapted for the burning of refuse which incorporates designs to overcome the difficulties which are inherent in the burning of refuse. These include the following.

(a) Refuse may be damp and may include up to 40% water.

(b) Refuse may be variable in size, moisture content and calorific value.

(c) Its density varies according to the composition and origin of the refuse. It may be bulky.

(d) The refuse is a mixture of soft and hard constituents and is likely to contain a mix of substances from dust to large furniture sections.

(e) It may include components with a low melting-point, such as glassware; this could lead to rapid clogging of the fire bars and other furnace parts.

(f) It may include noxious materials, such as plastics and rubber.

regain method of duct sizing An increase in static pressure due to expanding a fluid, e.g. the transfer of air through a transformation piece into a duct of larger dimensions. This is known as static regain.

regeneration Aerodynamic noise sources occurring in an attenuated air distribution system. It has the effect of increasing perceived noise levels in the conditioned areas.

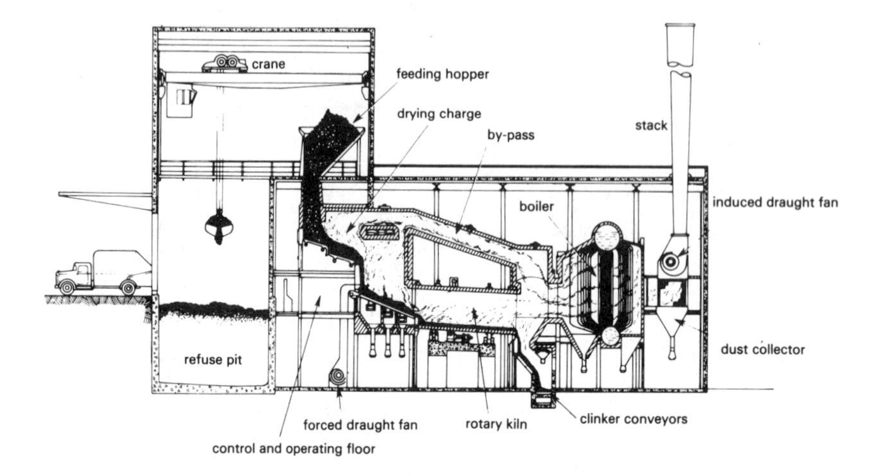

[Fig. 186] Layout of a large urban incinerator plant which operates on the Volund principle

Regional Electricity Companies The privatization of the electricity supply restructured the industry in England and Wales into twelve Regional Electricity Companies (RECs).

reheat Adds heat input to an air stream to increase its temperature above that achieved in the main 'air' heater or to make good transmission heat losses.

reheater, air An air heater which may be placed into an air-conditioning system to raise the air temperature following passage of the air through the cooling coils. It is also used to boost the temperature of the air supply to a higher temperature area or to compensate for heat losses in the ducts prior to the air reaching a particular air inlet.

reheater, air compressor An air heater that functions to reheat the compressed air to overcome the risk of freezing the residual moisture in the compressed air at the exhaust ports.

relative humidity See **humidity, relative**.

relay (electric) A device which is operated electromagnetically by the electric current in one circuit which causes electrical contacts to open or to close with the object of controlling the flow of current in another circuit.

relief valve A valve that automatically discharges fluid to relieve excess pressure. It is generally used on pipelines and for non-compressible fluids, e.g. liquids, such as water or oil, where immediate full-flow discharge is not essential since a small discharge flow reduces the pressure. Relief valves are available in various types and modes to suit particular applications, including full-lift and pilot operation.

remeasurement contract A contract which is based on, or largely on, provisional bills of quantities, bills of approximate quantities or a schedule of rates when the value of the works executed to the satisfaction of the supervising officer is ascertained by measurement and consequent valuation.

remote control and monitoring equipment A range of modules capable of data transmission enabling the remote control of heating plant, etc. subject to analogue or digital measurements.

Renoline system A system suitable for repairing a short section of drain by means of a fibreglass sleeve which is impregnated with a polyester resin. See **sectional in situ drain liner**.

resealing trap A type of trap specifically designed to retain an effective water seal after relieving excessive pressure variations at either the inlet or outlet of the trap. It is designed for use in unventilated small size discharge pipes fitted to appliances where, because of the arrangement of the pipework, siphonage would otherwise occur. Such traps do not protect the water seal from back pressure and they will become less efficient in resealing. If the criteria for the correct design of traps are not met, resealing traps must be routinely inspected and maintained.

resident engineer The engineer who represents the civil engineering consultant at site.

resistance, electrical The impediment offered by a conductor or appliance to the flow of current when an electromotive force is applied to it.

resistance thermometer A thermometer that utilizes the variation of the electrical resistance of a conductor with change of temperature. Resistance normally increases with a rise in temperature. The temperature is deduced from the resistance of a spiral of metal placed, for example, in the flow of boiler flue gases to measure its temperature on an appropriate scale. It is used extensively for measurement of the high temperatures in furnaces, hot exit gases, etc.

resistance welding The process of jointing metals by the electric arc welding method.

resonance The build up of excessive vibration in a resilient system. It occurs when the machine speed (disturbing frequency) coincides with the natural frequency, of the mounted machine or support system. The resonance of a system under forced vibration exists when any small increase or decrease in the frequency of excitation causes a decrease in the response of the system.

resonant frequency (Hz) The frequency at which resonance occurs in the resilient system.

rest bend See **duckfoot bend**.

retrofit To add or fit new components or assemblies to existing plant, usually to upgrade or modernize it. Also the provision of a service installation, or part thereof, into an established and previously serviced building.

reverberation Reflected sound that decays in a room after the sound source has stopped.

reverberation time The period of time required for the mean square sound pressure in an enclosure to decrease after the source is stopped to one-millionth of its initial value, i.e. by 60 dB.

reverse-acting float-operated valve A valve for use in large automatic flushing cisterns. It is open when the float is at the top water level and closed when the float is at the bottom water level. A pet cock on the supply initiates the operation.

reverse-acting intercepting trap A trap used when the water from a subsoil drainage system is connected into the main drainage system serving a building. A ventilating pipe is joined to the trap at the termination of the subsoil drain.

reverse cycle A sequence of operations which reverses a previous sequence of activity.

reverse osmosis water purification system A system used to overcome the problems of impure water in humidification systems. It removes unwanted minerals from the water, eliminating water-borne bacterial substances which may be harmful to air distributed by air-conditioning systems. Each unit is self-contained with microprocessor control designed for autonomous operation with an output range of 90–160 litres/h. Normal operation provides autoflush and drain-down cycles, continuous monitoring of water quality and a built-in ability to demand servicing when required.

reverse return system [Fig. 187] An arrangement that facilitates the balancing of a heating or cooling piped circulation by ensuring that the length of the circuit to each terminal unit, and back, is similar.

reversible cycle A cycle of operations which may be totally reversed. It usually requires additional energy input to compensate for energy losses in the original cycle.

reversing valve A valve that reverses the flow of refrigerant to perform either indoor heating or cooling.

Reynolds number (Re) A dimensionless quantity which is applied to a liquid flowing through a

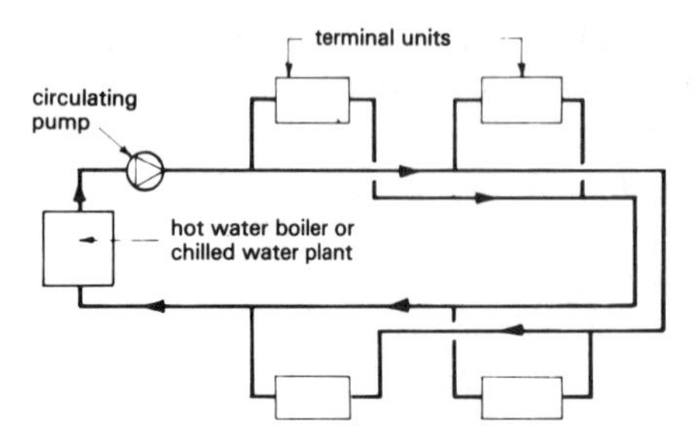

[Fig. 187] Reverse return system: diagrammatic arrangement

cylindrical tube, given by the relationship (Re) = $\mu\rho l/\eta$, where μ is the velocity of flow, ρ the density, l the diameter of the tube and η the coefficient of viscosity of the liquid. When flowing at low velocity, the flow of the fluid is streamlined. At a certain value of (Re) corresponding to a critical velocity μ_c, the flow changes to turbulent.

rheostat An adjustable resistance device for the regulation of current.

rhomboidal air controller An air controller suitable for fitting at air discharge terminals. It ensures a perfectly balanced distribution and comprises a series of vanes which are rhombic in cross-section. It is adjustable from fully open to fully closed positions by a regulating screw in each corner of the controller. By setting each screw individually, a variety of air flow patterns can be achieved to meet specified throw and air distribution parameters in the ventilated space. A rhomboidal air controller is also used on extract ducts where the differential control feature can establish an even exhaust velocity along the length of the grille.

riddlings Small unburnt pieces of solid fuel which tend to fall between the firebars of a furnace.

ring circuit An electric distribution circuit arranged in the form of a ring which starts from and returns to the subcircuit fuse or circuit breaker.

ring clip A pipe support fitting which comprises two halves of a circular steel clip which are placed over the pipe of the appropriate diameter and held together by means of screws. The top half incorporates a screwed socket into which the support is secured. This type of clip is used to support pipes off ceilings or off floors.

riser air relief valve The type of valve that is often fitted at the top of a fire-fighting dry riser to speed the filling of the pipe with water.

risk assessment Survey of relevant plant carried out by suitably qualified persons to ascertain the extent to which the owner or operator of a services installation is at risk through contamination caused by same to his staff and other persons. Currently this is particularly targeted at the risks arising from *Legionella* bacteria in water storage and distribution systems and in cooling towers due to stagnation of water flow and/or the presence of contaminants. The survey identifies all areas which offer a risk, reports on same, and usually embodies relevant recommendations. Risk assessment procedures usually include the preparation of guidelines for routine monitoring of the plant at risk and the issue of a logbook to record all relevant events.

rock fibre insulation Rock fibre in the form of slab, pipe section, flexible rolls, mattresses and quilts for thermal insulation to 950°C.

rock store An assembly of rocks or stones used as a heat accumulator. It stores excess heat fed into it and releases it when heat output is required under the action of fan-assisted air movement.

rocker (connection) A short length of drain-pipe placed immediately adjacent to a manhole or inspection chamber. It incorporates a flexible joint at each end to compensate for differential settlement.

rocking grate A grate supporting a solid fuel bed within a combustion chamber which is periodically moved (rocked) by an electric mechanism to move the fuel forwards along the grate and to dump the ash.

rodding eye A suitably located access fitting in a sanitary or drainage installation into which drain cleaning rods or other devices can be inserted for the clearance of blockages.

rods See **cleaning and descaling of sanitary installations**.

roll-type air filter A moving curtain air filter in which a roll of the filtering material is placed upon a frame and is advanced automatically (or sometimes manually) to continuously present a clean filter medium surface to the air flow. It operates under the action of a differential pressure switch and commonly incorporates an indicator light which warns when the filter roll has nearly run out. A stop prevents it from being fully run out of the assembly. The beginning of the new roll is attached to the end of the expired dirty filter and is cranked on to the frame whilst the dirty roll is run out. Filtration efficiency depends on the filter medium used and on the airtight construction of the frame assembly. It is often used in conjunction with a pre-filter.

roof-top air-conditioning unit [Figs 188, 189] A roof-located, weatherproofed and watertight assembly of all the essential components and controls required to achieve the air-conditioning functions of a space. It is usually brought to site in one piece (often using a crane for lifting the unit on to the roof) and requiring only the electric, water, drain and duct connections to become operational. It is commonly used by department stores and supermarkets.

room thermostat [Fig. 190] A device that accurately controls space heating temperatures. An integral accelerator heater can give minimum space temperature variation.

rotameter A device for measuring the rate of flow of fluids. It comprises a small float which is suspended by the fluid in a vertical calibrated tube. Under given conditions of flow, the height of the float is a measure of the rate of flow.

rotary burner A low-pressure air oil burner in which the fuel oil is thrown over the air blast from the lip of a rapidly rotating cup which is rotated by electric motor or air turbine.

rotary compressor A compressor that incorporates a rotor eccentric to the casing. As the rotor turns, it reduces the gas volume and increases its pressure. This offers the advantages of having relatively few moving parts, it is of simple construction and operates relatively quietly and free from vibration. The main field of use is in household refrigerators and window-type air-conditioners.

rotary converter An alternating electric current motor which is mechanically coupled to a direct

[Fig. 188] Roof-top air-conditioning comprehensive unit (Silentair Ltd.)

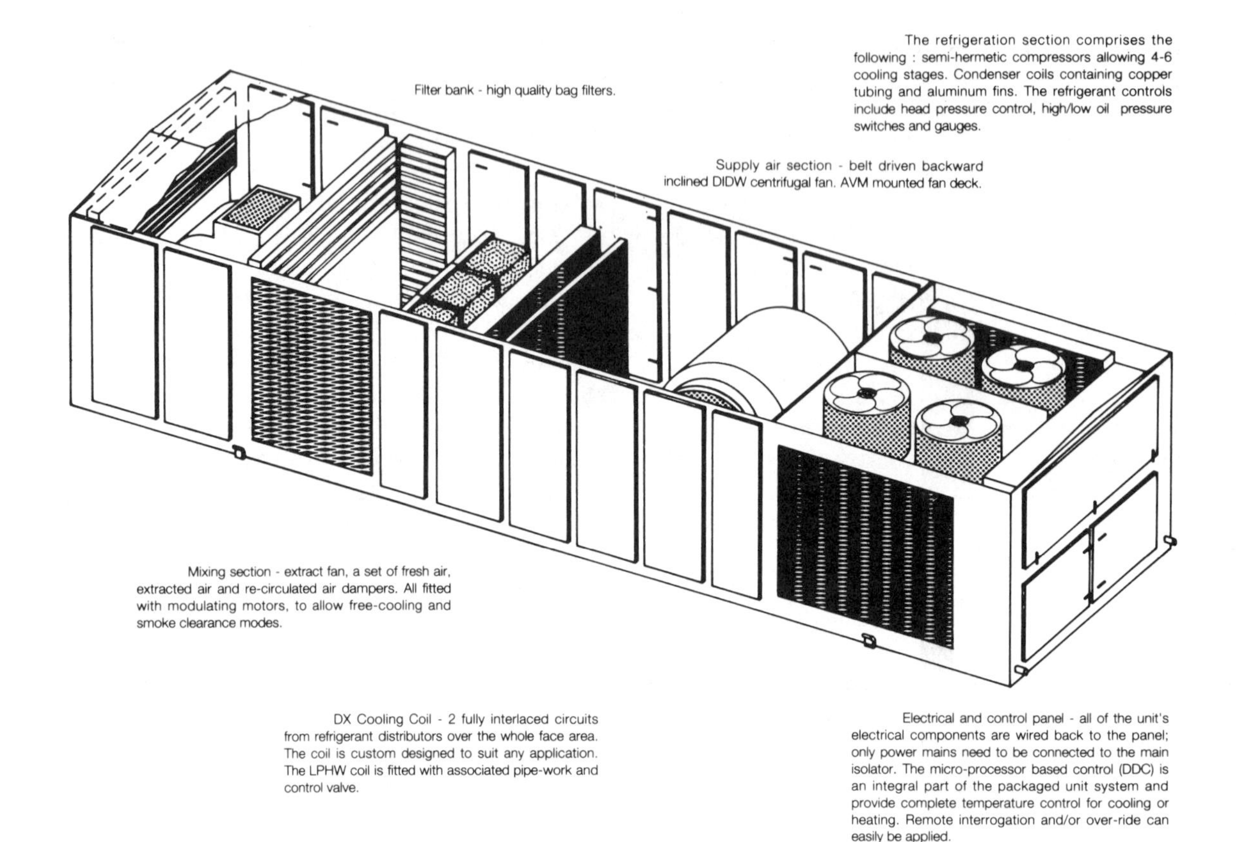

[Fig. 189] Roof-top air-conditioning comprehensive unit with top removed to show components (Silentair Ltd.)

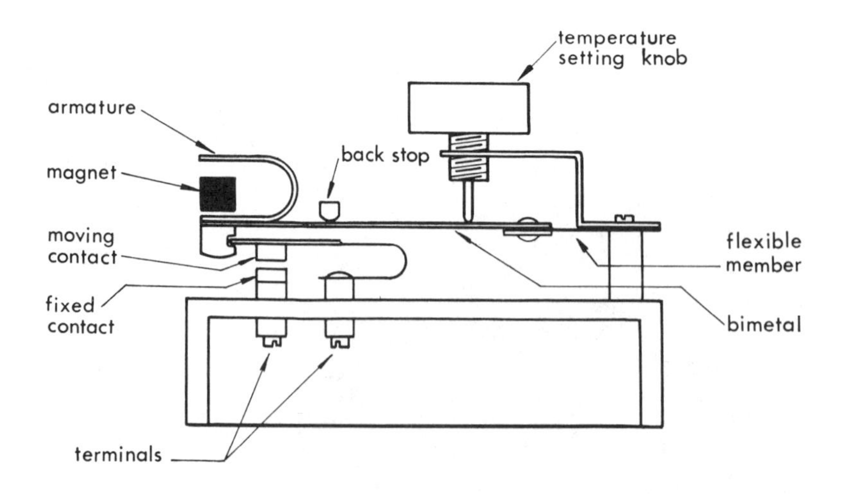

[Fig. 190] Diagrammatic arrangement of a room thermostat

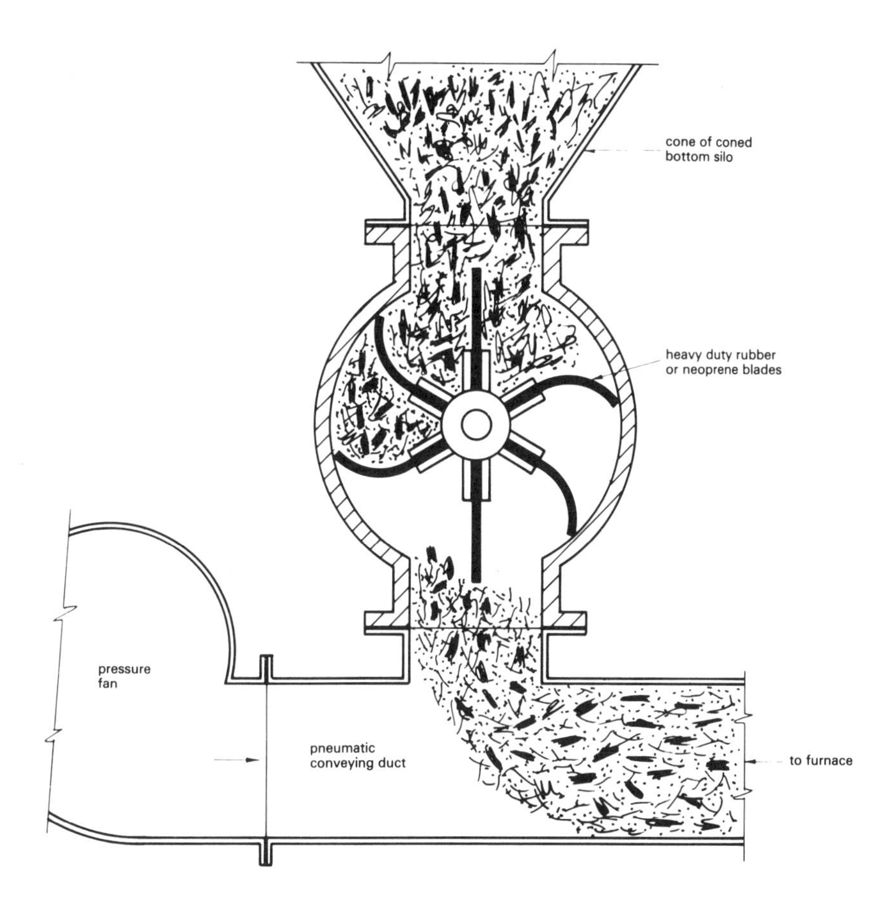

[Fig. 191] Rotary valve function

current generator. It converts an a.c. supply into direct current.

rotary flue gas regenerator The slowly rotating drum of a tightly packed heat transfer matrix which transfers heat from the hot flue gases to incoming cold air for combustion (or other) purposes. Radial and axial seals prevent excessive leakage and minimize cross-contamination. It is available for gas entering temperatures of up to 1400°C.

rotary regenerator See **heat wheel**.

rotary slide valve A valve in which the rotation of internal parts regulates the fluid flow by opening or closing a series of segmental ports.

rotary valve [Fig. 191] A valve that provides a suitable method of control over the quantity and rate of the waste fuel (wood chips, sawdust, etc.) fed from a silo (storage hopper) into a furnace. The speed of rotation of the valve controls the flow of the material via a direct current electric motor with non-overloading infinitely variable speed facilities.

The blades of a rotary valve are usually fabricated either from heavy gauge steel plate with heavy-duty rubber (or neoprene) tips or from wholly fire resistant heavy-duty rubber or neoprene to achieve safe and positive separation between the stored waste (which is at low pressure) and the high-pressure furnace feed line, to provide a fire-break.

rotary plug valve A spherical plug valve, in which the plug, which rotates through 90 degrees, incorporates a circular waterway to match the body and ports.

rotodynamic sewage lift pump A pump that most commonly comprises a centrifugal electric pump with open impeller, the pump being capable of passing without stalling a sphere of 87 mm diameter or 3 mm less than the diameter of the pump inlet, whichever is the greater, unless the blades of the impeller are capable of macerating solids or when a macerator is installed ahead of the pump. Such close-coupled pump and motor units are available for operation under submerged conditions.

rotor The rotating part of a generator, electric motor or turbine.

rubble drain A drain constructed by excavating a trench and filling it with selected rubble or stone through which water can percolate. It has a limited effective service life.

run-around heat recovery coil An arrangement with one coil in the exhaust air ductwork transferring a proportion of the heat to circulated water or water/glycol solution which is pumped through another coil in the supply air ductwork.

running cost The cost of actually operating a system or process.

running thread A pipe screwed thread which extends the whole length of a pipe or rod, used for radiator stays and for forming pipe connectors between adjacent screwed pipe joints which permit dismantling of a pipe section. The running thread permits the free movement of pipe sockets and backnuts.

running trap A tubular trap which has its inlet and outlet in horizontal alignment.

rust Hydrated oxide of iron, formed on the surface of ferrous materials when these are exposed to moisture or air.

Ruth's hot water accumulator A vessel used to meet peak demands of hot water by storing it. Surplus steam is fed into the accumulator and it discharges the hot water at times of peak demand. This permits the economic operation of steam boiler plant and reduces the required installed boiler capacity of a specific application.

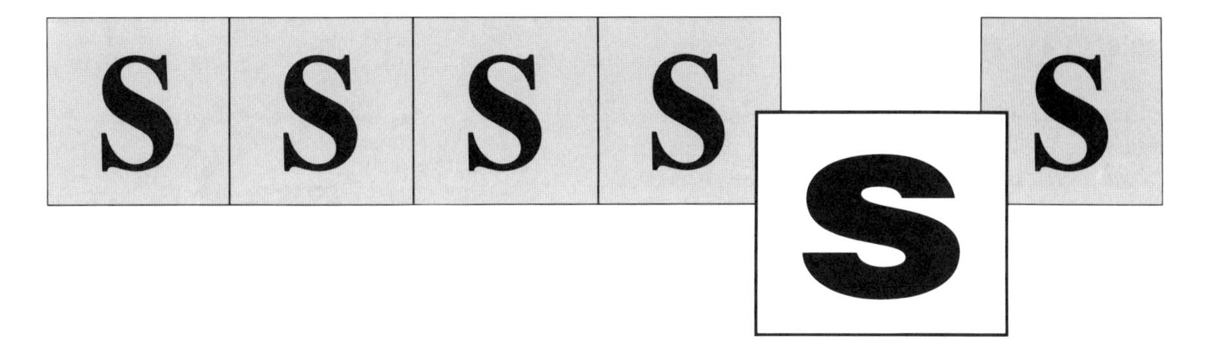

SR system A sewer renovation method (by Rice–Hayward) which uses a helically wound PVC strip to form a liner. See **uPVC spiral relining**.

sacrificial anode That electrode of an electrolytic or corrosion protection system at which oxidation occurs and the material of the anode goes into solution.

saddle [Fig. 192] An inverted U-form steel or plastic clip to secure a cable in position. Different types of saddle are available to suit the application. A space bar saddle keeps the cable some distance off the walls and allows cleaning and painting of the back of the cable.

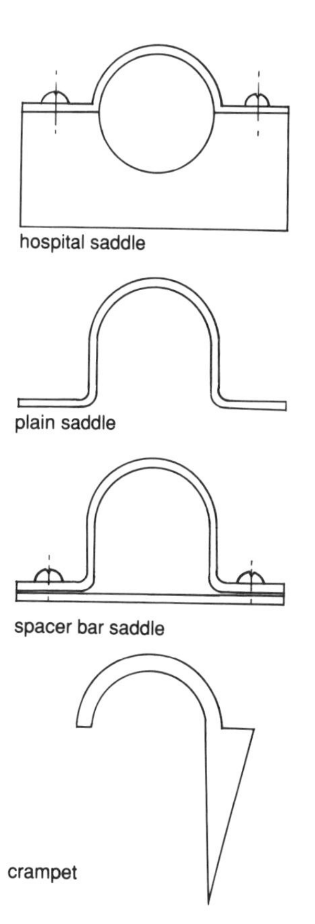

[Fig. 192] Saddles and crampet for fixing conduit

safety cut-out A device which switches off the power source to a system or process when dangerous conditions are about to arise.

safety factor A margin of performance or of structural strength built into a design to allow for exceptional operational factors which are not known at the design stage, e.g. excessively heavy snow loading on a roof, an earthquake, or accidental overloading.

safety valve A valve that automatically discharges fluid to atmosphere so as to prevent a predetermined safe pressure being exceeded. It is generally used for compressible fluids such as steam, air and other gases which require the speedy relief of overpressure and is available in different modes and types to suit particular applications. See also **relief valve**.

safety valve, pilot-operated See **pilot-operated safety valve**.

safety valve, pop type A safety valve used with air receivers, that is spring-loaded with a square valve seating which gives a tight shut off. It is usually complete with easing lever, adjusting screw and locknut, valve guard washer and shims for adjusting the clearance of the guard washer.

safety valve spring-loaded A safety valve in which the valve seat is held shut by the tension of a spring. Usually the pressure setting of the valve can be adjusted by changing the tension of the spring.

saline Containing salt, especially the salts of alkaline metals and magnesium. A saline solution is a solution of salts in water.

salinometer A form of hydrometer used for determining the concentration of salt solutions by measuring their density.

sander dust Finely powdered wood-waste collected off wood finishing or sanding machines. It is highly explosive, being spontaneously ignitable at about 400°C. It is usually collected in a separate dust collection system and cannot be safely incinerated in common types of incinerators without the risk of a dangerous explosion.

sand filter, pressure filter [Fig. 193] A filter contained within a steel pressure vessel. It can operate under pressure if the hydraulic conditions within the connected system require it to do so. It receives unfilterered water and after passage over the filter bed, discharges the filtered water to the process.

sanitary installation gradient This must be adequate to promote self-cleansing of the internal bore.

sanitary plumbing, modified single-stack system [Figs 194, 195] A system that provides ventilating pipework extended to atmosphere or connected to a ventilating stack and is installed where the disposition of some appliances on a branch discharge pipe could cause loss of their trap seals. The ventilating stack need not be connected directly to the discharge stack and can be smaller in diameter than that required for a ventilated stack system.

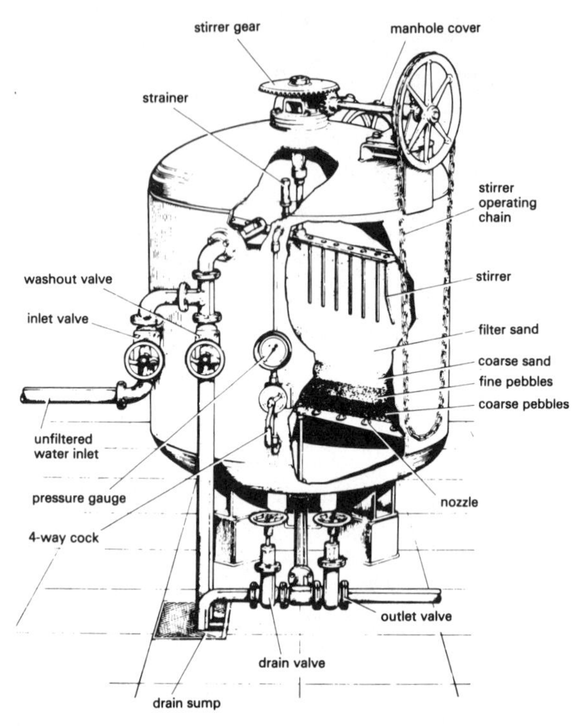

[Fig. 193] Cut-away view of an industrial sand filter

sanitary plumbing, single-stack system [Figs 196, 197] A system used in situations where it is practicable for the discharge stack to be installed in close proximity to the sanitary appliances, but only in cases where the diameter of the discharge stack is sufficiently large to limit pressure fluctuations without the need for a ventilating stack. All the sanitary appliances, including the WCs, are connected only to the discharge stack, no other stack being provided.

sanitary plumbing, stub ducts Ducts used in single-storey buildings and on the ground floor of taller buildings where positive pressures may cause problems if the sanitary appliance is connected to the main discharge stack. A short discharge stack

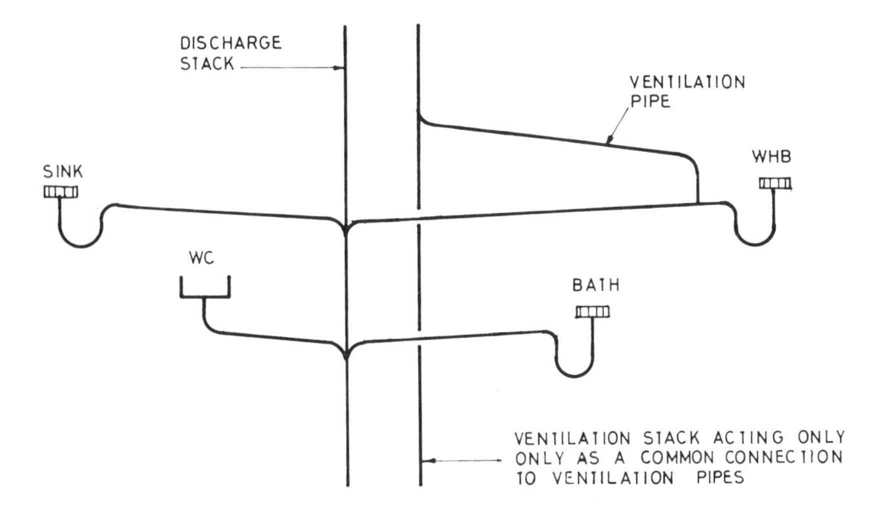

[Fig. 194] Sanitation: modified single stack system with single appliances

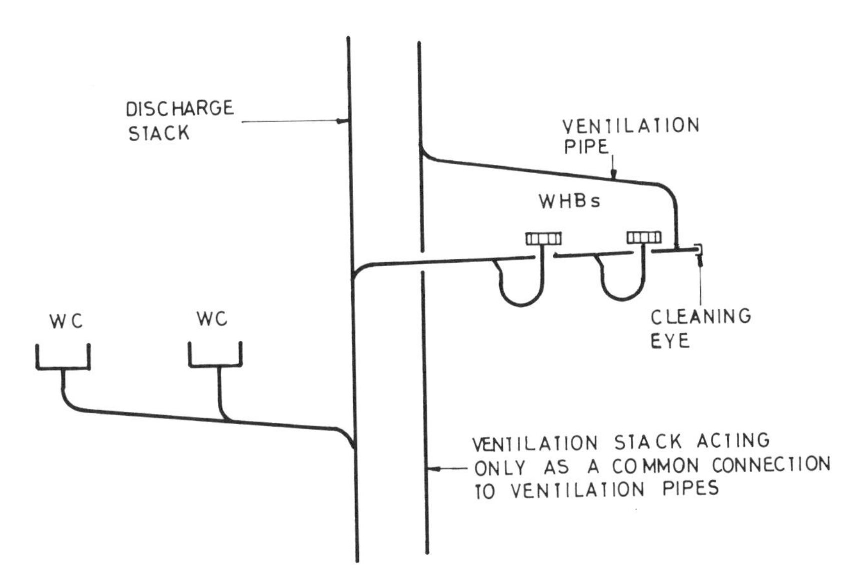

[Fig. 195] Sanitation: modified single stack system with multiple appliances

of 100 mm diameter with the top closed, preferably with an access fitting, may be used to connect each of the sanitary appliances (bath, shower, wash-basin, sink, washing machine and WC) directly to drain, providing that the depth to the drain invert level does not exceed 1.5 m from the crown of the WC trap and 2 m to the centre of the topmost connection to the stub stack. Drains which receive the connections of a number of such stub stacks must be adequately ventilated at the upstream end.

sanitary plumbing, ventilated stack system [Figs 198, 199, 200, 201] A system used in situations where the close grouping of sanitary appliances makes it practicable to install branch discharge pipes without also providing branch ventilating pipes. However, a separate ventilating stack is required. Trap seals are safeguarded by extending the two stacks to atmosphere and by cross-connecting the ventilating stack to the discharge stack. Single WCs or ranges of WCs are connected to the ventilating stack off the top of the horizontal connection.

sanitary plumbing, ventilated system A system used in situations where there are a large number of sanitary appliances installed in ranges or where these appliances are widely dispersed and close

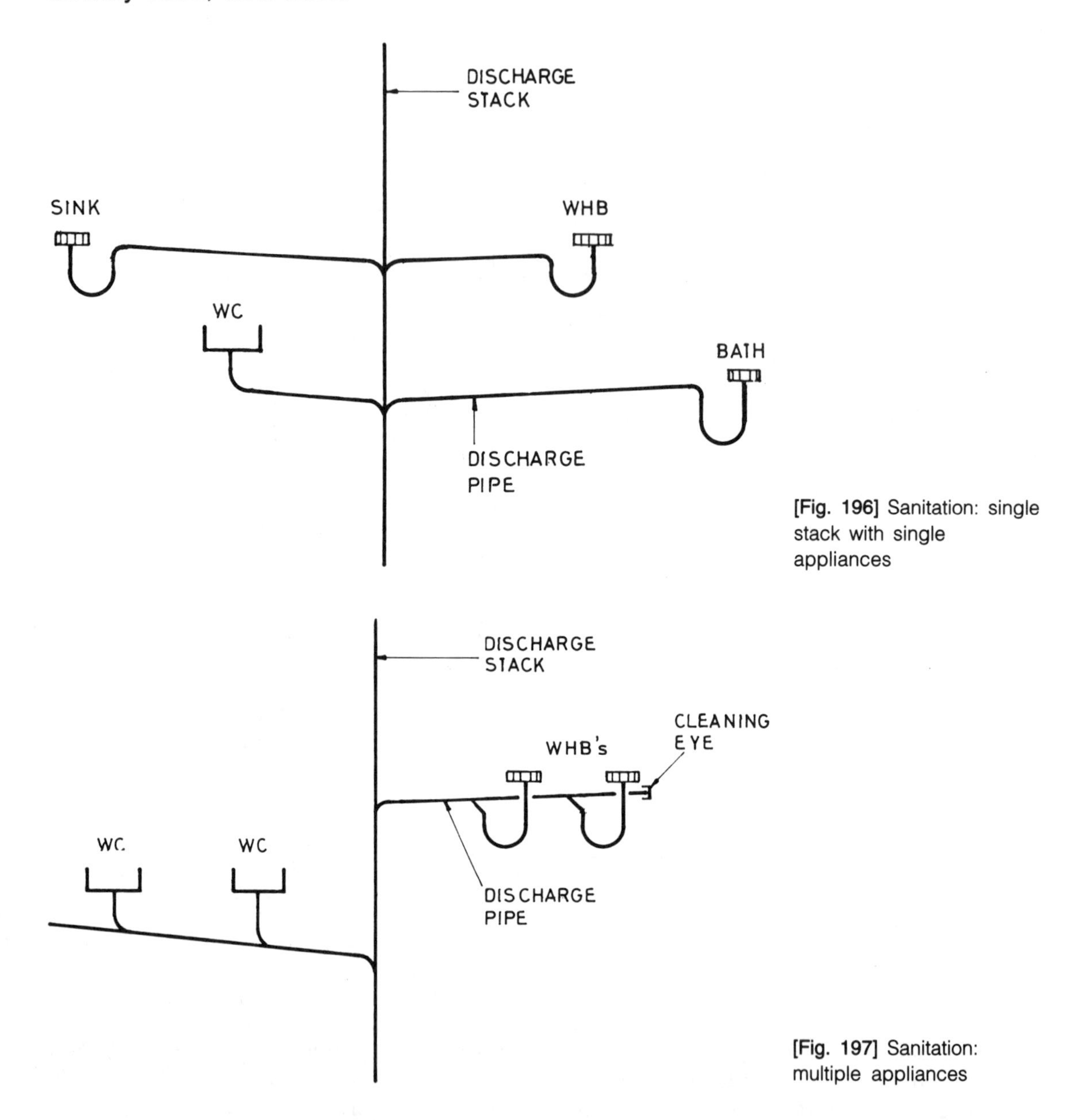

[Fig. 196] Sanitation: single stack with single appliances

[Fig. 197] Sanitation: multiple appliances

proximity to stacks is impracticable. The system comprises two separate stacks: a discharge stack and a ventilating stack. Trap seals are safeguarded from loss of water by extending each stack to atmosphere and, excepting ranges of WCs, providing individual branch ventilating pipes. Each WC, or range of WCs, is connected off the top of the horizontal WC connection into the ventilating stack.

sanitary stack, wind effect Wind blowing across roofs can induce pressure fluctuations in the vicinity of parapets and corners of a building. Discharge or ventilation stacks which terminate in such locations can become subject to unacceptable pressure fluctuations in the discharge system. The designer must exercise great care in the location of the stack terminals to avoid possible adverse wind effects.

sanitary ware A term that encompasses all the models, types and varieties of sanitary fittings, e.g. baths, basins, bidets, and sinks on a project.

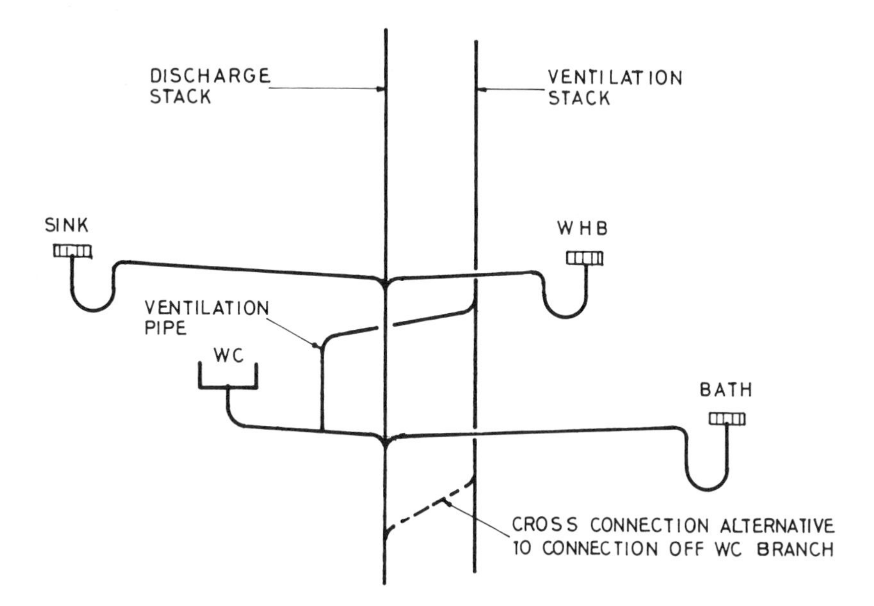

[Fig. 198] Sanitation: ventilated stack system with single appliances

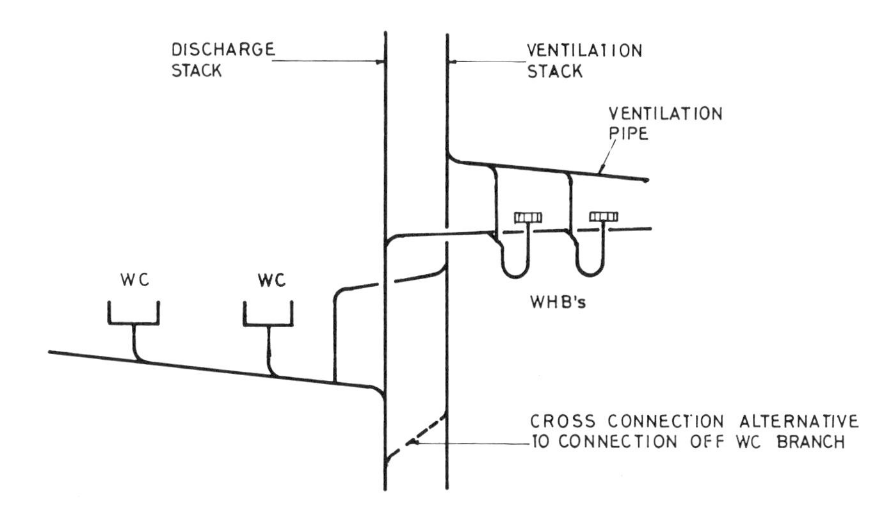

[Fig. 199] Sanitation: ventilated stack system with multiple appliances

sanitation, relief pipe An additional ventilating pipe connected to a soil or waste pipe at any particular point where excessive pressure fluctuation can be anticipated.

Sankey diagram [Fig. 202] A heat flow diagram which illustrates the destinations of the available heat by horizontal bands of appropriate widths.

saturated air Air that contains all the water vapour it can absorb at the prevailing temperature.

saturated steam Steam which has taken up its full capacity of latent heat and contains neither moisture nor suspended unevaporated water. It is also defined as dry saturated steam.

saturation The state of a substance in which it has absorbed as much of another substance as it can contain without spillage.

scale A solid encrustation derived from impurities in water. Accumulation of scale inhibits heat transfer and the free flow of fluid.

scale pocket An assembly comprising a small length of vertical pipe, a tee fitting and a drain plug, commonly fitted into a pipeline ahead of

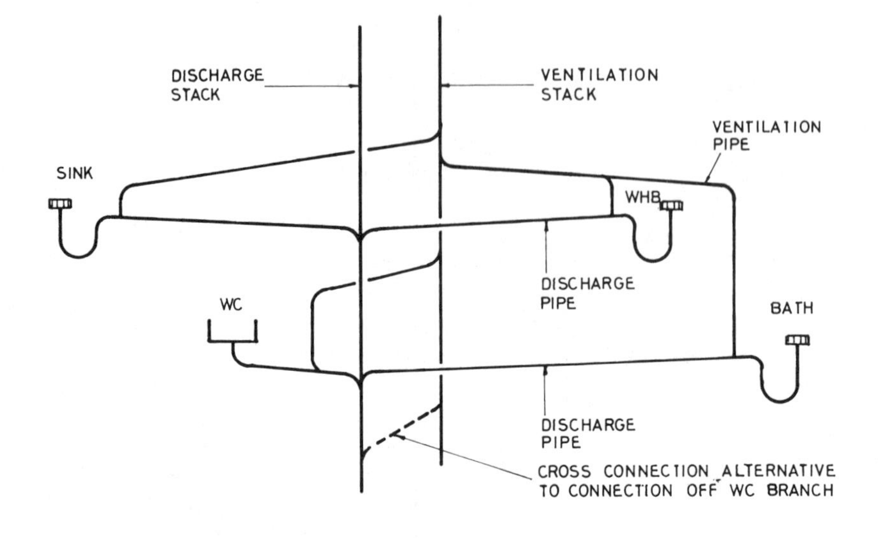

[Fig. 200] Sanitation: ventilated system with single appliances

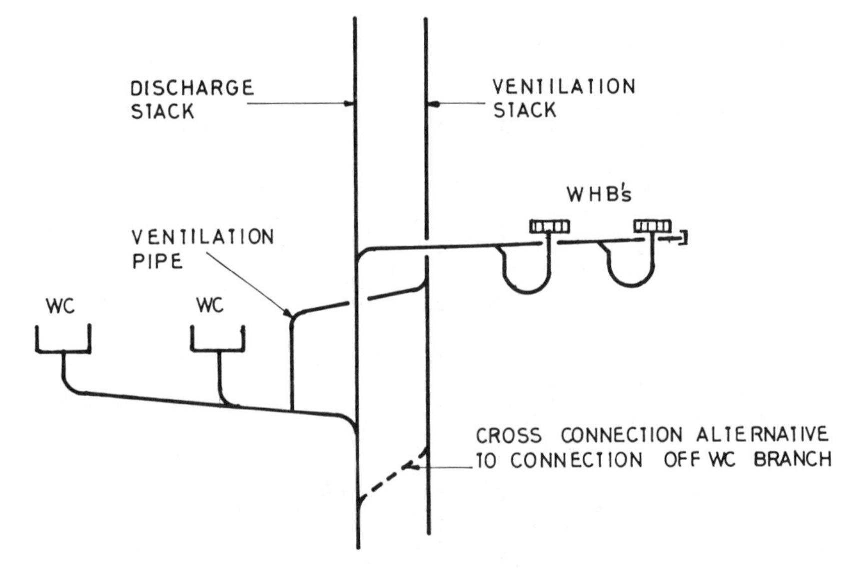

[Fig. 201] Sanitation: ventilated system with multiple appliances

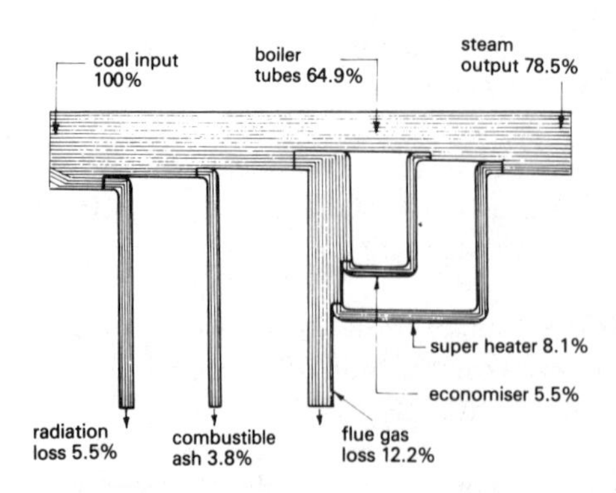

[Fig. 202] Typical Sankey diagram

stream traps, etc., to trap and collect dirt, scale and sludge.

schedule of rates A schedule prepared by the contractor on request indicating the itemized costs per unit of pipe, fitting, etc. for use in the subsequent agreed costing of the completed works.

Scottish Power one of two electricity supply companies in Scotland.

screened cable Cable that incorporates a protective shield which isolates it from undesirable electric or magnetic influences.

screen, electric A shield which isolates electrical equipment, circuits, etc. from the undesirable effects which could be caused by external electric or magnetic fields. It similarly applies to screened cable.

screw compressor A compressor in which two meshing helical-shaped screws rotate and compress the gas as the volume between the screws decreases towards the discharge end. It is generally used in the larger size ranges of positive displacement compressors, in capacities up to about 3000 kW. It is considered reliable, efficient and low-cost equipment.

screwed joint A method of joining pipes by screwing one component into another and securing tightness of soldering, compression or cementing.

screw-down and stop check valve A valve that incorporates a mechanism that can hold the disc in the closed position independently of flow or, alternatively, can restrict the lift of the disc. It differs from the globe valve only in that the disc is not attached to the valve stem. Optionally, it may embody a dashpot and a piston to cushion the valve movement.

screw-down stop valve A valve in which the disc is lifted from and lowered on to the body seat by a stem whose axis is at right angles to the face of the body seat. It may be operated by an inside screw or an outside screw. The actuating thread of the stem of the former type is engaged within the valve bonnet, whilst that of the latter is exterior to the bonnet.

screw pump A pump that incorporates a worm arrangement in place of the conventional pump impeller. It is used to pump viscous fluids, such as heavy fuel oil, and can develop high pressure.

sealed system In wet heating and hot water installations, this refers to a piped system which is not permanently vented to atmosphere and does not incorporate an open feed and expansion cistern to automatically replenish water lost from the system. Such systems incorporate a pressure vessel, appropriately sized for the water capacity and pressure of the particular installation. The pressure vessel has a diaphragm which separates its water space from the upper space which contains a pressurized air or nitrogen cushion. Water expands against the diaphragm during heating up.

Commercial-size sealed systems operate with a water break-tank and automatic controls which switch a boost pump to compensate for loss of pressure due to leaks at valves and pump glands. Domestic-size systems are usually provided with a flexible connection which is temporarily connected into the mains water supply to boost pressure. The satisfactory operation of such systems requires the provision of pressure gauges in accessible locations, so that the pressure may be closely monitored. Safety valves are required.

Sealed systems must be used with microbore pipe systems to exclude all particles of corrosion products. Sealed systems in general offer the advantages of exclusion of contact between the water and air and obviate the need for open vent pipes (particularly where such vent pipes are difficult or costly to accommodate). However, they present the system user with an added complication.

Caution: the correct selection and specification of the pressure and water capacity of the pressure vessel and the provision of pressure gauges are essential for the safe operation of the sealed system.

sealing compounds high temperature Compounds in ceramic or asbestos form with or without a hardening agent which are suitable for use on their own or with a gasket. They are used to effect an effective air/gas tight seal for flanges and joints which are subject to heat distortion over a period of time, e.g. boilers, furnaces, ovens or smelters.

seals mechanical Seals used with circulating pumps, replacing conventional pump glands. They do not leak water and do not require pump drains. Failure of the seal may cause a major water leak at the pump.

second law of thermodynamics The statement that no process is practicable whose sole result is the complete conversion of heat into work; there are always losses.

secondary air Air that supplements the primary air introduced to a combustion process to encourage complete combustion. It is usually introduced above the fuel bed in a controlled manner via adjustable air inlet ports and must be carefully monitored to ensure that only the correct quantity of secondary air is introduced; any excess will reduce combustion efficiency.

Applied to solid fuel combustion, it is introduced above the fuel bed to assist the process of complete combustion of the fine fuel particles in suspension above the bed. Secondary air must be closely regulated and monitored when combusting fine particle fuel, such as sawdust and sander dust because 75% or more of such fuel has to be burnt in suspension above the fuel bed.

Applied to the combustion of liquid or fuels, the secondary air is introduced around the fuel/primary air mixture to assist complete combustion and the establishment of the correct flame form or pattern.

secondary filter A filter installed behind the primary filter to arrest the main dust burden in a ventilation system. It acts as the main air filter.

secondary hot water circulation A pumped or gravity pipe circuit within which circulates the hot water being fed to draw-off points. It should be taken as close as practicable to the water-using taps or appliances to avoid cooling the water in deadleg pipe connections.

sectional in situ drain liner The Renoline system is suitable for repairing a short section of drain by means of a fibreglass sleeve which is impregnated at site with a chemically resistant polyester resin. It is applied by wrapping the sleeve, up to 1.2 m long, around a deflated air bag which is pulled into the defective section under the guidance of a closed-circuit television camera. The operation is completed by inflating the air bag and thereby expanding it tightly against the surface of the damaged pipe. It is left to cure for one hour, after which time the bag is deflated and removed leaving the drain ready for use.

security engineering Safeguards that restrict unauthorized entry into a building and discourage burglary.

sedimentation The process of separating an insoluble solid from a liquid in which it is suspended by allowing it to fall to the bottom of the containing vessel, with or without agitation of the liquid.

self-cleansing A feature of a correctly designed sanitary/drainage installation.

self-contained emergency luminaire A luminaire that is self-powered and which operates independently in an emergency. For most applications it must be capable of operating continuously for a period of 1–3 hours; 3 hours is the more common design basis. Routine maintenance of the units must be thorough to ensure performance in an emergency.

self-contained emergency luminaire, maintained In this mode the lamp is on all the time. Under normal conditions it is powered by the mains; under emergency conditions, it uses its own battery supply.

self-contained emergency luminaire, non-maintained The lamp is off when the mains power is available to charge the batteries; upon power failure, the lamp is energized from the battery pack.

self-contained emergency luminaire, sustained The lamp operates from the mains supply under normal conditions; in an emergency, a second lamp, powered from the battery pack, takes over. Such luminaires are used for exit signs.

self-exciting In a generator, having magnets which are excited by electric current drawn from the output of the generator.

self-priming pump Retains its water content when not pumping; hence starts pumping fluid immediately it is switched on regardless of the location of the pump relative to the pipeline to which the pump suction port is connected.

semiconductor A widely used component of an electronic circuit, tailored to conduct electric current within specified parameters. See also **thermistor**.

semi-hermetic compressor A compressor which is sealed into its casing and has a motor fitted outside the casing.

semi-rotary hand pump A handle-operated manual pump which draws liquid from one vessel and transfers it to another. It is suitable for small flows.

semi-silica A refractory material of limited application.

sensible cooling A reduction in the temperature of air without any change in its moisture content. It does not cool below the dewpoint of the air.

sensible heating An increase in the temperature of air without any change in its moisture content.

sensor A detector placed in a certain area or on a surface to measure some specific property. It is used in conjunction with a remotely located indicating or recording instrument.

separate drainage systems Foul sewage and surface water are collected separately. Regardless of the arrangement of the main sewer, drainage installations within buildings are most commonly designed as separate systems.

separate sewer systems Systems that collect and convey foul sewage and surface water within separate pipe installations.

sequence control The method by which a number or group of controlled pieces of equipment are operated within a preset sequence. For example, a number of boilers are brought on to line, one after the other, as the load increases and are taken off line in a similar manner as the load decreases.

sequence control for boilers See **boiler sequence control**.

series circuit A connection of two or more electrical circuits made in such a manner that the current traverses all the circuits in sequence without dividing.

service entry [Fig. 203] The arrangement at the boundary of a property where a piped service enters the building from a pipe duct or trench. Such special entry provision is usually required for electricity intake cable, gas pipe, telephone cable and water pipe.

service pipe, water The means of supplying water from a water main to any premises which are subject to water pressure from that main, or would be so subject but for the closing of some stop valve.

serving (electric) Layers of fibrous material, permeated with waterproof compound, applied to the exterior of an armoured or metal-sheathed cable for additional protection.

servo-control A method of amplifying, by electronic or other means, a small impulse from a detecting instrument or sensor, and thereby causing a larger force to operate a valve or other mechanism.

servo-element In automatic control applications, when a control impulse is received, the controlled element (e.g. valve, damper, etc.) is reset to return the controlled function to the control set point.

servo-mechanism A means of converting a small low-powered mechanical motion into a mechanical motion requiring considerable power. The output power is always proportional to the input power. The system may include a negative (usually electronic) feedback device.

sewage ejectors and pumps, selection The choice of selecting one or the other is subject to the following factors.

Location Where access to the sewage lifting apparatus is possible only from within the building, an ejector is preferred as the complete enclosure of the sewage lifting apparatus is practicable. A sump, normally required when using a lift pump, is not needed, structural provisions are minimized and hygiene safeguarded.

Flow rate Where the incoming sewage flow rate exceeds 7 l/s, an ejector is not commonly installed. The selected apparatus, whatever the type, should be capable of supporting a velocity in the delivery main that will be in the order of 9.7 m/s to 10 m/s, sufficient to prevent the settling out of solids.

Head Ejectors and pumps can lift sewage against

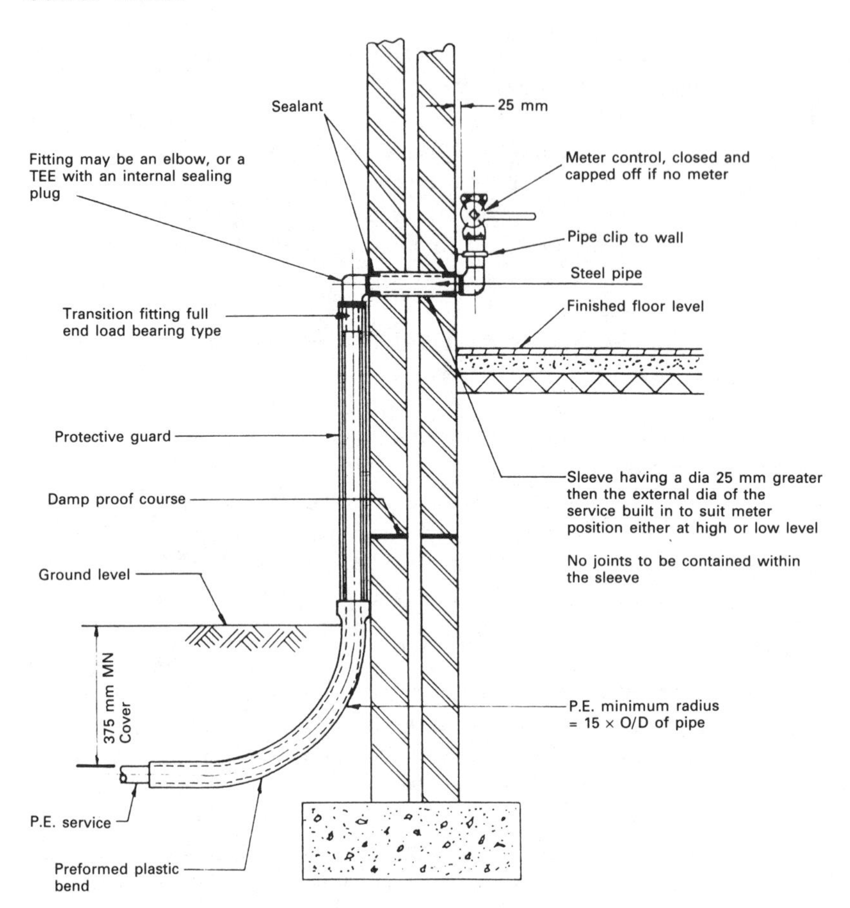

[Fig. 203] Standard detail of gas service entry outside riser (sectional elevation)

low heads; for total heads in excess of 25 m, pumps are preferred.

sewer invert See **invert**.

shade temperature This is measured or recorded by a thermometer which is placed in a location where there is no direct sunlight.

shade temperature air-conditioning An air-conditioning system without the capacity to compensate for the effect of solar energy heat gains.

shading The means of reducing solar heat transmission into a space. It utilizes internal, external or shades (blinds); it can also be located between double glazing panes. It is essential on a building face exposed to the Sun to obviate glare and to reduce the costs of air-conditioning.

shell and coil condenser A welded pressure vessel containing the condensing surface in a coil or grid formation, the cooling water flowing within the tubes. The water side may be cleaned only by chemical means. It is commonly used in packaged air-conditioners of up to 53 kW capacity.

shell and tube condenser A welded pressure vessel containing the condensing surface in a coil or grid formation. It operates with the cooling water in the tubes and refrigerant in the shell. The tubes are generally of copper construction and arranged for mechanical cleaning. Materials other than copper may be used to handle salt or corrosive water. These condensers are available in capacities of up to about 14 000 kW.

shell boiler A boiler in which the water being heated is contained within the shell of the boiler. The hot gases pass through furnace tubes which cross the water content in the shell.

sheradizing A rust-proofing process using a low-temperature heat treatment with zinc dust to produce a hard uniform coating or iron–zinc alloy suitable for the anti-corrosion protection of a wide range of ferrous components.

shop drawing A drawing that shows and details assemblies or items fabricated in the workshop, as opposed to fabricated on site.

short circuit (electric) A connection made accidentally or otherwise between two points in an electrical installation which have a difference of electric potential between them, the points of the short circuit connection being of sufficiently low resistance to allow a very much larger current than normal to flow through the circuit established by this connection. Fuses and other protective devices are fitted to protect circuits and subcircuits from the possibly harmful effects of short circuits.

short-circuiting The tendency in water circuits to divert the water via circuits having least resistance to the flow, e.g. circuits close to the circulating pump.

shower tray [Fig. 204] The base of a shower unit commonly manufactured from glazed fire-clay with a slip-resistant base and a grate-covered outlet with an overflow. The levels are arranged to provide a raised standing area adjacent to the shower with a slight fall into the tray to avoid water splashing outside the perimeter of the tray.

shrunk rubber joint ring A joint ring used with certain types of plastic pipes, such as PVC, employing an 'O' ring. Heat is applied and the socket is shrunk on to the spigot entrapping the rubber ring, thereby securing the joint.

shunt connection As used in electrics, the connection of two electric circuits in which the same e.m.f. is applied to both.

As used in mechanical engineering, a shunt connection is a bypass pipe or duct.

shunt duct A branch duct which rises parallel to the main duct to which it is connected.

shut-off damper An air flow control which completely prevents the passage of air through a duct fitted with this type of damper.

Sick Building Syndrome The chronic but typically low-level symptoms experienced when people are at work within a building and which disappear when they are away from the building for long periods, such as in the evenings and at weekends. The symptoms fall under four defined headings as follows.

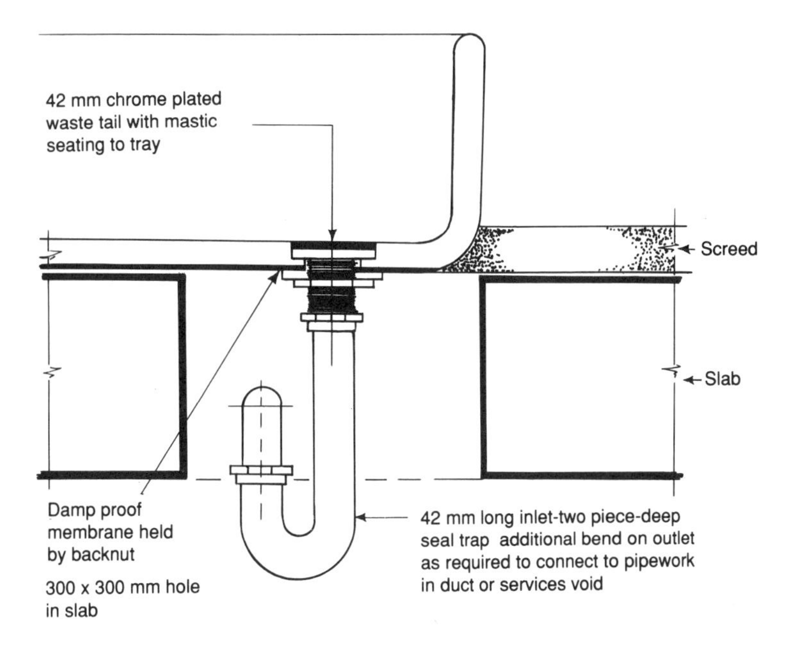

[Fig. 204] Water connection to shower tray (upper floors)

(a) Dry symptoms; stuffy nose, dry throat and dry skin.
(b) Allergic symptoms; runny or itchy nose, watering or itchy eyes.
(c) Asthmatic symptoms; tight chest.
(d) Undefined; undue lethargy, headache and similar.

It is accepted that good design and regular maintenance, together with health-oriented air movement, can prevent Sick Building Syndrome.

sight flow indicator A device installed in liquid pipelines to visually indicate fluid flow by means of a spinning disc or flap.

sight glass A device fitted within a steam trap assembly, following the trap which indicates via a transparent window whether the steam trap is functioning correctly, passing only condensate, or whether steam is allowed to pass the trap. A sight glass is valuable maintenance tool in steam installations. It may be single or double sided and with or without a flow indicator.

silencer A colloquialism for a sound attenuator. Device fitted to the exhaust gas discharge pipe of a diesel engine or generator to attenuate the noise emitted during operation.

silica A mineral compound found in various rocks. Its sensitivity to abrupt changes in temperature restricts its use as a refractory. It contains glass of high melting-point and retains its stiffness under load to within a few degrees of the melting-point and is, therefore, suitable for high furnace temperatures. Silica has a strong spalling tendency at low temperatures below dull redness.

silica gel [Fig. 205] A generic term which covers many forms of the material which can be obtained by acidifying sodium silicate solutions. It is used as a dehydrating agent. The material changes colour when saturated and can be reactivated by application of heat which drives off the moisture.

silicon carbide A refractory suitable for use with high-temperature applications, e.g. in cyclone furnaces in wood-waste incineration schemes.

silo [Fig. 206] An above-ground storage hopper, commonly constructed of steel or cast-iron sections. Two types are available for the storage of fuel, wastes, etc.: cone-bottomed and flat-bottomed. Cone-bottomed silos are designed to avoid bridg-

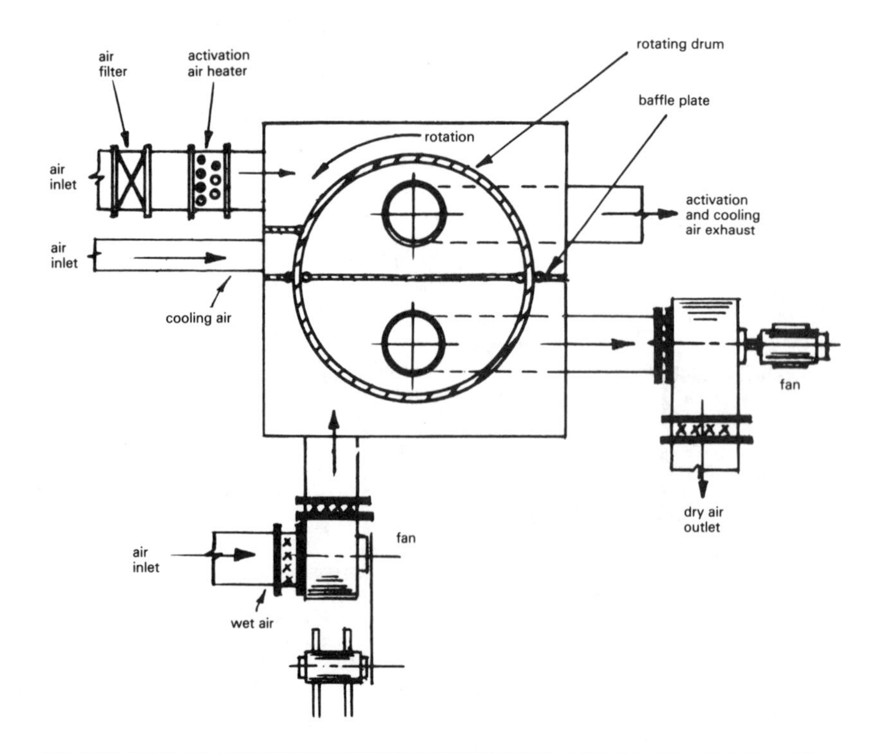

[Fig. 205] Silica gel rotary drier

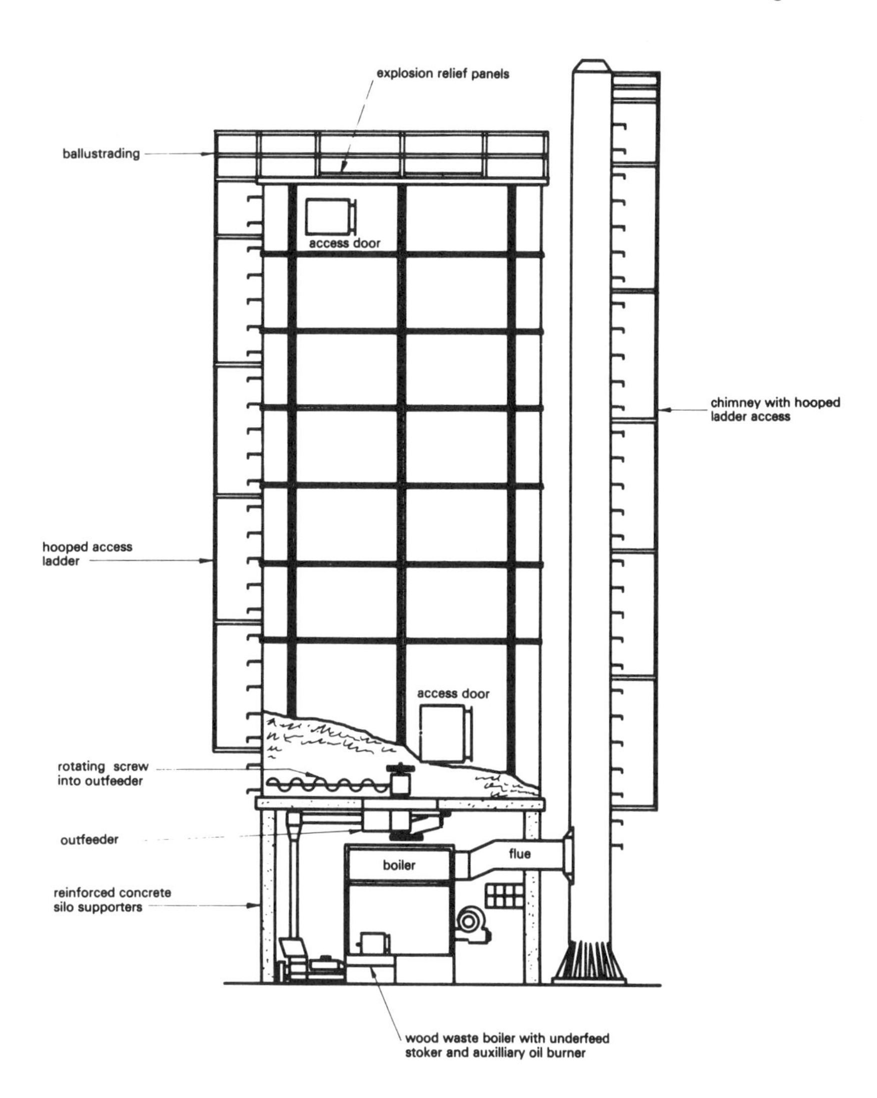

[Fig. 206] Flat-bottomed silo

ing within the stored material and to allow free flow towards the bottom outlet, usually via a rotary valve. Flat-bottomed silos depend on the positive displacement of the contents by means of a rotating arm or a screw-feeder fitted inside the silo. This offers the advantage of greater storage capacity for a silo of a given diameter.

silver-based brazing solder Comprises an alloy of copper, zinc and silver; cadmium is added sometimes for specific purposes. It melts between 600 and 750°C. The addition of cadmium improves fluidity and lowers the melting temperature. A suitable flux must be used in making the joint. Most material can be joined using this solder, including joints of dissimilar metals, providing that their melting point is above 750°C. The resultant joints are strong, often stronger than the parent metal, ductile and offer good resistance to corrosion. Brazed joints can be electro-plated. They are used widely in building services for the joining of copper pipes by butt-welding.

silver solder Hard solder suitable for harsh conditions of temperature, pressure and environment.

simulated cooling test A test based on the use of a simulated heat load with the object of testing the cooling capacity of a plant under designer conditions when these are not present e.g. testing an air conditioning plant during winter conditions.

simulated heat load A correctly sized heat source placed within an air-conditioned space for the purpose of testing the adequacy of the cooling facility.

simultaneous demand That demand generated at any one time by the in-use terminals of a system, e.g. water draw-off taps, or electric points. See also **maximum simultaneous demand**.

single-acting compressor A compressor that operates with only one stage of compression.

single-duct air-conditioning A high-velocity air-conditioning system in which warmed, humidified and filtered air is supplied from a central air handling plant into terminal induction cabinet units in which the cooling and distribution of the air is carried out locally.

single-leaf damper A type of damper which operates with one leaf, e.g. a fire damper. It is usually fitted with quadrant operating gear and an associated locking device.

single-phase circuit A circuit where a single alternating current is supplied by one pair of wires (conductors). This is generally used with domestic and small industrial or commercial systems.

single-pole switch A circuit breaker, cut out, fuse-switch and similar in which the circuit is broken in one pole only.

single-stack system See **sanitary plumbing, single-stack system**.

single-vane rotary compressor A rotary compressor operating with one rotary vane. See **rotary compressor**.

sinking fund A fund of money which is accumulated over a period of time by regular instalments to be eventually disbursed at a particular date in the future on the replacement of a physical asset, such as a new lift installation, boiler plant, computer, etc. The interest earned on the accumulated capital is incorporated into the fund.

sinks These are supplied for general cleaning duties or in areas such as flower handling rooms and laboratories. The shelf-back sink for heavy duties is popular for use in institutions and public buildings and is generally made of glazed fire-clay and supported on painted cast-iron legs. Taps would be the high neck 15 mm pillar type, the waste outlet would be a two-piece 37 mm with a 75 mm deep trap. Another type is a heavy-duty cleaner's sink with bucket grating manufactured of glazed fire-clay with a hardwood pad inset. Taps would be commonly 15 mm wall-mounted bib taps and the sink would be supported on metal legs as well as off the wall. These sinks are widely used in commercial, public and hospital buildings. Sinks manufactured of stainless steel are used where there is a risk of damage to the fire-clay alternative and where a high standard of cleanliness is required, such as in kitchens.

siphon A U-tube which has one leg longer than the other, used to transfer a liquid from one vessel to one at a lower elevation without recourse to pumping.

siphonage, induced The fast flow of liquid through a pipe may induce siphonage in a connected pipe, unless there is provision for an anti-siphonage facility.

siphoning The act of transferring a liquid from one vessel to another by means of a siphon. The shorter leg of the U-tube is placed over the rim of the upper vessel and is submerged in the liquid; the lower limb of the U-tube terminates inside the lower vessel. The siphoning process is started by applying a suction to the lower limb. Once the flow of liquid has been started, it will continue until the vessel has been emptied to the level of the limb of the upper U-tube.

site agent The main contractor's chief representative at the building site.

sketch scheme A plan prepared in the early stage of a project to outline the proposed scheme.

skirting heater A continuous radiant panel or convector casing heater of low height fitted to, or in lieu of, a skirting. See also **baseboard heater**.

slab urinal These are available in various lengths made up of components which are each about 600 mm long. The channels in these units are matched to provide continuous falls towards the outlet and they are fitted together with purpose-made components to form ends and divisions to give a close and watertight fit. In industrial and public lavatories, the channel must be set into a floor recess so that the channel edging tile is at the finished floor level, permitting the floor to be washed down into the urinal channel. The associated flushing cisterns may be exposed or concealed.

slaked lime injection flue gas cleaner [Fig. 207] A simplified method of cleaning flue gases with reduced energy consumption by means of a combination of a spray absorber and an electrostatic precipitator. The flue gases pass through the absorber into which slaked lime is continuously injected. The residual heat in the in-going gases to the water evaporate and the reacted components thereby form a free-flowing powder which settles out in the electrostatic filter and is periodically removed.

slave clock An individual clock connected to, and motivated by, a central clock system as directed by a master clock.

sleeve An over-sized short section of pipe which is built into a structure and allows a pipe to pass through it without making direct contact with the building fabric. It permits a pipe to move smoothly in conjunction with a thermal expansion system. End caps are available for sealing the ends of the sleeve over the pipe.

sleeve bearing A bearing that supports a rotating shaft, e.g. a fan shaft. It runs inside an oil reservoir which must be monitored and periodically replenished.

sleeved uPVC pipes for sewer or drain relining The Interline method (by Rice–Hayward) relies on the insertion into the defective drain of short lengths of unjointed uPVC pipes inside plastic sleeves. It offers structural strength, prevention of infiltration or ex-filtration, resistance to corrosion and chemical attack and satisfactory hydraulic characteristics. The method is suitable for sewers of any shape or size with severe infiltration defects and where man access is impracticable.

sleeve valve A free discharge valve in which the closing member is in the form of a sleeve.

slide rails Two parallel rails on which an electric motor is secured. The motor can be slid along the slide rails to vary the tension of the driving belts.

sliding expansion joint A joint that comprises two cylinders sliding inside a pipe gland to permit pipe

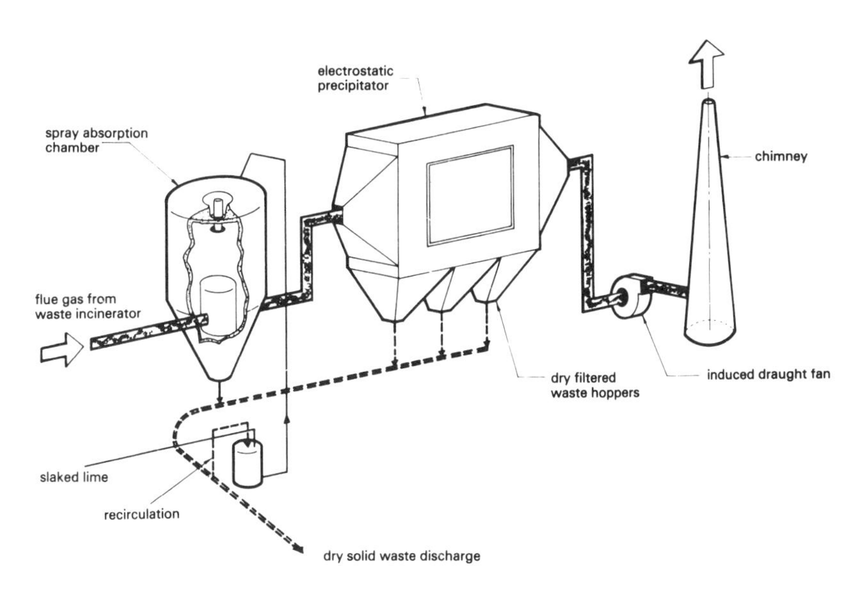

[Fig. 207] Slaked lime flue gas chamber

movement. It operates in conjunction with pipe anchors and pipe guides.

slip pipe coupling A coupling used to join plain-ended pipes. See **Viking Johnson slip coupling**.

slipper bend A short three-quarter section branch bend in a sanitation system.

slot diffuser An air distribution terminal in the form of a framed long slot, usually with a key-operated air flow adjustment damper fitted to the rear of the frame.

slow sand filter A filter in which the water is passed very slowly downwards through sand beds of about 75 cm thickness. After about a month in operation the surface has to be skimmed, but this period can vary widely depending on the condition of the raw water. Eventually the filter has to be taken out of service and refilled with washed sand.

sludge pot A collection vessel placed in the return connection to a hot water boiler to trap larger size impurities by gravity, or into pipelines ahead of sludge-sensitive equipment, e.g. small water content boilers, to arrest and intercept scale, sludge and dirt. It must be complete with a drain-off valve and plug.

sluice valve A solid wedge gate valve commonly specified for waterworks applications.

small-bore heating system A pump-assisted domestic-size installation employing 12.5 mm and 19 mm bore copper or thin bore special mild steel pipe distribution. The collecting headers at the boiler are usually of larger bore. It operates best as a closed system. It employs a canned rotor circulator to meet the head (resistance) requirement. Each sub-circuit must be valved to permit balancing of the circuits and to avoid the short-circuiting of water around the circuits with the least resistance.

smoke Visible evidence of incomplete combustion consisting of minute carbonaceous particles smaller than 5 microns. Emission of dark smoke is prohibited by legislation in certain 'smoke control or smokeless zones'. Smokes are relatively stable dispersions.

smoke cartridge A container of special chemicals used to generate smoke when ignited. This must be used with caution, taking great care that the ignited cartridge is not in direct contact with the air ducts and that the products of combustion have no harmful effect upon the materials of the system being tested.

Application: testing or commissioning of air distribution fittings and systems by observing the flow of smoke-laden air.

smoke control The division of a building into fire compartments and the incorporation of natural or mechanical means of smoke evacuation.

smoke density The intensity of smoke issuing from combustion equipment, measured on a scale of 1 to 10 on the Bacharach scale.

smoke density equipment [Fig. 208] A means of measuring the density of smoke emitted from industrial boilers and furnaces in terms of light obscuration.

smoke detector A device activated by smoke to give an audible alarm and/or start up fire-fighting equipment.

smoke hood A chamber at the flue outlet end of a boiler into which the flue gases are directed before passing via the flue pipe into the chimney.

smoke generator A device for generating smoke for the purposes of flow visualization. It is used for checking air movement across spaces, e.g. to locate and follow draught movements; indicating air flow patterns at air supply and extract grilles, slots, etc.; establishing air movement at windows and air curtains; monitoring fume and dust extraction, etc. At its simplest, the generator may consist of a smoke tube which is fitted into an aspirator. When both ends of the tube have been broken off, the smoke is aspirated. It is used for estimating the velocity of flow-moving ventilating currents by timing the travel of the smoke given off by the tube when air is aspirated through it. It can also be used to indicate the direction of air currents.

For research and more sophisticated smoke testing, electric mains-operated smoke generators are

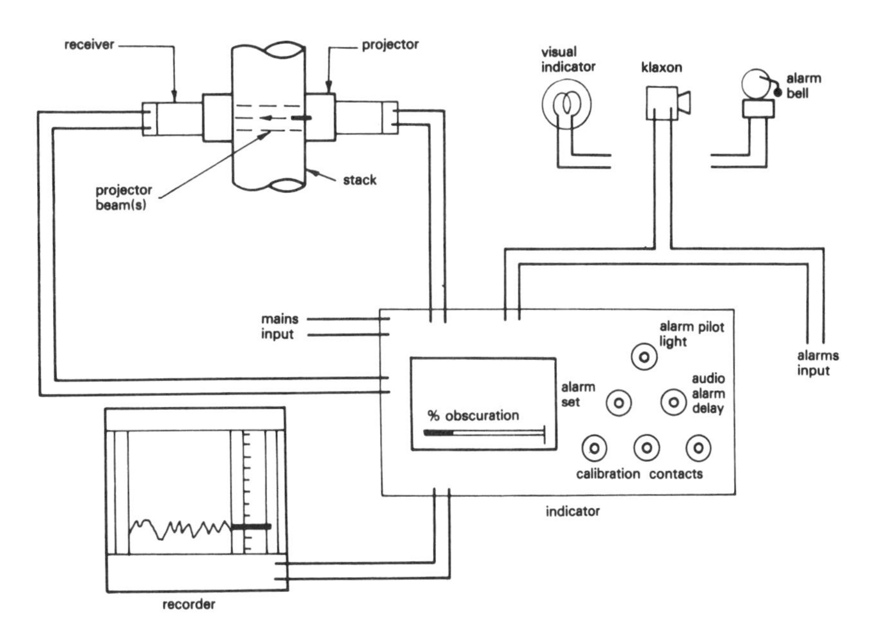

[Fig. 208] Smoke density measurement

available, such as the TEM Smoke Generator System developed by the National Physical Laboratory, Teddington, England. With this, the smoke probe operates in conjunction with an electric generator. The smoke is generated at the extreme tip of the probe by the vaporization of a special oil. Smoke is then emitted directly into the airstream from a compact screw-on vaporizer that is fed with oil and low voltage power via the thin stem of the probe. The standard probe is 368 mm long with a wake-minimizing curve into alignment along the airstream. Equipment is available for producing clearly visible smoke plumes at flow rates of up to 90 m/s.

Applications: for use in wind tunnels, industrial and architectural aerodynamics, air-conditioning, fume and dust extraction, heat exchanging, etc.

smoke test of sanitary installations A smoke generator introduces smoke under pressure into pipework suspected of being defective and leakages may be observed as the smoke escapes through defective joints. Smoke testing is unsuitable for plastic pipework due to naphtha having a detrimental effect on the plastic materials. Rubber jointing components can also be adversely affected.

smoke tube boiler See **fire tube boiler**.

smoke vent [Fig. 209] Framed multi-leaf louvres built into a roof, arranged to open automatically on a rise in space temperature to evacuate smoke.

smut emission The deposition of unburnt or partially burnt combustion products from chimneys into the surrounding atmosphere causing pollution. It is caused through inefficient combustion (which passes unburnt particles into the chimney) and through acid condensation brought about by cooling in the flue gas (or chimney) system below the acid dewpoint of the gas. Smut emission can be avoided by a good standard of combustion management and by suitable thermal insulation of the chimney system.

smuts Small particles of unburnt carbon carried in flue gas discharge from chimneys. Smuts tend to have an oily surface and cause a nuisance when settling on laundry, roofs, cars, etc.

Snap-lock system A method of internally lining sewers and drains. See **preformed new sewer replacement pipes**.

soakaway [Fig. 210] A borehole in the ground to which surface water is conveyed by drain and through which it percolates into the surrounding earth. A soakaway method of disposal relies on

GREENWOOD AIRVAC
SV-H SMOKE VENTILATOR

The SV-H is a highly insulated and sealed louvred smoke ventilator. This dual purpose ventilator is designed for day to day and smoke ventilation and can be fitted with a variety of controls.

DESIGN

The SV-H louvre and base is made from double skin marine quality aluminium fitted with insulation. All contact points are thermally insulated to avoid cold bridges.

The pivot points of the louvre turn in nylon bearing bushes and are housed in an aluminium frame outside the airstream. They require no maintenance.

The ventilator is totally insulated and air leakage free when closed.

OPTIONS:

SV-H TRANSLUCENT

Blades are triple skin translucent, extruded in U.V. stabilised, virtually unbreakable polycarbonate, in clear translucent (73% transp.) or opaque (40%) material.

BASE UNITS AND FLANGES

An insulated high base unit (150mm) is fitted as standard. The low base unit is not available on this model. Full details of fixing flanges are provided overleaf.

TECHNICAL SPECIFICATIONS (overleaf.)

In the table all SV-H ventilators are listed with their respective measured and aerodynamic free areas, weights and dimensions.

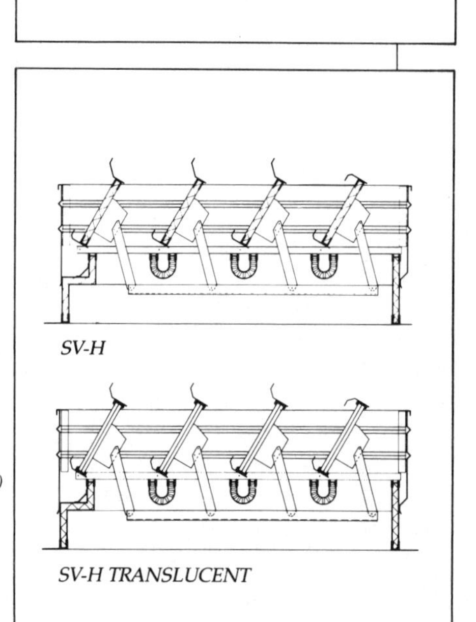

Greenwood Airvac reserve the right in view of their continuous programme of product development and improvement, to revise or alter their range of products and prices without prior notice.

[Fig. 209] Highly insulated and sealed double-skin marine quality aluminium smoke ventilator (Greenwood Airvac Ltd.)

the porosity of the ground to accept and disperse the water flow at the selected depth. Ground conditions must be investigated before deciding on the feasibility and/or depth of a soakaway.

soakaway, construction Such pits may be unlined and filled with hardcore for stability. Alternatively, surface water disposal may be via seepage trenches following convenient contours. Larger soakaway pits may be unfilled but lined with brickwork laid dry, joined honeycomb brickwork, perforated pre-cast concrete rings, segments laid dry or similar means. The lining should be surrounded with suitable granular material. An open pit must be safely roofed and a manhole cover be installed to permit access for inspection and maintenance.

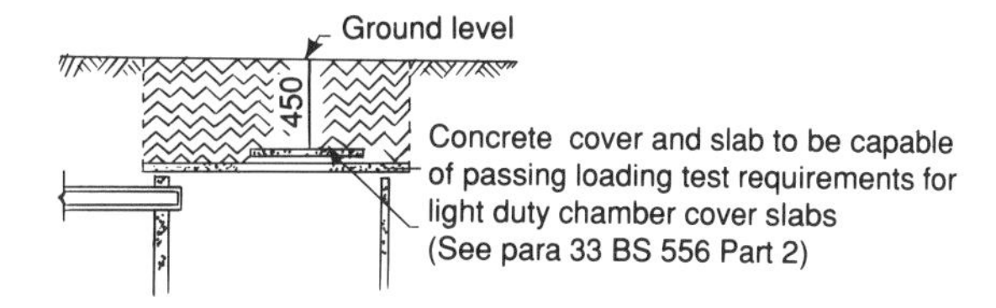

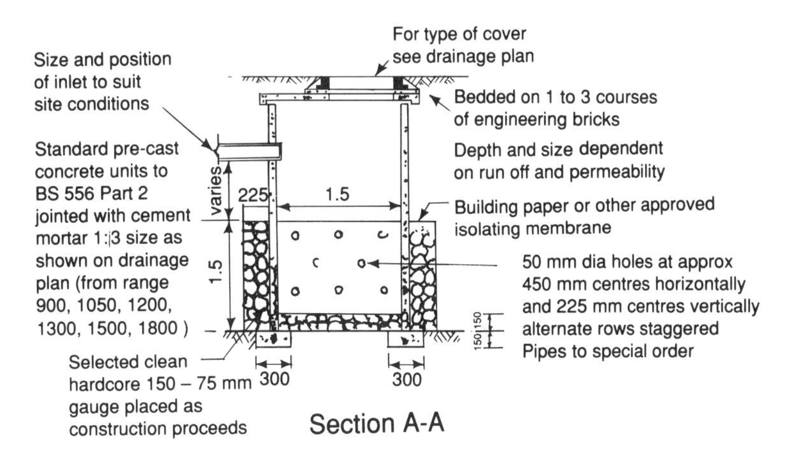

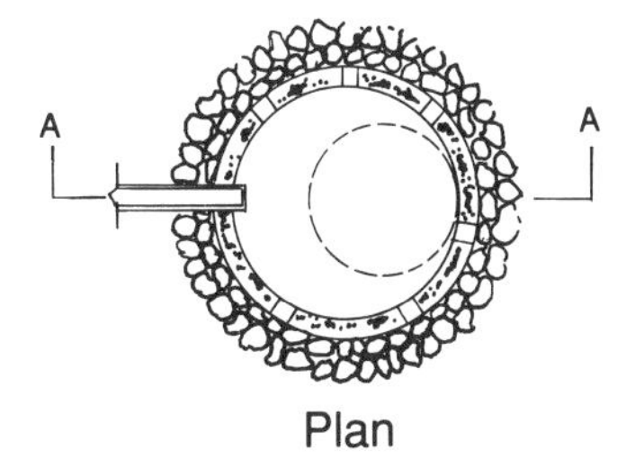

[Fig. 210] Soakaway

soap solution test of sanitary installations With the sanitary installation under air or smoke pressure, a soap solution is applied to the pipes and joints. Leakage can then be detected by observing the formation of soap bubbles at the defective parts.

soffit crown The highest point of the internal surface of a pipe or culvert at any cross-section.

soft coal Relatively friable coal which is subject to deterioration when being handled.

soft starter Primarily a means of alleviating the mechanical stresses to an electric motor during the starting sequence.

soft water Water which forms an immediate lather with soap.

software Programs used in a computer, especially the general programs supplied by the computer manufacturer. The actual equipment of the computer is termed the hardware.

soil stiffness This varies according to the nature of the subsoil material and the compaction of the bedding material.

solar-air temperature The external temperature of a hypothetical uniform environment which has the surroundings and the air at equal temperatures and which would provide the same rate of heat transfer through a building element as occurs under the actual prevailing conditions.

solar array A number of individual solar collectors arranged in a specific pattern to form a collector complex.

solar chimney A chimney that operates on the principle of guided convection. Air is drawn through the tall solar chimney, activating one or more turbines in its path to generate electricity. The essential components are: a central chimney exceeding a height of 200 m and a greenhouse environment formed of transparent plastic sheeting which is supported at a short distance above ground by a metal frame and which surrounds the base of the chimney.

In operation, solar incidence on the plastic screen causes the air which is trapped in the 'greenhouse' to heat up, thereby establishing a convection flow of air which is drawn up the chimney at a velocity of 20–60 m/s, sufficient to motivate low-pressure turbines which are placed within the circuit. The heated air is continuously replenished by cooler fresh air which flows through the 'greenhouse' to the chimney.

The area of the solar collector (greenhouse) extends over a considerable area, possibly a radius of 130 m, varying from 2 m at the circumference to 6 m high at the chimney inlet. To safeguard the investment, the specification of the plastic sheeting requires most rigorous appraisal.

solar collector [Figs 211, 212, 213, 214, 215] A device placed in the path of the Sun to gather incident solar energy for water heating, steam production or electricity generation. Designs range from a simple flat plate collector to the sophisticated computerized tracking concentrator type.

solar collector, concentrator This incorporates a reflector system to increase the intensity of sunlight on to the collecting area.

solar collector, concentration ratio The ratio of the heat flux within the image to the actual heat flux received on Earth at normal incidence.

solar collector efficiency This varies from sunrise to sunset especially when the collector is of the fixed-position non-tracking type. It can be indicated on a graph which incorporates all the important variables.

solar collector, evacuated tube A partial vacuum within transparent tubes arranged in parallel form over a reflector plate. Each tube usually houses one or more solar energy absorber tubes.

solar collector, flat plate A non-focusing flat-surfaced solar heat collecting device. It is the solar collector type most widely used for heating domestic hot water and is usually located on a roof or within the roof construction, in association with an adjacent hot water storage cylinder.

solar collector, parabolic A focusing type collector, usually arranged in trough form and having a line-focus.

solar collector, tracking A system of solar collectors which is arranged to follow or track the path of the Sun and to maximize the angle of incident radiation falling upon the collector surfaces. It may be computer guided.

solar constant The intensity of solar radiation outside the atmosphere of the Earth at the mean distance between the Earth and the Sun. The solar constant equals 1353 kW/m^2.

solar energy, active system A system that utilizes external energy (commonly electricity or sometimes gas) in the collection and/or utilization of solar energy. For example, a hot water system which incorporates a circulating pump, a warm air collector system which relies for its operation on fans and electric motors, a heat pump which incorporates a compressor, fan and automatic controls, etc.

solar energy, passive system This relates to a building in which solar energy input is stored and used without mechanical or electrical means, e.g. a trombe wall.

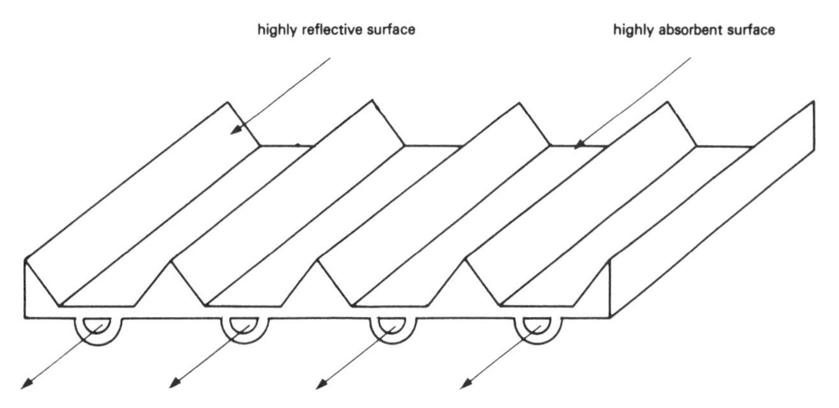

Moderately concentrating flat plate collector

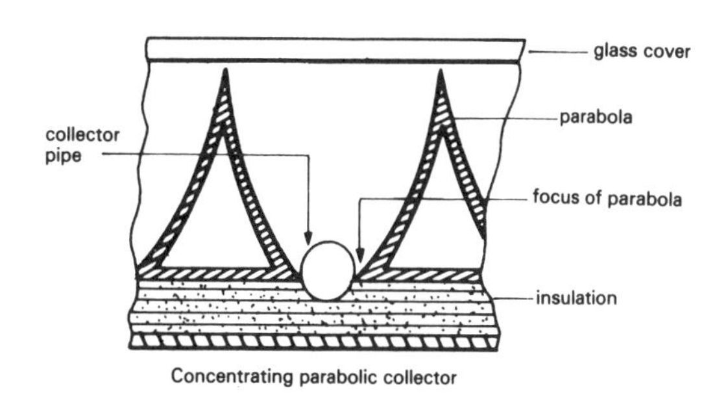

Concentrating parabolic collector

reflecting spiral curve

collector tube

Moderately concentrating trapezoidal collector

[Fig. 211] Concentrating solar collectors

solar heat gain Heat input into a building due to the incidence of solar energy. This occurs principally via windows and roofs.

solar screens system [Figs 216, 217] A means of reducing solar transmission through glazing. Various types and arrangements are available. When of modular construction, the system comprises a number of basic units assembled to form banks of horizontal, vertical or inclined (commonly aluminium) louvres with support clips of high quality engineering plastic to provide the required supporting strength and to promote silent operation. Louvres may be automated to follow the Sun cycle and incidence or they may be manually adjustable or be fixed in the optimum direction for solar energy exclusion. The louvres will be located externally to the building and thus deflect the solar energy before it reaches the building. Such duty imposes severe operating and weather conditions on all the components so that the selection of proven quality materials and workmanship

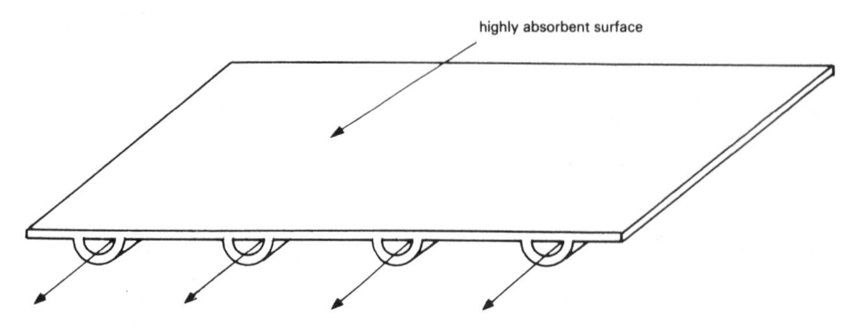

[Fig. 212] Conventional flat plate solar collector

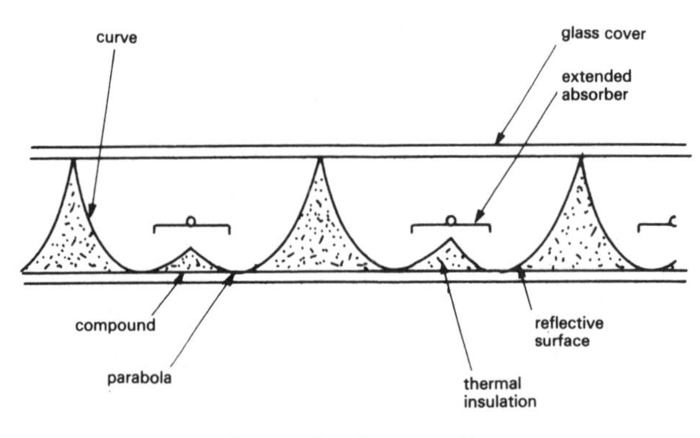

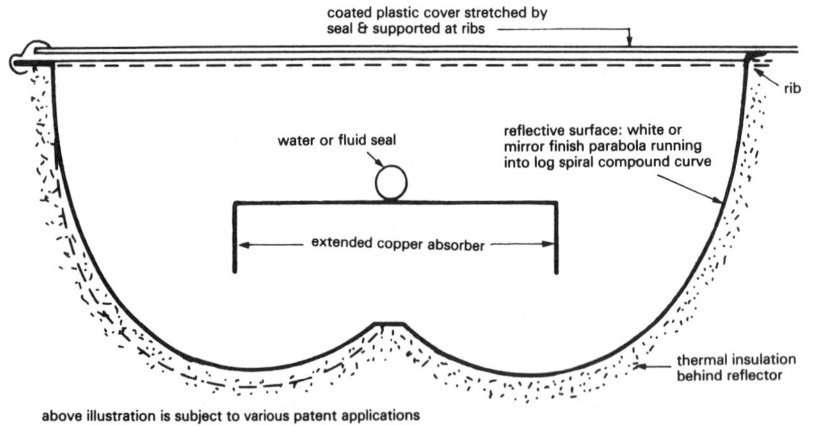

[Fig. 213] Parabolic focusing solar collector: general principles

is essential to the longer term success of such systems.

solar pond [Fig. 218] An artificially formed pond designed specifically to attract and trap solar energy for subsequent use. A typical pond has a depth of 1–2 m and the bottom of the pond is painted black. A salt solution is introduced to establish a density gradient from top to bottom. The bottom zone provides heat storage and extraction. Operating extraction temperatures of 90°C have been achieved. A solar pond can function in association

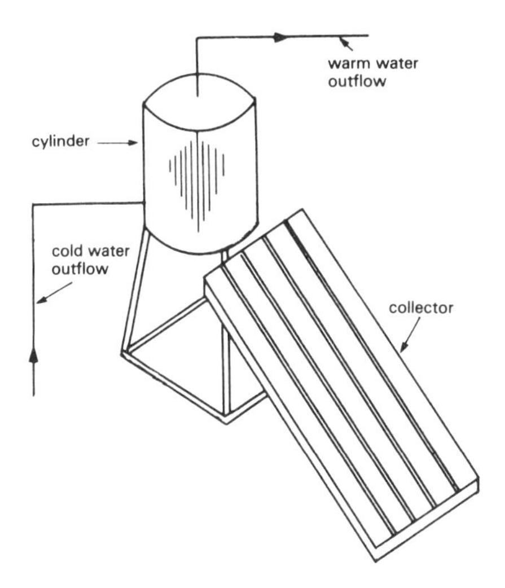

[Fig. 214] Single-panel solar collector with hot water storage tank on flat roof

with a low-pressure steam turbine to generate electricity suitable for heat recovery for water heating or to perform both functions.

solar window films These provide solar protection and thermal insulation to glazed areas. Available ranges include solar control films and thermal insulation films which combine solar insulation in summer and reduced heat loss in winter.

solder An alloy formulated for jointing metals. Soft solders are alloys of tin and lead in varying proportions to suit the required application; brazing solders are usually composed of copper and zinc.

soldering flux An agent used when jointing copper pipes. It ensures that the surfaces to be jointed are clean, or as free as is practicable from oxides, oxide by-products and other foreign matter, especially oil and grease, to ensure that the molten solder will flow freely and wet all surfaces of the joint. The flow of solder is obtained from the capillary forces of attraction.

solders, lead-free See **lead-free solders**.

solenoid A wound coil arranged for producing a magnetic field and used in conjunction with solenoid valves to control the flow of fluids. It is utilized to activate electrical equipment, such as motorized valves, dampers, etc.

solenoid valve An electrically actuated control or regulating valve with integral electric operating motor. It is actuated by the magnetic field generated in the solenoid attached to the valve and is widely used in control applications.

solid state (bottoming) In vibration isolation, the unwanted situation when a spring can be compressed no further and the coils are in contact.

In electronics, the description of circuitry using semiconductors instead of valves, e.g. transistors.

solubility The absorption of a gas by a liquid. The solubility of gases, such as oxygen and nitrogen, varies directly with the pressure and indirectly with the temperature.

solution The result of adding a dissolved substance to a solvent.

solution potential In a corrosion situation, the difference in potential between the anode part of the system and the cathode part. It has different values for different metals and under different conditions and has a major impact on the ongoing process of corrosion. In the natural corrosion process, it is always the anodic metal which corrodes.

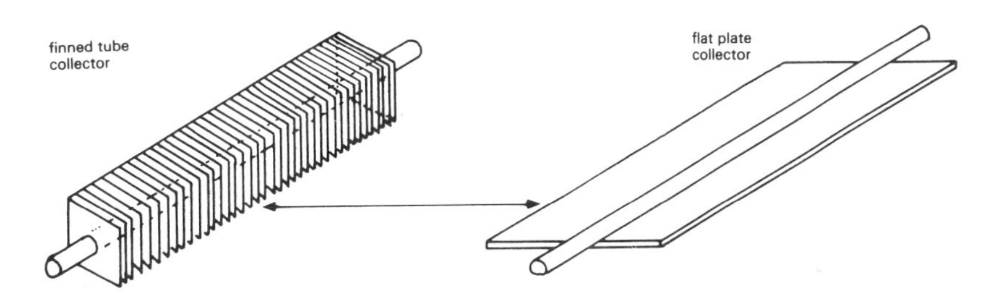

[Fig. 215] Solar collector: finned tube or flat plate

DESCRIPTION
Greenwood Airvac Solar Shade Louvres are assembled from extruded aluminium blades set on 45° or 25° support brackets, fixed to robust aluminium mullions. The assembly may be fixed horizontally, vertically, at an angle or in combination to suit the application. The blades are clipped into position ensuring no distortion due to expansion. Solar Shade louvres can be finished to specified colours to create an attractive feature on a building.

APPLICATIONS
Ideal for commercial and industrial buildings, Solar Shade Louvres shield occupants and equipment from direct sunlight, providing a more comfortable working environment and savings in air conditioning when applicable.
The open construction of a louvred Solar Shade allows natural air flow, preventing the build-up of pockets of warm air.

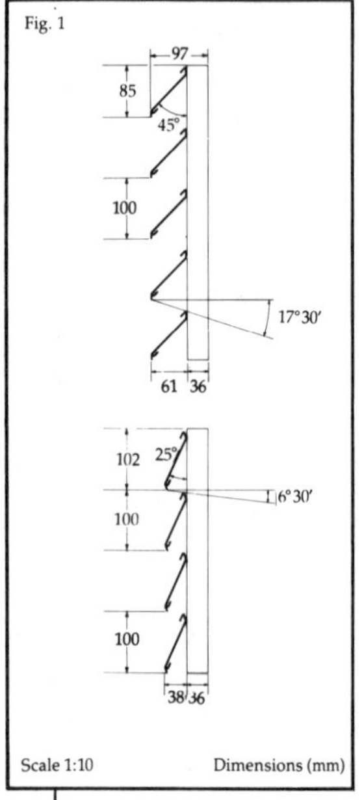

HORIZONTAL MOUNTING
*Solar Shade Louvres fixed to horizontally-mounted mullions are ideal for south-facing windows to provide shade from the overhead summer sun whilst allowing winter sun to penetrate into the building.**

VERTICAL MOUNTING
Vertically-mounted Solar Shade Louvres may be installed to provide shading to east and west-facing windows where the low sun angle would often make a horizontally mounted louvre impractical.*
**Northern hemisphere installations.*

ANGLED MOUNTING
Solar Shade Louvres, fixed to either 45° or 25° support brackets, may be inclined to any angle to suit the application. Support struts and blades may be affixed either from above or below to suit aesthetic requirements.

MANUFACTURE
Louvre blades and mullions — extruded aluminium sections to BS 1474:1972.
Assembly support — aluminium tubular sections and steel brackets.
Blade support brackets — injection moulded engineering quality nylon. 66.*
**Patented*

FINISHES
Mill finish: untreated aluminium extrusions.
Painted: stoved acrylic colours to BS 4800:1972 (suitable for external use) or polyester powder paint to available colours, matt or gloss finish.
Anodised: Available colours to Grade AA25, BS 3987:1974.

[Fig. 216] Section through externally mounted solar shade louvres (Greenwood Airvac Ltd.)

solvent A liquid organic chemical used in industry for dissolving fats, resins and similar materials. Most commonly used are the chlorinated solvents trichloroethylene, carbon tetrachloride and perchloroethylene.

soot Fireside deposits on the heating surfaces of boilers due to the combustion process. They must be removed periodically as their presence impedes heat transfer and may cause local overheating and failure of boiler tubes. Soot may be removed by brushing or vacuum cleaning and larger boiler units are generally equipped with automatically or manually operated soot blowers. Soot ejected from boiler chimneys can cause dirt nuisance to the neighbourhood.

soot blower A set of nozzles with related control valves, piping and programme controller for on-load cleaning of soot deposits, mainly used with

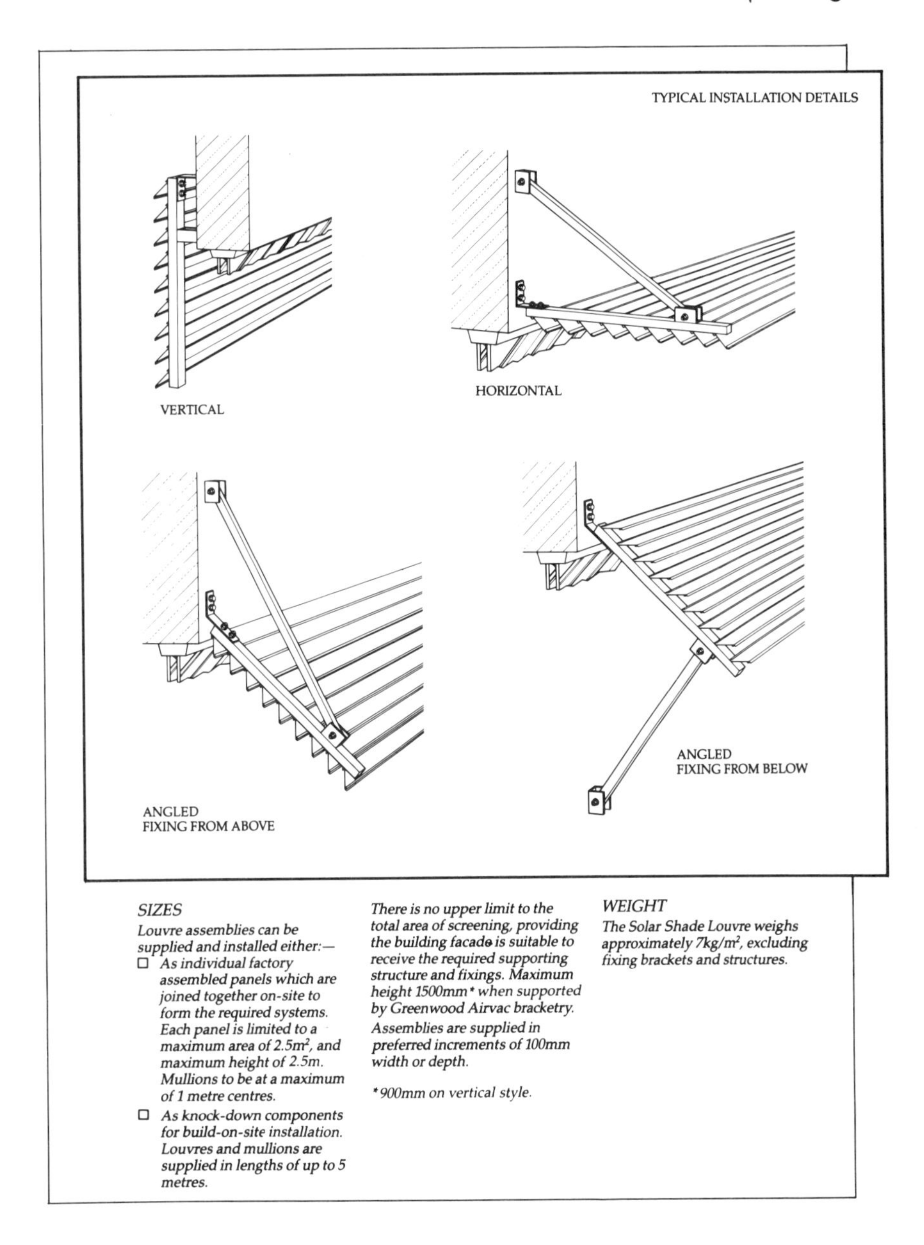

[Fig. 217] Assembly of solar shade louvres mounted externally to a building (Greenwood Airvac Ltd.)

water-tube boilers to clear hard deposits off the fireside surfaces of heat transfer tubes. A soot blower may be manually, pneumatically or electrically operated, using air, steam or water as the blowing medium at relatively high pressure: 13–35 bar. Correct positioning of the nozzles relative to the tube nest is critical. The distances between the nozzles and the tubes must be adjusted to suit the available blowing pressure (from close-up to up to 1 m).

Applications: essential for on-load cleaning of boilers, furnaces and heat transfer equipment in the power generation, marine, industrial and petrochemical service. It may be applied to water, tube and fire-tube boilers, superheaters, economizers, air pre-heaters and process furnaces. The effective programming of the soot blower operation is essential to avoid energy waste.

soot blowing The removal of fireside deposits from the boiler heating surfaces. The operation is used with the larger capacity water-tube boilers which

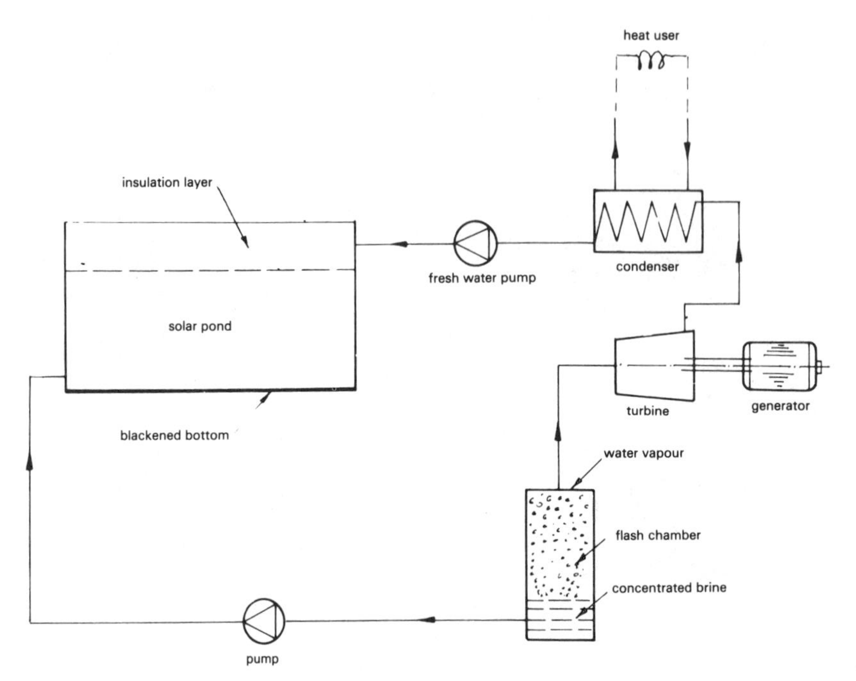

[Fig. 218] Extraction of heat and power from a solar pond

operate with high furnace temperatures causing the combustion deposits to bake hard on to the surfaces. The presence of soot deposits on the furnace and superheat tubes of water-tube boilers cannot be tolerated, as these reduce the heat transfer between hot gases and water and may cause local overheating leading to the fracture of the deposit-coated tube(s) and expensive down time and remedial works. The soot blowing nozzles must be carefully positioned to blow into the spaces between the tubes forming nests of tubes, avoiding direct impingement of the steam or air blast on to the thin water-tube walls of the boiler. In large boilers, the soot blowing facility is considered integrally with the design of the boiler, as correct soot blowing is critical to the boiler efficiency and reliability. A steam pressure of 13–35 bar is required for effective soot blowing.

The process may be manual (for smaller size boilers) or automatic (invariably so for large capacity boiler units). Excessive soot blowing is wasteful of energy. The fireside condition of the tubes must therefore be monitored to maintain optimum soot blowing effectiveness without wasting energy.

soot door A metal frame with a removable key-operated access door built into the base of a chimney to permit inspection and cleaning. It may incorporate a draught stabilizer.

sound absorption A method used to reduce undesired sound emission.

sound attenuation, inverse square law The reduction of perceived noise with distance from the noise source in terms of decibels. There is a decrease of 6 dB for each doubling of the distance from a point source when no reflective surfaces are apparent.

sound attenuator [Fig. 219] A device which provides sound attenuation. It may be a partition, a piece of lined duct or even the air itself, but the term is more commonly applied to describe factory-made purpose-designed sections of ductwork which incorporate sound-absorbent material.

sound attenuator, crosstalk See **crosstalk attenuator.**

sound conditioning Supplementary sound introduced into an area to reduce the variability of

[Fig. 219] Rectangular sound attenuator for attachment to flanged ductwork (Trox Brothers Ltd.)

fluctuating noise levels or the intelligibility of speech.

sound detector A device that senses the presence of a person in a room and activates space heating, air-conditioning, etc., for the occupant's comfort.

sound energy This is composed of pressure waves and measured in watts.

sound, flanking transmission The transmission of sound between two rooms by an indirect path of sound transfer.

sound insulation That property of a material which opposes the transmission of sound from one side to another.

sound level meter (noise meter) An instrument for measuring sound pressure levels. It can be fitted with electrical weighting networks for direct read-off in dBA, dBB, dBC, dBD and octave or third octave bands.

sound power A measure of sound energy in watts. It is a fixed property of an item of equipment irrespective of its environment.

sound pressure level (SPL) A measurable sound level that depends on the environment. It is a measure of the sound pressure at a point expressed in N/m^2, and is expressed in decibels of SPL at a specified distance and position.

sound pressure Since sound is composed of pressure waves, it is logical to measure its strength in terms of pressure (force per unit area).

sound pressure level (SPL), direct field Direct components of a sound level field are calculated from a given SWL by using the inverse square law and directivity, etc.

sound-proofing A method of lining a space with sound-absorbing material to reduce or eliminate the transmission of sound into or out of that space.

sound propagation The transmission of sound along a specific direction or path; e.g. an air duct or pipe.

sound reduction index The sound reduction index of a partition at a given frequency under specified conditions is 10 times the logarithm to the base 10 of the ratio of the sound energy incident upon the surface to that transmitted through and beyond the partition.

sound spectrum The separation of sound into its frequency components across the audible range of the human ear.

sound velocity The speed at which sound waves travel. Sound travels faster in materials having high elasticity and low density; below are indicated the approximate velocities of sound in different materials. The important parameters which determine the velocity of sound in a material are its modulus of elasticity and its density. The sound velocity in air is given with the air at a temperature of 20°C.

Air:	344 m/s
Lead:	1220 m/s
Water:	1410 m/s
Brick:	3000 m/s
Wood:	3400 m/s
Concrete:	3400 m/s
Glass:	4100 m/s
Aluminium:	5100 m/s
Steel:	5200 m/s

spa bath A bath maintained full of water and serving for therapy rather than washing. It has underwater water/air jets like the whirlpool but also incorporates heating and filtration equipment.

space cooling The cooling effected in a space for the purpose of providing comfort to the occupants of that space.

space factor The ratio, expressed as a percentage, of the sum of the overall cross-sectional area of cables (including their insulation and any sheath) to the internal cross-section of the conduit or other cable enclosure in which they are installed.

space heating Heat input into a space for the purpose of providing comfort to the occupants of that space.

space heating by floor warming For this type of heating, heating elements comprise serpentine pipe coils placed on top of the structural floor and screeded over. Pipes may be of steel or plastic. One method uses sleeved pipes to permit floor heating at full hot water flow temperature. Otherwise the temperature must be limited automatically to protect the floor finish from being damaged. The floor below the heating coils must be thermally insulated to direct heat only upwards. In view of the low operating temperature, heat output per unit area is restricted.

spalling A term used with refractories. It relates to the splintering or cracking of refractories where fragments of the brick are separated and fresh surfaces are exposed.

sparge pipe A pipe uniformly distributing steam or a liquid via integral holes or nozzles.

spark arrestor A device installed close to a chimney or furnace terminal to arrest and reduce the amount of incandescent flue gas components being discharged into the atmosphere. It takes the form of a metal screen.

specification A document listing the required parameters for a proposed installation or project. It may be accompanied by drawings, etc.

specific enthalpy The enthalpy per unit mass of a substance.

specific entropy The entropy per unit mass of a substance.

specific heat The quantity of heat required to raise the temperature of a unit mass of a substance through one degree of temperature. Experimental evidence shows that equal masses of different substances require unequal quantities of heat to raise their temperature through the same range. The number which expresses the specific heat of a substance is the same irrespective of the system of units which is employed. The specific heat of most substances varies somewhat over a range, depending on the temperature. For strict accuracy, a particular specific heat should be referred to the temperature at which it applies.

At a temperature of 100°C (212°F) the following are typical values of specific heat for common substances.

Aluminium	0.219
Copper	0.0936
Iron	0.119
Lead	0.0305
Tin	0.0552
Zinc	0.093
Ice (–20 to –1°C)	0.502
Water	1.00

Specific heat characterizes the thermal properties of a building, and is stated as the total heat loss rate per unit temperature difference (W/°C).

specific heat loss rate An index of the thermal properties of a building, given as the total heat loss rate per unit temperature difference (W/°C).

specific humidity The weight of water vapour per unit weight of dry air, in kg/kg dry air or g/kg dry air.

specific volume That volume, at a specified temperature and pressure, occupied by one unit mass (usually 1 kg) of the substance. It is the reciprocal of its density.

spectacle-eye valve A type of parallel slide valve in which the 'spectacle gate' incorporates one 'lens' of circular waterway and the other is of solid section.

specular reflection The reflection exhibited by shiny metal surfaces, such as chrome, silver or pure aluminium.

speculative building development A building constructed without a particular occupier in mind. This requires a flexible design approach to permit its subsequent adaptation to the occupier's needs.

spectrum colours Those colours visible in the continuous spectrum of white light, i.e. red, orange, yellow, green, blue and violet.

speech interference level The average of the octave band pressure levels of a noise centred on the frequencies 500 Hz, 1000 Hz and 2000 Hz, together with the frequency 250 Hz, if the level in this band exceeds the others by 10 dB or more.

speed of sound Wavelength × frequency = 335.3 m/s.

spigot The end of a pipe prepared to enter a socket to form a joint.

spigot and socket joint A joint that comprises a pipe end and a female entry connection on the pipe, to be joined together with appropriate jointing material or an 'O' ring. The joint is made by pushing the spigot into the socket and sealing it with jointing material. It is a common jointing method for cast-iron and some plastic pipes.

spikes The disturbance to consumers' electric supplies caused through the incidence of harmonic currents. See **harmonics (electrical connotation)**.

spinning cup oil burner See **rotary burner**.

spinning disc humidifier A type of mechanical humidifier in which the water flows as a fine film over the surface of a rapidly revolving (spinning) disc. It is thrown off by centrifugal action on to a toothed ring and atomized.

spiral fin A form of extended pipe heat transfer surface formed from a flat strip or by means of a process which results in the deliberate crimping of the fins at the base, care being taken to ensure that the strip of metal is wired on edge into a groove which has been previously formed within the thickness of the pipe wall. Such fins are easy to manufacture but, in the absence of metallic bonding or direct thermal contact, their effective heat transfer efficiency depends critically on the care taken in manufacture.

Application: in heat exchange coils, space heaters, etc.

spiral fin tube A tube or pipe with an extended heat transfer surface in the form of spirally wound fins.

spiral heat exchanger A heat exchanger comprising a spiral element with flow channels alternately open at one side and welded at the other to ensure the isolation of fluids. It is sealed between two gasketed cover plates. Hot fluid enters at the centre and flows spirally to the periphery, while cold fluid moves counter-currently.

spiral wound duct A wire helix spring-steel frame to provide flexibility, available with or without added thermal insulation in the form of glass fibre with a vinyl vapour barrier.

Spirovent micro-bubble air absorber (Spirotech Ltd.) A device used in central heating installations to deaerate water rapidly and completely, removing all air bubbles. It functions on the principle that water has a natural property whereby its ability to retain dissolved air decreases as the temperature rises. The reverse occurs on cooling, when air is absorbed in order to restore the balance. This absorption effect is utilized to capture all the released air in an installation and discharge it via the Spirovent. Air expulsion continues until finally only saturated and absorptive conditioned water remains.

This minimizes corrosion, pump wear and the need to bleed heaters, and reduces air noises generated within the system. It optimizes heat transfer at heat exchange surfaces and thereby enhances energy efficiency. The fitting embodies screwed, flanged or union connections for mounting to the pipe main from the boiler into the distribution mains. It is available in sizes from 22 mm to 300 mm in different models.

split air-conditioner This comprises two main assemblies: an air-conditioning cabinet which is located inside the space being cooled and a condenser located outside that space. Such an arrangement provides quieter operation as the condenser incorporates the (relatively noisy) heat dissipation fan. The cabinet and the condenser are linked by insulated refrigeration tubing.

spontaneous combustion Combustion of a substance possessing a low ignition point which results from the heat generated within the substance by its slow oxidation. It is particularly relevant to the bulk storage of coal which must be protected from becoming subject to spontaneous combustion.

spot heating The method of heating a limited area or operative selectively. It applies particularly to aircraft hangars, tall churches, factories, etc., where it can offer a high standard of comfort locally without the expense of heating large areas. Suitable equipment for spot heating includes: quartz linear electrical lamps, gas radiant panels, high-temperature hot water piped radiant panels and floor panels.

spot cooling When a cooling effect is applied to a small area, e.g. to give cooling relief to a clerk sitting close to an intensely lit display window.

spot welding When two faces are held together by applying the welding process to selected spots of the joint and not to the whole length. This is not suitable for the provision of water- or gas-tight joints.

spray chamber The section of an air washer in which water is injected into the passing airstream from a spray header with a nozzle or nozzles and which incorporates a sump for the collection of surplus water and a recirculating pump arrangement.

spray cooling tower [Fig. 220] A cooling tower that incorporates water spray nozzles to flood the tower fill. See also **cooling tower**.

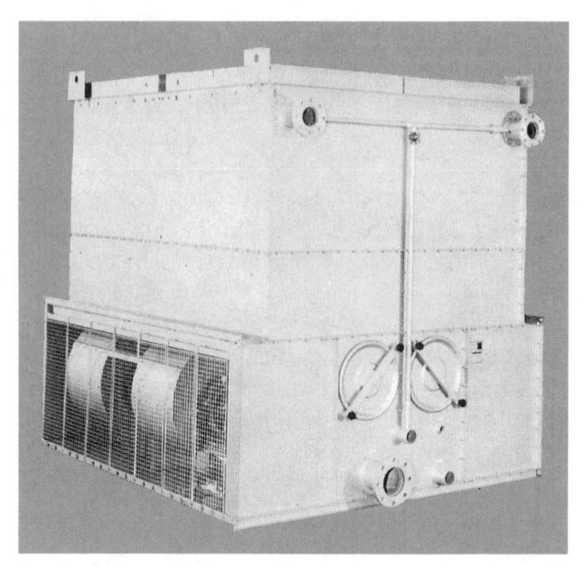

[Fig. 220] View of centrifugal fans and pipe connections

spray nozzle A fitting designed to discharge a spray of water into an airstream. It is usually one of a number screwed into a spray header and facing against the direction of the air flow.

spray recuperator [Fig. 221] A stainless steel device that condenses the water vapour of exhaust gases by the direct spraying of water. The process reclaims sensible heat (by lowering the exhaust gas temperature) and latent heat (through condensation).

spray tap A tap fitted with a purpose-made flow restrictor designed to minimize water consumption.

spray tap basin A wash-basin which has been designed to receive the discharges from a spray tap without splashing. It commonly incorporates only one tap hole, is without overflow and is fitted with a grated waste.

spray tap, drainage implication The rate of discharge from such spray tap basins is very low; in the order of 0.06 l/s. Given the pipe size that has to be used, self-cleaning velocities cannot be readily achieved and it is likely that soap residue, grease and hairs will accumulate in the discharge pipe. Such deposits can build up rapidly, especially in soft water areas. The rate of solids build-up can be

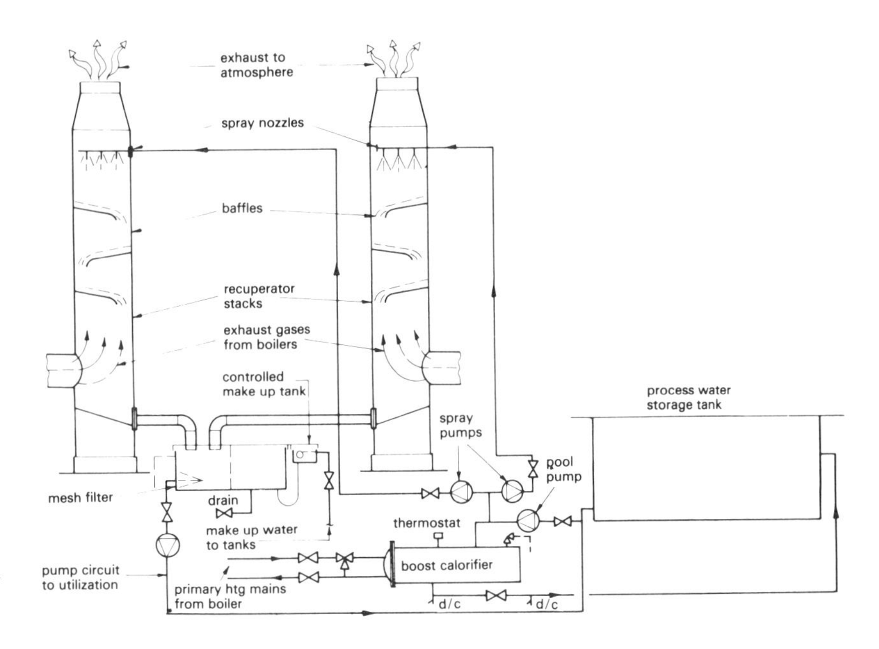

[Fig. 221] Multi-recuperator system

reduced by installing at the head of a group of spray tap basins an appliance, such as a cleaner's sink, which will give an occasional cleansing flush to the discharge pipe.

spray tree [Fig. 222] The arrangement of distribution headers, pipes and nozzles in a spray-type cooling tower.

[Fig. 222] Spray tree

sprayed coil condenser A condenser that rejects heat to atmosphere by spraying water on the heat exchange coils of the condenser, transferring some heat to the water as well as to the passing air and thereby increasing the capacity of the condenser. A pump, piping, spray nozzles and a collecting sump are required for the associated water circulating system. Fans are used to force the air through the unit.

spreader mechanical stoker A means of feeding a boiler or furnace with coal by spreading the coal evenly and thinly over most of the fuel bed, reducing the thickness near the end of the grate.

sprinkler grid Distance between sprinklers for a given coverage of a protected area is laid down by Fire Insurers.

spring hanger A pipe support which incorporate springs to reduce the transmission of the vibration from the pump into the pipe system and its associated structure.

sprinkler head Fitting screwed into sprinkler pipe system on a grid basis; incorporates fusible head which will melt in the presence of elevated temperature associated with a fire and discharge water into the protected area, at the same time actuating sprinkler main valve and audible alarm.

sprinkler heads, colour code Different sprinkler installations cater for different degrees of risk. Sprinkler heads are therefore coloured differently to indicate the temperature at which they will melt as follows. All temperatures are in °C.

Soldered and fusible chemicals type heads

Temperature	Colour
68–74	Uncoloured
93–100	White
141	Blue
182	Yellow
227	Red

Glass bulb type

Temperature	Colour
57	Orange
68	Red
79	Yellow
93	Green
141	Blue
182	Mauve
204–260	Black

Spares must be held of heads with the correct colour coding.

sprinkler, long-throw See **long-throw sprinkler**.

sprinkler mechanical stoker A means of feeding a boiler or furnace with coal, by sprinkling the coal evenly over most of the fuel bed, giving a thin and uniform bed which offers flexibility in the burning of a variety of coals as it reduces the effect of caking. This is claimed to result in improved steam raising and careful control leads to smokeless combustion. Since the volatile matter is released from the top of the fuel bed, this mode of combustion is termed 'over-feed'.

sprinkler protected Area or item of plant (e.g. silo) which is fitted with a sprinkler system.

sprinkler system, flushing connection This is required where an automatic pump draws water from a non-potable source, e.g. a canal, lake or river, located at extremities of the distribution pipes and used for periodic flushing of the main system.

sprinkler system, hazards The Fire Offices Committee identifies three classes of hazards for which sprinkler protection is required. The hydraulic design to provide an appropriate density of water discharge over an assumed maximum area of operation (the number of sprinklers likely to operate) in the highest and hydraulically most remote parts of a protected building depends on the perceived class of hazard as follows.

(a) Extra light hazard: density of discharge is 2.25 mm/min; maximum area of operation is 84 m^2.
(b) Ordinary hazard (four subgroups): density of discharge is 5.0 mm/min; maximum area of operation varies with group between 73 and 360 m^2.
(c) Extra high hazard: density of discharge, depending on the extent of the hazard, is 7.5 mm/min; maximum area of operation is 260 m^2.

sprinkler system, pipe distribution drainage All parts of a sprinkler system must be designed and laid to be capable of being drained completely via drain cocks fitted at all low points. The gradient in wet and dry systems should be at least 4 mm in each metre and in distribution pipes not less than 2 mm in each metre.

sprinkler system, priming tank An elevated water storage cistern provided to maintain the system pump(s) primed at all times under suction lift conditions. One priming tank is required for each pump in the system.

sprinkler system, pump priming Where automatic priming is required under conditions of suction lift, the design must ensure that the pumps are fully primed at all times. The adopted method is to provide an elevated cistern (priming tank) which is automatically maintained full at all times with an alternative water supply. The priming pipe is taken from the cistern to the pump delivery main header and fitted with a back pressure valve on the connection in close proximity to the pump. A separate priming tank is required for each pump installed in the system.

sprinkler system, pumps Pumps must be of a design such that the outlet pressure falls progressively with the rate of demand, so that, whilst being capable of providing the correct rate of flow and pressure at the highest and at the most remote parts of the protected premises, the pump output will be so controlled that there is not an excessive rate of discharge at the lowest level in areas close to the installation valves.

sprinkler system, risk of water damage Certain types of equipment and plant areas can suffer serious damage from sprinkler water discharge, whether accidental or in fighting a fire; such must not be sprinkler protected. Areas concerned are electrical equipment rooms, etc.

sprinkler system, water supplies Such systems must each be connected to two independent sources of water supply. Where two separate water supplies are not available from the water utility company, one source must be a sufficiently large capacity water tank or pond.

sprinkler valve Automatically actuates sprinklers discharge in the event of fire and also activates audible alarm by gong.

sprinkler valve chamber Small plant cupboard which houses sprinkler system isolating valve, actuating valve, test valve, gong, compressor and printed mounted instructions. Must be located for easy access by Fire Brigade.

spur A branch cable connected to a ring or radial final circuit.

squeeze valve A valve that operates with a vulcanized rubber sleeve reinforced with woven fabric. Larger size valves are reinforced with steel cords. The sleeve returns to its original shape when the valve reopens. Complete isolation is achieved when the valve is pinched shut. This type of valve can handle abrasive and corrosive fluids.

stable flame A flame that maintains its correct position relative to the combustion appliance.

stack A chimney. In plumbing a stack refers to a vertical discharge pipe.

stack effect The up-draught due to the stratification of the air temperature within a tall space.

stack loss That proportion of the heat input from the boiler fuel which is lost in the flue gases being discharged into the chimney.

stack, sanitary The main vertical discharge or ventilating pipe within a sanitary installation.

stack terminal, sanitary That point on a stack where gases vent to atmosphere. It must be suitably located to avoid adverse pressure effects caused by wind across the roof. Each terminal must be fitted with a guard or dome-shaped cage of durable metal or with some other type of cover which does not unduly restrict the flow of air, but which excludes birds and/or their droppings. Galvanized materials tend to corrode with time, particularly in an urban environment; their inspection should therefore be an item in any routine maintenance schedule for the building.

stage compression The process in which the compression of air or gas is performed in two or more stages (arranged in series) instead of only in one cylinder. This permits the cooling of the air in an external cooler on discharge from one cylinder and before entering the next.

stainless steel A class of chromium steels containing 70%–90% iron, 12%–29% chromium and 0.1%–0.7% carbon, affords good resistance to corrosion.

stall urinal Individual urinals with integral divisions and channels installed singly or in ranges. Whilst they are generally more expensive than slab urinals, this type offers the advantage of an integral channel and the absence of joints. The maximum length for stall urinal channels is about 2.4 m.

stand pipe A vertical valved pipe inserted into a water main, commonly provided for emergency use when pressure in the mains has been reduced and cannot fill high-level water cisterns.

standard air Concept of Standard Air is employed for the purpose of standardizing the performance of fans installed in ventilation and air-conditioning systems. Standard air is taken as having a density of 1.2 kg/m^3, which corresponds to atmospheric air at a temperature of 20°C, a pressure of 1013 bar and a relative humidity of 62%.

standby electrical generator Electricity generating equipment provided for use in the event of mains power failure. It is essential for ensuring the continuity of exhaust from underground garages, the operation of fire-fighters' lifts, etc.

standing charge That component of a utility tariff which applies whether or not there has been consumption.

standing overflow An overflow pipe which consists of a vertical tube standing within a cistern and passing through its base to the overflow outlet.

starter A device for switching on electrical apparatus, usually with green and red push buttons and overload protection. Other features may be incorporated to suit application. A starter may be suitable for wall fixing or for incorporation into a control panel.

starter, automatic A device that automatically starts up or stops controlled electrical equipment, usually incorporating a contactor coil.

starter, push-button A manual device that controls electrical equipment by the press of a button.

starting torque The resistance that must be overcome in getting a motor to start turning.

state-of-the-art Refers to items of equipment, assemblies or projects which incorporate all the relevant up-to-date technology.

static deflection The compression of vibration isolators under load.

static head The height of a column of water above any one point in a water system expressed in metres or millimetres.

static pressure That pressure exerted at the base of a column of water at any one point within a water system expressed in terms of newtons per square metre.

static rectifier A device widely used in electrostatic precipitators that converts an alternating supply of electricity into high-tension direct current electricity.

static regain See **regain method of duct sizing**.

static regain method of duct sizing The principle which underlies this method of duct sizing is that the reduction in velocity in the main duct consequent on the off-take of a volume of air through a branch duct results in a regain of static pressure. This regain is available to offset the friction loss in the succeeding section of ducting. The static pressure developed by a fan represents potential energy whilst the velocity pressure represents kinetic energy. If the size of a duct is decreased, thereby increasing the velocity of the air flow and the velocity pressure, a decrease in static pressure results. Conversely, enlargement of the duct reduces the velocity and the velocity pressure, resulting in an increase in static pressure.

It is commonly accepted that a regain of only about 75% is attainable across the duct section which comprises the branch duct, the short section of straight duct following the branch and the reducing section to the following section of duct. The remainder of energy which is theoretically available to be regained after a change in the duct size is lost due to turbulence following the take-off and because of the friction loss in the passage through the reducer. The static regain in any change of section of a duct can be calculated. This regain tends to lie between 60% and 120%, but for practical purposes an average static regain of 75% is considered reasonable. Charts are available to aid in the calculation of static regain in ducts. Systems designed on the basis of the static regain method have proved more successful and more easily amenable to regulation and balancing of air flow than similar systems sized by other methods.

stator That component of a dynamo or electric motor which incorporates windings and a core and does not revolve.

statutory requirements Published regulations to which an installation must conform in a particular locality. In the UK these would be the Public Health Acts, the Clean Air Act, Building Regulations, Building Standards, Regulations for Places of Public Entertainment, the Water Supply Act and others, including Local Authority by-laws and regulations.

steam Water in a gaseous state or water above its boiling point.

steam accumulator A vessel forming part of a system in which steam is stored at times of excess steaming capacity and discharged when demand exceeds boiler capacity.

steam blast burner An oil burner which consists essentially of a double concentric tube which directs steam at pressure to impinge on the oil being fed into the burner and atomizes it for combustion.

steam flashing See **flash steam**.

steam generator This term applies in general to all types of generators of steam but it specifically describes the smaller compact boiler assemblies which are designed for rapid steam raising using gas or oil as fuel.

A steam generator is also a pressure vessel into which high-temperature fluid is introduced via a heat transfer coil or battery and applied to heat a body of water to generate steam at some lower temperature.

steam heater battery [Fig. 223] An assembly of heat exchange pipes within one flanged casing for installation within a ducted air heating system. The pipes are usually finned to provide extended heat transfer surfaces or may take the form of heat pipes.

[Fig. 223] Steam heater battery coil showing steam inlets and condensate outlets

It incorporates steam inlet(s) and condensate outlet(s). It is usually fitted with a dirt pocket and a steam stop valve at the inlet and a steam trap set with an isolating valve at the outlet.

steam heating, direct See **steam injection**. This method utilizes the injection of live steam into a liquid whereby the steam yields its latent and some sensible heat. Some industrial processes use such primitive methods of heating which are wasteful in the use of energy, as the heat input cannot be closely controlled and the steam condenses into the liquid, diluting it and losing valuable treated boiler feed water.

steam heating, indirect See **indirect steam heating**. This method utilizes a heat exchanger to transfer the heat from the steam to the liquid being heated. Heat input can be controlled and the outlet of the exchanger is fitted with a steam trap which permits only condensate to pass into the return piping. The condensate is returned to the boiler plant.

steam heating, vapour system See **vapour steam heating system**.

steam heating vacuum pump A pump installed in a vacuum return system which must handle both air and hot condensate. It comprises, assembled on to one baseplate, a condensate receiver, a pump and associated controls. The condensate gravitates into the receiver and is returned to the boiler plant by the pump. Usually, there are two impellers on the same pump shaft, one handling air and the other water. The pump is switched on and off by an automatic control which is actuated by the subatmospheric pressure and by the water level in the receiver. A vacuum equivalent to 250–450 mm of mercury can be easily maintained so long as the whole system remains reasonably free from leakages. The steam traps must be routinely maintained in tiptop condition, as any leakage of steam into the return pipes will make it progressively more difficult (or even impossible) to maintain the essential vacuum.

steam humidifier Two different types are in use. The first is an arrangement of steam jets which discharge live steam into the moving air in ductwork systems. This is not recommended for installation in comfort air systems because of associated noise, odours and difficulty of control. The second is an automatic steam generator which uses lattice electrodes in a cylinder to heat water to boiling-point and discharges the low-pressure steam through a perforated distributor pipe. It is usually heated by electricity. This requires periodic maintenance to remove scale.

steam injection The direct injection of steam into a liquid for the purpose of heating it. The process may be under thermostatic control (crude) or under manual control (usually wasteful). Such injection is unsatisfactory for these reasons. The steam dilutes the solution being heated, condensate is not recoverable and thermostatic or manual control methods tend to be crude allowing an overrun of temperature and hence waste steam. When manual control is used, forgetfulness by the operator is not unknown when the steam is left on without need. See also **indirect steam heating**.

Application: temporary steam heating.

steam injector feed pump A device for feeding water into a boiler which is operating under pressure. It utilizes a steam jet and an arrangement of nozzles. It converts some of the kinetic energy in the steam into pressure energy. About 1 kg of steam is required for the injection of 10 kg of water into the boiler. This is suitable only for use with small capacity boilers; it cannot handle water at temperatures above 30°C.

steam jacketed valve A valve used in applications where it is necessary to maintain the line fluid at a specified high temperature. It is an alternative to an electrically traced valve.

steam jetting A method of cleaning articles or plant by the use of steam at an appropriate pressure. Generally, such cleaning should be discouraged, as it is likely to waste much steam. Other more economic cleaning methods should be considered.

steam-jet refrigeration [Fig. 224] A method of utilizing water vapour as a refrigerant for air-conditioning work. It employs a compression cycle which operates fundamentally like the other conventional compression systems, but uses a steam-jet aspirator as a compressor. The steam-jet compressor consists of one or more nozzles which discharge steam at a high velocity (1000–1500 m/s) and, by an aspirating effect, draw vapour from a cold tank and force it through a carefully designed throat into a surface condenser in which a vacuum of about 710 mm is maintained. The condenser water flows through the tubes of the condenser, leaving at an air temperature of about 37°C. It is necessary to incorporate a number of nozzles, as the working range of each individual nozzle is quite limited. The small amount of air which is entrained with the steam, due to inleakage and air present in the system on starting, is removed from the condenser by auxiliary ejectors, usually in two stages. These ejectors incorporate their own separate condensers. The subatmospheric pressure in the main condenser is necessary in order to decrease the

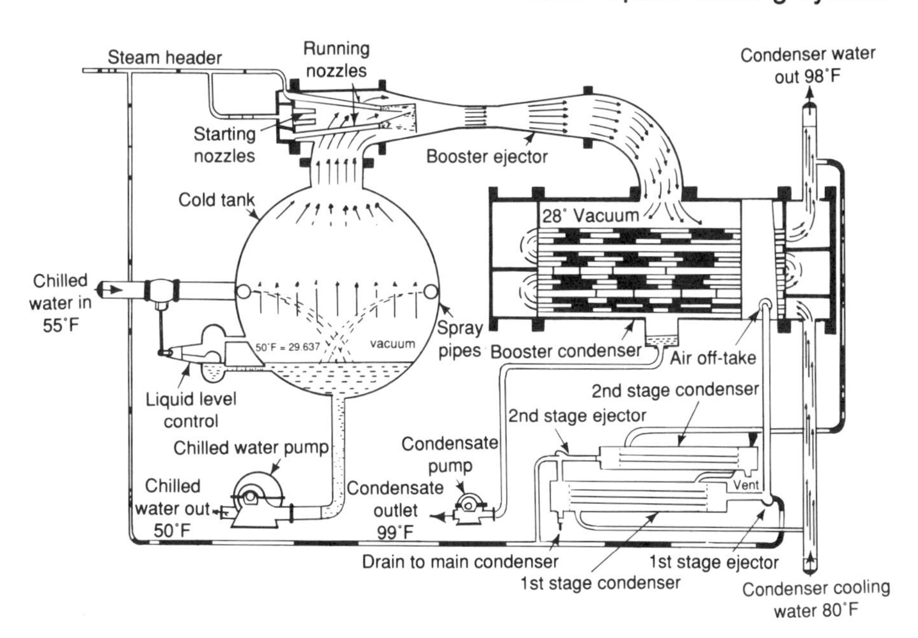

[Fig. 224] Steam jet refrigeration system

compression ratio and to minimize the steam consumption.

The steam-jet is simply one form of compressor which offers an economical method of compressing large volumes of gas. It is feasible to use it only when the refrigerant is water vapour, as it is necessary to mix the steam with the refrigerant. If an expensive refrigerant were used, it would be necessary to separate it from the steam after condensation. The economic justification for selecting a steam-jet system in preference to another system depends, in any specific case, on the relative costs of electricity, steam and water. A much greater quantity of water is required in the steam-jet system because it must condense the refrigerant as well as the steam discharged through the jets, so that the amount of water required is three to four times that used in mechanical compression refrigeration systems.

steam-jet refrigerator A refrigerator that operates on the principle of steam-jet refrigeration.

steam leak detector A handheld instrument for the detection of steam leaks at distances of up to 15 m from the source. It detects the ultrasonic noise generated by the leak and transmits it into an audible signal. The closer it is to the source, the louder the signal.

steam leak The loss of steam through leakages at pipe joints, fittings, valves, steam traps, process plant, etc. The magnitude depends on the size of the defect causing the leak, the operating pressure and the period of operation. It can cause a large loss of energy.

steam meter/condense meter A device that meters steam consumption by measuring the amount of condensate discharged from the steam appliance or system by rotating a paddle which is integral with a condensate container.

steam separator A means of draining water from a steam system incorporating a series of baffles.

steam space heating system A heating system that utilizes the liberation of the latent and some of the sensible heat in the steam as this is passed through heaters, such as radiators, unit heaters, pipe coils and convectors. The resultant condensate is usually piped back to the boiler plant to conserve treated feed water and its heat content. Each heater has an inlet valve and a steam trap set at the outlet. Thermostatic controls may be superimposed as required.

The steam pressure is maintained as low as practicable to safeguard the equipment (particularly cast-iron radiators and cast-iron boilers) and to optimize latent heat transfer. This will necessitate the provision of a steam pressure reducing valve in the pipe connection to the system from a steam boiler operating at higher pressure. Long runs of steam piping are sectionally trapped to maintain the dryness of the steam being conveyed. Because of the fairly high temperatures in the system, great care must be taken to compensate for thermal movement and to protect vulnerable persons from contact with the heaters and uninsulated steam and condensate pipes.

steam, superheated See **superheated steam.**

steam system with vacuum return line See **vacuum return line steam system.**

steam tables Tables that list the properties of steam at different pressures and conditions (e.g. saturated, superheated, etc.).

steam trap A device designed to remove condensate and air. Steam traps are available as thermodynamic, balanced-pressure, bimetallic, liquid expansion, ball float and bucket types.

steam trap efficiency monitor A device that checks whether a steam trap is wastefully blowing steam or not. A sensor chamber is close-coupled upstream of the trap with a cable running to a hand-held indicator. One such monitor, the Spira-Tec (Spirax–Sarco Ltd.) allows rapid checking without the need to shut down. Defective traps can be positively identified, limiting maintenance to only those that require it.

steam trap, float-operated type [Fig. 225] A steam trap that incorporates a substantial float and lever arm mechanism to control the flow of condensate and prevent steam from entering the condensate system. It can handle large volumes of condensate.

steam, trap, inverted bucket [Fig. 226] A steam trap that incorporates a lever arm and float in the shape of an inverted bucket to control the flow of condensate and prevent steam from entering the condensate system. This requires attention to air venting to prevent steam and air locking.

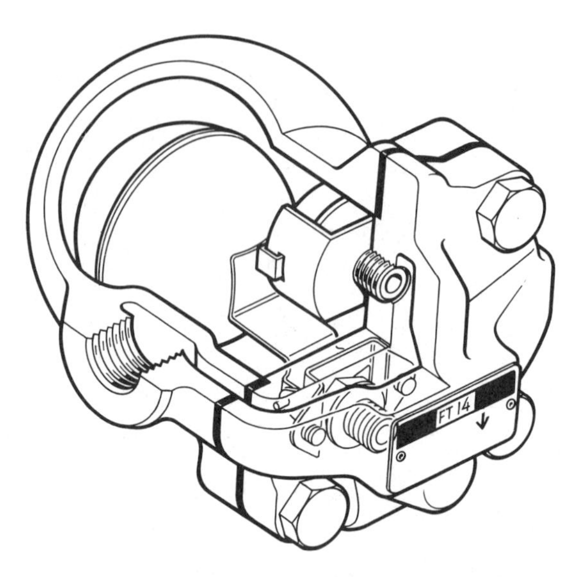

[Fig. 225] Steam float trap with thermal air vent (Spirax–Sarco)

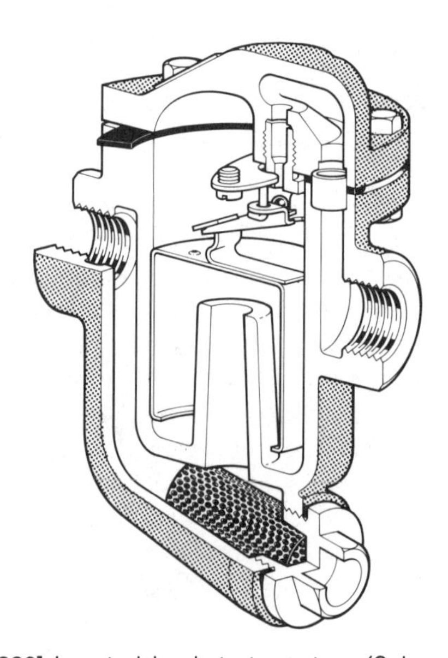

[Fig. 226] Inverted bucket steam trap (Spirax–Sarco)

steam trap set An assembly comprising a steam trap, sight glass, strainer, three-way test cock and associated isolating valves.

steam trap, thermodynamic See **thermodynamic steam trap.**

steam trap, thermostatic See **thermostatic steam trap.**

steam turbine A steam-driven prime mover in which jets of steam impinge upon hardened blades attached to one or more discs or wheels which are fixed on to a shaft supported on bearings. The turbine rotates to drive an electric generator or other machine.

steam whistle An audible signal conveyed by passing steam at pressure through a steam nozzle. Such a practice is likely to be wasteful of steam and should be discarded wherever practicable in favour of an electrically actuated sound.

Stefan-Boltzmann law A scientific law which states that a body will radiate heat in proportion to the fourth power of its absolute temperature.

stelvetite A material used for constructing air ducts. It comprises a coating of PVC (with a choice of colour) on galvanized mild steel. The PVC-coated surface forms the outside protective surface of the duct.

step control A method of controlling a device in two or more operational steps.

step controller An electrical mechanism for effecting a step control. See also **step control**.

step grate An inclined grate on to which solid fuel is fed in a combustion chamber which is located beneath the fire box of a boiler. It is particularly used with locomotive-type boilers.

step irons Individual treads constructed of steel and arranged for building into the walls of vertical shafts, chimneys or manholes to permit convenient ascent or descent by maintenance personnel. Where used inside manholes, they should be built into the wall at every fourth brick course (or otherwise at intervals of between 230 mm and 300 mm). Unless the step irons are of the straight bar corner type, they should be set staggered, in two vertical runs which should be constant at about 300 mm centres horizontally. The top step should offer direct access to the operative and be fixed no more than 750 mm below the ground surface. The lowest step iron should be fixed at no more than 300 mm above the manhole benching. Pre-cast manholes are usually constructed with step irons let into the walls.

sterile zone The area in a hospital which is maintained free from bacterial contamination, commonly an area associated with the operating theatre.

sterilizer See **autoclave**.

stochastic process A process which has some element of probability in its structure.

stoichiometric A term that describes any chemical process in which the ingredients are exactly balanced in quantity. Whenever a fuel is burnt, a certain amount of oxygen takes part in the process. The ratio of oxygen to fuel is stoichiometric when the oxygen supply is limited to that just required for complete combustion.

A stoichiometric compound is pure, and is one in which its component elements are present in the exact proportions represented in its chemical formula.

stoichiometric combustion Combustion achieved when the oxygen supply to the process is limited to exactly that required for the achievement of complete combustion.

stoke To feed a furnace with solid fuel by manual or automatic methods.

stoker A boiler attendant or the equipment for feeding solid fuel into a boiler or furnace.

stoker, automatic Equipment for feeding solid fuel automatically into a boiler or furnace. It comprises a feed mechanism and associated controls.

stoker, overfeed See **overfeed stoker**.

stoker, rotary grate A stoker that feeds solid fuel on to a moving grate system inside a boiler or furnace.

stoker, underfeed See **underfeed stoker**.

stokes A unit of kinematic viscosity in the metric system.

stopcock Usually a mains water isolating valve. It is also used as a general term for a fluid isolation valve.

strainer See **pipeline strainer**. It is a general term for equipment which is inserted into a fluid system to remove suspended particles. Its effectiveness depends on the configuration and size of the strainer mesh.

strainer basket A removable, perforated sheet metal mesh fitted inside a pipeline strainer to intercept impurities.

strainer check valve A pipeline strainer which incorporates a non-return valve to prevent back flow.

S trap A trap that connects to a drain-pipe vertically.

stratification A vertical temperature gradient in an air space or in a container of warm or hot liquids. It results in elevated air temperatures at the upper level of a space, unless the air is fan-circulated to avoid this. Stratification causes an increase in temperature across the height of a vessel containing warm or hot liquid, e.g. inside a hot water cylinder, whereby the hottest water accumulates in the upper part of the cylinder permitting hot water to be drawn off, even though part of the heated cylinder contains cold or cool water nearer the bottom.

streamline flow The type of flow, not often met in practice, where the velocity of flow is so small or the boundaries so close together that the flow is orderly. Contiguous layers of fluid flow parallel to each other and the ordered nature of the flow facilitates mathematical analysis.

stress A force per unit area. When a stress is applied to a body within its elastic limit, a corresponding strain is produced. The ratio of stress to strain is a characteristic constant of the body.

strip curtain A heat-retaining and draught-reducing curtain usually manufactured from clear strips of PVC. It allows vehicles and pedestrians to pass through whilst remaining in position when not in use.

Applications: in loading bays where doors are left open; for screening off sections of a factory; cold store doorways.

stroboscopic effect An effect that makes moving objects appear to be stationary or moving in a different direction to the actual movement, caused by the light output from most lamps fluctuating rapidly at twice the mains frequency. Where the effect is strong, it can have safety implications in factories and workshops or can mislead persons engaged in certain indoor sports where balls or racquets are used. Where a site is connected to a three-phase supply, the connection of luminaires to different phases may assist in reducing the stroboscopic effect.

structured wiring A backbone wiring system which utilizes risers or lift shafts to distribute electrical services to the various floors or areas of a building, together with a local access system for electrical distribution to the work areas. The concept derives from principles that have been followed for many years in low-voltage electric wiring. Structured wiring is usually planned for a service life of 10 to 15 years. Local area distribution may use 'star', 'ring' or 'bus' configurations.

stub stack See **sanitary plumbing, stub ducts**

stuffing box A device that seals a moving shaft against the leakage of fluid. It comprises an insert which is screwed into the receiving end of the device being sealed to press a jointing material firmly into place.

sub-circuit A branch of a main circuit supplied from a distribution fuseboard.

subcontractor A contractor appointed to carry out a selected package (or packages) of work under the overall direction of the main contractor.

sublimation The conversion of a solid directly into a vapour without melting and subsequent condensation, e.g. the transformation of ice into water vapour.

submerged combustion [Fig. 227] A method of gas-fired heating in which the flame is submerged and burns below the surface of the liquid being heated, the combustion products transferring directly to the contents of the tank in which the operation progresses. Combustion products take the form of numerous very small bubbles which present an

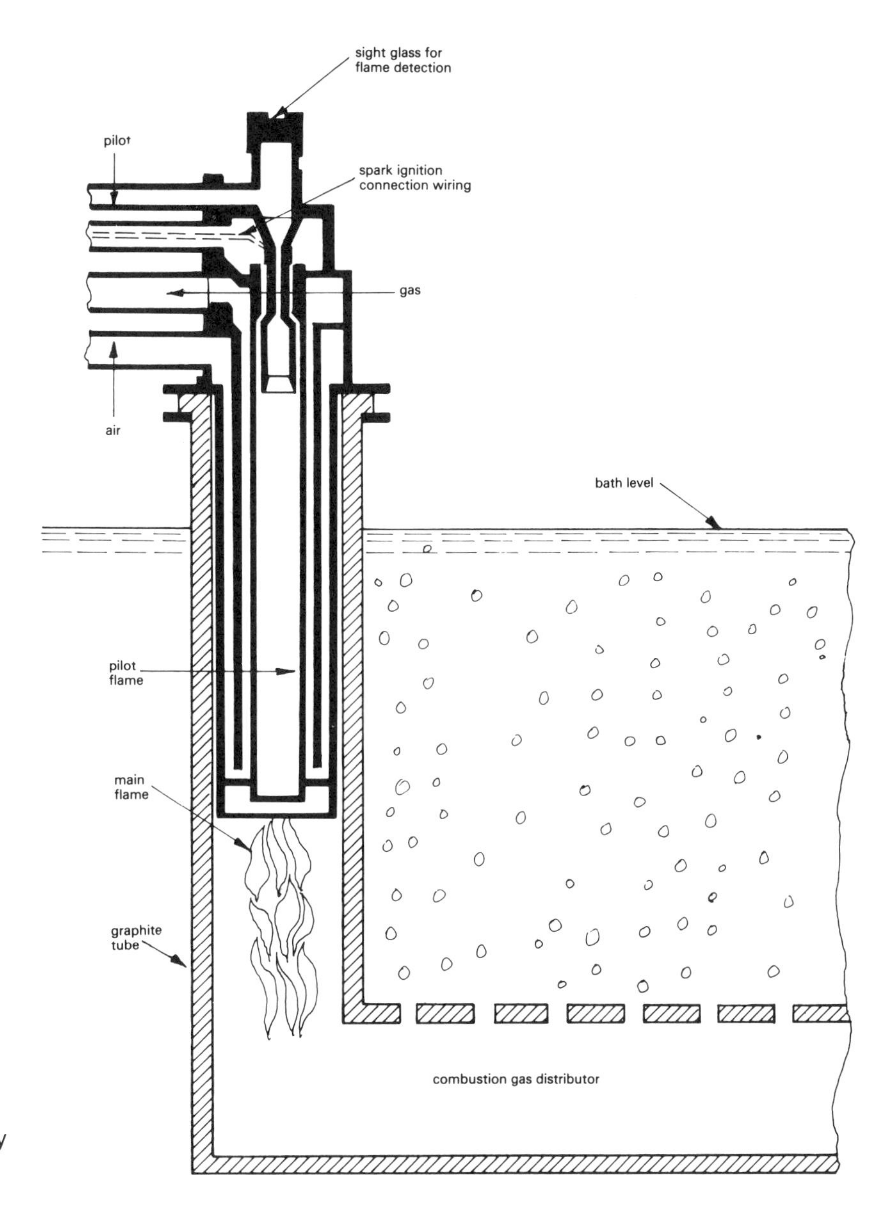

[Fig. 227] Application of submerged combustion by gas

extensive total heat transfer surface area, so that the combustion products cool quickly to the temperature of the liquid inside the tank. Combustion products and associated water vapour leave the surface together. Applications for such a process are in tank heating (especially of corrosive liquids) water and effluent treatment, concentration of dilute solutions and crystallization.

submersible pump See **pump, submersible**.

subsoil drain A drain that collects and conveys away subsoil water.

subsoil water drain A drain that disperses into the subsoil the effluents from septic tanks or from overflowing cesspools.

substation An enclosure containing apparatus for transforming or converting electricity from one voltage to another (stepping-up to a high voltage for transmission or stepping-down to a low voltage for distribution).

suction The raising or elevating of a liquid by atmospheric pressure which exceeds the pressure of the partial vacuum caused by the suction. It is a negative pressure which allows a fluid to be drawn into a pump, compressor, etc.

suction head Negative pressure at the inlet port of the device which creates the suction.

suction line accumulator A device that ensures the correct refrigerant control at a low ambient temperature.

sulphur dioxide A colourless pungent gas formed when sulphur burns in air. It is regarded as the most important air pollutant. Its presence is mainly due to the discharge into the atmosphere of the products of combustion of sulphur-bearing fuels. It is exceedingly corrosive to metals when combined with moisture, such as condensation formed within a chimney or stack.

sulphur trioxide A constituent of flue gases arising from the combustion of sulphur-bearing fuels in the presence of oxygen. It is a corrosion-promoting gas but its formation can be inhibited by strict control of the quantity of excess air supplied for combustion.

sump A reservoir in the lower portion of a device, e.g. a reservoir of lubricating oil in a compressor. It is also a pit or depression in a plant room floor into which water or oil leakages are directed via floor ducts and which is emptied periodically, generally by the use of a float-controlled pump or by a semi-rotary hand pump.

Superaqua method of renovating drainage system A system that relies on the chemical reaction between sodium silicate solution and dilute hydrochloric acid. See **two-part chemical grout for drainage renovation**.

superconductivity The loss of resistance in conductors of electricity to the flow of electric current when their temperatures are very low. Taken close to its limit, it should be possible to convey very large electric currents through thin conductors without appreciable loss of energy, an application of cryogenic engineering.

supereconomic boiler A highly efficient three-pass type self-contained economic boiler in which the second and third passes of the fire-tubes are placed below the main furnace tubes. The boiler employs the contraflow principle of heat exchange, and the design ensures optimum heat transfer and imparts a vigorous circulation to what is in other boiler designs a dead water zone. It also permits the effective use of a fusible plug.

superheat Sensible heat added to saturated steam in a superheater with the object of avoiding condensation during use, transmission and distribution of the steam.

superheated steam Steam at a temperature greater than 100°C which is obtained by heating water under a pressure greater than atmospheric.

superheater A heat exchanger for superheating steam comprising a set of inlet and outlet manifolds together with a housing box and multi-lap elements which can be of a welded or detachable design depending on the application.

superheating Heating a vapour above its boiling-point away from contact with its liquid of formation.

supersaturated air Air which contains more moisture than appropriate to its temperature.

supervisory data centre A programmable central control system of modular construction and incorporating microprocessors. It offers 24-hour and weekly programs to control the heating, ventilating and air-conditioning of buildings and building complexes such as shopping centres, hospitals, etc.

supplementary bonding This is necessary as a safety precaution where electrical continuity is interrupted by plastic (non-conducting) pipes, fittings, taps, etc. It is achieved by passing a bonding tape across the plastic component to maintain electrical continuity to earth.

supplementary insulation Insulation applied in addition to the basic insulation of an electrical installation in order to protect against electric shock in the event of failure of the basic insulation.

surcharge The flooding which occurs in drainage and sewer systems when the capacity of the system is exceeded during a storm.

surcharge, precautions Certain protective measures may be taken to avoid or mitigate the effects of back flooding caused by surcharge, apart from extra care taken in the design of buildings, hard standings and their drains in areas which are known to be the subject of occasional surcharge. The simplest form of protective device is the flap valve which may be installed at the outlet of a drain-pipe to protect it from back flooding. Build-up of head on the inlet side opens the flap valve under conditions of normal flow. (A generous upstream pipe gradient is necessary.) Such a valve requires frequent inspection as accumulated debris or stiffness due to lack of use is likely to cause the flap to stick and not function under conditions of surcharge.

More effective protection is afforded by an anti-flood valve of the flap and float pattern with a robust one-piece body in which a float operates within a side chamber which is ventilated to atmosphere, the flap and float being positioned at right angles to each other and attached to a horizontal spindle which is supported on a central pivot point. The most effective protection is afforded by the installation of an in-line valve which is manually or automatically operated when conditions of surcharge are anticipated. An associated audible or visual warning system should be provided and is essential with a manual valve. Where any type of anti-flood valve is installed, it may be necessary to provide additional ventilation of the protected drain to ensure the preservation of the trap seals.

surcharge, symptoms Surcharge causes back flooding and is most likely to occur in low-lying areas where the drainage system is laid at a shallow depth or with flat (inadequate) gradients. In a sealed system, the first indication of surcharge conditions in a drain may be the back-up and overflow of a WC pan. Back flooding only becomes a serious concern when the level reached causes overspill and flooding into the building via the openings of sanitary appliances; a most unpleasant occurrence.

surcharge, warning A gully or a rodding eye incorporating a warning valve may be located at a lower level external to the building being protected.

surface condenser A type of condenser in which the cooling water flows through a number of tubes and the steam condenses by being directed over the surface of these tubes.

surface heating A method of heating a fluid by applying a heating element to the outside of a vessel or pipe. These commonly comprise resistance heating elements usually mounted in a glass cloth carrier. Applied to the heating of electric vessels, tanks, road tankers etc., in order to maintain the temperature of the fluid being conveyed.

surface tension The molecular forces acting at a fluid interface. Because of unbalanced molecular cohesive forces and a slight pressure difference between the fluids on either side of the surface of a liquid, the surface appears to act as an elastic skin which is in tension in both directions. This tension is termed the surface tension. It is measured by the force which acts across the unit length of the surface in newtons/metre.

surface water [Fig. 228] Water run-off from roofs of buildings, paved areas or ground surfaces (e.g. roads, car parks, or hardstanding).

survey The investigation of an existing situation, e.g. ground conditions, buildings or utility services such as sewers, telephones, electric cables or water pipes.

suspended drainage system A collecting pipe which is supported off the soffit of a building. It is particularly suitable where the building incorporates basements and/or low-level car parks. It offers the major advantage of accessibility and visual

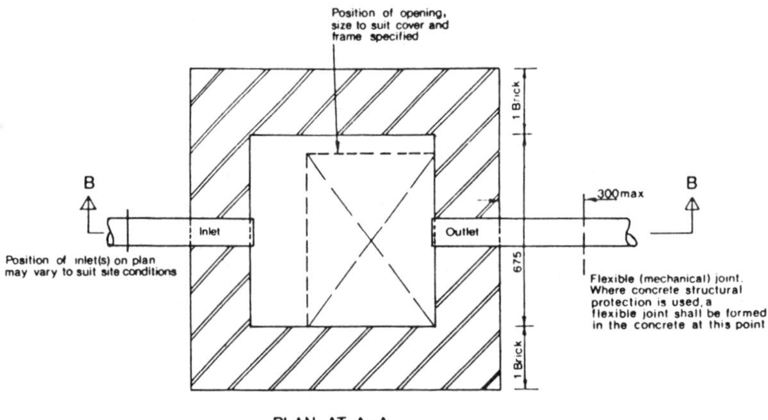

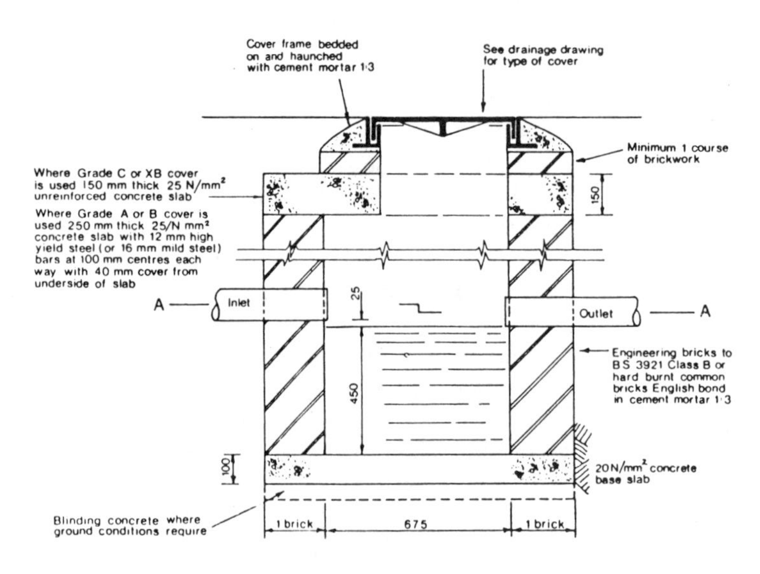

[Fig. 228] Surface water catchpit

indication of leaks. It is commonly constructed of cast-iron pipe.

suspended solids The solid particles in water. They can be filtered out, dried and weighed and are then expressed in terms of milligrams per litre.

suspension firing The method of firing of a fuel which burns whilst suspended in air. This method is used for combusting pulverized coal and certain types of wood-waste which are best burnt in suspension, together with the correct and well-directed quantities of secondary air.

sweep-up A fitting in an industrial exhaust system at floor level to collect dust off the floor. It must be complete with a grating to bar the entry of debris which could play havoc with the fan system.

swing check valve A valve in which the check mechanism incorporates a disc which swings on a hinge against a fixed stop.

swinging-type hose reel A hose reel that can be swung through 180 degrees. It is mounted at 900–1005 mm above floor level.

[Fig. 229] Swirl pattern air diffuser (Trox Brothers Ltd.)

swirl pattern air diffuser [Fig. 229] An air diffuser that discharges the supply air in a swirling motion.

switch A mechanical device installed for the purpose of making, carrying or breaking an electric current under normal circuit conditions.

switched socket A socket outlet which embodies an integral on/off switch.

switch-fuse A switch that embodies a suitable fuse carrier and fuse, and which is commonly applied to an isolating switch for plant. Its object is to fuse-protect the particular item of plant.

switchgear [Figs 230, 231] An assembly of switches, isolators, etc., mounted within the cubicle of a control panel or surface mounted on a wall or flat panel.

switch mode The operation of electrical equipment, e.g. computers, through frequent switchings.

[Fig. 230] Close-up view of electrical switchgear within control panel which serves a swimming pool complex (Contramec Installations Ltd.)

[Fig. 231] Central control panel serving a swimming pool complex: cover removed exposing switchgear and internal wiring (Contramec Installations Ltd.)

[Fig. 232] Cubicle-type switchpanel with switches and indicator lights (Contramec Installations Ltd.)

switchpanel [Fig. 232] An assembly on a backing board or in cubicles of switches and associated indicator lights, relays, etc. for the control of an installation.

symbiosis The relationship between two different types of organism which live together for their mutual benefit.

synchronization The commissioning of two or more generators of electric power to ensure that their power output is at the same frequency.

synchronous motor An alternating current machine whose speed of rotation is fixed according to the frequency of the electrical supply to which it is connected. See also **electric motor**.

T

tachometer An instrument for measuring the speed of rotation of shaft-driven machines.

tangential fan A fan that incorporates an impeller with blades shaped akin to those of a forward curved centrifugal fan impeller, but in this case both ends of the impeller are sealed. The fan is fitted into a casing in which the air enters at the periphery on one side, passes through the impeller, and leaves from the periphery at the other end. The flow of air takes place along a curved path, the axes of inlet and outlet begin roughly at right angles to each other. Most of the pressure development occurs in the form of velocity pressure and the fan efficiency is low. This type of fan is limited to small fan units, largely for use in domestic fan heaters.

tap (water) A discharge terminal on water systems where it is installed over basins, sinks, baths, etc., to turn the water flow on and off and to control the flow. It incorporates a renewable rubber washer and jumper.

tapped hole A hold drilled into a metal into which is superimposed an internal screw thread for the reception of a screwed fitting.

telemetering The transmission of signals from a measuring, monitoring or sensing instrument by radio or by telephone line to a remote recording station.

temperature, absolute See **absolute temperature**.

temperature deviation The movement of the temperature of a substance or gas away from a controlled or design set-point.

temperature indicating strip A strip which is attached to a flat surface and on which the indicators change from silver-grey to black as the temperature rises. It can check almost any surface temperature and is water- and oil-resistant. The temperature is indicated in degrees Celsius or degrees Fahrenheit or both.

temperature inversion This occurs when outside air is drawn from the top of a chimney around its periphery, whilst the hot gases rise up in the centre of the chimney. Inversion causes smut deposition around the chimney termination. It can occur at flue gas speeds below 5 m/s. Much higher gas discharge velocities are necessary to avoid inversion with certainty.

temperature regulating valve A valve that controls and maintains a predetermined temperature in a system.

temperature swing A variation of temperature around a set temperature point.

tempering of steel Imparting a definite degree of hardness to steel by heating it to a definite temperature and then quenching (cooling) in water or oil.

temporary hardness Hardness that settles out of the water on heating and/or boiling. It causes scale deposition.

tender The formal offer by a contractor for specified works commonly based on a specification and drawings. Tenders for building contracts commonly include a measured bill of quantities, as do tenders for larger building services contracts.

tender analysis A comparative statement of all tenders received.

tender documentation The specification, schedules, drawings, etc. issued with a tender invitation to permit the contractor to price the tender.

tender invitation A request to a contracting firm to submit a tender for works.

tendering Preparing a tender offer.

tensile strength Stress which has to be applied to a material in order to break it, measured as force per unit area (newtons per square metre).

terajoule A unit containing 1000 gigajoules or 10^{12} J.

terawatt A unit comprising 1000 gigawatts.

terminal valve A valve located at the draw-off (outflow) point of a distribution system.

terminal velocity The velocity of a flow at the outlet of a discharge or distribution system, e.g. the terminal velocity of chimney gases.

test cock A small-bore lever or key-operated valve which permits the convenient sampling or testing of the fluid flowing in a piped system.

test dust Dust prepared to standard specification for the testing of air filter materials, complete filters or an overall filter installation to assess their gravimetric efficiency. British Standard test dust is composed solely of aluminium oxide.

testing of sanitary installations See **air test of sanitary installations smoke test of sanitary installations** and **soap solution test of sanitary installations**.

test plug A plug fitted to a pipe or duct system for the purpose of connecting a test instrument.

tetrasilicates Mineral compounds used as a medium for the transfer of heat at high temperature at atmospheric pressure.

thermal comfort The comfort of the environment on the basis of air temperature, radiant temperature, humidity and air speed.

thermal conductivity (*k*) A measure of the ability of a material to transmit heat, expressed as W/m°C.

thermal contraction The reverse of thermal expansion.

thermal expansion The extension of a substance due to increases in its temperature. For example, pipes carrying hot fluids expand as the temperature is increased, with resultant extension in their length. In order to avoid damage to components of the pipe installation and its supports, steps must be taken to effect thermal compensation. Different substances have different coefficients of expansion. See also **coefficient of linear expansion**.

thermal fluid heater A heater used in process applications where the heat transfer fluid is heated to temperatures of up to 350°C.

thermal imaging A technique of resolving the different emissions of infra-red radiation from different temperatures and surfaces to give a meaningful and interpretative picture.

thermal index The temperature recorded by a thermometer placed at the centre of a 100 mm diameter blackened globe (globe thermometer). It is related to the radiant temperature of the air and the air velocity.

thermal insulation, corkboard See **corkboard thermal insulation**.

thermal insulation, loose fill Insulation that takes the form of granules which can be poured into space to be insulated, e.g. a loft space or a cavity wall.

thermal insulation, mineral fibre products Insulation that comprises rock fibre slabs, mats and blankets of various different densities and thicknesses bonded to produce flexible and semi-rigid materials having only a low proportion of phenolic resin. It is also made into wired and shaped insulation products. Suitable for thermal, acoustic and fire insulation use in a range of temperatures up to 1000°C.

thermal insulation, Northlite glazing An insulation unit which offers a second strong skin formed in PVC, typically about 2 mm thick, which is fitted to the exterior of roof-lights to form a double-glazed unit. Application is by special adhesive. One such unit Northlite I.V. 40/60 from Northlite Insulation Services Ltd.) is said to improve the U-value of the glazing from 7.54 $W/m^2/°C$ to 2.53 $W/m^2/°C$. For factory premises in use 12 hours a day, five days a week and with a 36 weeks heating season and heated by oil, a payback period of about two years is quoted.

thermal pipe insulation, protection A cement bandage applied over the thermal insulation material. One such product is Rok-Rap (Evode Ltd.). To use, soak in water and wrap over the insulation to obtain a rock-hard mechanical protection. It is suitable for all types of insulation material, including asbestos and is claimed to resist corrosion, fire and chemical attack.

thermal resistance (*R*) The product of thermal resistivity (1/*k*) and thickness of the material (*m*); expressed as m^2 °C/W.

thermal resistivity (1/*k*) The reciprocal of thermal conductivity, expressed as m °C/W.

thermal shock The reaction to a sudden increase or decrease in temperature of an object or system. For example, the refractory lining of a furnace must be gradually raised in temperature to avoid it being damaged by thermal shock.

thermal storage A method of smoothing out fluctuations between heat generation and demand by storing surplus heat in hot water and discharging it at times of excess demand or shut-down of the heat generator.

It can also utilize cheap off-peak electricity by heating a heat storage medium during off-peak time and discharging the heat at times of peak tariff.

thermal storage, discharging capacity See **discharging capacity, thermal storage**.

thermal storage electric system An electric system that operates with lower-priced off-peak electricity by heating water using electrode boilers during the off-peak period and storing it for use during 'working periods'. The required storage vessels are generally relatively bulky and they must be thermally insulated to the highest practicable efficiency. In use, the storage vessels connect to a pumped circulation which distributes the heat conventionally.

thermal storage vessel A container incorporated within an off-peak thermal storage system. It stores the hot water which is bled off to the system via thermostatic mixing valves.

thermal transmittance (*U*) The property of an element of a building fabric (or structure) which consists of given thicknesses of material. It is a measure of its ability to transmit heat under conditions of steady flow. It is defined as the quantity of heat that will flow through a unit area, in a unit time, per unit difference in temperature between the inside and outside environment. It is calculated as the reciprocal of the sum of the resistances of each layer of the construction and the resistances of the inner and outer surfaces and of any air space or cavity, expressed as W/m^2 °C.

thermistor A type of ceramic temperature-resistant resistor. It is a component of an electronic circuit. These semiconductors possess resistance values which may vary by the very large ratio of 10 000 000 to 1 and operate at temperatures from

−100°C to +450°C. (For special applications, the temperature range can be extended to −180°C–+500°C, or more.) Thermistors are available in a resistance range from ohms to megohms. Their thermosensitive characteristics, coupled with stability and high sensitivity, make them a most valuable tool for the measurement of temperature.

Among the semiconducting materials of which thermistors are manufactured are a number of metal oxides and their mixtures, including oxides of cobalt, copper, iron, magnesium, manganese, nickel, titanium, uranium and zinc. These oxides are usually compressed into the desired shape, are heat treated to recrystallize them, the process resulting in a dense ceramic body. Electric contact is made by one of various means; e.g. wire embedded before firing the material, plating or baked-on metal ceramic coatings. Forms in use are very small diameter beads, discs and rods. Flakes of a few microns in thickness are employed as infra-red radiation detectors or bolometers.

thermo-anemometer A hand-held air measuring instrument which combines air velocity and temperature sensing probes. It reads temperature and velocity on different knob-controlled scales. It is suitable for measuring low air velocities.

thermocouples Two wires of dissimilar metal joined together at both ends to form two junctions and used to detect, measure and relay temperature readings, particularly from remote locations. The junction at the point of the hot temperature measurement is termed the hot junction; that at the lower temperature is the cold junction. They must be so located that a temperature difference exists between them at the operative time which causes the generation of an electromotive force (e.m.f.) in the loop which actuates an appropriately calibrated associated measurement apparatus.

Thermodeck (by Straengbetong of Sweden) A method of integrating mechanical ventilation into the concrete structure utilizing the flywheel heat storage effect in buildings. It was pioneered in Sweden and uses extruded hollow-core floor slabs and thereby permits higher heat transfer efficiency due to turbulent air flow within the ducts and offers an increased surface area for heat storage purposes within the core of the slab, combined with improved control of the heat flow to and from the concrete.

Hollow-core concrete is particularly suited to Swedish construction methods. The façade of a Swedish Thermodeck building typically consists of load-bearing pre-cast wall elements which reduce air leakage to a minimum and provide a high degree of thermal storage capacity. The slabs are generally longitudinal structural members of about 1.2 m wide, commonly 0.3 m deep and contain five smooth-faced cores. Thermodeck requires only one distribution duct for supply air. This must be placed centrally and perpendicularly to the slabs. Every hollow-core slab in a room is connected to this duct. The supply air then travels back and forth through the slab until it reaches a simple ceiling diffuser which is usually placed near the perimeter of the ventilated space. During the summer, cool outdoor air is circulated overnight to cool the slab; in the daytime, the warmer outside air is cooled as it passes through the cool slab. Assuming ventilation air flows of around 5–9m^3/m^2 of floor area, the thermal storage available in the pre-cooled concrete can absorb between 10 and 40 W/m^3 of the internal heat gains generated from lighting, occupancy and solar radiation during the day. During the heating season, the temperature of the slab must be raised. This is achieved through the use of off-peak electric heating which warms the incoming night air to about 45°C and then passes this air into the slab to raise its temperature.

Since the Thermodeck system covers the entire horizontal surface area of a space, it then becomes possible to install the complete ventilation system before deciding on the location of partitions – a facility which is particularly useful for speculative office buildings. The system is very quiet in operation as the concrete cores function as very efficient attenuators against fan noise and the movement of dampers.

thermodynamics The science and application of heat, work, temperature and related matters, applied especially to heat engines and the behaviour of gases. See also **first law of thermodynamics** and **second law of thermodynamics**.

thermodynamic steam trap [Fig. 233] A steam trap that relies on the phenomenon that flash steam

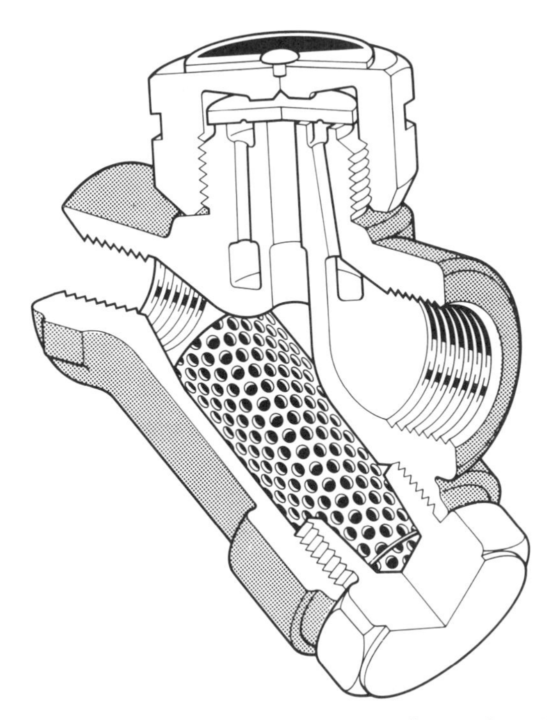

[Fig. 233] Thermodynamic steam trap (Spirax–Sarco)

is generated when condensate at a temperature close to its related steam temperature is reduced in pressure at an orifice and which utilizes the increased velocity of the flash steam to close a disc against a seating. The cycle is repeated when the steam pressure within the control chamber has been dissipated through heat emission; the disc lifts and condensate is discharged.

thermoelectric pyrometer A temperature-measuring instrument which uses the Peltier effect generated at a hot junction of two different metals. Two wires of dissimilar conductors jointed together at the hot end are enclosed within a heat-resistant protective sheath connected between the thermocouple and the measuring or recording instrument where the temperature is indicated to an appropriate scale.

thermograph A self-registering thermometer which records temperature variations on a graph during a period of time.

thermomechanical Describes the process of thermal expansion or contraction of a temperature-sensitive metal or fluid employed to actuate the mechanism of a control device on a change of temperature.

thermometer An instrument for the measurement of temperature. Any physical property of a substance which varies with temperature change can be used to measure it.

thermometer, maximum and minimum A thermometer that records the highest and lowest temperatures reached during a given period of time. It comprises a bulb filled with alcohol which, by its expansion, pushes a mercury thread along a fine tube which is graduated in degrees of temperature. At each end of the mercury thread is a small steel index which is pushed by the mercury. One index is left at the farthest point reached by the mercury thread corresponding to the maximum temperature and the other at the lowest point which corresponds to the minimum temperature.

thermometer pocket An open-top sheath filled with an appropriate heat-conducting liquid which is screwed or inserted into a pipe or vessel and serves for the insertion of a thermometer.

thermostat A sensor, electrical or non-electrical, measuring temperature fluctuations and activating a related control function.

thermostat, direct See **direct-acting thermostat**.

thermostatic air vent for steam systems [Fig. 234] A device that incorporates metal bellows which expand and contract when in contact with steam or condensate. It functions to release accumulated air which might cause air blanketing of heating surfaces (and thereby reduce output and efficiency) and/or water hammer. It is fitted at the high points of a steam pipe or of steam-fed machinery.

thermostatic expansion valve A valve that is widely used in dry expansion evaporator systems in which a small opening between the valve seat and the valve disc results in the required pressure drop. It also regulates the flow of refrigerant according to need. A bulb filled with a fluid is strapped to the suction line and thus senses the gas suction

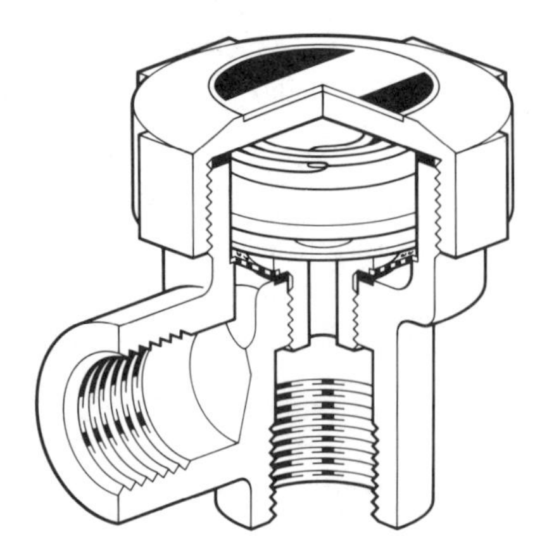

[Fig. 234] Thermostatic air vent for steam systems (Spirax–Sarco)

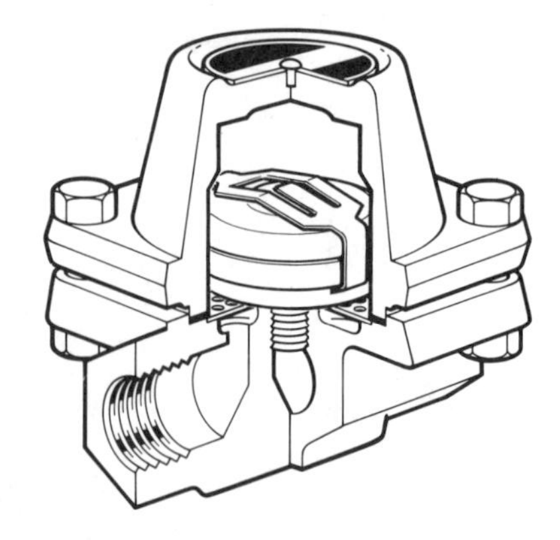

[Fig. 235] Balanced-pressure thermostatic steam trap (Spirax–Sarco)

temperature. This bulb is connected to the valve by a tube in a manner which ensures that the pressure of the fluid in the bulb tends to open the valve against a closing spring pressure. If the load in the system increases, the refrigerant in the evaporator picks up more heat and the suction gas temperature rises, causing the pressure in the bulb to increase also and it opens the valve more. This increases the refrigerant flow needed to handle the increased load. The reverse occurs when the refrigerant load decreases.

thermostatic mixing valve A valve that combines temperature selection and control over fluid flow.

thermostatic radiator valve A self-acting liquid-filled radiator valve sensitive to ambient temperature and in sizes to fit most pipe diameters for gravity-circulated flow, pumped water and steam systems. Models are available with integral and remote sensors.

thermostatic steam trap [Fig. 235] A steam trap that relies on bimetallic devices or liquid-filled bellows which sense the difference in temperature between steam and condensate. The trap remains open to condensate flow and closes when the higher temperature steam enters it and bars the steam from entering the condensate system. When the steam has cooled and condensed, the trap reopens. Such traps are suitable for the trapping of steam-fed appliances which carry a small load such as radiators, for air venting and for draining of steam lines.

thermosyphon A hydraulic system in which a fluid circulation is caused by temperature (and thus density) differences within a body of fluid.

thermosyphon cooling A method of providing chilled water for large plants at elevated temperatures, taking advantage of low ambient temperatures to provide 'free cooling' throughout the year by allowing natural circulation of the refrigerant without running the refrigeration compressors.

thermosyphon cooling effect Describes the principle whereby water is cooled by the evaporation of the refrigerant which, under suitable ambient conditions, is allowed to pass to a condenser, bypassing the compressor by means of the pressure difference generated by the difference in temperature between the water being chilled and the ambient temperature. This relies for its efficiency on sensing the wet bulb temperature. Where this cannot be easily sensed, it may base itself on the dry bulb temperature reading which is subjected to an appropriate correction factor.

thermosyphon cooling system Depending on the external (wet bulb or corrected dry bulb) temperature, the control system decides whether or not the thermosyphon effect is utilized. If free cooling is possible, the system will not commerce mechanical cooling at that point but will allow a cascading effect to extract heat using the thermosyphon principle.

In one system which was closely monitored over one year, with diurnal variations also taken into account, free cooling met 55% of the cooling load, the remaining 45% being satisfied by the mechanical system. Projected energy savings in running costs are stated to show up to 70% for a typical plant compared with standard water chillers and up to 25% compared with other free-cooling systems, with a corresponding increase in the COP.

threshold of audibility The minimum sound levels at each frequency that a person can hear.

threshold of pain The sound level at which the average person experiences physical pain (typically 120 dB).

three-phase circuit An electrical installation where the alternating current is divided into three branches or phases. It is generally used with non-domestic and the larger commercial/industrial/lift installations.

three-pipe system A system that comprises two flow pipes and one common return. It is used with induction and air-conditioning systems in which hot water and chilled water are separately piped to each terminal unit by various patented arrangements. The principal advantages are the flexibility and the rapid response compared with more conventional two-pipe systems, due to a separate source of heating and of cooling being available to each unit at all times. Zone change-over control is eliminated and variable internal and external heat gains can be effectively compensated.

three-way boiler escape valve A valve that enables a common open vent system to be used in multiple boiler plants. It ensures that the boiler is always safely vented to atmosphere.

three-way cock A lever-operated three-way valve arrangement.

three-way valve An isolating or shut-off valve with three ports at right angles to each other. It is used, for example, on a multiple boiler installation connected to a common open vent pipe to ensure that the boiler on line is always safely vented.

throttle valve A non-tight closing butterfly valve which incorporates a centrally hinged flap which can be locked in any desired position to regulate fluid flow.

through-the-wall air-conditioner An air-conditioning unit which is built into a wall, usually below a window. All the components, including the evaporator and the condenser, are mounted within one assembly.

throw-away-type air filter [Fig. 236] An air filter made of glass fibre, coated paper, fabric or another relatively inexpensive filtration medium in the form of a framed filter pad which may be used singly or in banks. The medium cannot be cleaned when dirty and must be replaced (thrown away). The assembly should include a differential pressure indicator, e.g. an inclined manometer, to indicate when the filter pad has become dirty and requires replacement. The filtration efficiency depends on the nature of the filtration medium and on the construction of the assembly.

tie rod A steel rod which is threaded at each end and is often used in pairs to pull up sections of an assembly, e.g. the sections of a cast-iron or steel sectional boiler.

timber, moisture content The water contained within a particular sample of timber is calculated as a percentage by weight of the dry timber. This does not relate to the total weight of wood substance and water within the sample. For example, a piece of timber of 2.0 kg at a moisture content of 100% contains 1.0 kg of wood and 1.0 kg of water.

timber, moisture content, effect of [Fig. 237] Wood-waste is considered to be relatively dry when the moisture content is below 20% (by weight) and it can then be incinerated without undue difficulty. Conventional incinerators cannot successfully combust very wet wood; it must be pre-dried before being fed into the incinerator. A cyclone furnace can effectively handle wet timber as there is generally adequate dwell time within the furnace before combustion.

Certain modern underfeed stoker-type furnaces can combust wood-waste with a moisture content

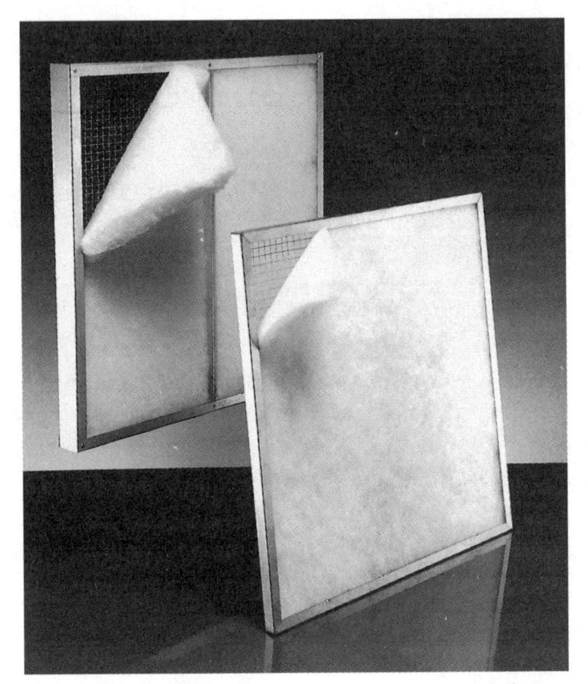

[Fig. 236] Fibreglass mat air filter module (Waterloo-Ozonair Ltd.)

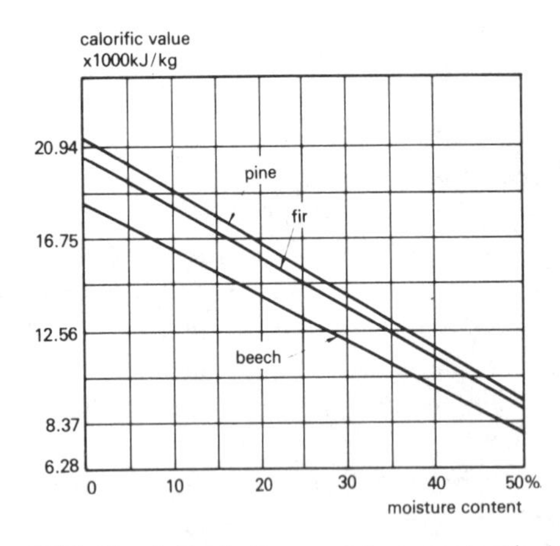

[Fig. 237] Chart illustrating moisture content/calorific value variation for wood

of up to 50% but wetter material is progressively more difficult to burn in such equipment. The burning of wet waste cannot be as efficient as burning dry material as a proportion of the heat being generated is absorbed in drying the material, i.e. converting the moisture into steam.

time clock A device for switching electricity on or off at predetermined times. It may include a day-omitting feature and may allow numerous switching operations during a 24-hour cycle. It is commonly driven by an electric motor but is also available with a hand-wound spring action which requires periodic manual rewinding. May incorporate an emergency spring reserve to maintain time clock operative over periods of power failure.

time delay The time interval before a process or function can be activated.

time of concentration The longest time taken for the rain falling on an area to reach the associated drain, plus that time taken by it to travel to the point under consideration which may be expressed as:

$$t = \text{Time of entry} + \frac{\text{Length of drain}}{\text{Full bore velocity of flow}}$$

timer time switch [Fig. 238] This may be manually-operated, clockwork or motor driven, programme or function controlled.

tin–lead soft solder This type of solder is typically suitable for a melting point of up to 65°C and 10 bar. It is available in sizes 12–28 mm.

tin–silver solder The type of solder is typically suitable for up to 100°C and 16 bar. It is available in sizes 12–28 mm.

ton of refrigeration (TR) A term used to measure the cooling capacity of refrigeration plant. One ton is equivalent to a capacity of 3.15 kW.

toroidal combustion A combustion method which utilizes the principle of the vortex to achieve the continuous intimate mixing of fuel and air. It increases the dwell time of the fuel particles within the combustion zone and is suitable for the burning of difficult fuels and wastes in equipment such as the cyclone furnace.

torque The moment of a force or of a system of forces which tends to produce rotation.

Torrent (Terrain Ltd.) [Figs 239, 240] A proprietary system of uPVC rainwater collection fittings and of jointing them.

[Fig. 238] Control cubicle incorporating a range of time switches and associated switchgear (Contramec Installations Ltd.)

[Fig. 239] The high-capacity Terrain Torrent plastic rainwater system installed externally on a school building

torsion The act of twisting about an axis produced by the action of two opposing couples acting in parallel planes.

total energy See **energy, total**.

total energy system [Fig. 241] A system that supplies all the energy requirements of a particular premises, site or area by on-site generation of electricity using a prime mover. It includes waste heat recovery with the added provision of any required supplementary heat and may take the form of a combined heat and power system which optimizes energy usage.

total head The sum of all the pressures which act upon an item of equipment. For example, the total head on a hot water boiler would be the imposed head due to the height of the feed and expansion tank or pressurization vessel and the pressure (positive or negative) exerted by the system pump(s).

total heat See **heat, total**.

towel radiator Rows of serpentine hot water piping from which towels may be hung. Alternatively it may comprise piping rails which extend from flat-fronted radiators.

towel rail, electric A bathroom towel airer incorporating an electric heating element.

towel rail, hot water A bathroom towel airer, commonly connected to the primary hot water system; when it is connected to the direct hot water circulation it is likely to be prone to airlocking. It may comprise a number of chromium-

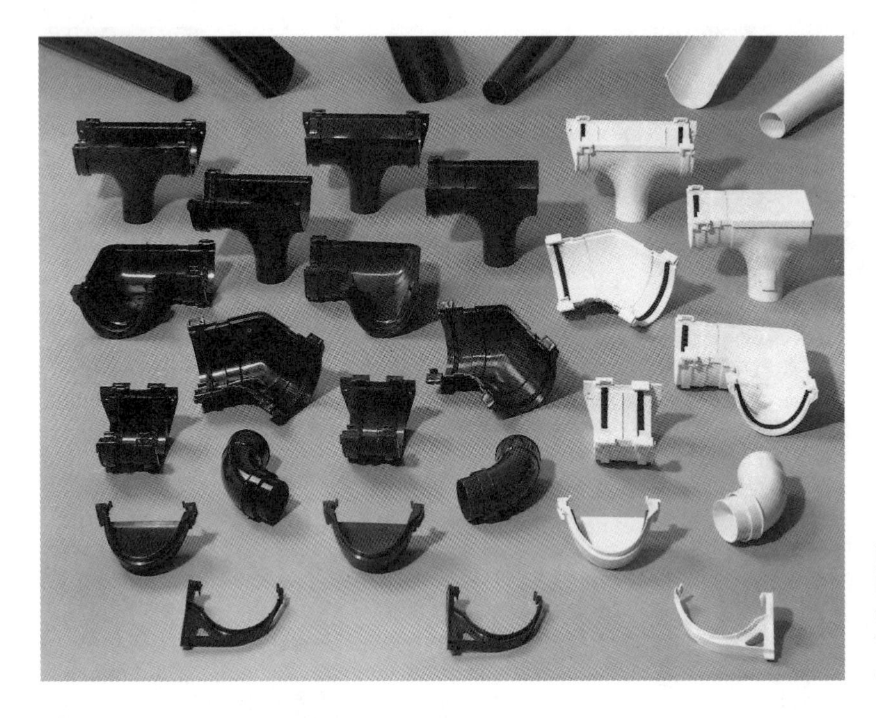

[Fig. 240] Selection of MuPVC plastic guttering fittings of the Terrain Torrent range with colour options

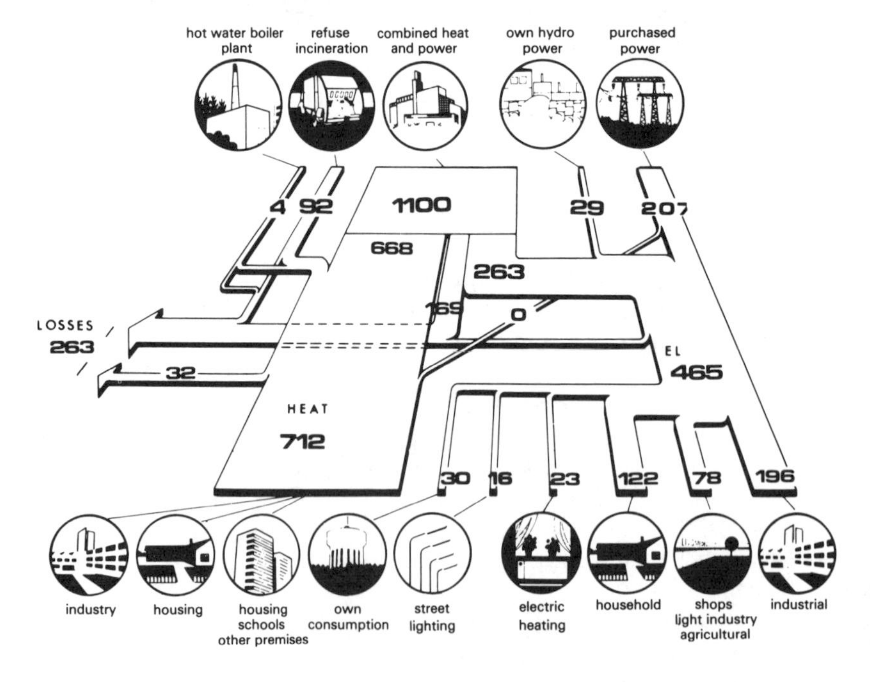

[Fig. 241] Total energy balance LINKOPING district

plated heated pipes or incorporate a panel radiator section.

towel rail primary circuit A circuit that connects the towel rail to the closed heating circuit and which is piped independently of the space heating so that the circuit remains live throughout the year.

town gas Gas manufactured from coal, usually by the process of carbonization in gas retorts, and distributed by public utilities for heating and the fuelling of prime movers. Gross energy as supplied to consumers is in the order of 20 MJ/m^3. It has a distinctive smell which assists to alert the public to leaking gas.

toxic Poisonous.

toxicology The study of poisons.

trade effluent Waste product in liquid form discharged by a commercial or industrial establishment. Certain pollutant and/or poisonous wastes may either not be discharged into the public sewers, may require prior treatment before discharge or be permitted to be discharged only in specified limited quantities over a period of time.

traffic loads Such loads imposed upon drainage systems comprise a combination of static wheel loads and impact factors. Main road traffic load also includes an impact factor.

transfer grille A grille that transfers air from one directly ventilated area to another or to exhaust; usually fitted in partition walls. Vision-proof grilles are available to promote privacy.

transfer pump A pump that conveys a liquid from one receiver to another. For example, an oil transfer pump will transfer oil from the main storage to a smaller daily service tank serving an oil-fired boiler or electricity generator. Drinking-water pump will transfer water from a basement main water storage to the drinking-water header on the roof of the building.

transformer Equipment for the stepping up or stepping down of the voltage of an electric supply. It essentially comprises two coils of wire which are wound on to a common soft iron or mild steel core. The supply passes through the primary winding where the alternating current sets up an alternating flux in the core generating an induced electromotive force which results in a current flowing through the secondary winding. In a step-down transformer (a common type within a transmission system), the voltage in the secondary windings is less than the supply voltage; in a step-up transformer, the secondary current is at the higher voltage. The relative number of turns in primary and secondary windings determines the voltage ratio between the primary and secondary circuits. The process of transformation generates heat which is dissipated via an oil, water or oil-filled cooling system.

transient heat flow The irregular or inconstant flow of heat, e.g. the transmission of solar heat through a particular face of a building during the period of time solar energy falls on to that face.

transistor A semiconducting device which incorporates three or more electrodes. It is used with electronic (control) equipment.

translucent Permitting the passage of light in such manner that an object cannot be seen clearly through the substance, e.g. frosted glass.

transmissibility The amount of vibratory force that is transferred to the structure through an isolator, expressed as a percentage of the total force applied.

transmission loss A set of values measured by a specific test method to establish the actual amount of noise that will be stopped by the material, partition or panel when placed between two reverberation rooms.

transparent reflective surface A surface coating which allows short-wave radiation to pass through it while infra-red (thermal radiation) is reflected.

transparent thermal insulation (TI) This insulating material is placed on the external surfaces of a building and utilizes the fact that TI is as good an insulator as conventional opaque insulation materials. Since it is transparent to solar radiation, it allows the direct solar gain to warm the wall behind the TI modules. It is anticipated that the development and wider use of transparent insulation will have a major beneficial impact on the energy efficiency of buildings.

trap [Fig. 242] A fitting or part of an appliance within a sanitary system or pipe arranged to retain

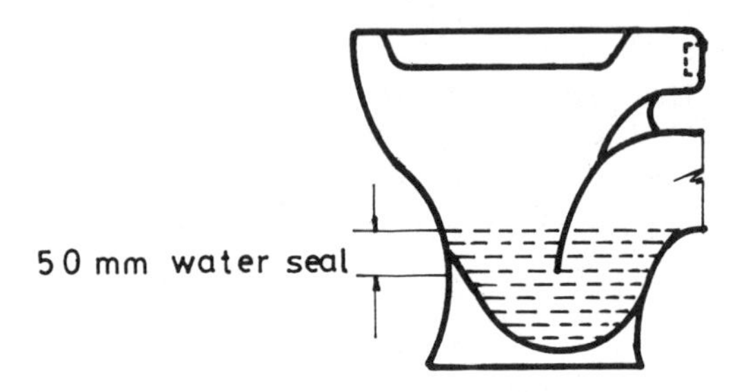

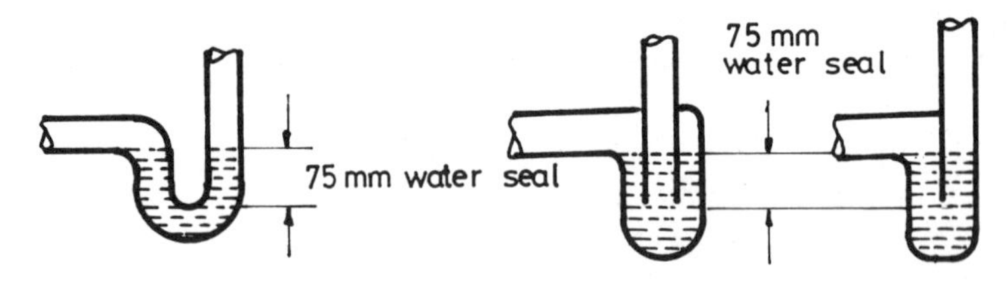

[Fig. 242] Sanitation: types of sanitary trap

water or other fluid with the object of preventing the passage or escape of foul air.

trap, catchpot See **catchpot trap**.

trap, deep seal See **deep seal trap**.

trap, depth of seal Traps with outlets for pipes up to and including 50 mm size require a minimum water seal of 75 mm. Those with outlets for pipes above 50 mm size require a minimum seal of 50 mm. Traps installed on appliances with a trailing waste discharge installed on ground floors and discharging into an external gully may have a reduced water seal of not less than 40 mm.

trap design and location Traps should be designed so that deposits do not accumulate therein. A trap which is not an integral part of an appliance should be attached to, and be immediately beneath, its outlet and be self-cleansing and its internal surface must be smooth throughout.

trapping bend A unsocketed 90 degree cast-iron bend which incorporates one long arm for use as a dip pipe. It is used in a petrol interceptor.

trap, P-type See **P trap**.

trap, Q-type See **Q trap**.

trap, resealing See **resealing trap**.

trap, running See **running trap**.

trap, S-type See **S trap**.

travelling grate Part of a system for the combustion of solid fuel or waste in which the fuel is moved through the furnace by a moving grate operated by an electric motor, so that there is an intake of fuel at the front of the furnace and the discharge of ash and clinker at the rear. It is widely used in refuse incinerators which handle a large throughput of waste.

tribology The science and technology of friction, wear, lubrication and associated aspects.

trickle ventilator A ventilator that introduces into a room small quantities of air by natural ventilation. It commonly takes the form of a slot incorporating a removable shutter fitted above a window frame. It is designed to meet the requirement for natural ventilation in domestic premises.

trombe wall [Fig. 243] A passive solar collection method which uses a thermally massive wall construction for the storage of solar energy. This is placed between the living space of a dwelling and an external glazed sun-trap. The wall collects solar radiation and stores the heat energy within its mass; this is then reradiated through the back of the wall and/or by convection air movement of the cooler room air along the warm face of the trombe wall by the natural buoyancy of the air. This thermosyphonic effect is generated on the face of the wall which is exposed to the Sun by locating a glazed screen some 50 mm distant from the sun-facing side of the wall, encouraging the desired air circulation by forming openings near the top and close to the bottom of the trombe wall. The air will then circulate through the openings and space between the trombe wall and the glazed screen, absorbing heat and thereby being raised in temperature. Thermal control over the heat flow is by selecting the mass of the trombe wall to give sufficient time lag before reradiation occurs and by installing manually or automatically operated dampers to regulate the air flow through the openings. Sun blinds are fitted and manipulated to prevent overheating during the summer.

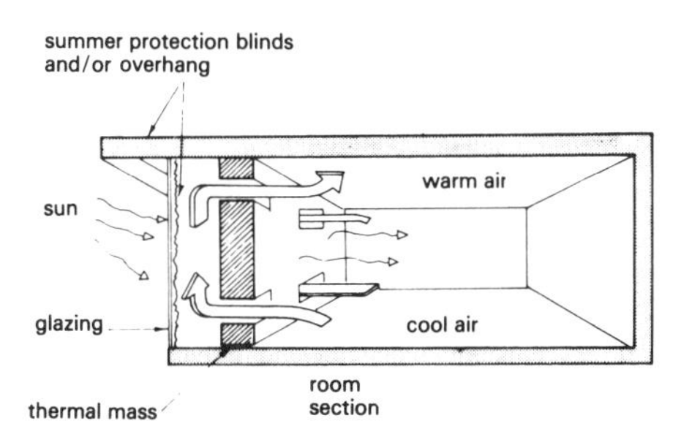

[Fig. 243] Trombe (thermal) wall application

trunk duct A main run of ducting.

tube An alternative term to pipe, commonly applied to the heat transfer pipes within boilers and furnaces.

tube bundle An assembly of pipes which constitutes a heat transfer battery, condenser, etc. or an assembly of water tubes in a boiler which is contained by the tube plates.

tube expander A tool used to expand the ends of tubes to secure them tightly into collection plates within boilers and furnaces.

tube plate A steel plate which contains and seals the smoke, fire or water tubes in a boiler. Usually, one tube plate is fitted at each end of the tube assembly.

tubular trap A trap formed as a simple U-bend of uniform cross-section.

tumble drier A clothes drier that typically consists of a rotating drum with a non-recirculatory heated fan-driven air flow as the drying medium. Most industrial tumble driers consume large amounts of energy as a large proportion of the drying cycle discharge comprises exhaust air at high temperature. Such driers are commonly steam-heated using multi-row batteries with the leaving air temperature approaching that of the steam. At the start, the discharge air is relatively cool and moist and this becomes hotter and dryer close to the end of the cycle. Filters are fitted on the discharge from the drier to avoid excessive lint deposits on the surrounding areas. When the lint on the filters builds up to a level where this restricts the air flow, the lint deposits are usually removed by hand.

tumble drier, energy efficiency measures Most industrial tumble driers consume greatly excessive amounts of energy. The main options for reducing their energy consumption are:

(a) the attainment of optimized and constant rate of air flow;
(b) the introduction of air recirculation towards the end of drying cycle;
(c) pre-heating the incoming fresh air supply via the leaving exhaust air.

Constant air flow cannot be achieved when conventional lint filters are used, as the air flow must then vary due to the added resistance caused by a build-up of lint on the filtration surfaces. A suitable alternative is offered by a spray wash system. Recirculation may prove impracticable in situations where the high fibre content of the exhaust air is likely to excessively clog the equipment in the recirculation circuit. Preheating the incoming air is often the preferred option, commonly using flat place cross-flow recuperators with wide plate spacing, smooth air flow passages to minimize the build-up of fibrous material and convenient access for cleaning.

Experience has shown that a vertically upward exhaust from the drier up through the recuperator minimizes the clogging problem and eases cleaning by allowing washing of the recuperator with the aid of gravity. The advantage of a vertical configuration is that some of the fibres that are caught on the recuperator fall to the plenum below when the fans are turned off at the end of each drying cycle. When the fans are turned on for the next cycle, these fibres are picked up by the air stream and a large proportion then pass through the recuperator plates.

tumbling manhole See **back-drop manhole**.

tungsten lamp A lamp that incorporates a metallic element used as a filament. In use, tungsten lamps are inefficient relative to fluorescent lighting. They are now used mainly for domestic and special effects lighting.

turbine A machine that directs the flow of a liquid or gas at high pressure to provide rotational movement. A turbine is equipped with blades, vanes, nozzles or similar to effect efficient transformation of energy. Steam turbines are employed for electricity generation in major power stations. Gas turbines are used for smaller size peak-load or localized electricity generation.

turbine, back-pressure See **back-pressure turbine**.

turbine, closed-cycle gas See **closed-cycle gas turbine**.

turbine, condensing A turbine that is operated with the object of achieving the highest possible electric power generation efficiency by employing a vacuum on the outlet side. It is practicable to achieve a pressure of 4 kPa on the condensing side of a turbine.

turbine, gas See **gas turbine**.

turbine, open-cycle gas See **open-cycle gas turbine**.

turbine, pass-out See **pass-out turbine**.

turbine, steam See **steam turbine**.

turbo-compressor A machine very closely resembling a multi-stage centrifugal pump, in that its working components consist of impellers and diffusers, rather than discs and diaphragms, as in turbines. It converts kinetic energy efficiently into pressure energy. Turbo-compressors are capable of very large outputs of compressed air.

turbo-compressor diffuser A component of a turbo-compressor comprising an annular chamber which is fitted around a corresponding impeller. It carries blades which guide the air within the compressor in all directions and provides a smooth and short divergent passage in which the kinetic energy in the air is converted into pressure energy.

turbo-generator A turbine which is coupled to an electric generator for the generation of electric power. It is driven by steam in conventional power stations and by water in hydraulic power schemes.

turbulators Metal strips which are formed into opposing bands. They are inserted into boiler flueways to create turbulence of the hot gases, thereby increasing the rate of heat transfer within the boiler.

turbulent flow A flow of fluid which is disturbed and flows in eddies, or a confused state of air flow that may cause noise to be generated inside, for example, a ductwork system.

turning vanes [Fig. 244] Shaped metal baffles attached to the inside of an air duct at changes of direction to induce a smooth flow in the desired direction and to minimize friction loss.

turnkey project A project where the main contractor is responsible for all aspects of the project and the employer is not involved until the project is handed over.

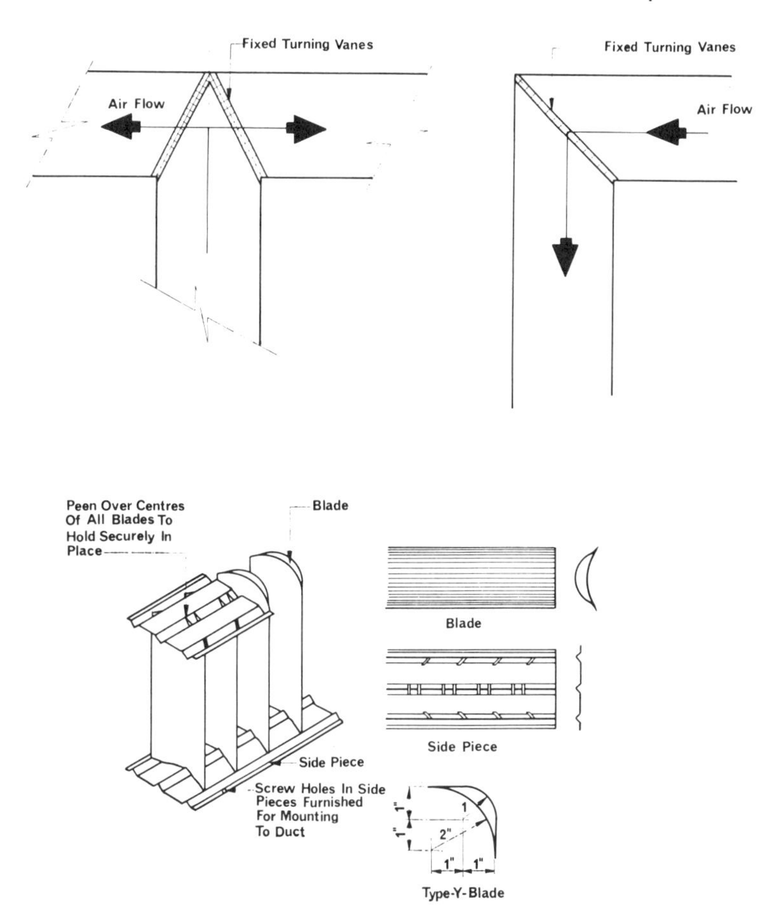

[Fig. 244] Turning vanes

twin-compartment pipeline strainer A strainer that incorporates two strainer compartments which are brought into use alternately, enabling one strainer to be cleaned whilst the other remains operational. Change-over is usually effected by the operation of a lever.

twin-cylinder compressor A compressor that incorporates two stages of compression to achieve higher output air pressures. Capacity is partially controlled by unloading one cylinder.

two-duct air-conditioning system A system that utilizes two separate air supply duct systems, one carrying warm air and the other cooled air into a mixing box from where the correctly blended air is carried to the conditioned space. This system is now generally out of favour due to high cost and space requirements.

two-part chemical grout for drainage renovation This method (Superaqua) uses a solution of sodium silicate followed by diluted hydrochloric acid and flushing of the drain to grout up cracks and gaps. It is applied by filling a cleaned and plugged-off drain run with the sodium silicate solution and then quickly pumping this back into the tanker. The diluted hydrochloric acid is then introduced into the sewer and allowed to stand whilst a chemical reaction takes place between the two in the ground surrounding the drainpipe, effectively grouting up cracks, holes and gaps. The acid is pumped back into the tanker and the pipe flushed with water; it should then be free of leaks.

two-part tariff A tariff that comprises a fixed charge element levied per week/month/quarter/annum proportional to the maximum electrical demand or to the total connected load, plus a charge per unit consumed. This is applied in district or group heating schemes and comprises a fixed charge in respect of the common charges such as depreciation of central plant and distribution mains, plus a variable charge which depends on the metered use of heat by the individual consumer. Fixed charges may also include the cost of standing losses from central and common plant.

two-pipe heating A heating system in which each heater is connected at its inlet to a flow pipe taken off the boiler or boiler plant flow header and at its outlet to a separate return pipe which recirculates the water back to the boiler or return manifold of the boiler plant.

two-wire circuit An electric circuit which functions with live and neutral wiring only, there being no earth wire.

U

UIO Universal information outlet. See **premises distribution system**.

UPS Uninterruptible power supply. This is used as a back-up to computers and data communication systems.

U bolt An inverted U-shaped metal rod with a screw thread at the terminations. It secures a pipe firmly in position over a bracket where the nut and bolt fastenings are applied from underneath the bracket. It is unsuitable for guiding pipes subject to thermal movement.

ultimate stress That load required to fracture a material divided by its original area of cross-section at the point of the fracture. Ultimate stress is divided by the factor of safety to obtain the acceptable working stress.

ultrasonic Describes inaudible (to the human ear) pulsations of high frequency sound, typically 28 000 cycles/second.

ultrasonic air and steam leak detector A device designed to detect faulty steam traps, steam leaks and air and gas leaks from both pressure and vacuum systems. It has audio and visual indicators and is battery-operated for portability.

ultrasonic boiler cleaning The ultrasonic pulses are transmitted to the boiler water through a length of pipe. The metal disc transmitter operates at 28 000 cycles/s frequency and pulsates about three times a second for a duration of 1/1000 s. Caution is essential in that ultrasonic power must be limited to prevent the destruction of the protective mill scale on the boiler steel plates.

Application: Satisfactory results have been reportedly achieved with shell boilers steaming on a supply of chemically untreated water of average hardness at pressures up to 30 bar. Normal operation permits the scale to build up to eggshell thickness. It is then broken off in patches which drop to a scale collection pocket.

ultrasonic frequency Very high frequency in excess of 20 000 Hz.

ultrasonic generator A device for the generation of pressure waves of ultrasonic frequency.

ultrasonic scanning A way of detecting leaks that are almost impossible to find by any other method. Gas escaping through a restricted orifice passes from a pressurized laminar flow to low-pressure turbulent flow, generating a broad spectrum of sound with strong ultrasonic components. Because the ultrasounds are loudest at the site of the leak, a sharp rise in ultrasound during a scan gives a clear indication of leakage. Ultrasonic sound waves are extremely short and travel in straight lines; they cannot penetrate solids, although they filter through the tiniest of openings or cracks. These properties leave ultrasonic detectors largely unaffected even in the noisiest environment.

ultrasound Sound created by high-frequency vibrations above the threshold of hearing. The phenomenon is exploited in the technology of ultrasonics.

ultraviolet That band of electromagnetic wavelengths adjacent to the visible violet (0.10 nm + 0.38 nm) of particular interest in solar energy applications since various unprotected materials change in colour due to the reception of ultraviolet radiation. If the materials are unstable, they are then likely to suffer failure due to decay or fracture.

underfeed stoker [Fig. 245] A type of mechanical stoker in which the coal is supplied by a Archimedean screw or by a reciprocating ram, the fuel being fed into the bottom of a retort below the fire and gradually forced up into the combustion zone of the furnace. The combustion air is injected through ports or tuyeres into the fuel bed just below the combustion level. It is not normally necessary to supply secondary air through the fire door. As the coal rises in the retort, the released volatiles pass upwards through the burning fuel where they mix with the incoming combustion air and burn at the top of the fire with a short flame. Under the correct combustion conditions only coke remains when the fuel reaches the surface of the fuel bed. The coke burns as it moves outwards towards the perimeter of the fire; combustion is completed and ash collects around the retort from where it must be removed periodically. This type of stoker is fitted to central heating boilers but is prone to emission of smoke when there is intermittent operation or whenever the fire is disturbed. This can be largely overcome by the use of overfire jets.

under-floor heating See **floor heating**.

underground drainage renovation See **closed-circuit television (CCTV)**, **cement grouting with pistons**, **liquid resin grouting**, **two-part chemical grout for drainage renovation**, **drain liners**, **in situ drain liners**, **uPVC spiral relining**, **sleeved uPVC pipes for sewer or drain relining**, **preformed new sewer replacement pipes**, and **pipe bursting**.

union [Fig. 246] A pipe fitting consisting of two parts, each of which is screwed or otherwise jointed to a pipe end. One part incorporates a nut which is passed over the mating screwed fitting on the other part and is then tightened to secure the joint. A union is fitted in locations where it may be necessary to easily disconnect one section of pipe from another or from an item of equipment, such as for maintenance, inspection or replacement.

unitary air-conditioner A packaged air-conditioner which consists of factory-fabricated assemblies designed to provide all the required air treatment functions. It is made operational by being located in the required position and connected to the electric supply.

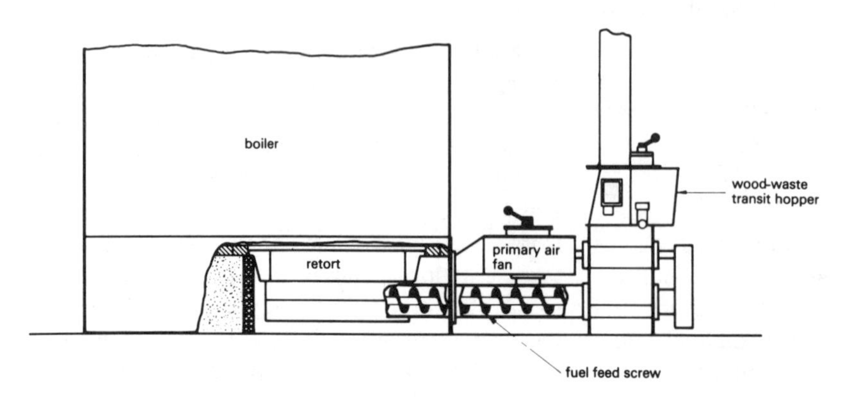

[Fig. 245] Underfeed worm stoker

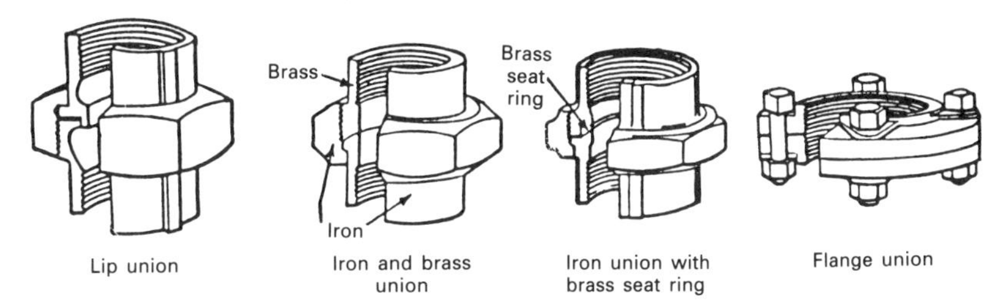

[Fig. 246] A selection of pipe unions

unit heater The assembly of a fan heater battery and an adjustable discharge grille within a common casing. It may be mounted to discharge horizontally or vertically.

unloading equipment The automatic air governor of an air compressor. When the air in the air receiver of the compressor has reached a predetermined pressure, the automatic air governor passes compressed air into the unloading cylinders which are fitted above the compressor suction valves. When the pressure falls to the predetermined lower limit of the air governor adjustment, the air supply to the unloading cylinders is cut off by the air governor.

unsaturated air Air that is capable of absorbing further moisture at the prevailing temperature.

unstable flame A flame that does not maintain its position relative to the combustion appliance.

unvented hot water storage system, installation To meet with the terms of the BBA certificate, the unit or package should be installed by an approved installer, who must also install the discharge pipe taken from any safety valve releasing hot water.

unvented hot water storage system, manual resetting thermal cut-out This should be connected to the direct or indirect heat source in accordance with the current IEE Regulations.

unvented hot water storage system, package A hot water storage system which incorporates the comprehensive range of required safety devices supplied by the manufacturer with a kit containing any other operating devices to prevent back flow, control working pressure, relieve excess pressure and accommodate the expansion facility which is to be fitted by the installer.

unvented hot water storage system, safety cut-out In a directly heated system, the thermal cut-out should be on the storage system. In an indirectly heated system, the heating coil should only be connected to an energy supply fitted with a temperature-actuated energy cut-out.

unvented hot water storage system, safety device discharge pipe The discharge pipe must be of a suitable metal and its pipe size should be the same as the discharge outlet on the safety device. The discharge should exit via an air break to a tundish; the pipe should be laid to continuous fall and must not exceed 9 m in length, unless the pipe bore is increased. The final discharge should terminate in a visible but safe location, such as over a gully, where there is no risk of contact with the hot water being discharged and persons using or passing the building.

unvented hot water storage system, safety requirements The BBA certificate requires a minimum of two temperature-activated devices operating in sequence, a non-resetting thermal cut-out and a temperature relief valve. The two devices are additional to any thermostatic control which is fitted to maintain the temperature of the stored water.

unvented hot water storage system, unit This incorporates a comprehensive set of safety devices and other operating devices to prevent back flow, control working pressure, relieve excess pressure and accommodate an expansion facility fitted on the unit by the manufacturer.

unvented hot water system An arrangement of hot water generation and supply which is connected to mains water and is not fitted with an open vent.

uplighter A luminaire which directs the light upwards towards the ceiling.

uPVC spiral relining A sewer renovation system (known as the SR system by Rice–Hayward) which comprises the *in situ* formation of a structural liner from a helically wound PVC strip which is fed into a winding machine located at the bottom of a manhole or other excavation. The edges of the strip interlock to give a positive joint between adjacent turns of the helix without the use of adhesives. With the liner full of water, annulus grouting is carried out to seal the system. The sizes and thicknesses of the PVC strip and the various grout formulations are available as required.

urea formaldehyde foam A widely-used insulating material for existing brick or timber cavity walls. It can be conveniently injected into such cavities through small holes on the inside or outside of the wall and sets without any appreciable heat change. It is unaffected by dilute acids, alkalis, oils or solvents and it neither rots nor shows any sign of ageing. It is combustible and simply shrivels up when exposed to high temperatures. It is greasy in nature, so that, while it presents no barrier to water vapour, it tends to repel liquid water by capillary action.

urinal Four types are available: wall-mounted bowl and trough, floor mounted, recessed slab and stall types. They are usually manufactured of fire-clay, stainless steel, vitreous enamel or GRP. The former are used more commonly in public lavatories.

urinal, bowl-type See **bowl and trough urinal**.

urinal, discharge stack Stacks which serve only urinals are likely to become encrusted with sediment leading to a corresponding reduction in bore. It is an advantage also to connect the discharges of other appliances (e.g. WCs or hot water using appliances) to a urinal stack to reduce such encrustation.

urinal, flushing Flushing is usually carried out by automatic cisterns. The capacity of the flush should not exceed 4.5 litres per hour for each wall urinal and 13.5 litres per hour per 700 mm of stall urinal. Water conservation is achieved by fitting each urinal or range of urinals with a time switch, Cisternmiser or an electronic device which activates the flushing mechanism when the urinal is being used.

urinal, slab-type See **slab urinal**.

urinal, stall-type See **stall urinal**.

urinal, trough-type See **bowl and trough urinal**.

U-value calculation The coefficient U (thermal transmittance) of a composite building structure or element is expressed as:

$U = 1/(R_{si} + R_{so} + R_{as} + R_1 + R_2 + R_3)$
R_{si} = internal surface resistance
R_{so} = external surface resisitance
R_{as} = resistance of air space or cavity forming part of the building element
R_1 = resistance of first layer of material
R_2 = resistance of second layer of material
R_3 = resistance of third layer of material (if present)

Example: Calculate the U-value of a wall comprising:
outer face – unrendered
brickwork – clay @ 1700 kg/m^3 – 100 mm thick
cavity (sealed – air space) – 50 mm wide
internal face formed in lightweight block – 500 kg/m^3 – 100 mm thick
plaster finish on internal wall – 10 mm

R_{si} = 0.123 m^2 °C/W
R_{so} = 0.550
R_{as} = 0.180
External wall leaf (R_1) = 0.10/0.84 = 0.119
Internal wall leaf (R_2) = 0.10/0.19 = 0.526
Plaster finish (R_3) = 0.01/0.19 = 0.053
U-value = 1/1.056 = 0.95 W/m^2 °C

(Tabulated information on the above parameters and coefficients is available from various publications, including those of the Chartered Institution of Building Services, Building Research Establishment (UK), as well as from numerous technical books.)

U-value definition A measure of the ability of the different building elements (walls, floors, roofs, windows) to conduct heat out of the building (thermal transmittance). The greater the U-value, the greater the heat loss through the building element. The total heat loss through the building fabric is the summation of all the elemental U-values and the related externally exposed areas, multiplied by the difference in temperature between the internal and external environment.

The U-value of a structure (or building element) varies to some extent about a mean value from one situation to another. It is affected by moisture content, wind speed, internal conditions, etc. The standard U-values are calculated from the resistances of the building components based, in turn, on standard assumptions concerning moisture contents of materials, rates of heat transfer to surfaces by convection and radiation and airflow rates in ventilated air spaces. Some allowance is also made for certain heat bridging through the structure. The standard assumptions are essentially based on practical conditions, although these cannot apply in all circumstances of building locations. Measured U-values are not accepted as standard, because the conditions of the measurement only seldom agree precisely with the standard assumptions.

UV irradiation A method of disinfection or removal of bacterial growth using ultraviolet lamps.

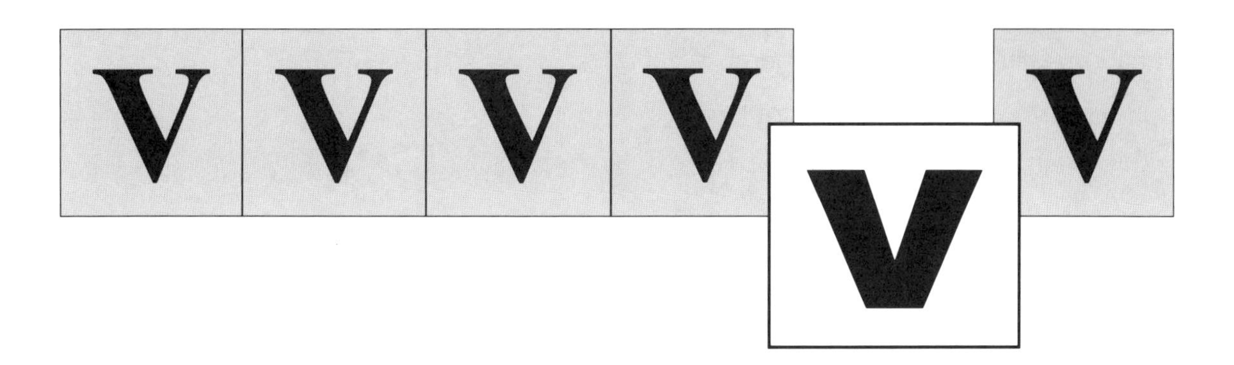

VDU Visual display unit or console; i.e. the terminal device of a computer on which text can be displayed and usually associated with a keyboard.

VAV Variable air volume. See **VAV fan-powered terminal, VAV system** and **VAV throttling-type terminal**.

VRI Vulcanized rubber insulated cable. See also **vulcanized rubber**. VRI has a limited working life due to hardening and eventual brittleness of the insulation. It has largely been superseded for current insulation purposes by other materials, such as PVC.

VRV Variable refrigerant volume system of air-conditioning which utilizes the latent heat of a refrigerant as a heat transfer medium. It is twenty times more efficient than air and ten times more efficient than water. Based on a modular zone-controllable arrangement, the standard VRV system comprises a 5 hp outdoor condensing unit which is connected via small-bore refrigerant piping to any combination of indoor cooling or heat pump units with a total capacity equal to the outdoor unit. Heating and/or cooling can be effected on a floor-by-floor or zone basis using either individual or central control.

vacuum breaker Essentially an automatic air vent operating in reverse in that it admits air to break a vacuum when the pressure external to the vessel exceeds that inside. When the pressure in a vessel drops below atmosphere (1 bar) a partial vacuum is created and this might lead to dangerous conditions, such as implosion of a copper hot water tank or cylinder. A vacuum breaker is fitted to protect the vessel in such circumstances.

vacuum pump and condensate receiver A packaged assembly of pump and electric motor(s) for the collection and delivery of condensate in a vacuum return line steam heating system.

vacuum relief valve A valve that prevents a vacuum falling below a safe level.

vacuum return line steam system [Fig. 247] A system developed as a sophistication of vapour steam heating to adapt it to the heating of larger buildings. Whilst the return pipes of a vapour system convey the air and the condensate by gravity flow, the vacuum system relies on a suction pump to maintain a subatmospheric pressure in the return piping to create a positive flow of air and condensate. Rapid filling of the radiators with steam is assured because of the greater pressure differential across the system.

Another advantage offered by this system is that radiators may be located below the level of the return main or below the boiler. Some degree of

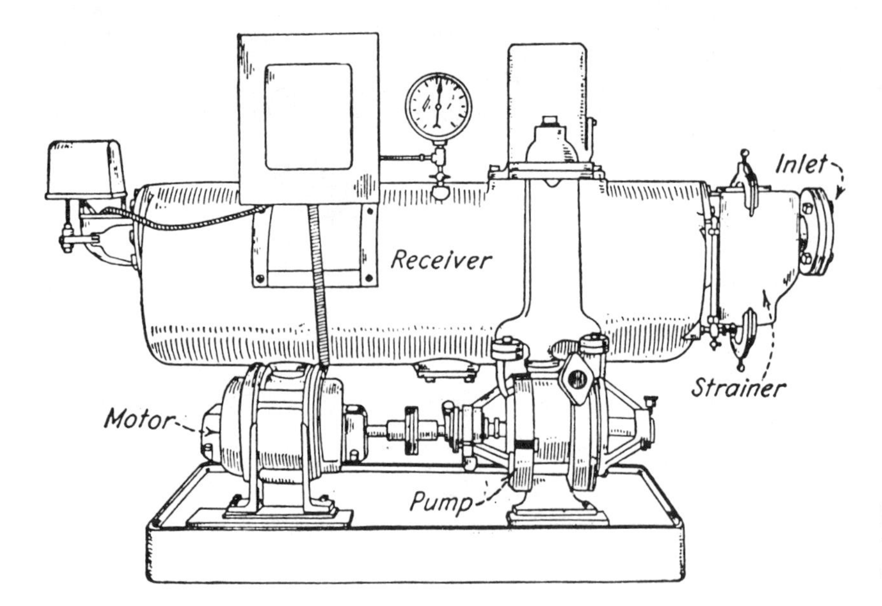

[Fig. 247] Motor-driven vacuum pump and condensate receiver

temperature control can be achieved, as the steam temperature, and hence the heat emission from the radiators, can be regulated by reducing the steam supply and operating the vacuum pump. When exhaust steam is used for space heating, a vacuum return system permits a lower back pressure on the engines or turbines and thereby improves the overall economy of the energy system. Operational difficulties arise in the maintenance of the required vacuum, as this will be impaired by leaks within the system and at defective steam traps.

vacuum sanitary pipework, general Stacks are designed to ensure that the velocity of flow does not exceed 10 m/s during peak loading and the pipe sizes for larger installations are calculated on the quantities of air required for the removal of foul water. Branch pipes from sinks, wash-basins, etc., can be 32 mm or 35 mm; the floor gully outlet is 40 mm. The related floor gully incorporates an odour seal in the form of a diaphragm pressure valve in lieu of the conventional trap with a water seal. Pressure valves similar to those in the floor gully are fitted to other appliances served by the system.

vacuum sanitary pipework system This system is based on the principle of using air instead of water to convey foul water from water closets. Vacuum flushing of WC contents requires only one litre of water and pipelines need not be laid to the demanding gradients required for gravity water-borne installations.

value engineering An organized approach to providing the necessary functions of a project at the required quality and at minimum cost.

valve, air See **air valve**.

valve, angle See **angle valve**.

valve, anti-vacuum See **anti-vacuum valve**.

valve, balancing See **balancing valve system**.

valve, ball See **ball valve**.

valve bounce Water tap washer oscillation caused by a defective washer and unsuitable washer material. It can be cured by changing to a washer of a suitable material.

valve, butterfly See **butterfly valve**.

valve, conduit See **conduit valve**.

valve, diaphragm-operated See **diaphragm-operated valve**.

valve, double regulating A valve that incorporates a facility for locking the required setting whilst also acting as an isolating valve without disturbing the valve setting.

valve, extended spindle A rod extension fitted to a valve stem or wheel to permit convenient remote access.

valve, fail-safe See **fail-safe valve**.

valve, fire hydrant See **fire hydrant valve**.

valve, flush-fitting tank See **flush-fitting tank valve**.

valve flutter Uncontrolled oscillation leading to unstable conditions.

valve, free discharge See **free discharge valve**.

valve, gate See **gate valve**.

valve, globe See **globe valve**.

valve, hollow jet See **hollow jet valve**.

valve, hydraulic See **hydraulic valve**.

valve, magnetic See **magnetic valve**.

valve, mixing See **mixing valve**.

valve, multi-port See **multi-port valve**.

valve, needle See **needle valve**.

valve, oblique See **oblique valve**.

valve, packless gland flow control See **packless gland flow control valve**.

valve, pinch See **pinch valve**.

valve pit [Fig. 248] A brick-built or concrete chamber which gives access to control or isolating valves in an underground pipeline.

valve, plate See **plate valve**.

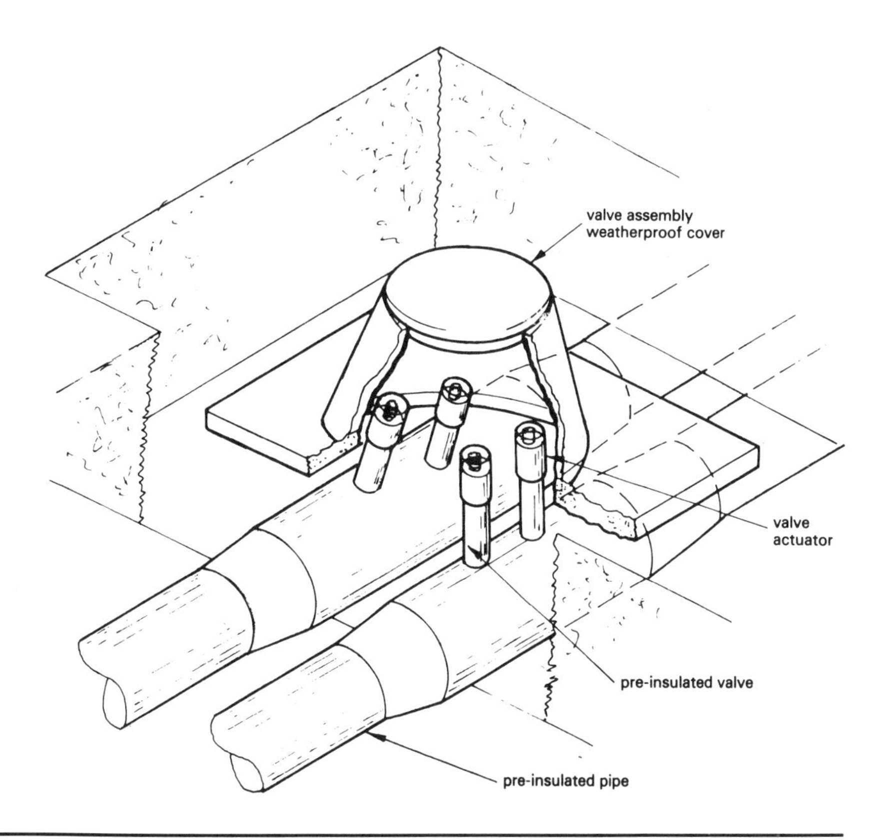

[Fig. 248] Pre-insulated pipe system: valve pit

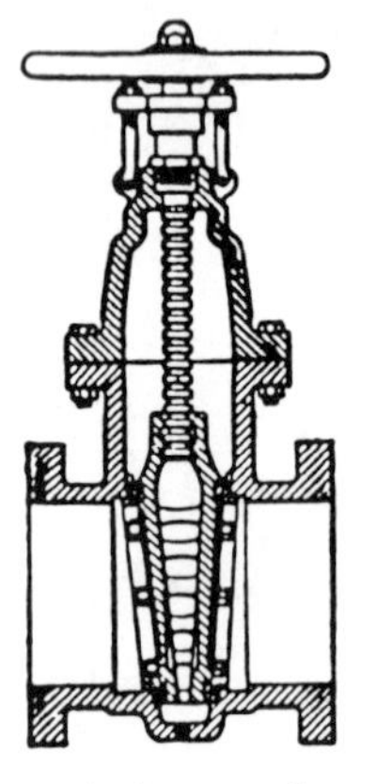
Iron-body gate valve, non-rising stem

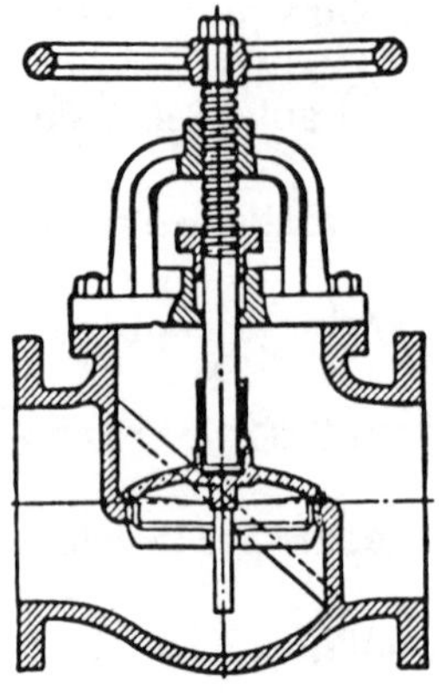
Iron-body globe valve, rising stem

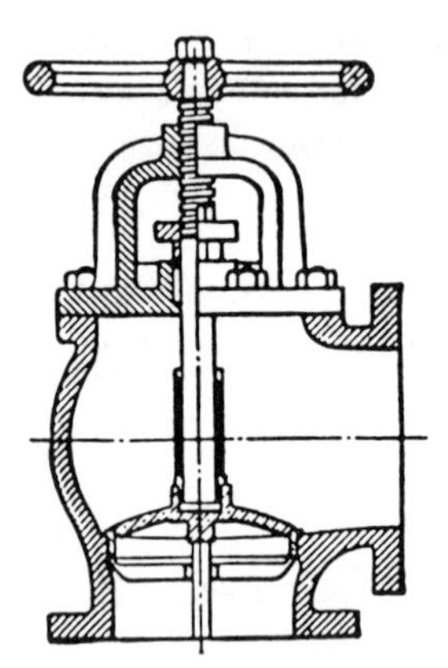
Angle valve

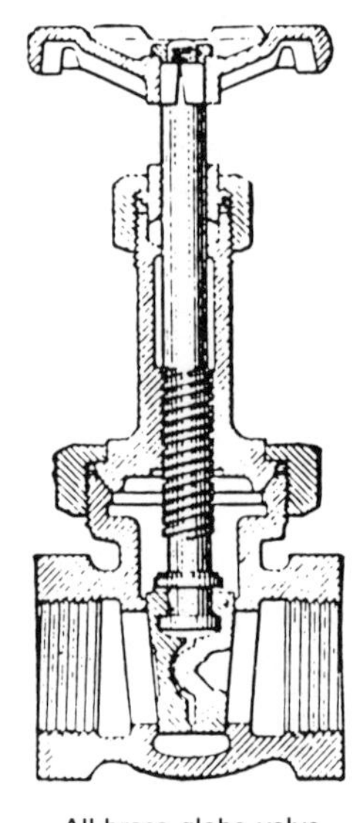
All-brass globe valve

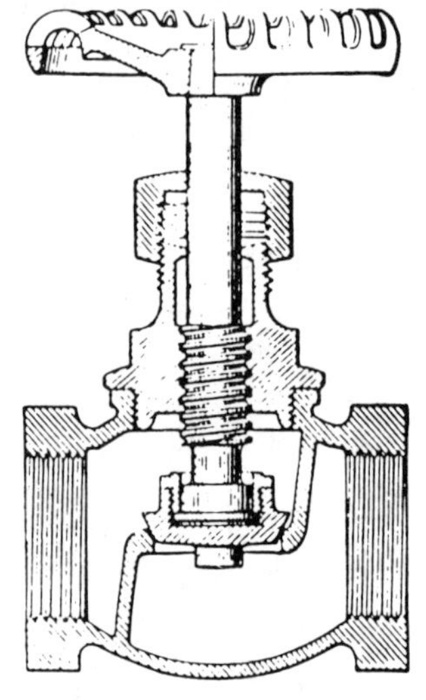
All-brass globe valve

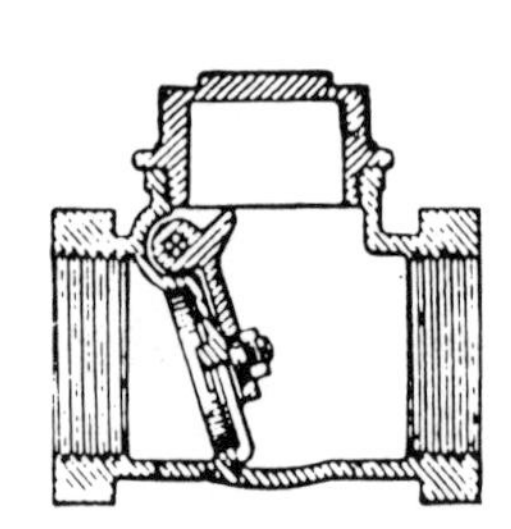
Swing cheek valve

[Fig. 249] Typical brass valves

valve, plug See **plug valve.**

valve, relief See **relief valve.**

valve, rotary See **rotary valve.**

valve, rotary slide See **rotary slide valve.**

valve, safety See **safety valve.**

valve, screw-down stop [Fig. 249] See **screw-down stop valve.**

valve, sleeve See **sleeve valve.**

valve, sluice See **sluice valve.**

valve, spectacle eye See **spectacle eye valve.**

valve, steam jacketed See **steam jacketed valve.**

valve, swing check See **swing check valve.**

valve, thermostatic mixing See **thermostatic mixing valve.**

valve, throttle See **throttle valve.**

valve, vacuum relief See **vacuum relief valve.**

valve, Y-type A valve that has its spindle inclined at an angle into the direction of flow. It offers the minimum resistance to liquid flow, consistent with good stop-valve design. It is available as a simple or double-regulating valve pattern and also with drain/vent cocks or with blank plugs. It is also available with a backseating machined feature on the valve stem which enables the valve to be packed under pressure when fully open.

vane anemometer A small, compact and portable air measurement instrument. It is essentially a windmill which is free to rotate inside a metal guard and a centrally located dial which is graduated in metres per second, over which moves a pointer gear-driven by the windmill. It may incorporate a stop-and-start and set-to-zero mechanism.

Application: for measurement of air flow at air registers and other locations where the air velocity is relatively low.

vane control The adjustment of the output of a fan by an arrangement of vanes at the suction port.

vapour A substance in the gaseous state which may be liquefied by increasing pressure and reducing temperature.

vapour barrier A continuous layer of material that is impervious to moisture. The most effective materials are polythene (heavy gauge), metal foil and metal sheet. A seal is placed across the warm side of the insulation (the inner face will be the cold face). The inner wall coating should incorporate some porosity to permit small amounts of moisture to pass through small defects in the insulation rather than be trapped and cause a breakdown of the insulation properties.

vapour steam heating system A two-pipe system in which the steam circulates at very low pressure – just above atmospheric. The steam enters the radiators, flows along the top part and then downwards into the various sections, driving the air through the return pipe. The supply of steam (and hence the heat emission) can be regulated by only partially closing the inlet valve. The condensate passes from the radiator via a steam trap and gravitates back to the boiler plant.

The temperature of the top surface of the radiator which is filled with steam is close to that of the steam, whilst the remainder of the radiator is warmed by the condensation which trickles down the inside surfaces, so that the temperature of the discharged condensate is thus materially lowered. The advantages claimed for this system are: the improved method of air removal which eliminates noise, smell and drips; greater facility for output control; and more even heating and greater economy of operation.

vaporization The process of converting a substance from the liquid state to a gaseous state by the addition of the latent heat of vaporization at constant pressure and without a change in temperature.

vaporizing burner An oil burner, also known as a pot burner, used in conjunction with domestic heating and hot water boilers. The oil in the pot is vaporized by heat and the vapour mixes with the correct amount of air to achieve combustion. It is designed to operate with the lightest grade of fuel oil. The burner may rely solely on the chimney draught for its operation or be fan-assisted, a fan supplying the combustion air.

variable refrigerant volume air-conditioning See **VRV**.

variable speed drive A variable frequency inverter for driving standard induction motors at variable speed. Such an operation of fan, pump and compressor systems improves efficiency and control instead of damping or throttling. For example, a 100 hp pump running at 80% throughput will absorb approximately 85 hp if throttled back or approximately 50 hp if run at variable speed. Two years is common payback period but it may be less for the larger systems.

variation order A formal instruction to the contractor to vary some aspect of the specified works.

vat and tank, surface insulation Hollow plastic balls which float and cover surface tanks, baths, etc., to minimize fumes or retain heat.

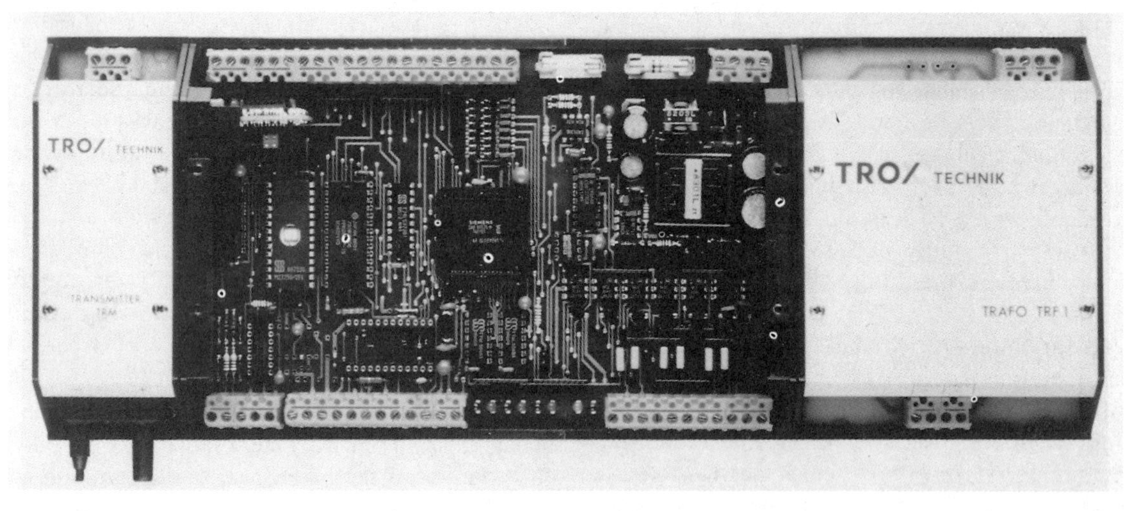

[Fig. 250] Varycontrol controller for use with single- or dual-duct terminal units (Trox Brothers Ltd.)

[Fig. 251] Series fan VAV terminal (Trox Brothers Ltd.)

VAV fan-powered terminal [Figs 250, 251] This replaces the conventional VAV room terminal with one which incorporates a controllable fan. By boosting the air flow this provides an improved alternative to conventional throttling terminals. It mixes the cooler primary air with recirculated room air, thereby providing a near-constant volume of supply air to the room with variable mixed air temperature to satisfy the cooling load.

VAV system Variable air volume system of air-conditioning in which a single main duct is run from the central air handling unit and the quantity of air supplied to each branch duct is varied. A room

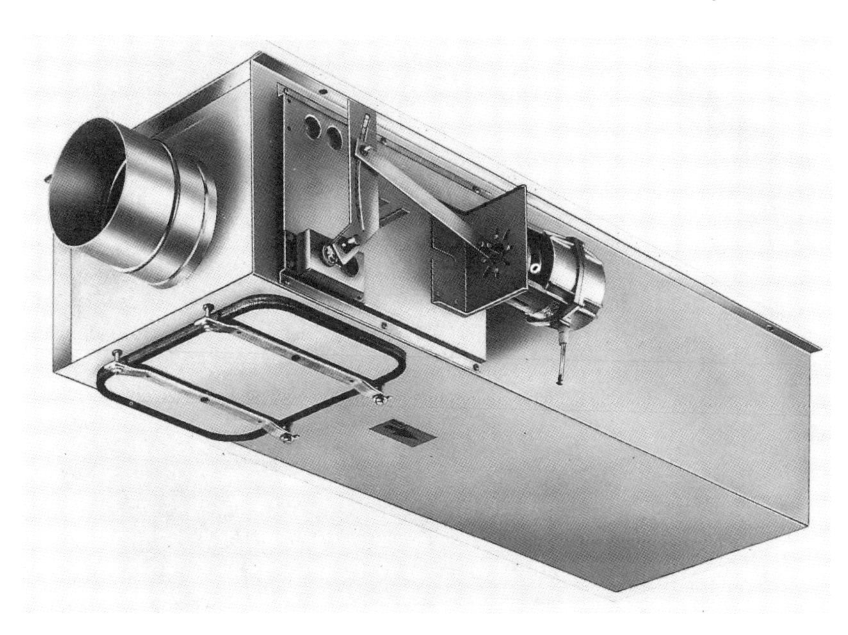

[Fig. 252] Single-duct variable air volume terminal box (Trox Brothers Ltd.)

thermostat operates a damper or other throttling device that regulates the rate of air flow to a specific zone or room in response to load, but usually subject to a minimum air supply setting. Thus, when a room is subject to full load (occupants and equipment) and needs its full ration of treated air, the throttling device will open to admit maximum air. At lower load conditions the throttling effect reduces the air supply down to the minimum level. Various types of VAV terminals are available; see **VAV fan-powered terminal** and **VAV throttling-type terminal**.

Figure 253 illustrates a Varofoil axial flow fan with pneumatic pressure control which permits variation of the blade pitch in motion. Suitable for VAV systems; functions to trim the power taken by the fan to the system load. Offers a most favourable power saving curve.

VAV throttling-type terminal This terminal relies for its performance on the progressive throttling of the supply air quantity as the cooling load reduces. When throttling to turn down below 40% of design air flow, the koanda effect is lost and the air will tend to dump into the space leading to poor distribution.

vee belt A driving belt manufactured in a vee section. It provides a greater friction surface and grip than a flat belt.

vee belt drive A connection between a driving pulley and a driven pulley. These are specifically manufactured to accept vee belts. The complete drive arrangement generally comprises a number of vee belts engaging with corresponding grooves in the pulleys.

Vee-mat-type air filter An air filter in which the filtration medium is placed across the frame in a vee-type formation to offer to the air flow a greater filtration area. It can be used in a roll-type air filter. One type of medium is acrylic fibre mat. The minimum particle size handled is 5 microns at a face velocity of 2 m/s and efficiency of up to 95%, depending on the medium and assembly construction. It is also suitable as a pre-filter. The medium can be self-extinguishing and fire retardant, non-clogging and with constant resistance to air flow.

velocity The rate of motion in a given direction, measured as distance traversed per unit of time.

velocity contour This is established by measurement of the velocity of the pattern of air flow at different locations in the vicinity of air supply or exhaust fittings. Knowledge of the velocity contour permits the selection to be made of the correct terminal fittings for a particular application. It is of particular importance in the design of exhaust

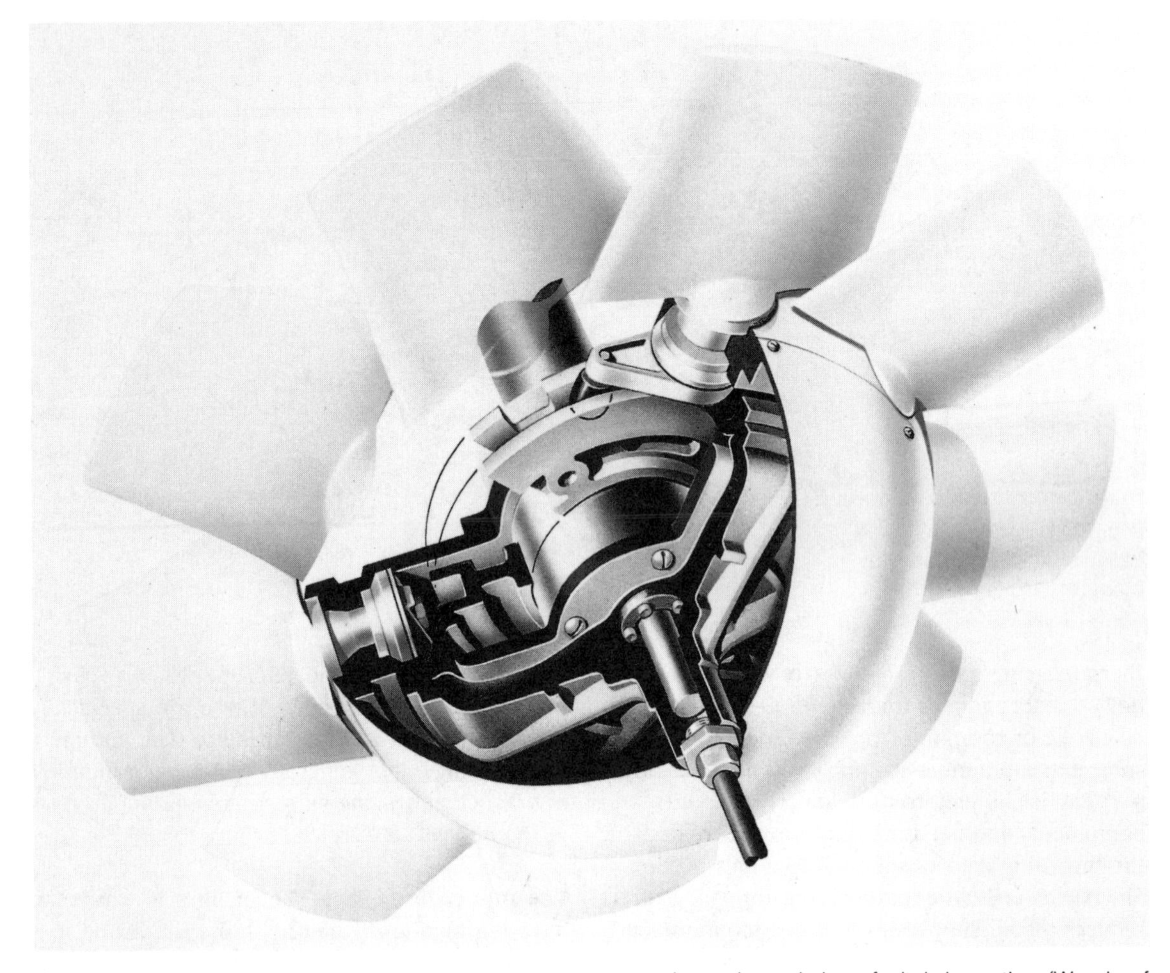

[Fig. 253] Varofoil axial flow fan: the pneumatic pressure control permits variation of pitch in motion (Woods of Colchester Ltd.)

canopies for dust and fume extraction in order to achieve the optimum capture of particles.

velocity head A measure of the kinetic energy of a moving fluid. It is the pressure caused by and related to the velocity of flow of a fluid.

velocity reduction method of duct sizing In this method of sizing ductwork, the initial step is to select a suitable velocity for the first section of the ducting after the fan, this velocity being determined by either the limitation of noise generation or by the permissible friction loss. The velocity for this first section of ductwork will determine the remaining velocities throughout the ducted system. The procedure then followed is for the air velocity in the main ductwork to be decreased after each off-take of a branch duct by 30–100 m/min, depending on the air volumes being conveyed. The velocities in the branch duct should be about 130–160 m/min less than in the main duct to which the branch is connected. The difference between the velocities in the branch and main ducts decreases as the index duct is approached.

Having decided on the rate of velocity reduction, the velocity and air volume being conveyed are marked on a diagram of the ducting installation against each section duct. The required cross-sectional area of the duct is established by dividing the air volume in the section concerned by its design velocity. This area is then checked against a duct sizing table and the equivalent circular diameter or the square or rectangular dimensions for this duct are determined. Adding the resistance of all ducts making up the index run to that offered by the fittings and main plant items establishes the total resistance which the supply fan must overcome.

venetian blind An adjustable blind with horizontal louvre blades, usually operated by some type of string which permits the manual operation of opening and closing the louvres. It is fitted to the room side of a window to reduce glare but is difficult to clean and maintain.

vent pipe A pipe fitted to a hot water appliance for the escape of air and for the safe discharge of any steam generated due to the malfunctioning of the system controls.

vented hot water system An arrangement of hot water generation and supply which incorporates an open vent pipe.

ventilated ceiling A purpose-made (most commonly) metal ceiling suspended below the soffit of the structural ceiling of the space adapted for the downward displacement of ventilation air. Air outlets may be perforations or adjustable slots located in module configuration. The ceiling incorporates acoustic and thermal insulation. For the best effect, it is integrated with flush moulded light fittings.

Applications: clean rooms, computer suites, etc.

ventilated stack system See **sanitary plumbing, ventilated stack system**.

ventilated system See **sanitary plumbing, ventilated system**.

ventilating pipe, sanitary A pipe provided in a sanitary installation to limit the pressure fluctuations within the discharge pipe systems.

ventilation The supply and extraction of air to and from an environment. It may include provision for humidification, heating, infiltration and air distribution.

ventilator, trickle See **trickle ventilator**.

Versatemp multi-room total air-conditioning system [Fig. 254] A proprietary system (by Temperature Ltd.), which operates on the reverse cycle heat pump principle. Each room or zone is provided with its own Versatemp unit which is connected to a water circuit maintained at room temperature. The unit is controlled to draw heat from the circuit (for space heating) and to reject heat into the water circuit (when cooling). At times when the total demand for heat exceeds that available in the circulating water, a top-up heat source (e.g. a small boiler or heat pump) makes up the difference. Conversely, any excess heat is dissipated by means of a heat rejector (e.g. a closed-circuit cooling tower, dry cooler or heat pump).

The Versatemp range of individual units encompasses wall, floor, overhead or underfloor mountings as appropriate to the application. Each individual unit incorporates a refrigeration circuit with a hermetic compressor, finned refrigerant-to-air heat exchanger, reversing valve, protection devices, a multi-speed centrifugal fan which circulates the room air through the Versatemp, and a withdrawable air filter. A two-pipe water circuit connects the units with the heat source and with the heat rejector, forming a closed circuit which usually operates at a temperature of 27°C. This system provides energy reclaim, as for a large proportion of the operating time neither the heat source nor the heat rejector are activated because of the thermal balance of the overall system. For example, where a building faces north-to-south there will be periods, particularly during spring and autumn, when surplus heat is available for transfer from the Versatemps on the south face to those calling for heating on the north face.

In applications where a number of Versatemp units are connected to one circuit, it is good practice to incorporate random start facilities so that all the units will not draw maximum starting current simultaneously when switched on by a time switch, or similar device. The system generally offers only recirculation of room air, though kits are also available for the direct introduction of outside air through the unit; this has obvious limitations in dusty and noisy areas. It is therefore necessary to install a back-up fresh air ventilation system which will introduce heated, filtered and humidified air and extract the exhaust air. Commonly, the back-up system is sized to handle two air changes per hour.

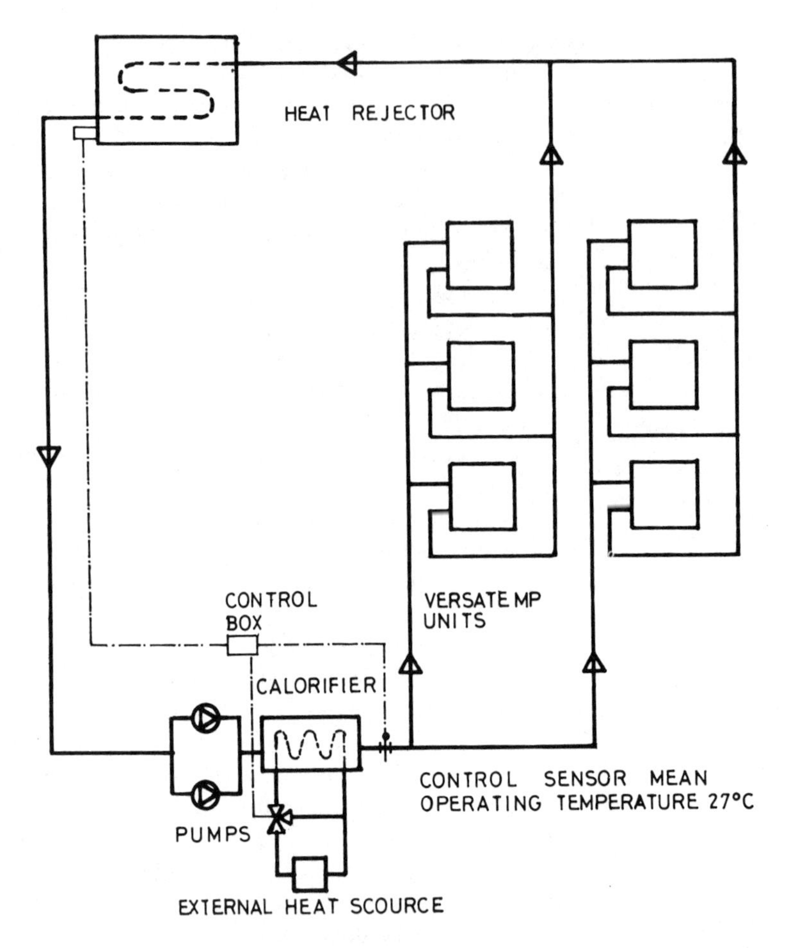

SCHEMATIC ILLUSTRATION OF THE VERSATEMP SYSTEM

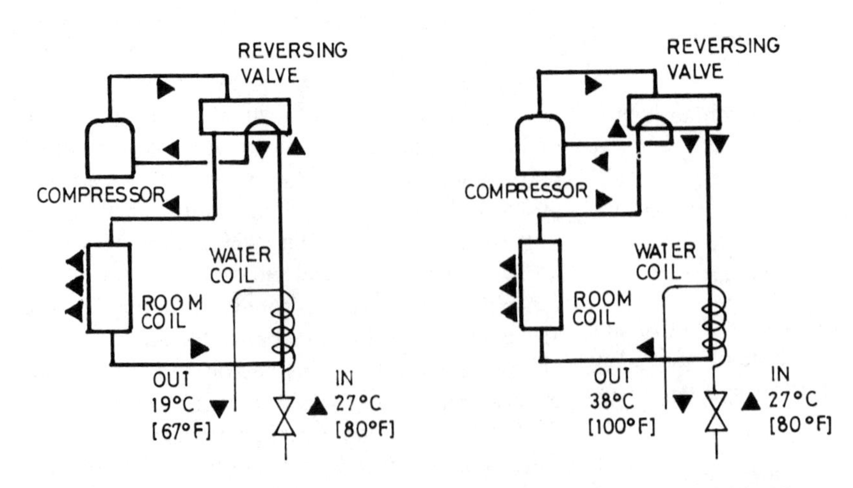

[Fig. 254] Versatemp air-conditioning system

The Versatemp air-conditioning offers important advantages over conventional air-conditioning alternatives: circulating pipes are not thermally insulated or vapour-sealed; there is no central chiller and associated plant area; heat is reclaimed; and there is a major reduction in the associated builders' work.

vertical check valve A valve used in applications where a non-return valve is required to be installed in a vertical pipe, allowing flow in an upwards direction only.

vertical compressor A reciprocating compressor which operates with a vertical piston movement.

venturi meter [Fig. 255] A short length of pipe which tapers to a narrow throat in the middle for measuring rate of flow of a fluid. Detector tubes are connected at the throat and at the ends and these measure the pressure of the fluid at these points. As the liquid traverses the throat, its velocity must increase due to the reduced area; consequentially, the pressure will also be reduced. Applying Bernoulli's theorem of measurement to the readings taken at the enlarged end of the throat, the quantity of liquid flowing through the meter can be measured and calculated. Various types of meters for the measurement of fluid flow are based on the venturi principle. Since inevitably there will be a loss of head (pressure) in the meter between the enlarged end and the throat, significant errors can be introduced at low velocities of flow. Hence, this type of meter is unsuited to the accurate measurement of fluids flowing at low velocity.

vermiculite The geological name given to a group of hydrated laminar minerals which are aluminum magnesium silicates and resemble mica in appearance. The sintering temperature of vermiculite is about 1260°C (2300°F) and its melting-point is 1315°C (2400°F). The material is found in many parts of the world, one of the main supply regions being the Palabora deposits in the north-east Transvaal where it is mined by open-cast methods. Rock and other impurities are removed and the crude ore is crushed and graded.

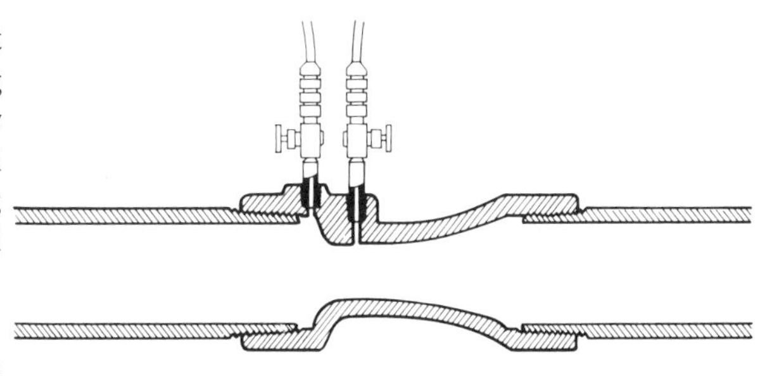

[Fig. 255] Barco-Aeroquip venturi system for measuring water through its velocities

Vermiculite is also a thermal insulation material used in granular form for loft insulation. It can be embedded within floor and roof screeds.

vibration frequency Vibrations are caused by pressure waves which may have many causes, such as moving machinery, e.g. pumps, fans, compressors, boosters or wind effects on structures. The vibration frequency is measured in hertz. The frequencies of vibration encountered in buildings lie mostly within the range of 5–50 Hz and when these exceed about 20–30 Hz, they pass into the audible range. If sufficient vibrational energy is present, the excitation will be perceived by the human ear as sound.

vibrating grate A type of grate that supports a solid fuel fire bed. It is vibrated by an electric motor to free ash and riddlings from the burning fuel.

vibration isolation Any of several means of reducing the transfer of vibrational force from the mounted equipment to the supporting structure, or vice versa.

vibration isolation efficiency The amount of the vibration force absorbed by an isolator and thereby prevented from entering the supporting structure. It is expressed as a percentage of the total force of vibration applied to the isolator.

Victaulic grooved joint, pipe end preparation The formation of circumferential grooves close to the pipe ends. The housing segments are secured in these grooves and thereby securely lock the pipe ends together, whilst still permitting some flexibility, thermal movement and deflection. Grooves may be formed by one of two methods, roll grooving or cut grooving, the method used depending on the particular application. Light- and medium-weight pipes are normally roll grooved, heavy-weight pipes are cut grooved. The manufacturers of the coupling provide equipment for the grooving operation. The depth of material removed in the grooving process corresponds to the depth of one pipe thread and does not significantly reduce the strength of the pipe.

Victaulic grooved pipe jointing system A system where the joint is constructed by grooving the ends of the pipes to be jointed (using special techniques) and stretching an elastometric sealing ring across the grooves, then placing two rolled steel segments of the housing over the sealing rings and securing the two segments with heat-treated bolts designed to permit tightening of the nuts from one side using a single spanner. This is suitable for working pressures of up to 34 bar. See **Victaulic joint**.

Victaulic joint [Fig. 256] A widely-used proprietary coupling for the jointing and subsequent easy dismantling of sections of pipes. Its use is approved by the major relevant authorities for a wide variety of installations, including fire protection sprinkler systems, mechanical services, chilled water and condensate water, air-conditioning systems, fire hydrants, compressed air, etc.

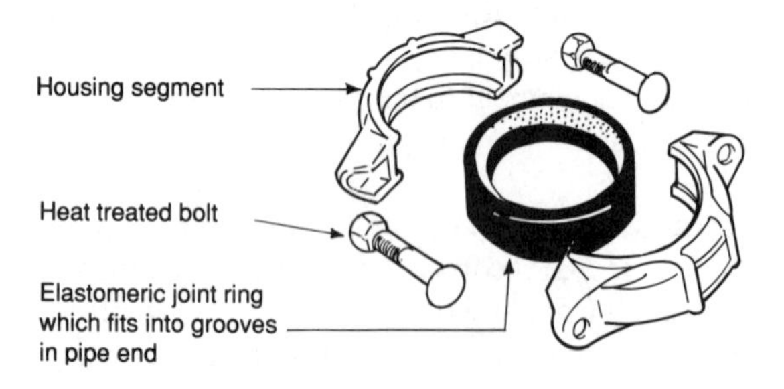

[Fig. 256] Components of Victaulic grooved pipe joint

Victaulic joint, sealing rings [Fig. 257] The standard ring is Grade E made of ethylene propylene, identified by a green colour flash. It is suitable for most building services and process fluid, but not for petroleum products, oil, compressed air or hydrocarbons. Other non-standard sealing ring materials are Nitriline, natural rubber, polyacrylic and others, depending on the fluid being conveyed and the required temperature range. Each specification is identified by a specific colour flash.

An increase in line pressure automatically strengthens the seal by acting internally on the lips of the sealing ring; the higher the pressure (within the permitted range) the tighter the seal. The sealing rings must be lubricated with a vegetable-based lubricant prior to assembly over the pipe ends. The absence of lubrication risks the rings being pinched and thereby causing the joint to leak. The manufacturers market a suitable lubricant.

Victaulic jointed pipe anchorage The Victaulic joint is designed to prevent pipe separation due to forces acting in an axial direction only. Where the pipes follow a tortuous route, particularly under high pressure, such a system only partially or improperly restrained, could impose unacceptable bending moments on the grooves and housings, leading to a possibility of pipe separation (the pipe pulling out of its coupling). Restraints to such systems must take into account the forces due to thermal movement and the pressure and velocity of the medium being conveyed. It is essential that such joints are not subjected to non-axial forces as would arise with inadequate pipe supports and anchorages, particularly where the jointed section of pipe is close to a bend and screwed joint(s) which would permit torsional movement of the pipe. The coupling should be anchored to a substantial support which is attached to, or built into, a solid wall, floor or ceiling and which holds the coupling firmly in line.

Viking Johnson slip coupling Probably the simplest form of pipe joint used for plain-ended pipes as special preparation of pipe ends (no grooving is required) is entirely eliminated and the laying of pipe mains is reduced to a simple and speedy operation.

Viking Johnson slip coupling, anchorage Pipelines under pressure are subject to forces which tend to separate the component parts of the coupling. Flexible pipelines, such as those jointed with Viking Johnson slip couplings, are particularly

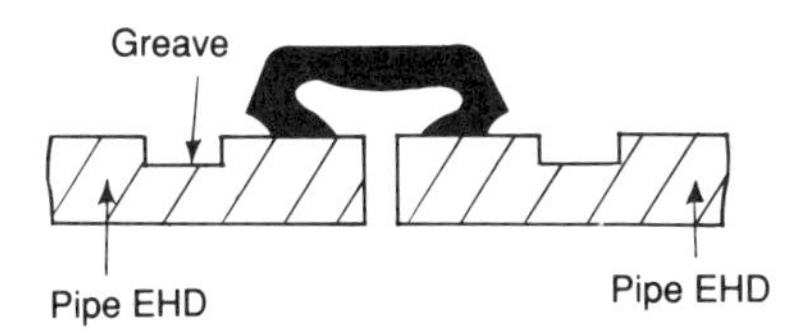

1. Stretching the Victaulic ring over pipe ends puts the angular lips in immediate and automatic sealing tension

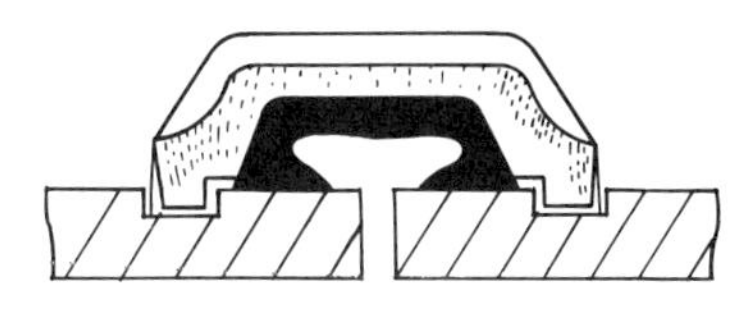

2. Application of the Victaulic half housings automatically centralizes the Victaulic ring and seats it on the pipe ends firmly

3. Line pressure automatically strengthens the seal by acting internally on the ring lips. The higher the pressure the tighter the seal

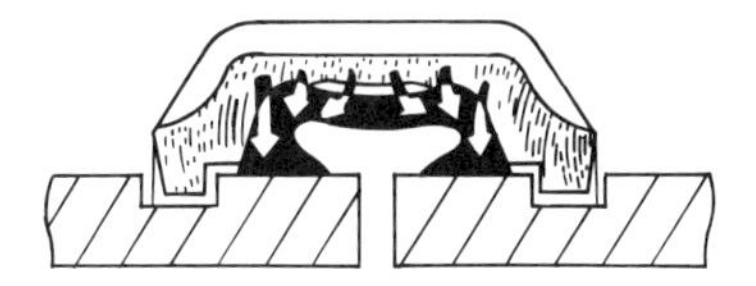

4. Suction or vacuum also automatically strengthens the seal. The higher outside atmospheric pressure acts as a compressant on the walls of the Victaulic ring

[Fig. 257] Victaulic grooved pipe joint: action of sealing rings

subject to forces acting on bends, tee junctions, blank ends, etc. With suspended pipelines it is essential to take full account of the thrusts produced by the internal pressure and to restrain them with thrust blocks, anchorage or tie bars. Buried pipelines can generally be restrained by means of anchor blocks at abrupt changes of direction, the minor forces at each straight joint being restrained by the soil friction.

One common method of preventing pipe separation in steel pipes is the use of a harness assembly, consisting of lugs for welding to steel pipes and connecting tie rods tightened up between the lugs. Another method is to weld brackets to the centre sleeve of the coupling and these are then bolted directly to the supporting structure. Cast-iron pipes require special attention; they can sometimes be harnessed through the use of additional flanges cast on the pipe ends. The use of locating plugs is recommended with anchored slip couplings to assist in resisting pipe movement. The manufacturers of these couplings provide detailed installation instructions and it is essential that these are followed to avoid the risk of pipe separation with the potential risk of possibly major consequent damage because of leakages.

Viking Johnson slip coupling, construction The joint comprises a centre sleeve, two wedge-section packing rings and two end flanges which hold the packing rings in contact with the sleeves and pipe ends. The sleeve normally incorporates a centre register which acts as a locating stop. Couplings can be supplied without the register if so required. The length of the centre sleeve permits normal longitudinal movement of the pipes without causing leakage. The joint provides flexibility and lengthening of the pipeline laid in the ground where subsidence is likely to occur. Standard couplings are suitable for hydraulic test pressures of up to 57 bar; heavier quality couplings are supplied for much higher pressures.

viscosity The resistance of a fluid to flow, sometimes called dynamic or absolute viscosity, η. The dynamic viscosity divided by the density of the fluid is called the kinematic viscosity, ν.

viscous The property of a liquid having a high viscosity. Such liquid flows slowly like treacle.

viscous air filter An air filter that incorporates a filtering medium of coarse fibres which are coated with a viscous adhesive. Two commonly used such materials are glass fibres and metal screens. The air filter will remove large dust particles but not small particles. It is low in cost and is often used as a pre-filter. Resistance to air flow increases with progressive accumulation of dirt. Resistance must be monitored and the filter cleaned and recoated when the maximum permitted resistance has been reached.

vitreous Describes a surface finish which is composed of or resembles glass.

vitrification A term used with refractories which relates to the conversion of a material under heat into a glass or glass-like substance with increased hardness and brittleness. This particular characteristic of a refractory brick is of great importance in its development of high strength and hardness.

volt A unit of electric potential. It is defined as the difference of potential between two points along a conducting wire which carries a constant current of one ampere when the power dissipated between these two points is one watt.

voltage The potential difference or electromotive force of a supply of electricity as measured in volts.

voltage drop Wherever there is a flow of electricity, there must be a drop in the voltage of the supply reaching a particular point. This depends on the current carried, the size of cable and on its length. IEE Regulations require that the voltage drop between the supply position in any building and any outlet must not exceed $2\frac{1}{2}\%$ of the nominal voltage.

voltmeter A hand-held or panel mounted instrument for measuring the potential difference between two points.

volumetric run-off coefficient The proportion of rainfall on the paved areas which appears as surface run-off in a storm water drainage system. This varies from 0.6 on catchments with rapidly draining soils to about 0.9 on catchments with heavy soils.

vortex The rotating mass of a fluid. It is relevant to fluid flow in fans which set up such a motion in the air which passes through the impeller. It is the basis of the operation of the cyclone dust separator.

vortex flow Rotational flow, e.g. in a dust separating cyclone or in a cyclone furnace.

vortex, forced See **forced vortex**.

vortex, free See **free vortex**.

VRV air-conditioning Variable refrigerant volume air-conditioning (concept by Daikin Ltd.) which utilizes refrigerant as a heat transfer medium which is conveyed through small-bore pipes. It relies on the superior heat transfer characteristics of refrigerants over air and water. (Under typical conditions the heat transfer capacity of air is 2.4 joules/kg, that of water 5 joules/kg, whilst refrigerant transfers 49 joules/kg.) Additionally, comparison of the energy required to transfer 100 000 kcal/h of heat indicates that air requires 7.4 kW, water 47 kW and refrigerant only 2.5 kW.

The Daikin system permits the connection of up to eight indoor distribution units of varying outputs to one single outdoor condensing unit. An inverter-driven compressor enables the output of the condensing unit to be modulated to match exactly the cooling or heating demands of the zone which it controls and each indoor unit can be individually or centrally regulated to optimize flexibility and minimize energy use. The system can function with refrigerant piping runs of up to 100 m in length, with 50 m vertical distance between indoor and outdoor units and with a 15 m vertical distance between the first and last indoor units on any one circuit. Utilizing high efficiency oil separators, it can be installed without oil traps in buildings of up to 15 storeys high. Vertical pipe runs can be extended over four or five storeys. Indoor distribution units may be built-in or suspended ceiling cassettes, concealed or free-standing wall- or floor-mounted cabinets, whichever is best suited to the particular heating/cooling applications. The system is suitable for retrofits.

vulcanized rubber The substance obtained by heating rubber in the presence of sulphur. It used to be widely employed for insulating electric conductors.

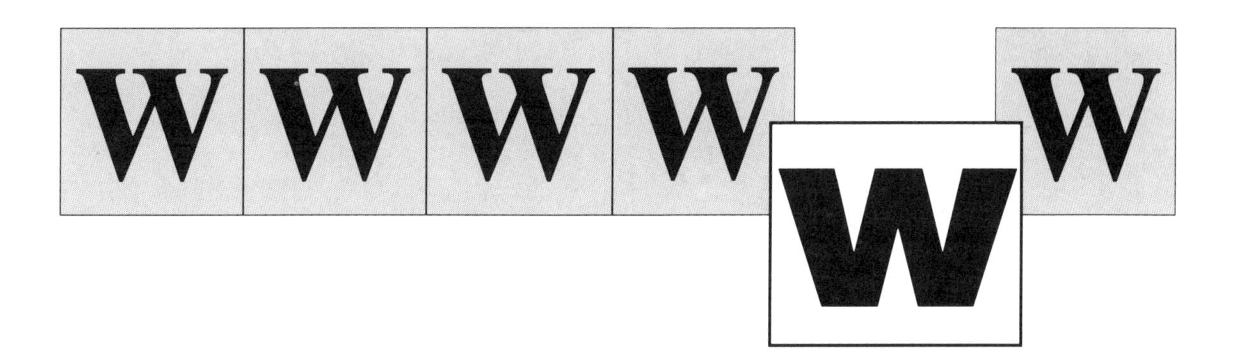

WC Water closet.

WWP Water waste preventor. This is a storage facility attached to a water closet incorporating a ball valve and float and which controls the quantity of water discharged when flushing into the water closet.

walkthrough duct A corridor formed for the routeing of services. Its dimensions permit a person of average height to walk upright through the duct.

warm air curtain A blanket of moving warm air across a doorway or other opening to counteract incoming cold air/draught. It comprises a high-velocity duct terminal at each side (or for narrow openings from top to bottom), designed to fully blanket the opening together with a fan, a heater plant and controls which may include a thermostat and/or a door switch which activates and deactivates the warm air curtain operation. The air velocity within a curtain at public entrances must be restricted to avoid discomfort to people entering.

warm air curtain, photoelectric control A photoelectric beam system which automatically actuates the fan plant when entrance doors are opened or closed. It is particularly useful for warm air curtain control of entrances to warehouses and workshops which are infrequently used.

warm air heater [Fig. 258] The complete assembly within one casing of a fuel burner (oil or gas), a combustion chamber and flue outlet, an air distribution fan, a discharge head with spigots for the connection of air distribution ducts or a header with integral discharge grilles, and a return air inlet and controls.

warning pipe An overflow pipe installed and located so that its outlet, whether inside or outside the building, discharges water where it can readily be seen and be a nuisance.

wash-basin Wash-basins are manufactured in a wide range of designs and configurations ranging from small cloakroom basins for hand rinsing to large double-bowl fittings and ablution troughs for industrial and sports ground use and may incorporate special anti-vandal features. Specialized designs are manufactured for hospital/medical uses. Most basins are made in vitreous china, though heavy-duty and vandal-resistant models are made of fireclay or stainless steel. Other materials used are enamelled pressed steel, acrylic and GRP.

wash-basin, vanity unit Vanity units are often used in public buildings and as special features. They come complete with a self-rimming design which can be set into a counter top. The arrangement provides secure fixing and also permits

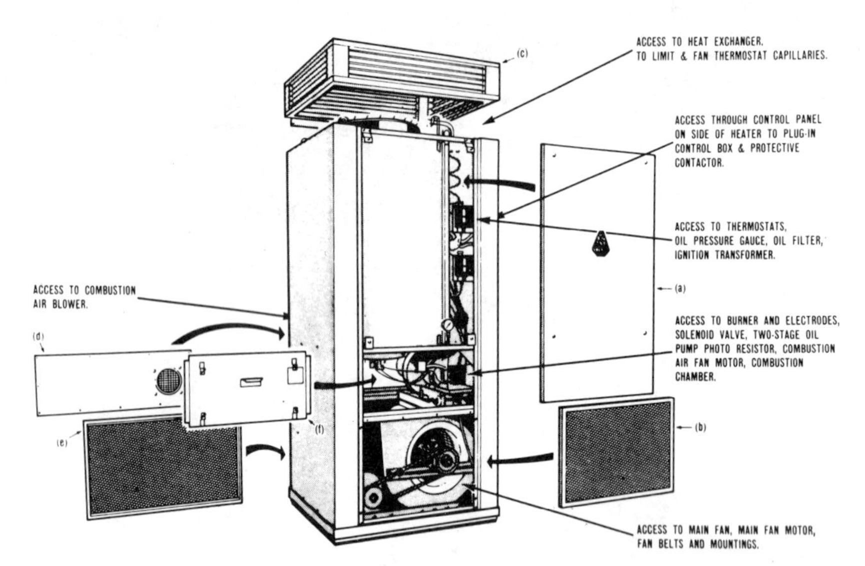

[Fig. 258] Self-contained warm air heater

concealment of the connecting pipework, thereby giving a flush and neat appearance. A flexible waterproof sealant should be applied to the joint between the basin and the counter into which it is set. Vanity tops are also available with integral basins made from plastic materials such as GRP acrylic, etc., in one-piece mouldings.

waste audit The identification of the components in the waste stream of a business with the object of establishing an optimum scheme (or schemes) for handling it and also for reducing its volume.

waste disposal authorities Generally, the County Councils or their authorized agents.

waste heat boiler [Fig. 259] A steam or hot water boiler, usually adjunct to an incinerator, which uses the combustion gases from it. Since the combustion gases issuing from incinerators tend to include a fairly high proportion of particles, waste heat boilers must be equipped with large-bore fire tubes and an easy means of cleaning them.

waste oil burner A burner that can usually burn all grades of oil, including waste oil from manufacturing and vehicle sumps. The burner operates on the principle of emulsification of the oil/air mix.

waste power station A power station that serves the dual purpose of incinerating municipal or large-scale industrial waste and then utilizing the resultant energy to generate electricity.

waste regulation authorities Generally, the County Councils or their authorized agents.

waste ventilating pipe A ventilating pipe which is part of a waste pipe system.

water absorption The property of certain materials to absorb water readily when exposed to a moist atmosphere.

waterbar A channel section (commonly of PVC) placed inside the shuttering of a concrete wall adjacent to a manhole or inspection chamber to waterseal the junction. It is fitted in conjunction with a kicker joint.

water-cistern, byelaw 30 (2) requirement The following are mandatory requirements:

(a) a screened air inlet;
(b) a feed and expansion pipe (vent pipe) grommet;
(c) an access cover;
(d) insulation;
(e) a screened overflow chamber;
(f) a sealed cover.

water chiller [Figs 260, 261] A refrigeration compressor and associated equipment and controls for

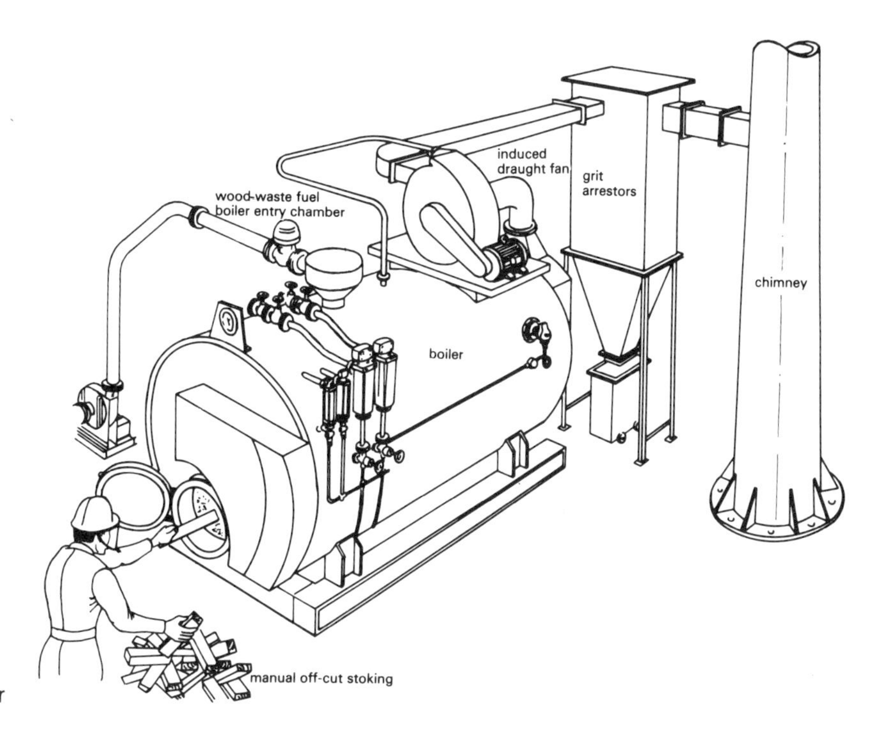

[Fig. 259] Waste heat boiler

[Fig. 260] Water-cooled water chiller of 40 TR capacity

the cooling (chilling) of water for circulation through the cooler batteries of air-conditioning systems.

[Fig. 261] Water-cooled water chiller of 22 TR capacity

water closet configuration, back-to-wall pan
This type of pan has shrouded connections, is neat in appearance and is easily cleaned. It can be used with medium- or high-level cisterns, but preferably with a concealed arrangement.

water closet configuration, close-coupled
Usually of similar material to the pan, the close-coupled cistern can be used with either syphonic or washdown pans, but it is less effective with the latter due to the reduced velocity of the water. It

is bolted to the pan and is commonly screwed to the wall as well for extra strength of support. With some designs the cistern may be concealed within the duct.

water closet configuration, low-level washdown pan and cistern This is the most widely used arrangement as it is simple to maintain and to replace. It can be converted into a side outlet 'P' or 'S' trap with a waste adaptor. The introduction of the horizontal, or convertible outlet has eliminated the need for the basic 'P' or 'S' trap.

water closet configuration, wall-hung pan This is fixed on concealed built-in brackets, partly in the wall and partly in the floor and is generally used with concealed cisterns. Some patterns of brackets have to be cast into the floor screed or be fixed before the floor finish is laid.

water closet, syphonic This type of water closet incorporates either a double or single trap. It is quiet in operation, but there is some risk of blockage if it is misused. It is installed with a close-coupled cistern which is normally exposed.

water closet, vacuum suite This differs in construction from the conventional water closet in that the bowl contains no trap, but incorporates an air-tight valve and flushing apparatus. The branch pipes from WC stacks and main sanitary pipework need only be of 40–50 mm diameter.

water closet, washdown A water closet that is cleared by the volume and force of the flush water. This type is usually in vitreous china, but is also available in fire-clay and stainless steel where greater strength is needed. This is the simplest design and offers the minimum risk of blockage. It can be used with a close-coupled medium- or high-level cistern and has a 50 mm water seal.

water-cooled condenser A condenser that utilizes water as the condensing medium. It is usually of shell and tube construction. Water is drawn from storage systems, rivers, lakes or wells. More commonly, for reasons of economy, the water must be recirculated through a cooling tower.

water cooling tower A means of recovering water for reuse in a process to save the cost of the water and of its water treatment. It relies on evaporative cooling in which the hot water is recirculated and is brought into contact with a stream of fan-propelled air.

water display An arrangement of waterfalls, fountains, artistically designed water sculptures and cascades for display purposes, often combined with lighting effects and computerized sequencing.

water equivalent (heat capacity) The mass of water which requires the same quantity of heat to raise the temperature through one degree as the body requires. For example, the water equivalent of 9 grams of iron (specific heat 0.119) is about 1 gram (specific heat of water taken as 1.0). If M is the mass of the body, s is the specific heat of the material (body), then the water equivalent (or heat capacity) is Ms. See also **heat capacity**.

water flow indicator A device used to monitor the water flow to equipment which is reliant on constant water circulation; e.g. an air compressor and receiver. It may be magnetic or incorporate an impeller. The dial indicates the water flow visually.

water flow restrictor A device fitted within the body of a water draw-off tap which reduces the flow to the required level. It is used in hot and cold water installations in schools, public buildings, institutions, factory washrooms, offices, etc. and will reduce the water flow by 50–75%.

water hammer elimination Water hammer can be avoided by the careful design of systems and, in particular, by ensuring that all steam and condensate pipelines are correctly graded to permit the flow of water in the direction of the steam flow, the elimination of all unvented high points where air might collect, efficient routine maintenance of all steam trap sets and strainers, the fitting of a condensate drain, having dirt pockets and trap sets at low points and in long runs of steam piping and ensuring that all reducing fittings in the system are of the eccentric type. In water systems, water hammer can generally be avoided by the elimination of air pockets and the effective maintenance of ball valves and taps.

water hammer, general Vibration (usually accompanied by loud banging noises) in a water or steam system. It can be caused in a variety of ways, such as the presence of air pockets in pipes and/or equipment, chattering valve parts, a loose ball valve, etc. Water hammer effect can damage pipe installations.

water hammer, in steam systems This is usually caused by contact between live steam and water (mainly condensate) setting up a high-velocity wave motion. Shock waves generally occur at obstructions and changes of direction when a slug of water collides with the metal and is brought to rest with considerable shock. Consequent destructive effect may result in cracked fittings, the fracture or collapse of thermostatic and compensator bellows, etc.

water hardness A term used to measure the extent of water supply scaling impurity commonly caused by calcium and magnesium salts. Derived carbonates and bicarbonates are generally responsible for 'temporary' hardness; chlorides, sulphates and nitrates contribute to 'permanent' hardness of water.

water heater battery [Fig. 262] An assembly of heat exchange pipes within one flanged casing for installation within a ducted air heating system. The pipes are usually finned to provide extended heat transfer surfaces or they may take the form of heat pipes. Isolating/regulating valves are usually fitted at the flow and return pipe entries and an air vent is required when served with pipes from below.

water leakage detector A system of probes placed into spaces vulnerable to water damage (e.g. under suspended computer floors) in conjunction with a monitor which sounds an alarm in the event of leakage being detected.

water meter An appliance for measuring and recording the quantity of water passing through it to the consumer.

water, scale The solid encrustation caused by water impurities sometimes found inside water pipes and on heat exchanger surfaces, also called 'furring'. It can cause water flow restriction and blockage.

[Fig. 262] Water heat exchange coil showing internal connections

water softener, base-exchange (zeolite) process Equipment for the treatment of hard water in which the water containing the hardness forming dissolved calcium and magnesium salts is passed through a bed of insoluble sodium zeolite to exchange the calcium and magnesium ions for those of sodium. The sodium zeolite is thus gradually converted into an insoluble calcium and/or magnesium zeolite, whilst the water now contains sodium ions equivalent to the calcium and magnesium salts which have been removed (the atoms of calcium and magnesium present throughout the water before treatment are termed ions). The name zeolite refers to a group of minerals which are essentially hydrated silicates of aluminum, calcium, sodium, potassium or iron. The term zeolite, as used in water softening practice, is restricted to both the natural and the artificial sodium-alumino silicates which have base-exchange properties.

The active sodium zeolite can be regenerated by washing the treatment bed with a strong salt solution (brine) for some 5–10 minutes, during which period the relatively high concentration of sodium ions in the brine replaces the calcium and magnesium ions in the zeolite and reforms the sodium zeolite. The calcium and magnesium ions pass into the solution and out to waste when the brine is run off; the regenerated bed is then washed free from brine by an upward flow of water before more raw water is admitted into the softener. The full cycle of operations is therefore: softening (i.e. removal of the hardness salts), back-washing, regeneration, rinsing and then again softening. In some plants this cycle is performed manually using the readings of the raw water meter as a guide; in others, this operation can be programmed to take place automatically after a measured quantity of water has been softened. The regenerated zeolite will now function again and remove the calcium and magnesium ions. By alternate partial exhaustion, regeneration with brine and rinsing, the zeolite may be used almost indefinitely, though some fresh zeolite must be added from time to time to compensate for the slight disintegration of the original bed which does occur, particularly if hot water is softened or if the zeolite is allowed to become completely exhausted before regeneration. Such softeners are effective over a wide range of flow rates and they can deal efficiently with waters of widely and rapidly varying hardness. They can produce treated water of near-zero hardness; the average effluent may be assumed to have a residual hardness of 0.5 parts of calcium carbonate per 100 000 parts of water. However, softened water is rarely required at near-zero hardness as very soft water may, under certain conditions, promote corrosion. Thus, for domestic hot water use, the soft water leaving the softener is usually blended with a quantity of untreated raw water and by-passed around the softener to give a final hardness to the consumer of about 7 parts of calcium carbonate per 100 000 parts of water. In order to accommodate the salt required to make up the brine for regeneration a brine vessel and a salt store is required as part of the softener installation. The salt must be stored in dry conditions to avoid deterioration.

water softening The process of treating hard waters to eliminate hardness salts. This is desirable in hard water areas to obviate the formation of scale in hot water pipes and equipment, to avoid the formation of stains in wash basins and baths and the furring up of kettles and cooking utensils. See also **water treatment**.

water storage, baffle plates These are installed within the water storage tank/cistern to prevent waves on the surface of the water affecting the float and causing oscillations and attendant water hammer. Alternatively, they are installed where the surface waves are magnified when the length of the tank is in sympathy with the wave length of the surface water movement and cause oscillations with attendant water hammer. The baffle plates are fixed across the width of the storage tank and will usually eliminate the oscillations.

water storage cistern, actual capacity The capacity of the cistern measured up to the water line.

water storage tank A tank that stores water pending consumption.

water storage tank, access manhole Larger size tanks require to be fitted with a hinged access cover adjacent to the ball valve to permit internal inspection and ball valve repair.

water storage tank, bird flap This is fitted to the termination of larger overflow pipes to prevent birds from entering.

water storage tank, ball valve A float and arm operated sliding valve through which water enters the tank. The valve incorporates a solid rubber washer against which the valve closes and this requires periodic replacement.

water storage tank, insect screen Fitted to termination of overflow pipe to prevent insects from entering.

water storage tank, lining A plastic lining applied to the inside of tank surfaces to reinstate them. These are useful in locations where it is difficult to replace a defective tank.

water storage tank, overflow pipe A pipe connected to the overflow boss on a tank and piped to discharge to the outside in a location where attention is drawn to the overflow and where this constitutes a nuisance.

water storage tank, plastic sectional A tank assembled from standard flanged plastic sections bolted together. It is essential that the tank is correctly supported according to the manufacturer's instructions.

water storage tank, steel sectional A tank assembled from standard size cast-iron or steel flanged plates commonly one metre square and bolted together. It may be assembled with the flanges facing to the outside of the tank or to the inside. The latter arrangement is particularly useful when there is a lack of space for the assembly of the sections.

water storage tank, warning pipe A 25 mm bore pipe fitted below the level of the overflow to give advance warning of the tank overflowing. It applies to large size tanks.

water supply by-laws By-laws issued by water utilities and supervisory bodies to safeguard public water supplies. These by-laws govern all water-using installations. It is necessary to be familiar with those applicable to the area of the works.

water supply, down service See **down service water supply**.

water supply, mains See **mains water supply**.

water supply, potable See **potable water supply**.

water treatment A process for rendering harmless the adverse impurities in water.

water treatment, lime soda A precipitation process in which lime and soda ash (if required) are added to waters containing temporary hardness salts. Non-soluble carbonates are formed which settle out and are removed from the water system.

water treatment, magnetic Apparatus which incorporates permanent magnets between which water flows. Its passage through the magnetic field breaks down the scale crystals which are subsequently discharged in the water outflow. It operates without an external electric supply and prevents heat exchangers, boilers, pipes, etc., from being encrusted with scale. The apparatus does not contribute towards softening hard water. It is used widely with once-through district heating systems.

water-tube boiler [Figs 263, 264] A boiler in which the water being heated circulates in tubes which are located above the combustion chamber of the boiler. The hot (flue) gases pass over the tubes, thereby heating the water.

water vapour Water in a gaseous or vaporous state. It is present in the Earth's atmosphere in varying amounts.

watt A unit of electric power. It is that power which gives rise to the production of energy at the rate of 1 joule per second.

wattage Electric power measured in watts.

wattmeter An instrument for the direct measurement of power in watts flowing in an electric circuit. It can be hand-held or panel mounted.

wavelength The distance between two like points on a wave shape, such as the distance from crest to crest. For example, it may be the distance between the succeeding layers of compression of a sound, depending on the speed of the variations.

way-leave The permission to route pipes, drains, cables, etc. through land not under the control of the contractor.

water heating compensator control A device employed to reduce fuel consumption and to provide correct conditions in space heating systems. A compensator comprises external and pipe located detector thermostats which act through a control box to adjust the heat input to the system. It may be self-acting or electrically motivated. The control may act directly on to the firing equipment (adjusting the rate of firing to the weather conditions) or actuate a mixing valve in the heat supply to the building or zone (essential when boilers provide heating and domestic hot water).

weather factor Used in the theoretical estimation of the heat and fuel consumption of a space heating system, based on its heating capacity, to allow for the periods during which the system operates below peak capacity due to seasonal weather variations. This varies usually in the UK between 0.6 and 0.7.

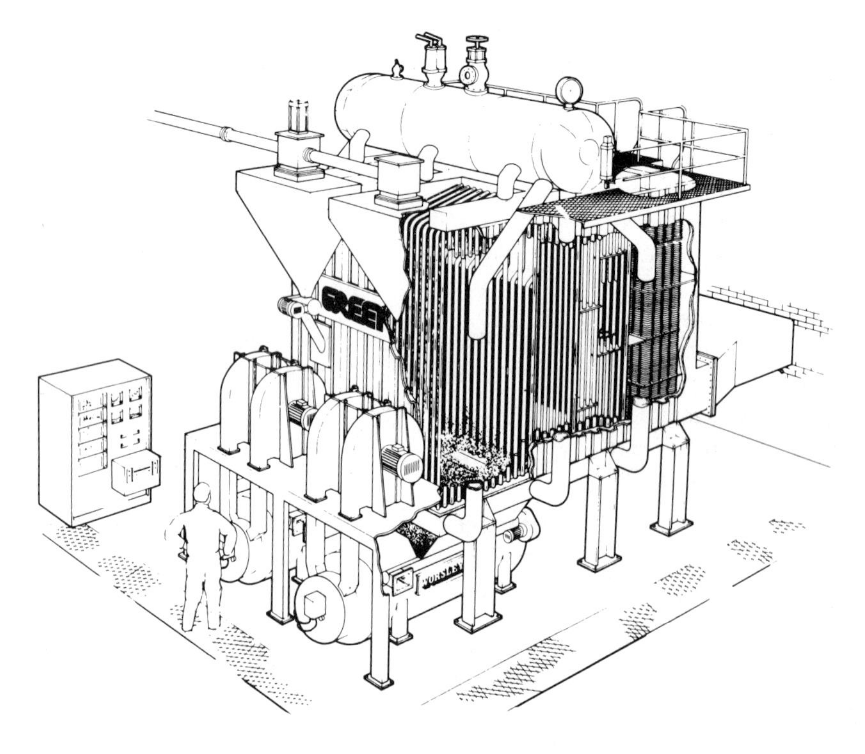

[Fig. 263] Green's packaged coal-fired water-tube boiler with Worsley fluidized bed combustor

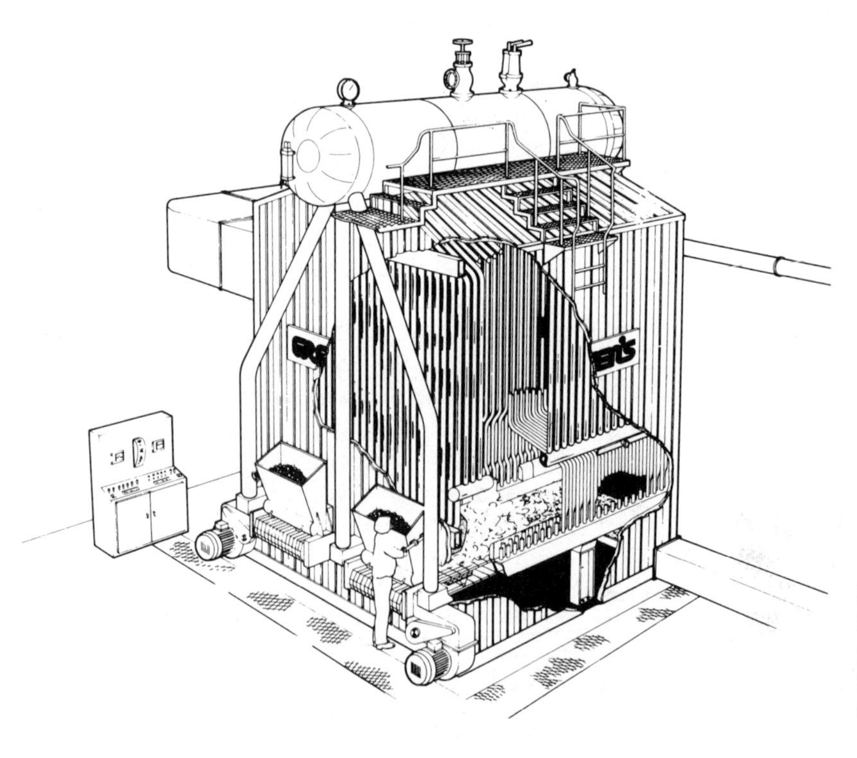

[Fig. 264] Green's 'A' frame water-tube packaged boiler for coal firing with chain grate stokers

weathering A method of making watertight at points and areas where pipes, flues, louvres, etc. pass through the external building envelope.

weathering, buildings A means of preventing water or moisture penetrating into a building. Lead and zinc are common roofing materials used as 'flashing' or weathering strip between a solar collector and roof tiles. Other materials used are synthetic compounds, aluminium and rubber-based strips.

wedge gate valve A valve in which the closure is effected by the wedge action between the gate and the body seat. This may incorporate a solid wedge in which the gate is in one piece, solid or cored, or a split wedge in which the gate is in two pieces.

weight-loaded valve A type of valve, usually a safety valve, which is held closed by a weight fixed to one end of a bar in the top of the valve. The valve will lift when the pressure exerted on it exceeds that of the weight, thereby relieving excess pressure.

welding fitting A plain (unscrewed) pipe fitting designed for butt welding to pipes.

welding rod A rod of appropriate metal held in one hand by the welder and melted to form a seam or joint in an oxyacetylene or electric arc welding operation.

wet air filter See **capillary air washer**.

wet and dry thermometer An instrument for the measurement of relative humidity. It comprises two conventional mercury-in-glass thermometers mounted on a common frame with a handle. One of the two thermometers has its bulb immersed in a wet wick which must remain wetted. The instrument is whirled by its handle for several minutes when the air movement over the wet wick causes some evaporation at the wet wick and thereby depresses the temperature. This is the wet bulb temperature. Readings of the dry bulb and the wet bulb thermometers are noted and checked against values in a table of humidities from which the relative humidity of the air being sampled can be established. More sophisticated versions of wet and dry bulb thermometers incorporate an aspirating fan which draws air over the wet wick (wet bulb).

wet bulb depression The difference between readings of a dry bulb thermometer and a wet bulb (covered with wet muslin). The wet bulb reading is always the lower.

wet bulb temperature The temperature of air as indicated by an ordinary thermometer which has its bulb covered with a wet wick. Evaporation occurs at the wick and the wet bulb thermometer actually measures the temperature at which the water is evaporating. The rate at which the water evaporates at the wick relates to the prevailing relative humidity.

wet riser, fire-fighting A pipe riser in a building to which a permanent water supply is connected. It does not depend on using the fire-fighters' water supply.

wet steam Saturated steam which contains entrained water droplets.

Wheatstone's bridge An accurate and convenient method of comparing electrical resistances by the principle of the divided circuit. It is incorporated into electrical control and measurement devices.

whirling hygrometer A hand-held instrument for measuring humidity. It comprises two mercury-in-glass thermometers in one frame, one with its bulb covered by wetted muslin. After whirling the instrument vigorously by its handle the difference between the readings of the two thermometers indicates the humidity which is read off an associated chart.

white meter tariff A tariff related to the time of day or night when electricity is consumed.

wind farm The assembly of a number of aerogenerators on one site to feed into a common electricity off-take. The location must be carefully selected to ensure wind availability. It usually operates at wind speeds of 7–27 m/s (15–60 miles per hour).

window fan See **fan, window**.

wing nut A fastening device which incorporates two wings (or ears) by which it can be turned by hand on to a matching screw. It is used in locations where the fastening will be repeatedly secured and unscrewed.

wiped joint A method of joining lead pipes in which the two ends are cleaned of dirt and grease and soldered together. The solder is wiped across the joint to give a neat raised closure.

wire drawing Damage caused to a pipe fitting and/or valve through excessive velocity of fluid flow.

wiring diagram A diagram illustrating how an electrical installation is connected. Such a diagram must be provided with each completed electrical installation to permit fault checking and modifications.

working drawing A drawing that is used as a guide for the construction of an item of equipment or a project.

working fluid That fluid which carries the energy into, around or out of a heating or cooling system; for example, water heated in a gas-fired boiler which circulates heat to radiators and/or an indirect hot water storage cylinder; refrigerant which alternately evaporates (absorbing heat) and condenses (liberating heat); air warmed by a furnace, heat pump, heat exchanger, etc. circulated to a drying process or space heating; or steam piped to heat exchangers or turbines. A system may have more than one working fluid; e.g. the operation of a heat pump, an absorption refrigerator or an air-conditioner.

working pressure That pressure at which a pipe, vessel, pump, boiler, etc. is designed to work.

worm feed An arrangement for feeding solid fuel into a boiler by means of a steel worm or screw device. It is used for moving coal from a fuel storage bunker into a boiler fuel hopper.

wrap-around heat recovery coil See **run-around heat recovery coil**.

wrist action tap A quarter-turn tap with a short lever handle which can be operated by the wrist. It is suitable for use by medical personnel.

wrought iron The purest commercial form of iron, nearly free from carbon. It is tough, fibrous, ductile and can be welded.

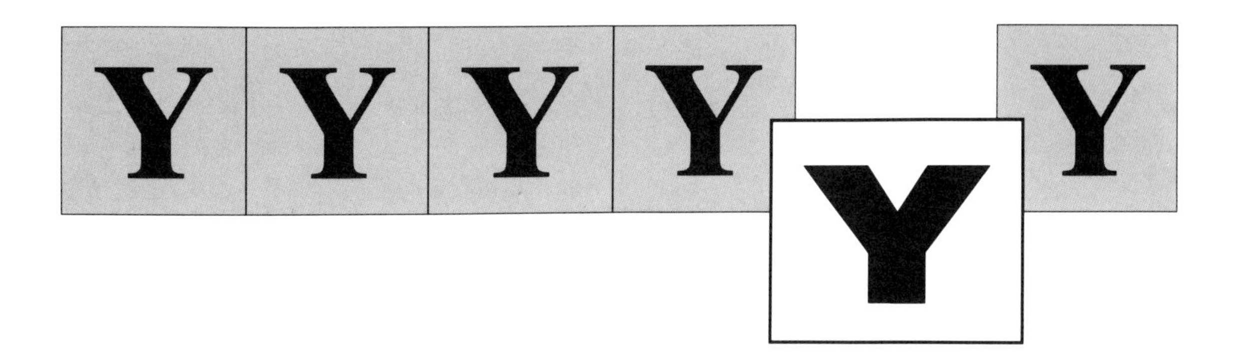

yoke vent A short relief pipe fitted between the main soil or waste pipe and a main ventilating pipe.

Yorkasal pipe fitting A proprietary make of zinc-free fitting for copper pipes which cannot be affected by dezincification.

Y-type valve See **valve, Y-type**.

Z

zig-zag air filter A type of air filter in which the filter medium is formed into a zig-zag shape over a wire framework. The filter medium may be of fabric, glassfibre, paper or plastic. It offers a large area to the air flow in a relatively small space and can have a high collecting efficiency, depending on the type of filter medium.

zoning The separation of a system (heating, ventilation, air-conditioning) into different areas for purposes of improved control of the zonal environment. Zoning may be relative to solar and weather exposure, building usage, occupancy, etc.